Methods in Cell Biology

Methods for Analysis of Golgi Complex Function

Volume 118

Series Editors

Methods in Cell Biology

Methods for Analysis of Golgi Complex Function

Volume 118

Edited by

Franck Perez
Institut Curie, CNRS UMR144
26 rue d'Ulm
Paris Cedex, France

David J. Stephens
School of Biochemistry
University of Bristol
Medical Sciences Building
University Walk
Bristol, UK

ELSEVIER

AMSTERDAM • BOSTON • HEIDELBERG • LONDON
NEW YORK • OXFORD • PARIS • SAN DIEGO
SAN FRANCISCO • SINGAPORE • SYDNEY • TOKYO
Academic Press is an imprint of Elsevier

Academic Press is an imprint of Elsevier
525 B Street, Suite 1800, San Diego, CA 92101-4495, USA
225 Wyman Street, Waltham, MA 02451, USA
The Boulevard, Langford Lane, Kidlington, Oxford, OX5 1GB, UK
32 Jamestown Road, London NW1 7BY, UK
Radarweg 29, PO Box 211, 1000 AE Amsterdam, The Netherlands

First edition 2013

ISBN: 978-0-12-417164-0
ISSN: 0091-679X

For information on all Academic Press publications visit our website at store.elsevier.com

Printed and bound in USA
13 14 15 16 11 10 9 8 7 6 5 4 3 2 1

Contents

Contributors xiii
Preface xix

PART 1 MODEL SYSTEMS AND FUNCTIONAL STUDIES

CHAPTER 1 Analysis of Golgi Complex Functions: *In Vitro* Reconstitution Systems 3
Frank Adolf, Felix T. Wieland
Introduction 4
1.1 Materials 5
1.2 Methods 5
1.3 Discussion and Perspective 9
Acknowledgments 11
References 11

CHAPTER 2 RNA Interference Approaches to Examine Golgi Function in Animal Cell Culture 15
Victoria J. Miller, Caroline M. McKinnon, Harry Mellor, David J. Stephens
Overview of Gene Depletion Methodologies 16
2.1 Key Considerations for the Analysis of Organelle Function 17
2.2 Strategies for RNAi Experiments 18
2.3 Choosing Between siRNA and shRNA 18
2.4 Sources of Reagents for RNAi 19
2.5 Controls 19
2.6 Methods 20
2.7 Discussion 31
Summary 32
Acknowledgments 32
References 32

CHAPTER 3 Trafficking Along the Secretory Pathway in Drosophila Cell Line and Tissues: A Light and Electron Microscopy Approach 35
Margarita Zacharogianni, Catherine Rabouille
Introduction and Rationale 36
3.1 Materials 37
3.2 Methods 38

Conclusions/Perspectives 47
Acknowledgments 47
References 47

CHAPTER 4 Studying Membrane Trafficking in the Worm *C. elegans* by RNA Interference 51
Zita Balklava, Elizabeth Sztul
Introduction and Rationale 52
4.1 Materials 55
4.2 Methods 56
4.3 Discussion 64
Summary 65
Acknowledgment 65
References 65

CHAPTER 5 Endomembrane and Golgi Traffic in Plant Cells 69
Giovanni Stefano, Anne Osterrieder, Chris Hawes, Federica Brandizzi
Introduction 70
5.1 Leaves as a Model System for Analyzing the Plant Golgi and Secretory Pathway 71
5.2 Mutant Screens for Trafficking Proteins 73
5.3 Photobleaching for Transport Studies 75
5.4 Förster Resonance Energy Transfer for Protein Interactions 77
Summary 81
Acknowledgments 81
References 81

CHAPTER 6 Biochemical Analysis of Secretory Trafficking in Mammalian Cells 85
Peristera Roboti, Tomasz M. Witkos, Martin Lowe
Introduction 86
6.1 Materials 89
6.2 Methods 91
6.3 Discussion 99
Summary 101
Acknowledgments 101
References 102

CHAPTER 7 Methods for Analysis of Apical Lumen Trafficking Using Micropatterned 3D Systems 105
Alejo E. Rodríguez-Fraticelli, Fernando Martín-Belmonte
Introduction 106
7.1 3D Culture of MDCK Cysts on Micropatterns 107
7.2 Discussion 120
Acknowledgments 120
References 121

CHAPTER 8 Using Replication Defective Viruses to Analyze Membrane Trafficking in Polarized Epithelial Cells 125
Lucy Pigati, Richard S. Kang, Heike Fölsch
Introduction and Rationale 126
8.1 Materials 129
8.2 Methods 130
8.3 Discussion 135
Summary 136
Acknowledgments 136
References 136

CHAPTER 9 SNAP-Tagging the Retrograde Route 139
Ludger Johannes, Massiullah Shafaq-Zadah
Introduction 140
9.1 Materials 141
9.2 Methods 142
9.3 Conclusion/Discussion 152
Summary 153
Acknowledgments 153
References 153

CHAPTER 10 Assessing ER and Golgi N-Glycosylation Process Using Metabolic Labeling in Mammalian Cultured Cells 157
Romain Péanne, Jorick Vanbeselaere, Dorothée Vicogne, Anne-Marie Mir, Christophe Biot, Gert Matthijs, Yann Guérardel, François Foulquier
Introduction and Rationale 158
10.1 Materials 162
10.2 Methods 166
10.3 Discussion 172

Summary ... 175
Acknowledgments ... 175
References ... 175

PART 2 IMAGING-BASED APPROACHES

CHAPTER 11 Fluorescence-Based Analysis of Trafficking in Mammalian Cells ... 179
Gaelle Boncompain, Franck Perez
Introduction ... 180
11.1 Methods ... 184
11.2 Discussion ... 190
Summary ... 192
Acknowledgments ... 193
References ... 193

CHAPTER 12 Photobleaching Methods to Study Golgi Complex Dynamics in Living Cells ... 195
Erik Lee Snapp
Introduction and Rationale ... 196
12.1 Materials ... 197
12.2 Methods ... 198
12.3 Discussion ... 213
Summary ... 214
Acknowledgments ... 214
References ... 214

CHAPTER 13 Photohighlighting Approaches to Access Membrane Dynamics of the Golgi Apparatus ... 217
Prabuddha Sengupta, Jennifer Lippincott-Schwartz
Introduction ... 218
13.1 Objectives and Rationale ... 219
13.2 Materials and Instruments ... 219
13.3 Methods ... 220
Summary ... 232
References ... 233

CHAPTER 14 **Live Cell Visualization of Golgi Membrane Dynamics by Super-resolution Confocal Live Imaging Microscopy** **235**

Kazuo Kurokawa, Midori Ishii, Yasuyuki Suda, Akira Ichihara, Akihiko Nakano

Introduction 236

14.1 Instrumentation 237

14.2 Materials 237

14.3 Methods 238

14.4 Representative Results and Discussion 239

Acknowledgments 241

References 242

CHAPTER 15 **Analysis of Golgi Complex Function Using Correlative Light-Electron Microscopy** **243**

Elena V. Polishchuk, Roman S. Polishchuk

Introduction 244

15.1 Materials 247

15.2 Methods 249

15.3 Discussion 255

Acknowledgments 256

References 256

CHAPTER 16 **Viewing Golgi Structure and Function from a Different Perspective—Insights from Electron Tomography** **259**

Brad J. Marsh, Margit Pavelka

Introduction 260

16.1 New Insights Into Golgi Architecture and Functional Organization Afforded by Electron Tomography 268

Summary 276

Acknowledgments 276

References 276

CHAPTER 17 **High-Content Screening and Analysis of the Golgi Complex** **281**

George Galea, Jeremy C. Simpson

Introduction 282

17.1 Materials 284

17.2 Methods ... 285
17.3 Discussion ... 292
Summary ... 293
Acknowledgments ... 293
References ... 293

PART 3 EMERGING STUDIES [NEW FRONTIERS]

CHAPTER 18 (Re)Modeling the Golgi ... 299
Pierre Sens, Madan Rao
Introduction: Coarse-Grained Modeling Based on Nonequilibrium Physics ... 300
18.1 Organelle Biogenesis and the Mechanistic Origins of Compartmentalization ... 300
18.2 Protein Transport of Golgi and Maintenance of Cisternal Identity ... 302
18.3 Kinetics of Budding and Cisternal Morphology ... 305
18.4 Future Outlook ... 307
Acknowledgments ... 309
References ... 309

CHAPTER 19 Golgi Depletion from Living Cells with Laser Nanosurgery ... 311
Paolo Ronchi, Rainer Pepperkok
Introduction and Rationale ... 312
19.1 Theory ... 312
19.2 Equipment ... 313
19.3 Materials ... 315
19.4 Methods ... 315
19.5 Discussion ... 322
Acknowledgment ... 323
References ... 323

CHAPTER 20 Studying Intracellular Trafficking Pathways with Probabilistic Density Maps ... 325
Kristine Schauer, Tarn Duong, Carina S. Gomes-Santos, Bruno Goud
Introduction ... 326
20.1 Material and Methods ... 330
20.2 Results and Discussion ... 339

Summary 341
Acknowledgments 341
References 341

CHAPTER 21 Methods to Study Signaling at the Golgi Apparatus 345
Veronika Reitere, Francesco Baschieri, Valentina Millarte, Hesso Farhan
Introduction and Rationale 346
21.1 Materials 346
21.2 Methods 347
Concluding Remarks 356
Acknowledgments 356
References 356

CHAPTER 22 Golgi-Dependent Signaling: Self-Coordination of Membrane Trafficking 359
Jorge Cancino, Anita Capalbo, Alberto Luini
Introduction 361
22.1 Generation of Traffic-Dependent Golgi Signaling: ER-to-Golgi Traffic Synchronization Protocols 364
22.2 Morphological and Quantitative Analysis of Traffic-Generated Golgi-Based Signaling 368
22.3 Analysis of KDELR-Dependent Signaling 371
22.4 The KDELR Transduction Machinery 375
22.5 Discussion 378
Summary 379
Acknowledgment 379
References 379

CHAPTER 23 Signaling at the Golgi During Mitosis 383
Antonino Colanzi, Christine Sütterlin
Introduction 384
23.1 The Organization of Mammalian Golgi Membranes in Interphase 385
23.2 Mechanism of Mitotic Golgi Disassembly 386
23.3 Mechanism of Postmitotic Golgi Reassembly 390
23.4 The Significance of Golgi Reorganization in Mitosis 390
23.5 Assays to Study Mitotic Golgi Reorganization 393
Concluding Remarks 395
Acknowledgments 395
References 395

CHAPTER 24 Ice Recovery Assay for Detection of Golgi-Derived Microtubules **401**
Ashley D. Grimaldi, Maria Fomicheva, Irina Kaverina
Introduction and Rationale 402
24.1 Materials 403
24.2 Methods 405
24.3 Discussion 412
Summary 413
Acknowledgments 413
References 414

Index **417**

Contributors

Frank Adolf
Heidelberg University Biochemistry Center (BZH), University of Heidelberg, Im Neuenheimer Feld 328, Heidelberg, Germany

Zita Balklava
School of Life and Health Sciences, Aston University, Aston Triangle, Birmingham, United Kingdom

Francesco Baschieri
Biotechnology Institute Thurgau, Kreuzlingen, Switzerland, and Department of Biology, University of Konstanz, Konstanz, Germany

Christophe Biot
Structural and Functional Glycobiology Unit, UMR CNRS/USTL 8576, IFR 147, Avenue Mendeleïev, University of Lille 1, Villeneuve d'Ascq, France

Gaelle Boncompain
Institut Curie, and CNRS UMR144, Paris, France

Federica Brandizzi
Department of Energy Plant Research Laboratory, and Department of Plant Biology, Michigan State University, East Lansing, Michigan, USA

Jorge Cancino
Telethon Institute of Genetics and Medicine, and Istituto di Biochimica delle Proteine (IBP), Consiglio Nazionale delle Richerce (CNR), Napoli, Italy

Anita Capalbo
Telethon Institute of Genetics and Medicine, and Istituto di Biochimica delle Proteine (IBP), Consiglio Nazionale delle Richerce (CNR), Napoli, Italy

Antonino Colanzi
Institute of Protein Biochemistry, CNR, Naples, Italy

Tarn Duong
Theoretical and Applied Statistics Laboratory (LSTA), University Pierre and Marie Curie—Paris, and Institute of Translational Neurosciences (IHU-A-ICM), Pitié-Salpêtrière Hospital, Paris, France

Hesso Farhan
Biotechnology Institute Thurgau, Kreuzlingen, Switzerland, and Department of Biology, University of Konstanz, Konstanz, Germany

Heike Fölsch
Department of Cell and Molecular Biology, Northwestern University Feinberg School of Medicine, Chicago, Illinois, USA

Maria Fomicheva
Department of Cell and Developmental Biology, Vanderbilt University Medical Center, Nashville, Tennessee, USA, and Department of Biology, Lomonosov Moscow State University, Moscow, Russia

François Foulquier
Structural and Functional Glycobiology Unit, UMR CNRS/USTL 8576, IFR 147, Avenue Mendeleïev, University of Lille 1, Villeneuve d'Ascq, France

George Galea
School of Biology and Environmental Science, Conway Institute of Biomolecular and Biomedical Research, University College Dublin (UCD), Dublin, Ireland

Carina S. Gomes-Santos
Molecular Mechanisms of Intracellular Transport, Unité Mixte de Recherche 144 Centre National de la Recherche Scientifique/Institut Curie, Paris cedex, France

Bruno Goud
Molecular Mechanisms of Intracellular Transport, Unité Mixte de Recherche 144 Centre National de la Recherche Scientifique/Institut Curie, Paris cedex, France

Ashley D. Grimaldi
Department of Cell and Developmental Biology, Vanderbilt University Medical Center, Nashville, Tennessee, USA

Yann Guérardel
Structural and Functional Glycobiology Unit, UMR CNRS/USTL 8576, IFR 147, Avenue Mendeleïev, University of Lille 1, Villeneuve d'Ascq, France

Chris Hawes
Department of Biological and Medical Sciences, Oxford Brookes University, Oxford, United Kingdom

Akira Ichihara
Live Cell Molecular Imaging Research Team, Extreme Photonics Research Group, RIKEN Center for Advanced Photonics, Wako, Saitama, Japan

Midori Ishii
Live Cell Molecular Imaging Research Team, Extreme Photonics Research Group, RIKEN Center for Advanced Photonics, Wako, Saitama, and Department of Biological Sciences, Graduate School of Science, The University of Tokyo, Hongo, Bunkyo-ku, Tokyo, Japan

Ludger Johannes
Institut Curie—Centre de Recherche, Trafic, Signaling and Delivery group, Paris Cedex 05, France

Richard S. Kang
Department of Cell and Molecular Biology, Northwestern University Feinberg School of Medicine, Chicago, Illinois, USA

Irina Kaverina
Department of Cell and Developmental Biology, Vanderbilt University Medical Center, Nashville, Tennessee, USA

Kazuo Kurokawa
Live Cell Molecular Imaging Research Team, Extreme Photonics Research Group, RIKEN Center for Advanced Photonics, Wako, Saitama, Japan

Jennifer Lippincott-Schwartz
The Eunice Kennedy Shriver National Institute of Child Health and Human Development, National Institutes of Health, Bethesda, Maryland, USA

Martin Lowe
Faculty of Life Sciences, University of Manchester, Manchester, United Kingdom

Alberto Luini
Telethon Institute of Genetics and Medicine, and Istituto di Biochimica delle Proteine (IBP), Consiglio Nazionale delle Richerce (CNR), Napoli, Italy

Brad J. Marsh
Institute for Molecular Bioscience, Centre for Microscopy and Microanalysis, The University of Queensland, Brisbane, Queensland, Australia

Fernando Martín-Belmonte
Centro de Biología Molecular "Severo Ochoa", Madrid, Spain

Gert Matthijs
Center for Human Genetics, KU Leuven University, Herestraat 49, Leuven, Belgium

Caroline M. McKinnon
Cell Biology Laboratories, School of Biochemistry, University of Bristol, Bristol, United Kingdom

Harry Mellor
Cell Biology Laboratories, School of Biochemistry, University of Bristol, Bristol, United Kingdom

Valentina Millarte
Biotechnology Institute Thurgau, Kreuzlingen, Switzerland, and Department of Biology, University of Konstanz, Konstanz, Germany

Victoria J. Miller
Cell Biology Laboratories, School of Biochemistry, University of Bristol, Bristol, United Kingdom

Anne-Marie Mir
Structural and Functional Glycobiology Unit, UMR CNRS/USTL 8576, IFR 147, Avenue Mendeleïev, University of Lille 1, Villeneuve d'Ascq, France

Akihiko Nakano
Live Cell Molecular Imaging Research Team, Extreme Photonics Research Group, RIKEN Center for Advanced Photonics, Wako, Saitama, and Department of Biological Sciences, Graduate School of Science, The University of Tokyo, Hongo, Bunkyo-ku, Tokyo, Japan

Anne Osterrieder
Department of Plant Biology, Michigan State University, East Lansing, Michigan, USA

Romain Péanne
Center for Human Genetics, KU Leuven University, Herestraat 49, Leuven, Belgium

Margit Pavelka
Department of Cell Biology and Ultrastructure Research, Center for Anatomy and Cell Biology, Medical University of Vienna, Schwarzspanierstrasse 17, Vienna, Austria

Rainer Pepperkok
Cell Biology and Biophysics Unit, European Molecular Biology Laboratory, Heidelberg, Germany

Franck Perez
Institut Curie, and CNRS UMR144, Paris, France

Lucy Pigati
Department of Cell and Molecular Biology, Northwestern University Feinberg School of Medicine, Chicago, Illinois, USA

Elena V. Polishchuk
Telethon Institute of Genetics and Medicine, Naples, Italy

Roman S. Polishchuk
Telethon Institute of Genetics and Medicine, Naples, Italy

Catherine Rabouille
Hubrecht Institute for Developmental Biology and Stem Cell Research, and The Department of Cell Biology, UMC Utrecht, Utrecht, The Netherlands

Madan Rao
Raman Research Institute and National Centre for Biological Sciences (Tata Institute of Fundamental Research), Bangalore, India

Veronika Reitere
Biotechnology Institute Thurgau, Kreuzlingen, Switzerland, and Department of Biology, University of Konstanz, Konstanz, Germany

Peristera Roboti
Faculty of Life Sciences, University of Manchester, Manchester, United Kingdom

Alejo E. Rodríguez-Fraticelli
Centro de Biología Molecular "Severo Ochoa", Madrid, Spain

Paolo Ronchi
Cell Biology and Biophysics Unit, European Molecular Biology Laboratory, Heidelberg, Germany

Kristine Schauer
Molecular Mechanisms of Intracellular Transport, Unité Mixte de Recherche 144 Centre National de la Recherche Scientifique/Institut Curie, Paris cedex, France

Prabuddha Sengupta
The Eunice Kennedy Shriver National Institute of Child Health and Human Development, National Institutes of Health, Bethesda, Maryland, USA

Pierre Sens
Laboratoire Gulliver, CNRS-ESPCI, UMR 7083, 75231 Paris, France

Massiullah Shafaq-Zadah
UMR144 CNRS, Paris, France

Jeremy C. Simpson
School of Biology and Environmental Science, Conway Institute of Biomolecular and Biomedical Research, University College Dublin (UCD), Dublin, Ireland

Erik Lee Snapp
Department of Anatomy and Structural Biology, Albert Einstein College of Medicine, New York, USA

Giovanni Stefano
Department of Energy Plant Research Laboratory, and Department of Plant Biology, Michigan State University, East Lansing, Michigan, USA

David J. Stephens
Cell Biology Laboratories, School of Biochemistry, University of Bristol, Bristol, United Kingdom

Christine Sütterlin
Department of Developmental and Cell Biology, University of California, Irvine, California, USA

Yasuyuki Suda
Live Cell Molecular Imaging Research Team, Extreme Photonics Research Group, RIKEN Center for Advanced Photonics, Wako, Saitama, Japan

Elizabeth Sztul
Department of Cell, Developmental and Integrative Biology, University of Alabama at Birmingham, Birmingham, Alabama, USA

Jorick Vanbeselaere
Structural and Functional Glycobiology Unit, UMR CNRS/USTL 8576, IFR 147, Avenue Mendeleïev, University of Lille 1, Villeneuve d'Ascq, France

Dorothée Vicogne
Structural and Functional Glycobiology Unit, UMR CNRS/USTL 8576, IFR 147, Avenue Mendeleïev, University of Lille 1, Villeneuve d'Ascq, France

Felix T. Wieland
Heidelberg University Biochemistry Center (BZH), University of Heidelberg, Im Neuenheimer Feld 328, Heidelberg, Germany

Tomasz M. Witkos
Faculty of Life Sciences, University of Manchester, Manchester, United Kingdom

Margarita Zacharogianni
Hubrecht Institute for Developmental Biology and Stem Cell Research, Utrecht, The Netherlands

Preface

Understanding Golgi apparatus organization and function continues to be the subject of intense research. Its unique structure and central role in both anterograde and retrograde membrane traffic, along with its conservation across most eukaryotes, place the Golgi apparatus at the center of many studies. Despite knowledge of its existence since its discovery by Camillo Golgi in 1897, the Golgi apparatus continues to be the subject of research that exploits the very latest in cell biological approaches. Collecting protocols in a single volume such as this not only serves multiple purposes in terms of dissemination of new techniques among researchers and sparking ideas but also emphasizes the key role that members of the Golgi community have had in developing these methods and refining the protocols in the first place. These range from *in vitro* reconstitution to whole animal experiments. In this volume, we have sought to bring together examples of state-of-the-art tools that are applied to study Golgi biology. The contributors are truly expert in their arena both in terms of their knowledge of Golgi biology and the techniques described. Many of these techniques are of course equally applicable to many other fields of biology, and this volume provides a snapshot of some of this diversity.

We have organized the volume into three core sections covering Model Systems and Functional Studies, Imaging Studies, and New Frontiers. This organization serves to provide some context for systems that have proven, and continue to prove, immensely valuable to the study of Golgi function across many years.

The section on Model Systems includes examples that span *in vitro* and *in vivo* contexts. *In vitro* reconstitution provides the opportunity to derive functional insight using controlled addition of purified components. The use of model organisms, notably yeast, nematodes, plants, and flies, provides important functional insight in a more physiological context. Bridging these studies are the *in vitro* cell culture models. Here, more advanced modes such as polarized systems and 3D cell cultures are proving of great benefit in providing a link to tissue and organism biology. As it is the case in most cell biology fields, cellular genetics, through the use of RNA interference, is now a common approach to improve our understanding of Golgi organization and function. Many of the specialized functions of the Golgi are also given space in this section—dedicated methods provide clear insight into processes including retrograde traffic through the Golgi (perhaps the true interface of the secretory and endocytic pathways) and in the complex yet tightly controlled glycosylation events. The addition of complex polysaccharides to newly synthesized proteins is of great importance to normal human health and development, and this is evidenced by the complex array of congenital disorders of glycosylation that have been mapped to mutations in genes controlling Golgi function.

The past 20 years have seen an explosion in the application of bioimaging to the study of many processes in cell biology, perhaps none more so than membrane trafficking to and from the Golgi. The juxtanuclear location of the Golgi in many commonly used cell lines provides a clear end point for imaging experiments in the early secretory pathway or a starting point for post-Golgi trafficking. What are now relatively trivial live-cell imaging experiments based on GFP tagging are augmented by complex protocols enabling controlled trafficking of select components, applications of photobleaching to study and quantify protein dynamics, and more recently the broader adoption of higher resolution light and electron microscopy technologies including superresolution light microscopy and electron tomography. All of these techniques are included in the Imaging Studies section. These technological developments also now enable wider application of high-content screening methods to define the cohort of genes involved in specialist aspects of Golgi function and processes that involve it.

Recent years have seen the introduction of new tools and approaches to this field; these are described in our New Frontiers section. The ordered structure yet complex function of the Golgi provides opportunities for the development of mathematical models to better understand and in future predict key aspects of Golgi function. Nanosurgery methods have been applied to try to define the origin of the Golgi in the absence of a preexisting structure; micropatterning provides opportunities to encourage cells to adopt predefined and more uniform shape, imposing particular intracellular organization. This will likely benefit analysis of those processes for which Golgi orientation and intracellular location are of key importance. We also cannot, of course, understate the importance of the Golgi as an integrator of cellular signals and processes. The core examples we have chosen here develop newly emerging themes of Golgi-dependent signaling pathways and the role of protein phosphorylation in more general Golgi function. This, of course, also links directly to the unique role of the Golgi in mitosis where the ordered disassembly of this single copy organelle is a prerequisite for mitotic entry. The structure, organization, and function of the Golgi are all inherently connected to the cytoskeleton. We focus here on the role of microtubules, yet many of the core concepts can equally be applied to other cytoskeletal networks.

This collection of protocols showcases the intense research activity that is focused on the Golgi itself as well as the importance of this organelle in diverse processes in cell biology. We are very grateful to the contributors for providing these carefully detailed methods, most of which contain the very lab-specific information that is almost impossible to glean without normally visiting the lab concerned. We also would like to acknowledge the help of people at Elsevier, and in particular Sarah Lay and Zoe Kruze.

David Stephens and Franck Perez

PART

1

Model Systems and Functional Studies

CHAPTER

Analysis of Golgi Complex Functions: *In Vitro* Reconstitution Systems

1

Frank Adolf and Felix T. Wieland

Heidelberg University Biochemistry Center (BZH), University of Heidelberg, Im Neuenheimer Feld 328, Heidelberg, Germany

CHAPTER OUTLINE

Introduction **4**
1.1 Materials **5**
1.1.1 COPI Coat Proteins 5
1.1.2 Antibodies 5
1.1.3 Reagents 5
1.2 Methods **5**
1.2.1 Preparation of Semi-intact Cells 5
1.2.2 COPI Vesicle Formation from Semi-intact Cells 6
1.2.3 Ultarstructural Analysis of COPI Vesicles 8
1.3 Discussion and Perspective **9**
Acknowledgments **11**
References **11**

Abstract

In vitro reconstitution is prerequisite to investigate complex cellular functions at the molecular level. Reconstitution systems range from combining complete cellular cytosol with organelle-enriched membrane fractions to liposomal systems where all components are chemically defined and can be chosen at will. Here, we describe the *in vitro* reconstitution of COPI-coated vesicles from semi-intact cells. Efficient vesicle formation is achieved by simple incubation of permeabilized cells with the minimal set of coat proteins Arf1 and coatomer, and guanosine trinucleotides. GTP hydrolysis or any mechanical manipulations are not required for efficient COPI vesicle release.

Methods in Cell Biology, Volume 118

ISSN 0091-679X
http://dx.doi.org/10.1016/B978-0-12-417164-0.00001-X

INTRODUCTION

Transport vesicles functionally link organelles of the endomembrane system. In the early secretory pathway, COPI and COPII vesicles mediate transport of newly synthesized proteins and lipids between the endoplasmic reticulum (ER) and the Golgi apparatus, whereas clathrin-coated vesicles with various adaptor proteins mediate transport in the late secretory system, including traffic between the late Golgi, endosomes and the plasma membrane, and endocytosis (reviewed in Edeling, Smith, & Owen, 2006; Popoff, Adolf, Brugger, & Wieland, 2011; Szul & Sztul, 2011).

The identification of COPI-coated vesicles, their characterization in terms of coat components, and finally the minimal requirements for the generation of this class of vesicular carriers were significantly facilitated by various *in vitro* reconstitution systems. COPI vesicles were initially identified by electron microscopic analysis of isolated, enriched Golgi membranes incubated with cytosol and ATP (Orci, Glick, & Rothman, 1986). Large-scale incubations of Golgi-enriched membranes with cytosol and the poorly hydrolyzable GTP analogue GTPγS allowed purification of *in vitro* generated COPI vesicles by sucrose gradient centrifugation (Malhotra, Serafini, Orci, Shepherd, & Rothman, 1989). This *in vitro* reconstitution system subsequently allowed to identify the heptameric coat complex coatomer (Waters, Serafini, & Rothman, 1991) and the small GTPase Arf1 (Serafini et al., 1991) as cytosolic components involved in the formation of COPI vesicles. *In vitro* vesicle reconstitution from chemically defined liposomes revealed Arf1 and coatomer as cytosolic components required and sufficient for the formation of COPI vesicles (Bremser et al., 1999; Spang, Matsuoka, Hamamoto, Schekman, & Orci, 1998).

Likewise, functions at the molecular level of a variety of proteins involved in the formation of such transport vesicles have been deduced capitalizing on liposomal *in vitro* reconstitution systems. The discovery and characterization of curvature-sensitive domains in ArfGAP1 is one example for the power of such reductionist systems. ArfGAP1, in addition to its catalytical domain that induces hydrolysis of GTP in the small COPI coat GTPase Arf1, possess ArfGAP1 lipid-packing sensor motifs that confer recognition of and increasingly efficient binding to increasing membrane curvature (Bigay, Casella, Drin, Mesmin, & Antonny, 2005; Bigay, Gounon, Robineau, & Antonny, 2003). This mechanism was established by quantitative binding studies using defined recombinant protein domains in combination with liposomes of defined lipid composition and defined curvatures.

Such defined liposomal systems have revealed a function of the small GTPases Sar1 (Lee et al., 2005) and, combined with cryoelectron tomography, of Arf1in the scission reaction of COP vesicles (Beck et al., 2011, 2008) and have allowed first insight into the structural organization of a coat protein on a membrane, using reconstituted liposomal COPI vesicles (Faini et al., 2012).

Several questions (e.g., mechanisms that underlie the sorting of cargo) in the process of COPI vesicle biogenesis, however, require the use of native membranes. To meet these requirements and to allow quantitatively comparative studies, semi-intact cells can be used to reconstitute various intracellular transport processes. More recently, permeabilized cells were also used to reconstitute the formation of COPII-coated vesicles with purified Sar1 and the coat protein complexes Sec23/24 and Sec13/31 (Kim et al., 2007; Mancias & Goldberg, 2007). Here, we describe how this assay serves the *in vitro* formation of COPI carriers that can be isolated as coated vesicles from semi-intact cells (donor membranes) by simple differential centrifugation. Advantages of the assay are discussed here.

1.1 MATERIALS

1.1.1 COPI coat proteins

Full-length human Arf1 wild type was co-expressed with yeast *N*-myristoyl-transferase in *Escherichia coli* BL21 Star™ (Invitrogen) and purified as described previously (Franco, Chardin, Chabre, & Paris, 1995). Coatomer from rabbit liver was purified as described (Pavel, Harter, & Wieland, 1998). More detailed descriptions about the preparation of myristoylated human Arf1 and rabbit liver coatomer can be found in Nickel and Wieland (2001). Recombinant mouse coatomer isotype γ1ζ1 (CMγ1ζ1) was produced in Sf9 insect cells and purified as described (Sahlmuller et al., 2011). After thawing of stored aliquots, protein aggregates of coat protein were removed by ultracentrifugation at 100,000 × *g* for 15 min at 4 °C.

1.1.2 Antibodies

The following antibodies were used: β′-COP 899 (Harter, Draken, Lottspeich, & Wieland, 1993), δ-COP 877 (Faulstich et al., 1996), and p24 (Gommel et al., 1999), Calnexin (Abcam), ERGIC53 (C6) (Santa Cruz Biotechnology), Sec22b (29-F7) (Santa Cruz Biotechnology), and GM130 (B10) (Santa Cruz Biotechnology).

1.1.3 Reagents

Alpha MEM, digitonin, trypsin–EDTA, trypsin inhibitor, nucleotides, and creatine phosphokinase were all purchased from Sigma-Aldrich. Creatine phosphate was purchased from Roche.

1.2 METHODS

1.2.1 Preparation of semi-intact cells

HeLa cells (ACC57) were cultured in alpha MEM (Sigma-Aldrich) supplemented with 10% FCS, 100 μg/ml penicillin, 100 μg/ml streptomycin, 2 mM L-glutamine, at 5% CO_2 at 37 °C. In general, the preparation of semi-intact cells was essentially carried out as described previously (Mancias & Goldberg, 2007). Cells were split 48 h prior permeabilization to reach approximately 80–90% confluence when used.

Cells were washed 1 × with PBS (preheated to 37 °C) and detached by incubation with 1 ml trypsin–EDTA (Sigma-Aldrich) for 1–2 min at 37 °C. Per dish 300 μl of a (1 mg/ml) trypsin inhibitor (Sigma-Aldrich) stock solution prepared in PBS was added, and cells were resuspended in 25 ml ice-cold PBS. All following steps were carried out either on ice or at 4 °C. Cells from two 15 cm dishes were combined into a 50-ml centrifuge tube and sedimented by centrifugation at 300 × *g* for 5 min at 4 °C. The supernatant was discarded, and cell pellets were resuspended in 20 ml ice-cold PBS, carefully mixed by inverting after addition of 20 μl of a 40 mg/ml digitonin (Sigma-Aldrich) stock solution prepared in DMSO, and subsequently incubated for 5 min on ice. To stop permeabilization/remove the digitonin, 30 ml ice-cold PBS was added and cells were sedimented by centrifugation at 300 × *g* for 5 min at 4 °C. The supernatants were discarded, and the permeabilized cells were carefully resuspended in 50 ml ice-cold assay buffer and incubated for 10 min on ice. Finally, semi-intact cells were harvested by centrifugation at 300 × *g* for 5 min at 4 °C. Cell pellets prepared from four 15 cm dishes were resuspended in 600 μl assay buffer (see below). Concentrations of the semi-intact cell suspensions were determined quantification of protein by the Bradford assay (Bradford, 1976). For COPI vesicle budding assays, typically semi-intact cells corresponding to 100 μg membrane protein were used per sample.

1.2.2 COPI vesicle formation from semi-intact cells

For the generation of COPI-coated vesicles from semi-intact cells, permeabilized HeLa cells corresponding to 100 μg membrane protein were incubated for 30 min at 30 °C in 1.6 ml low binding reaction tubes (Biozym Scientific) with 2 μg Arf1, 10 μg coatomer (either purified rabbit liver coatomer or recombinant mouse coatomer isoform γ1ζ1), and 0.5 mM GTP in a total volume of 200 μl, containing 25 mM HEPES, pH 7.2 (KOH), 150 mM KOAc, and 2 mM MgOAc. Addition of an ATP-regenerating system to the addition of an ATP-regenerating system (40 mM creatine phosphate (Roche), 0.2 mg/ml creatine phosphokinase (Sigma), 1 mM ATP (Sigma)) to the reaction mix did not enhance the yield of released vesicles produced (data not shown). After incubation, newly formed COPI-coated vesicles were separated from donor semi-intact HeLa cells by differential centrifugation. Permeabilized cells are pelleted by medium-speed centrifugation at 10,000 × *g* for 10 min at 4 °C. Supernatants are carefully transferred into 1.5 ml ultracentrifugation reaction tubes, and COPI-coated vesicles in the medium-speed supernatant were harvested by ultracentrifugation at 100,000 × *g* for 30 min at 4 °C. Supernatants from ultracentrifugation were carefully aspirated and discarded. Vesicle-containing pellets as well as samples of semi-intact cells (input) were solubilized in SDS gel electrophoresis sample buffer and denatured for 10 min at 95 °C. Seventy-five percentage of the ultracentrifugation pellets (vesicle fractions) and 5 μg semi-intact cells (5% input) were separated by SDS-PAGE, transferred to PVDF membranes, and analyzed with the antibodies indicated. For analysis of efficiency of vesiculation, Western blot signals of membrane vesicle markers and, and as a control, of nonvesicle markers were quantified with the LI-COR Odyssey system equipped with the Image Studio (Version 2.1.10) software and blotted in percentage of total input.

Examples of COPI vesicles generated from semi-intact cells separated from the donor membranes by differential centrifugation are shown in Fig. 1.1A and B. COPI vesicle formation, as judged by the presence of the vesicle membrane marker p24 and

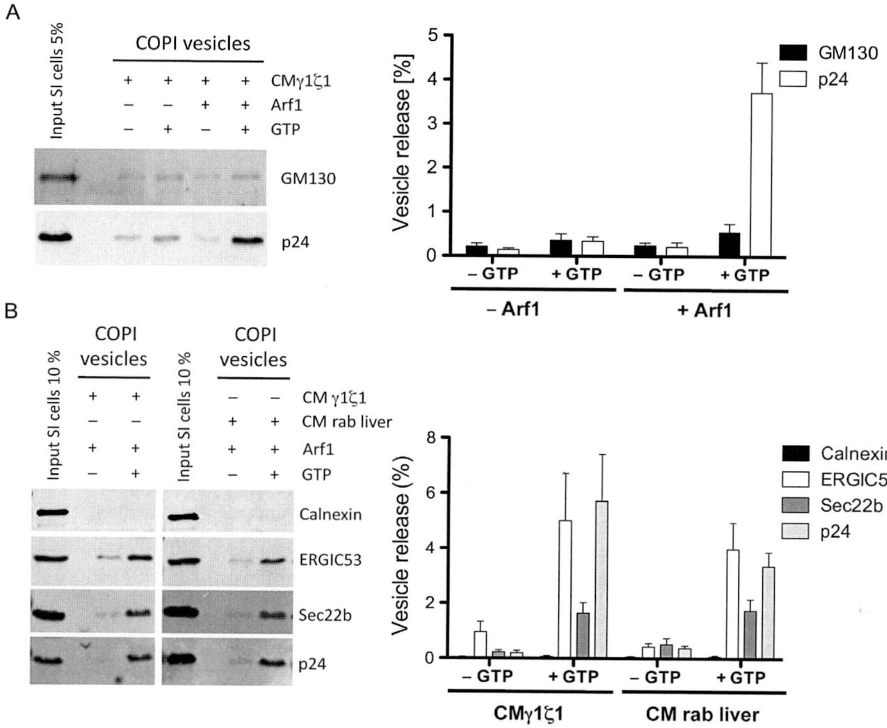

FIGURE 1.1

COPI vesicle formation from semi-intact. (A) Semi-intact cells were incubated with the small GTPases Arf1, mouse coatomer isoform γ1ζ1 (CMγ1ζ1), and GTP (in the presence of an ATP-regenerating system) as indicated. COPI vesicles were separated from donor membranes by differential centrifugation (see text) and analyzed by Western blotting for the presence of the membrane markers indicated. COPI budding efficiency of the vesicle membrane marker p24 and the nonvesicle, Golgi membrane marker GM130 was quantified with Li-Cor Image Studio (means ± SEM, $n=4$). (B) Semi-intact cells were incubated with the small GTPases Arf1, GTP (in the presence of an ATP-regenerating system) and either mouse coatomer isoform γ1ζ1 (CMγ1ζ1) or purified rabbit liver coatomer (CM rab liver), as indicated. COPI vesicles were separated from donor membranes by differential centrifugation and analyzed by Western blotting for the presence of the membrane markers indicated. COPI budding efficiency of the vesicles membrane marker ERGIC53, Sec22b, and p24 and the nonvesicle membrane marker Calnexin was quantified with Li-Cor Image Studio (means ± SEM, $n=4$). (See color plate.)

the absence of the nonvesicle marker GM130 that was previously reported to be excluded from COPI vesicle preparations generated from Golgi-enriched membranes (Sonnichsen, Watson, Clausen, Misteli, & Warren, 1996), was dependent on addition of Arf1, coatomer, and GTP (Fig. 1.1A). Vesicles generated in a nucleotide-dependent manner with either purified rabbit liver coatomer or recombinant mouse coatomer isoform γ1ζ1 exclude the ER marker calnexin and contain the well-established COPI vesicular marker proteins ERCIC53, Sec22b, and p24, further demonstrating efficient COPI vesicle formation under the conditions applied (Fig. 1.1B).

For analysis of the coat proteins on vesicles, it was necessary to further separate COPI vesicles released into the medium-speed centrifugation supernatant from excess free soluble coat proteins because the large protein complex coatomer is partially pelleted under the conditions of centrifugation to harvest the vesicles from the medium-speed supernatant. The buoyant density of COPI-coated vesicles derived from enriched Golgi membranes corresponds to 40–43% (w/v) sucrose (Malhotra et al., 1989). Here, we separate free soluble coat proteins from COPI vesicles by density centrifugation with a single sucrose cushion. The vesicle-containing medium-speed supernatant was layered on top of a 200 μl 37% (w/v) sucrose cushion and centrifuged at 100,000 × *g* for 60 min at 4 °C. Western blot analysis of COPI vesicles separated from the donor membranes and from soluble coatomer as described above is shown in Fig. 1.2. Here, the COPI budding assay was conducted with the poorly hydrolyzable GTP analogues GTPγS or GMP-PNP. Nucleotide-dependent signals for coatomer (subunits β′-COP and δ-COP) as well as for the COPI vesicle membrane markers ERGIC53 and p24 were observed regardless of whether GTP or the poorly hydrolyzable GTP analogues GTPγS or GMP-PNP were utilized to initiate COPI vesicle formation (Adolf et al., 2013), indicating that COPI vesicles remained coated under all conditions applied.

1.2.3 Ultarstructural analysis of COPI vesicles

The presence of regular bona fide vesicles in these vesicle formation assays was probed by electron microscopic analysis. For this purpose, the assay described above was scaled up three times. After separation of semi-intact cells from newly formed COPI vesicles by medium-speed centrifugation at 10,000 × *g* for 10 min at 4 °C, samples were fixed by addition of glutaraldehyde at a final concentration of 2% and incubated for 30 min on ice. Subsequently, fixed vesicles were harvested by ultracentrifugation at 100,000 × *g* for 30 min at 4 °C. Supernatants were carefully aspirated and discarded. The COPI vesicle-containing pellets were further fixed with 2% glutaraldehyde for 75 min and subsequently treated with 0.5% tannic acid for 60 min both in assay buffer. After washing 2 × 10 min with assay buffer and 1 × 10 min with 0.1 M sodium phosphate, pH 7.4, the samples were postfixed with 2% osmium tetroxide/1.5% potassium ferrocyanide in 0.1 M sodium phosphate, pH 7.4, for 1 h on ice. After postfixation, samples were washed, contrasted en bloc with uranyl acetate, dehydrated with an ascending series of ethanol, and embedded in glycid ether 100-based resin. Ultrathin sections were cut with a Reichert ULTRACUT S

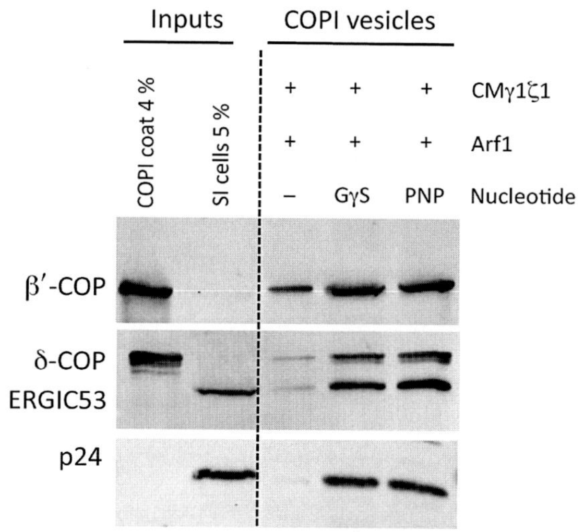

FIGURE 1.2

Characterization of coat proteins found on COPI vesicles generated from semi-intact. (A) Semi-intact cells were incubated with the small GTPases Arf1, mouse coatomer isoform γ1ζ1, and nucleotides (GTP, GTPγS (GγS), or GMP-PNP (PNP)) (in the presence of an ATP-regenerating system) as indicated. After medium-speed centrifugation to separate newly formed vesicles from donor membranes, the vesicle containing medium-speed supernatant was layered on top of a 37% (w/v) sucrose cushion to further separate COPI vesicles from excess soluble coatomer. COPI vesicles were harvested by ultracentrifugation and pellet fractions were analyzed by Western blotting for the coatomer subunits as well as for the vesicle membrane markers indicated.

ultramicrotome (Leica Microsystems). After contrasting with uranyl acetate and lead citrate samples were inspected with an EM 10 CR electron microscope (Carl Zeiss). Sections of embedded COPI vesicles generated from semi-intact cells with Arf1, mouse coatomer isoform γ1ζ1 and GTP, or the poorly hydrolyzable GTP analogues GTPγS or GMP-PNP are depicted in Fig. 1.3.

1.3 DISCUSSION AND PERSPECTIVE

Semi-intact cells have served in various investigations to reconstitute various intracellular transport processes. More recently, permeabilized cells have also been used as donor membranes to reconstitute COPII vesicle formation with purified mammalian COPII coat proteins (Kim et al., 2007; Mancias & Goldberg, 2007).

We have recently shown that semi-intact cells can be used as donor membranes for the formation of COPI-coated vesicles with Arf1 and coatomer as the minimal set cytosolic coat proteins (Adolf et al., 2013). It is important to note that no mechanical manipulations (e.g., shearing by pipetting or centrifugation) are required for efficient

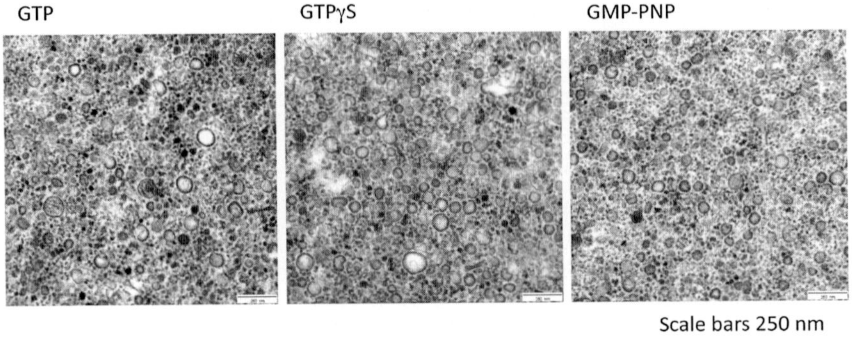

FIGURE 1.3

Ultra-structural analysis of COPI vesicles generated from semi-intact cells. Ultrathin sections of resin-embedded COPI vesicle fractions generated from semi-intact cells with purified Arf1, mouse coatomer isoform γ1ζ1, and the nucleoside triphosphate derivatives (GTP, GTPγS, or GMP-PNP), as indicated, were analyzed by electron microscopy (scale bars 250 nm).

release of COPI vesicles from semi-intact cells. In agreement with various other studies, which have reported COPI vesicle formation with Arf1 and coatomer as the minimal set of cytosolic proteins from enriched Golgi membranes (Orcl, Palmer, Amherdt, & Rothman, 1993) and chemically defined liposomes (Bremser et al., 1999; Spang et al., 1998), we did not observe the requirement of an additional cytosolic factor as a stochiometric COPI coat component. Involvement of additional cytosolic factors that might be required to control COPI vesicle biogenesis *in vivo* cannot be excluded. At least one catalytically required factor is the cytosolic protein GBF1, the ArfGEF involved in COPI vesicle formation (Kawamoto et al., 2002; Spang, Herrmann, Hamamoto, & Schekman, 2001; Zhao, Lasell, & Melancon, 2002). Addition of GBF1 is not required for COPI vesicle formation from Golgi-enriched membranes with purified coat proteins, even when membrane fractions were washed more stringently as semi-intact cells (Beck, Adolf, Weimer, Bruegger, & Wieland, 2009). Thus GBF1, which cycles between a cytosolic and membrane bound from, most likely is present on the membranes of preparations of semi-intact cells in amounts sufficient to catalyzed Arf1 activation and thereby COPI vesicle formation.

The use of semi-intact cells as donor membranes opens new opportunities to study various aspects in COPI vesicle formation and cargo sorting. Mammalian coatomer exists in various isoforms (Futatsumori, Kasai, Takatsu, Shin, & Nakayama, 2000; Wegmann, Hess, Baier, Wieland, & Reinhard, 2004) that all are available as recombinant protein complexes from insect cells. Differential localization of coatomer isoforms within the Golgi stack points to distinct functions similar to the various roles the different tetrameric clathrin adaptor protein complexes have in the late secretory and endocytic pathways (Moelleken et al., 2007; Wegmann et al., 2004). This possibility is in agreement with identification of different subpopulations of

COPI vesicles containing distinct sets of cargo proteins in living cells (Orci et al., 1997) and *in vitro* (Lanoix et al., 2001; Malsam, Satoh, Pelletier, & Warren, 2005). It is unknown whether COPI vesicles containing different sets of cargo proteins are formed by distinct (combinations of) coatomer isoforms, or whether homogeneously coated vesicles exist *in vivo*. As the intracellular membranes of semi-intact cells have an ordered organization more closely resembling the organization of the organelle membranes (e.g., the stacked organization of the mammalian Golgi apparatus) in living cells, they might well offer a significant advantage over Golgi-enriched membranes when applied to study roles of individual coatomer isoforms by comparative proteomics and lipidomics of isoformic COPI vesicles.

Coatomer promotes sorting of targeting proteins (e.g., SNARE proteins) and cycling transmembrane cargo adaptors (e.g., KDEL receptors, proteins of the p24 family and the ERGIC53 family) into COPI vesicles by direct interaction with cytoplasmic exposed domains. Soluble cargo proteins containing the sorting signal KDEL are dependent on the KDEL receptor for incorporation into COPI vesicles. More recently, it was reported that Golgi enzymes require Vps74p/GOLPH3 for proper localization in a COPI-dependent mechanism (Schmitz et al., 2008; Tu, Chen, & Banfield, 2012; Tu, Tai, Chen, & Banfield, 2008). Combination of siRNA-based knock-down with *in vitro* COPI budding assays in semi-intact cells has the potential to help identify or verify new transmembrane cargo adaptors or membrane associated cytosolic cargo adaptor proteins.

Last but not least, the semi-intact cell system does not only offer a high efficiency in vesicle formation but, as precise amounts of membranes defined by numbers of cells can be used to generate different vesicles in parallel, will also allow quantitative comparison of the biogenesis of various coated vesicles, including clathrin-coated vesicles.

Acknowledgments

We are grateful to Andrea Hellwig for her electron microscopic analysis conducted in the lab of Hilmar Bading (IZN, University of Heidelberg). This work was supported by the German Research Council SFB638 and the Graduiertenkolleg GRK1188.

References

Adolf, F., Herrmann, A., Hellwig, A., Beck, R., Brugger, B., & Wieland, F. T. (2013). Scission of COPI and COPII vesicles is independent of GTP hydrolysis. *Traffic*, *14*, 922–932.

Beck, R., Adolf, F., Weimer, C., Bruegger, B., & Wieland, F. T. (2009). ArfGAP1 activity and COPI vesicle biogenesis. *Traffic*, *10*, 307–315.

Beck, R., Prinz, S., Diestelkotter-Bachert, P., Rohling, S., Adolf, F., Hoehner, K., et al. (2011). Coatomer and dimeric ADP ribosylation factor 1 promote distinct steps in membrane scission. *The Journal of Cell Biology*, *194*, 765–777.

Beck, R., Sun, Z., Adolf, F., Rutz, C., Bassler, J., Wild, K., et al. (2008). Membrane curvature induced by Arf1-GTP is essential for vesicle formation. *Proceedings of the National Academy of Sciences of the United States of America*, *105*, 11731–11736.

Bigay, J., Casella, J. F., Drin, G., Mesmin, B., & Antonny, B. (2005). ArfGAP1 responds to membrane curvature through the folding of a lipid packing sensor motif. *The EMBO Journal*, *24*, 2244–2253.

Bigay, J., Gounon, P., Robineau, S., & Antonny, B. (2003). Lipid packing sensed by ArfGAP1 couples COPI coat disassembly to membrane bilayer curvature. *Nature*, *426*, 563–566.

Bradford, M. M. (1976). A rapid and sensitive method for the quantitation of microgram quantities of protein utilizing the principle of protein-dye binding. *Analytical Biochemistry*, *72*, 248–254.

Bremser, M., Nickel, W., Schweikert, M., Ravazzola, M., Amherdt, M., Hughes, C. A., et al. (1999). Coupling of coat assembly and vesicle budding to packaging of putative cargo receptors. *Cell*, *96*, 495–506.

Edeling, M. A., Smith, C., & Owen, D. (2006). Life of a clathrin coat: Insights from clathrin and AP structures. *Nature Reviews. Molecular Cell Biology*, *7*, 32–44.

Faini, M., Prinz, S., Beck, R., Schorb, M., Riches, J. D., Bacia, K., et al. (2012). The structures of COPI-coated vesicles reveal alternate coatomer conformations and interactions. *Science*, *336*, 1451–1454.

Faulstich, D., Auerbach, S., Orci, L., Ravazzola, M., Wegchingel, S., Lottspeich, F., et al. (1996). Architecture of coatomer: Molecular characterization of delta-COP and protein interactions within the complex. *The Journal of Cell Biology*, *135*, 53–61.

Franco, M., Chardin, P., Chabre, M., & Paris, S. (1995). Myristoylation of ADP-ribosylation factor 1 facilitates nucleotide exchange at physiological Mg2+ levels. *The Journal of Biological Chemistry*, *270*, 1337–1341.

Futatsumori, M., Kasai, K., Takatsu, H., Shin, H. W., & Nakayama, K. (2000). Identification and characterization of novel isoforms of COP I subunits. *Journal of Biochemistry*, *128*, 793–801.

Gommel, D., Orci, L., Emig, E. M., Hannah, M. J., Ravazzola, M., Nickel, W., et al. (1999). p24 and p23, the major transmembrane proteins of COPI-coated transport vesicles, form hetero-oligomeric complexes and cycle between the organelles of the early secretory pathway. *FEBS Letters*, *447*, 179–185.

Harter, C., Draken, E., Lottspeich, F., & Wieland, F. T. (1993). Yeast coatomer contains a subunit homologous to mammalian beta'-COP. *FEBS Letters*, *332*, 71–73.

Kawamoto, K., Yoshida, Y., Tamaki, H., Torii, S., Shinotsuka, C., Yamashina, S., et al. (2002). GBF1, a guanine nucleotide exchange factor for ADP-ribosylation factors, is localized to the cis-Golgi and involved in membrane association of the COPI coat. *Traffic*, *3*, 483–495.

Kim, J., Kleizen, B., Choy, R., Thinakaran, G., Sisodia, S. S., & Schekman, R. W. (2007). Biogenesis of gamma-secretase early in the secretory pathway. *The Journal of Cell Biology*, *179*, 951–963.

Lanoix, J., Ouwendijk, J., Stark, A., Szafer, E., Cassel, D., Dejgaard, K., et al. (2001). Sorting of Golgi resident proteins into different subpopulations of COPI vesicles: A role for ArfGAP1. *The Journal of Cell Biology*, *155*, 1199–1212.

Lee, M. C., Orci, L., Hamamoto, S., Futai, E., Ravazzola, M., & Schekman, R. (2005). Sar1p N-terminal helix initiates membrane curvature and completes the fission of a COPII vesicle. *Cell*, *122*, 605–617.

Malhotra, V., Serafini, T., Orci, L., Shepherd, J. C., & Rothman, J. E. (1989). Purification of a novel class of coated vesicles mediating biosynthetic protein transport through the Golgi stack. *Cell*, *58*, 329–336.

Malsam, J., Satoh, A., Pelletier, L., & Warren, G. (2005). Golgin tethers define subpopulations of COPI vesicles. *Science*, *307*, 1095–1098.

Mancias, J. D., & Goldberg, J. (2007). The transport signal on Sec22 for packaging into COPII-coated vesicles is a conformational epitope. *Molecular Cell*, *26*, 403–414.

Moelleken, J., Malsam, J., Betts, M. J., Movafeghi, A., Reckmann, I., Meissner, I., et al. (2007). Differential localization of coatomer complex isoforms within the Golgi apparatus. *Proceedings of the National Academy of Sciences of the United States of America*, *104*, 4425–4430.

Nickel, W., & Wieland, F. T. (2001). Receptor-dependent formation of COPI-coated vesicles from chemically defined donor liposomes. *Methods in Enzymology*, *329*, 388–404.

Orci, L., Glick, B. S., & Rothman, J. E. (1986). A new type of coated vesicular carrier that appears not to contain clathrin: Its possible role in protein transport within the Golgi stack. *Cell*, *46*, 171–184.

Orci, L., Stamnes, M., Ravazzola, M., Amherdt, M., Perrelet, A., Sollner, T. H., et al. (1997). Bidirectional transport by distinct populations of COPI-coated vesicles. *Cell*, *90*, 335–349.

Orcl, L., Palmer, D. J., Amherdt, M., & Rothman, J. E. (1993). Coated vesicle assembly in the Golgi requires only coatomer and ARF proteins from the cytosol. *Nature*, *364*, 732–734.

Pavel, J., Harter, C., & Wieland, F. T. (1998). Reversible dissociation of coatomer: Functional characterization of a beta/delta-coat protein subcomplex. *Proceedings of the National Academy of Sciences of the United States of America*, *95*, 2140–2145.

Popoff, V., Adolf, F., Brugger, B., & Wieland, F. (2011). COPI budding within the Golgi stack. *Cold Spring Harbor Perspectives in Biology*, *3*, a005231.

Sahlmuller, M. C., Strating, J. R., Beck, R., Eckert, P., Popoff, V., Haag, M., et al. (2011). Recombinant heptameric coatomer complexes: Novel tools to study isoform-specific functions. *Traffic*, *12*, 682–692.

Schmitz, K. R., Liu, J., Li, S., Setty, T. G., Wood, C. S., Burd, C. G., et al. (2008). Golgi localization of glycosyltransferases requires a Vps74p oligomer. *Developmental Cell*, *14*, 523–534.

Serafini, T., Orci, L., Amherdt, M., Brunner, M., Kahn, R. A., & Rothman, J. E. (1991). ADP-ribosylation factor is a subunit of the coat of Golgi-derived COP-coated vesicles: A novel role for a GTP-binding protein. *Cell*, *67*, 239–253.

Sonnichsen, B., Watson, R., Clausen, H., Misteli, T., & Warren, G. (1996). Sorting by COP I-coated vesicles under interphase and mitotic conditions. *The Journal of Cell Biology*, *134*, 1411–1425.

Spang, A., Herrmann, J. M., Hamamoto, S., & Schekman, R. (2001). The ADP ribosylation factor-nucleotide exchange factors Gea1p and Gea2p have overlapping, but not redundant functions in retrograde transport from the Golgi to the endoplasmic reticulum. *Molecular Biology of the Cell*, *12*, 1035–1045.

Spang, A., Matsuoka, K., Hamamoto, S., Schekman, R., & Orci, L. (1998). Coatomer, Arf1p, and nucleotide are required to bud coat protein complex I-coated vesicles from large synthetic liposomes. *Proceedings of the National Academy of Sciences of the United States of America*, *95*, 11199–11204.

Szul, T., & Sztul, E. (2011). COPII and COPI traffic at the ER-Golgi interface. *Physiology (Bethesda, MD)*, *26*, 348–364.

Tu, L., Chen, L., & Banfield, D. K. (2012). A conserved N-terminal arginine-motif in GOLPH3-family proteins mediates binding to coatomer. *Traffic*, *13*, 1496–1507.

Tu, L., Tai, W. C., Chen, L., & Banfield, D. K. (2008). Signal-mediated dynamic retention of glycosyltransferases in the Golgi. *Science*, *321*, 404–407.

Waters, M. G., Serafini, T., & Rothman, J. E. (1991). 'Coatomer': A cytosolic protein complex containing subunits of non-clathrin-coated Golgi transport vesicles. *Nature*, *349*, 248–251.

Wegmann, D., Hess, P., Baier, C., Wieland, F. T., & Reinhard, C. (2004). Novel isotypic gamma/zeta subunits reveal three coatomer complexes in mammals. *Molecular and Cellular Biology*, *24*, 1070–1080.

Zhao, X., Lasell, T. K., & Melancon, P. (2002). Localization of large ADP-ribosylation factor-guanine nucleotide exchange factors to different Golgi compartments: Evidence for distinct functions in protein traffic. *Molecular Biology of the Cell*, *13*, 119–133.

CHAPTER

RNA Interference Approaches to Examine Golgi Function in Animal Cell Culture

2

Victoria J. Miller, Caroline M. McKinnon, Harry Mellor and David J. Stephens
Cell Biology Laboratories, School of Biochemistry, University of Bristol, Bristol, United Kingdom

CHAPTER OUTLINE

Overview of Gene Depletion Methodologies **16**
2.1 Key Considerations for the Analysis of Organelle Function **17**
2.2 Strategies for RNAi Experiments **18**
2.3 Choosing Between siRNA and shRNA **18**
2.4 Sources of Reagents for RNAi **19**
2.5 Controls **19**
2.6 Methods **20**
2.6.1 Gene Depletion Using siRNA 20
2.6.1.1 Calcium Phosphate-mediated Delivery 20
2.6.1.2 Lipid-based Transfection 21
2.6.2 Gene Depletion Using shRNA 22
2.6.2.1 Lentivirus Production 22
2.6.2.2 Stable Cell Line Production 23
2.6.3 Validation of Suppression 24
2.6.3.1 Immunoblotting 24
2.6.3.2 Immunofluorescence 26
2.6.3.3 Quantitative PCR 27
2.6.4 Rescue Experiments 30
2.7 Discussion **31**
Summary **32**
Acknowledgments **32**
References **32**

Methods in Cell Biology, Volume 118

ISSN 0091-679X
http://dx.doi.org/10.1016/B978-0-12-417164-0.00002-1

Abstract

The ability to deplete specific proteins from cells has transformed cell biology. Targeting of gene transcripts using RNA interference has allowed for a highly refined approach to the analysis of gene function that has been applied to all aspects of cell biology. Developments of the technology have reached a point where it is now a relatively trivial task to assess the role of an individual protein in a particular cell function. RNAi also allows for genome-wide screening as a discovery step toward the identification of new components of cellular pathways and machines. The technique has been applied extensively to the analysis of Golgi complex function, leading to significant insight into the biology of this complex organelle. Here, we describe the commonly used options for targeting individual genes for both transient and stable knockdown. We consider the alternative methods for introducing these reagents into cells and outline methods that we and others have used widely for validation of specificity and efficacy of gene targeting.

OVERVIEW OF GENE DEPLETION METHODOLOGIES

The molecular mechanisms by which small RNA molecules regulate gene expression were first defined in *Caenorhabditis elegans* by Andrew Fire and Craig Mello for which they shared the 2006 Nobel Prize for Physiology or Medicine. The application of this technology to experimental work exploits the ability of short-interfering RNAs to target individual transcripts. The application to mammalian cells was described first by Elbashir et al. (2001). We do not intend to provide an extensive history of the methods nor a discussion of microRNAs and other noncoding RNAs that can be used for selective depletion of gene products. Instead, we provide methodologies and examples that relate more specifically to the analysis of Golgi function.

There are several ways in which one can achieve specific gene depletion. These include the use of short-interfering RNAs, which feed directly into the RNAi pathway; the use of longer RNA templates that are cleaved by Dicer to effect gene depletion; and the use of short hairpin RNAs (shRNAs), which provide a similar template. These three tools each have their advantages. It has been described that longer RNA templates induce an interferon response in many cell lines and so short-interfering RNAs are often preferred. Conceptually, one might also be concerned about a great array of off-target effects using a longer template. Long RNAs and short hairpins have the advantage of enabling expression from vector and viral systems which open up the possibility of stable and inducible expression.

Perhaps the biggest challenges in using these tools are ensuring efficient and effective delivery into the cells of choice and validating that one does indeed obtain specific gene depletion and that the experiment does not suffer from "off-target" effects. We describe these approaches below and provide some context based on our own experience.

2.1 KEY CONSIDERATIONS FOR THE ANALYSIS OF ORGANELLE FUNCTION

When designing an experiment to probe the mechanisms that define organelle structure or function, there are some fundamental points that one must consider. First, in metazoans, especially in mammals, there is often some overlap of function between closely related proteins. This is often considered to be "redundancy" but typically a specific function can be assigned to each gene product involved. If one looks, for example, at the process of transport from the ER to the Golgi, there are multiple isoforms of nearly all subunits that comprise the COPII (Coat Protomer II) complex that mediates this step of the secretory pathway. Often these redundant functions might not be uncovered using RNAi in cultured cells but animal models frequently highlight such distinctions. One recent example would be the four human Sec24 isoforms (Tang, Kausalya, Low, Lock, & Hong, 1999) that comprise the primary cargo-selective component of the COPII coat (Zanctti, Pahuja, Studer, Shim, & Schekman, 2011). *In vitro* (i.e., cultured cells) individual depletion of these subunits fails to define any phenotype in a widely used assay of membrane traffic (Simpson et al., 2012). Indications of specific functions were suggested by RNAi experiments in the parasite *Trypanosoma brucei* (Demmel et al., 2011). Recently, a mouse knockout of Sec24A defined both a selective role in packaging of PCSK9 and other functions that were redundant with Sec24B (Chen et al., 2013). In this case, if one were to wish to perturb Sec24 function in cultured cells, one might need to deplete multiple isoforms, possibly all four. Solutions to such issues can be found through targeting subunits of a multipartite complex for which there is only one mammalian form. We applied this approach ourselves to the COPII complex in our work targeting the Sec13 subunit (Townley et al., 2008).

Similar considerations are required with regard to the products of a single gene: in this case splice forms. Many trafficking components are defined by multiple splice forms within the same cell type. As with the issue of isoforms, this requires significant data mining to define RNAi sequences that will target all isoforms (if desired). In some cases, this sequence specificity also allows for the development of reagents selective for only one particular isoform; this of course would allow more refined analysis of splice form function.

Given our extensive knowledge of the core machinery of trafficking systems, much of which comes from yeast genetics, one might also need to consider the possibility of overlapping function of closely related proteins from the same functional family (where they might not be isoforms *per se*). Examples would include members of the Rab protein family where one might expect overlapping roles between closely related isoforms. An essential component of any project seeking to define specific gene function is therefore to characterize the cell type to be used in terms of the cohort of isoforms and splice forms as well as related family members that are expressed. Fortunately, the advent of whole genome sequencing projects makes this much simpler than before. Classical RT-PCR approaches can be used to supplement knowledge from available datasets.

There has been extensive use of RNAi-based screens for high-content screening of membrane trafficking pathways including of traffic to and through the Golgi. Examples include analysis of Golgi morphology, transport to and through the Golgi (Bard et al., 2006; Simpson et al., 2012), and the kinases and phosphatases that control Golgi function (Chia et al., 2012; Farhan et al., 2010). Such approaches have identified molecules of key importance as well as identifying additional regulatory control points for Golgi function. We do not deal with such screens here and instead focus on analysis of individual gene function in mammalian cells.

2.2 STRATEGIES FOR RNAi EXPERIMENTS

Entire volumes have been written regarding the design of sequences for siRNA, and the reader is referred to other recent texts for an in-depth discussion (Petri & Meister, 2013). Many companies have invested heavily in siRNA design tools, and some now also offer prevalidated sequences where the efficacy has been tested using quantitative PCR (qPCR). In most cell biology labs, these custom algorithms or predesigned sequences are routinely used.

Two approaches are commonly used in gene depletion experiments. Either one targets a gene using multiple individual duplexes independently or one pools multiple duplexes to form pools. The latter approach has been used for both siRNAs and shRNAs. It is often argued that this has the advantage of one being able to use these pools at lower overall concentrations and that as a result the chances of off-target effects are reduced. However, such pools are also less "controllable" with one being less able to define off-target effects and "rescue" experiments (which we will describe later) being more complicated. Screening approaches are often easier using pools of reagents and such experiments often involve multiplexing not only duplexes to target a single gene but also multiple targets to reduce the volume of screening required.

2.3 CHOOSING BETWEEN siRNA AND shRNA

siRNA reagents are cheap and highly effective. Delivery methods are simple, widely used, and require little specialized equipment. As such, for transient depletion, siRNAs are the method of preference. shRNAs provide additional refinements. Since they are plasmid based, they facilitate stable suppression of gene expression. The plasmid backbone also allows incorporation of additional control elements, notably for inducible expression, for example, using tetracycline. The plasmid backbones can also be modified to allow generation of lentiviral delivery systems for these shRNAs, enhancing the efficiency of delivery and facilitating stable expression. We discuss these options in some more detail below.

Fundamentally, one must consider the goal of a gene depletion experiment. In many cases, transient suppression using a set of single siRNAs will prove most useful; in others, a small-scale screen involving pools of shRNAs will be required. These decisions are entirely dependent on the experiment to be undertaken and little further advice can be given.

2.4 SOURCES OF REAGENTS FOR RNAi

In the early days of RNAi, reagents themselves were often prohibitively expensive. Innovative solutions were developed to overcome this including *in vitro* synthesis of siRNA duplexes from plasmid DNA synthesis using T7 RNA polymerase (Donze & Picard, 2002). Costs have reduced significantly to the point where ordering from one of the many suppliers is a simple and cost-effective solution. Similarly, extensive work has been done on chemical modification of synthesized siRNAs, which lead to increase stability and activity while reducing toxicity (Bramsen et al., 2009). Many such modifications are vendor specific, which necessitate individual exploration of the benefits of such modifications.

Similarly, while construction of custom shRNA is straightforward, predesigned vectors can be purchased from many suppliers at reasonable cost, notably when multiple reagents are ordered together. Much of this has been driven by the work of the RNAi Consortium which has developed a series of genome-wide shRNA libraries that are in many cases prevalidated and are widely available (Root, Hacohen, Hahn, Lander, & Sabatini, 2006).

2.5 CONTROLS

Choice of controls is a complex task in siRNA experiments. Conflicting opinions support the options of using nontargeting sequences (that will not cause depletion of any cellular target) versus specific targeting of a protein not involved in the process of interest. Two options for the former are the use of a completely irrelevant siRNA sequence such as that targeting a gene sequence from another species. An siRNA duplex termed GL2 is widely used; this targets firefly (*Photinus pyralis*) luciferase and was originally used by Tuschl and colleagues in their proof-of-principle paper (Elbashir et al., 2001). In other early work, lamin A/C was described as a nonessential gene in cell culture (Harborth, Elbashir, Bechert, Tuschl, & Weber, 2001). This led to its widespread use as a control in siRNA experiments. A Golgi-related caveat to its use is that Gundersen and colleagues have, in fact, shown that it plays a key role in nuclear migration during directed cell migration in response to scratch wounding (Folker, Ostlund, Luxton, Worman, & Gundersen, 2011). Polarization of the Golgi toward the wound edge is a common marker in these assays which could preclude the use of lamin A/C in Golgi-based assays. The use of a targeting control has the advantage that it will engage the full RNAi machinery and therefore has the potential to act as a more comprehensive control in these experiments.

A further common target is glyceraldehyde-3-phosphate dehydrogenase owing to its ubiquitous expression. However, its involvement in processes from ER-to-Golgi transport (Tisdale, Kelly, & Artalejo, 2004) to fast axonal transport (Zala et al., 2013) raises a note of caution in its use as well.

One solution to this is to ensure that phenotypes are validated against both targeting and nontargeting controls.

2.6 METHODS

2.6.1 Gene depletion using siRNA

Three commonly applied methods for siRNA delivery are described below. In our hands, the modified calcium phosphate method remains the protocol of choice. We find in many cell lines including HeLa, RPE1, MRC-5, and LLC-PK1 that this method leads to high efficiency of transfection (the proportion of cells that receive the siRNA) and also good uniformity across a culture. We have previously validated this method using fluorescent siRNAs and immunofluorescence labeling of the depleted gene (in this case the microtubule plus end binding protein EB1) to show 98.8% efficiency in terms of transfection and 98.0% in terms of effective depletion (Watson & Stephens, 2006). In contrast, lipid-mediated transfection can work better on some cell lines but tends to lead to regions of the culture that are very effectively depleted of the target gene and large areas that are not. This "clumping" can be useful in terms of defining phenotypes as the nonsuppressed cells effectively act as an internal control. A caveat of course is that it is harder to know that one is looking at an effectively suppressed cell unless one can and does label for the target gene or has a robust quantifiable phenotype. Reverse transfection methods are very popular for screening applications owing to the uniformity and ability to prepare multi-well plates in advance. We have also had some success with cell types that are harder to transiently transfect using the Neon electroporation system (Life Technologies). This has enabled efficient knockdown of genes in primary fibroblasts to a much greater degree than other methods we have used.

2.6.1.1 Calcium phosphate-mediated delivery

2.6.1.1.1 Materials

1. 0.25 M $CaCl_2$, filtered-sterilized through a 0.2 μm filter
2. 2 × BBS buffer (50 mM BES, pH 6.85, 280 mM NaCl, 1.5 mM Na_2HPO_4), filtered-sterilized through a 0.2 μm filter

2.6.1.1.2 Method

The protocol is adapted from the method of Chen and Okayama (1988), which gives very high efficiency of transfection compared to traditional calcium phosphate methods. The protocol relies on slow formation of the calcium phosphate particles as the pH of the cell media changes at the lower CO_2 level. The day prior to transfection, the cells should be plated at 1×10^5 cells/ml. The aim is to have the cells at approximately 80% confluence on the day of transfection. The transfection should be set up in the next afternoon as follows (volumes given are for a 10 cm dish). Mix 1000 pmol of siRNA with 500 μl of 0.25 M $CaCl_2$, then add 500 μl of filter-sterilized $2\times$ BBS. Allow the complexes to form for 20 min at room temperature, and then add the mixture dropwise to the plate, with gentle swirling. Place the cells in a 37 °C incubator humidified with 3% CO_2 overnight. The following morning, it will be possible to see fine grains of calcium phosphate around the cells and a slick of calcium phosphate on the surface of the media. Remove the medium by washing once with phosphate-buffered saline (PBS) and replacing with fresh media and returning the cells to a 37 °C, 5% CO_2 incubator for 1 h. After 1 h, wash the cells again with PBS, replace with fresh media, and incubate the cells overnight at 37 °C, 5% CO_2. The cells can be harvested the following day although longer incubation time may be necessary depending on the target protein to be depleted.

Success depends on accurate concentrations of calcium and phosphate in the two buffers and on the CO_2 level in the incubator. It is worth calibrating the incubator CO_2 level regularly. The protocol is optimized for cells grown at 5% CO_2 in the appropriate bicarbonate-buffered media. HEPES-buffered media is not recommended as the protocol relies on a pH change that HEPES will prevent.

2.6.1.2 Lipid-based transfection

2.6.1.2.1 Materials

1. Lipofectamine® 2000 (Life Technologies) or similar transfection reagent
2. Opti-MEM® (Life Technologies) reduced serum media

2.6.1.2.2 Method

This method has been adapted from that provided by Life Technologies for Lipofectamine® 2000.

Lipid-based transfection incubates nucleic acids with cationic lipids to facilitate entry into the cell. Positively charged liposomes form and are thought to enter the cell by endocytosis. A wide range of transfection reagents are in common use; however, not all are recommended for transfection of siRNA. The transfection protocol is similar to the calcium phosphate method, but the materials required are commercially available and a 3% CO_2 incubator is not required. For each reagent, the manufacturer's own instructions should be followed.

Plate cells the day prior to transfection to be at approximately 70% confluence the following day. Volumes given are for one 35 mm diameter dish. Take two tubes and in each place 100 μl Opti-MEM®. To the first tube add 75 pmol siRNA. To the second tube, add 7.5 μl Lipofectamine® 2000. Add the contents of tube 1 to tube 2, mix gently then incubate for 5 min at room temperature. Add the RNA-reagent mixture dropwise to the cells, gently rocking the plate from side to side to mix—efficient liposome formation is promoted by surface impact with the media, so this step must not be omitted. Incubate the cells at 37 °C, 5% CO_2 overnight and replace the media the next morning. This incubation can be shortened to as little as 4 h if required to reduce cytotoxicity. Knockdown efficiency should be assessed 24–72 h after transfection (see later protocols), allowing time for protein levels to decrease.

Transfection efficiency varies with cell type and may be increased by optimizing cell confluence at the time of transfection, the amount of nucleic acid, and/or transfection reagent used or by using an alternative transfection reagent. Co-transfection of plasmid DNA and siRNA is possible using lipid-based transfection (e.g., for rescue experiments). Consideration might also be given to the use of reverse transfection where lipid–nucleotide mixtures are first spotted on substrates prior to the addition of cells (Ziauddin & Sabatini, 2001). This can have advantages for some cell types and is widely used in high-throughput experiments (Erfle et al., 2007).

2.6.2 Gene depletion using shRNA

Previous protocols have described siRNA transfection methods. These create a transient reduction in protein level, the extent of which will vary across the population of cells. The creation of stable cell lines using shRNAs overcomes these difficulties.

The RNAi Consortium (TRC), based at the Broad Institute, is a private–public consortium founded to generate libraries of shRNAs for scientists to use. Over 80,000 constructs are now available to target over 20,000 human and mouse genes, with more sequences constantly being generated. Constructs consist of a 21-base base stem and a 6-base loop in the pLKO.1 vector. Sigma-Aldrich distributes the TRC library in a variety of formats including bacterial glycerol stocks, plasmid DNA, and lentivirus particles. Other sources for these or similar collection are also available (see, e.g., Thermo/OpenBiosystem). As these sequences have not yet been validated for effective knockdown, it is best to order a clone set of three to five constructs and experimentally determine the most effective for the gene of interest. A negative control should also be included (suitable constructs are available from Sigma-Aldrich). A wealth of further information can be found at http://www.broadinstitute.org/rnai/public/.

2.6.2.1 Lentivirus production

shRNAs are commonly introduced into cells using lentivirus infection. The shRNA plasmid of interest is packaged into lentiviruses and released into the cell culture medium. This is used to infect the desired cell line. Stable clones form by random integration into the genome and are then isolated using antibiotic selection.

As lentiviruses infect human cells, appropriate safety procedures *must* be followed. Typically, these involve working in a designated tissue-culture hood, use of a suitable viricidal disinfectant (Virkon or a similar bleach), use of filtered tips and serological pipettes, and separate collection and disposal of all liquid waste and plastic ware. The biosafety classification will vary depending on country and host institution. Consult your local biological safety officer to ensure you are compliant with all local and national guidelines.

2.6.2.1.1 Materials

1. HEK-293 T cell line
2. Lentiviral packaging system such as Lenti-X (Clontech)
3. Mission shRNA of interest (Sigma-Aldrich)

2.6.2.1.2 Method

Seed 4×10^6 HEK-293 T cells in a 10 cm dish for each lentivirus to be generated, to be 80% confluent the next day. Transfect the cells using the provided transfection reagent to cotransfect cells with the Mission shRNA plasmid and plasmids expressing the viral packaging proteins. These packaging genes are provided on separate plasmids to prevent spontaneous lentivirus formation. In the Lenti-X system, expression of the viral proteins is under control of a tetracycline-dependent promoter, so tetracycline-free fetal bovine serum should be used in the culture medium to ensure high expression. Incubate the cells at 37 °C, 5% CO_2 overnight and replace the media the next morning. The virus will be released into the culture media. Collect media after 24–48 h. We do not find it necessary to filter or centrifuge the samples to remove cell debris. One milliliter aliquot can be stored at −80 °C for future use; however, freeze–thaw cycles will significantly impair virus titers.

2.6.2.2 Stable cell line production

2.6.2.2.1 Materials

1. Cell line of interest (e.g., MRC 5 cells)
2. Lentiviral supernatant (see above)
3. Puromycin
4. *Optional* polybrene (30 mg/ml stock). Caution, polybrene is toxic to some cell lines

2.6.2.2.2 Method

To infect the desired cell line with lentivirus, seed cells in a 10 cm dish to be 70% confluent the following day. Add fresh media and add 1 ml of the virus-containing media. Incubate the cells at 37 °C, 5% CO_2 overnight and replace the media the next morning. Polybrene may be added to media at 8 μg/ml prior to addition of virus to increase infection as its positive charge neutralizes repulsion between virus particles and the cell surface.

Wait at least 24 h before adding puromycin or other antibiotics. For each cell line used, you must first establish the appropriate concentration of antibiotic to use. A useful list of conditions used for previous stable cell line construction to act as a starting point can be found at http://www.sigmaaldrich.com/life-science/functional-genomics-and-rnai/shrna/learning-center/lentivirus-cell-line.html. Test a range and use the lowest concentration that kills all nontransfected cells within 24 h. We use 3 μg/ml puromycin for MRC 5 cells. Change to fresh media with puromycin every 2–3 days. After a few days, widespread cell death will occur, leaving a few surviving stable colonies. Continue to change media while waiting for clones to form colonies large enough to isolate by transferring into individual wells using a 200 μl pipette tip (do not use trypsin at this stage). Protein knockdown can then be assayed after clone amplification. Include a low concentration of puromycin at all times while passaging stable cell lines to maintain the integration.

2.6.3 Validation of suppression

The choice of technique here largely depends on the availability of reagents. Immunoblotting and immunofluorescence are of course only possible in experiments where suitable antibodies are available. qPCR provides an alternative where no suitable antibody is available.

2.6.3.1 Immunoblotting

A simple method for defining the level of suppression of gene expression across an entire population of cells is immunoblotting. Here, one assesses the level of depletion of an entire population. This can give a clear indication of both transfection efficiency and efficacy of depletion. The caveat here is of course that one loses the ability to correlate suppression with phenotype on a cell-by-cell level. Modern approaches of immunoblotting that employ quantitative imaging techniques also enable one to accurately quantify suppression. Proteins are first separated by polyacrylamide gel electrophoresis, then transferred or "blotted" to membranes where they can be probed with specific antibodies to report levels. Here Odyssey imaging with fluorescent secondary antibodies is used to allow quantitative analysis. Such analysis is not possible with the traditional enhanced chemiluminescence-film technique, which is only semiquantitative due to the nonlinear nature of film.

2.6.3.1.1 Materials

1. Lysis buffer (50 mM Tris–HCl, pH 7.4, 150 mM $NaCl_2$, 1% Triton X-100)
2. Cell scrapers with pivoted blades (#3010, Corning)
3. BCA protein assay kit (Pierce)
4. Sample buffer (NuPAGE LDS Sample buffer with reducing agent)
5. Protein ladder
6. Acrylamide gel and equipment (gel, tank, running buffer)
7. Western blotting transfer apparatus
8. PBS with 0.05% Tween-20
9. Primary antibodies against protein of interest and tubulin or other loading control
10. IRDye® 680 and 800 fluorescent secondary antibodies (Licor)
11. Odyssey imager

2.6.3.1.2 Method

Allow two 35 mm wells of cells for each sample. Following knockdown as above, wash cells in PBS and then place dishes on ice. Add 200 μl of chilled lysis buffer to first well and incubate for 2 min. Scrape cells with scraper to recover lysed cells from the dish, add buffer to second well and repeat. Transfer to a microfuge tube. Centrifuge for 10 min at 10,000 × *g* at 4 °C to pellet cell debris. Recover supernatant to a separate tube. Retain some of the supernatant to perform a BCA comparing to known standards prepared to determine protein concentration; mix the rest with sample buffer with reducing agent and heat at 95 °C for 5 min or 80 °C for 15 min. Samples can be stored at −20 °C.

Load approximately 30 μg of protein on a standard acrylamide gel. Equalize loading volumes between samples with loading buffer. The appropriate acrylamide percentage of the gel will depend on the molecular weight of the protein of interest. In one lane load 5 μl protein ladder; we recommend using a pre-stained ladder so that protein position can be readily tracked while running the gel and protein transfer can be easily confirmed after Western blotting. Run the gel according to manufacturer's instructions until the dye front has reached the end of the gel and run off. We use NuPAGE® gels and associated equipment for both gel electrophoresis and Western blotting.

Remove gel from the running tank and prepare a "sandwich" of sponge pads, Whatman paper, gel, and membrane. Pads and Whatman paper should be presoaked in transfer buffer. Either nitrocellulose or PVDF membranes can be used; PVDF membrane must be wetted briefly with methanol prior to use. Care should be taken when assembling the transfer sandwich to avoid trapping air bubbles which will impede transfer—rolling over the surface with a serological pipette using firm pressure will remove any bubbles that are present. Do not touch the membrane with bare hands, to prevent contamination with proteins from the skin and in general only handle the edges. Blunt-ended tweezers are ideal for handling membranes.

Set transfer device to blot (e.g., for 1 h at 30 V when using the NuPAGE® transfer system). Both wet and semidry transfer machines are available and give very similar results. If working with very large proteins (200 kDa and over), blotting time can be extended to improve transfer, although at the risk that smaller proteins may pass through the membrane. Transfer can be confirmed by Ponceau staining of the membrane if desired.

Block remaining exposed binding sites on the membrane using Odyssey blocking buffer or 5% nonfat dried milk in PBS. Carry out all incubations for 1 h at room temperature on a shaker. Then incubate with primary antibodies diluted at appropriate concentration in blocking buffer. Wash 3 × 10 min with PBS/0.05% Tween-20 to remove excess. Incubate with fluorescent secondary antibodies diluted 1:1000 in blocking buffer. Protect from light during this incubation using foil or similar. Wash 3 × 10 min with PBS/0.05% Tween-20 to remove excess. Rinse with PBS alone. Place membrane (can be wet or dry) on to imaging platform and scan at desired wavelengths. Adjust scanning intensity if necessary. Save images and analyze as desired.

It is advisable to probe for a second protein as a loading control (frequently tubulin, as shown in Fig. 2.1), to show that any reduction in the protein of interest is a consequence of the RNAi. Imaging with the Odyssey system has the advantage that a single blot can be double-probed, even if the proteins are of similar sizes, as the fluorescent secondary antibodies used are individually excited and can be functionally separated when imaging. When using conventional ECL, it is necessary to either cut the membrane or run duplicate blots.

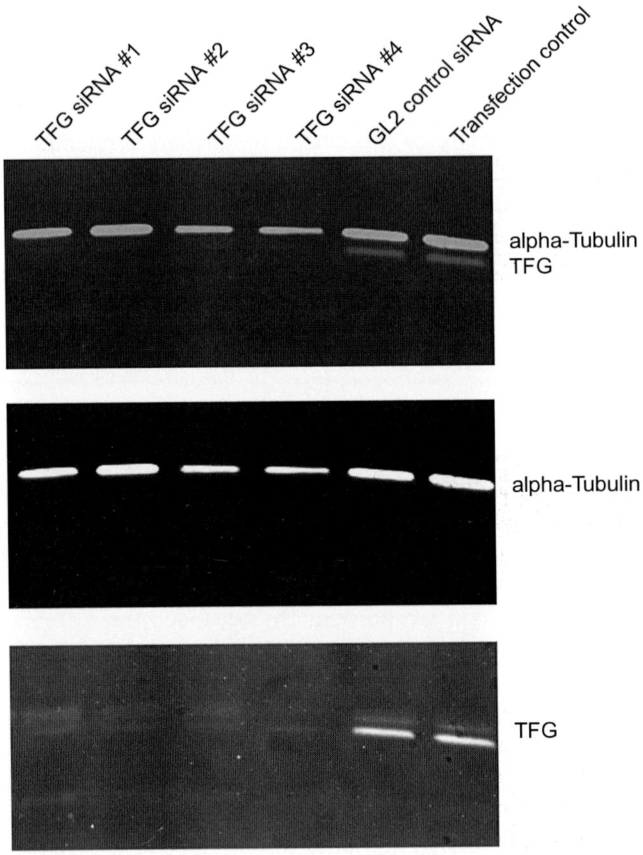

FIGURE 2.1

Confirmation of RNAi knockdown by immunoblotting using the Odyssey system. Western blot showing expression levels of endoplasmic reticulum protein TFG1 involved in ER-Golgi trafficking. RPE1 cells were transfected with siRNAs (#TFG1–4) against TFG or GL2 as a negative control (see text for discussion of negative control selection) using calcium phosphate-mediated delivery. Extract from nontransfected cells is also included. Samples were prepared as described in the text and 38 μg protein loaded for each sample. The blot was probed with α-TFG (Imgenex)/800 IRDye secondary (red) and α-tubulin (Sigma)/680 IRDye secondary (green). Clear depletion of TFG can be seen with all four siRNAs tested, while no reduction is seen with the GL2 siRNA when compared to the nontransfected cells. (See color plate.)

2.6.3.2 Immunofluorescence

In the case of the Golgi, one can preserve morphology using methanol fixation. Since siRNA-mediated suppression is rarely 100% effective, one can use nontransfected cells as an internal control. Such an approach can also be integrated into an experimental workflow to allow one to correlate the efficacy of gene suppression with the penetrance of a particular phenotype. Such quantitative approaches are often highly informative in the context of mechanism.

2.6.3.2.1 Materials

1. Methanol (stored at −20 °C)
2. Phosphate buffered saline (PBS)
3. Bovine serum albumin (BSA)
4. Antigen-specific primary antibodies
5. Species-specific secondary antibodies (Jackson Immunoresearch)
6. ProLong® Gold Antifade Reagent (Invitrogen)
7. Fluorescence microscope with CCD camera

2.6.3.2.2 Method

Note that fixation with methanol permeabilizes the cells and there is no need to add additional detergent. Note also that some proteins cannot be efficiently detected after methanol fixation and that paraformaldehyde fixation followed by permeabilization may be preferable in this case.

Aspirate medium from cells that have been depleted using RNAi and fix with cold methanol for a maximum of 4 min at −20 °C (cells should be placed in the freezer during this time). Aspirate the methanol and replace immediately with PBS. Block nonspecific binding by incubating cells for 30 min in PBS containing 3% BSA (PBS-BSA). This step is not always necessary but we find that it aids reproducibility. Incubate the sample with primary antibody at a suitable dilution in PBS-BSA for 1 h. Wash the cells carefully by aspiration (3 times, 5 min each wash). Incubate cells for 1 h with the relevant species-specific secondary antibody diluted to an appropriate concentration in PBS-BSA. Repeat washes (3 times, 5 min each). Include one final wash and mount the sample using ProLong® Gold Antifade Reagent or equivalent. After curing of the mounting medium (typically overnight), image cells using appropriate fluorescence filters and ensure image acquisition for controls is within the linear range without saturation. ImageJ (we use the Fiji implementation at www.fiji.sc) or equivalent software can then be used to measure fluorescence intensities within a region of interest (e.g., the Golgi apparatus) to quantitate the level of target suppression. Typically, we would repeat this process on three independent occasions to generate a histogram and use a statistical test to compare samples as appropriate (Fig. 2.2).

2.6.3.3 Quantitative PCR

One can readily assess the efficacy of RNAi by monitoring the levels of messenger RNA remaining using qPCR. An important caveat is that the levels of messenger RNA do not always correlate with levels of protein. This is particularly true for many trafficking machinery components that have long half-lives and therefore turn over with relatively slow kinetics.

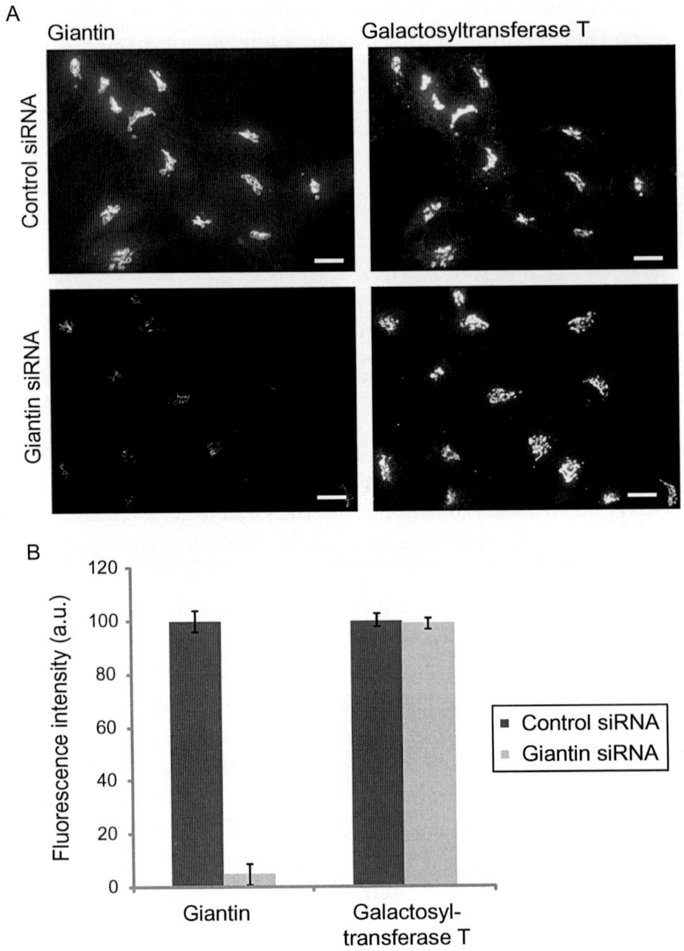

FIGURE 2.2

Validation of RNAi knockdown by immunofluorescence. (A) Immunofluorescence of cells labeled for the Golgi proteins giantin and galactosyltransferase T as indicated. Cells were transfected with either control (GL2) siRNA duplexes or siRNA to specifically deplete giantin. Here, the use of an additional Golgi marker allows one to evaluate the level of suppression at an organelle level as well as across the entire cell. Bar = 10 μm. (B) Quantitation of giantin and galactosyltransferase T fluorescence at the Golgi in control and giantin-depleted cells.

2.6.3.3.1 Materials

1. TRIzol reagent (Life Technologies, #15596-026)
2. Chloroform
3. Isopropanol
4. Ethanol

5. RNAse-free water (e.g., Life Technologies, #AM9939)
6. A UV spectrophotometer with appropriate cuvette
7. Omniscript RT reverse transcription kit (Qiagen, #205110)
8. Oligonucleotide primers for qPCR
9. SYBR green qPCR mastermix (Thermo, #F-410L)
10. Random hexamer mix (Life Technologies, #N8080127)
11. RNAsin (Promega, #N2511)

2.6.3.3.2 Protocol

2.6.3.3.2.1 RNA isolation. Add 1 ml TRIzol per well of a 6-well plate and leave to stand at room temperature for 5 min. Pass through a 19G needle three times and transfer to a 1.5 ml microfuge tube. Add 0.2 ml chloroform, shake vigorously for 15 s, and then allow to stand for 2–15 min at room temperature. Centrifuge at 12,000 × *g* for 15 min at 4 °C. The samples separate into three phases: the red, lower, organic phase contains protein; the interphase contains DNA; and the colorless, upper, aqueous phase contains the RNA. Transfer the aqueous phase to a fresh 1.5 ml microfuge tube. Add 0.5 ml isopropanol; invert to mix then allow to stand for 5–10 min at room temperature. Centrifuge at 12,000 × *g* for 10 min at 4 °C. The RNA precipitate forms a pellet on the side and bottom of the tube. Remove supernatant and wash the pellet by adding 1 ml 75% ethanol. Vortex sample and centrifuge at 7500 × *g* for 5 min at 4 °C. Briefly dry the RNA pellet for 5–10 min by air-drying or under vacuum. Add an appropriate volume of water or 0.5% SDS solution to the RNA pellet (typically 40 μl RNAse-free water). Mix by pipetting, at 55–60 °C, for 10–15 min. Quantify the RNA by reading the absorbance at 260 and 280 nm. The A_{260} of 40 μg/ml RNA is 1. The A_{260}/A_{280} ratio of pure RNA should be 1.8–2.0 and lower readings indicate contamination with protein. Purified RNA can be stored at −20 °C for a year.

2.6.3.3.2.2 Reverse transcription. Perform all preparatory steps of the reverse transcription reaction on ice using the Omniscript RT kit. The reaction uses 2–4 μg of the isolated RNA. Per reaction, add the following reagents to a chilled 0.5 ml microfuge tube:

RT buffer	3 μl
dNTPs	3 μl
RNasin (10 units)	1.5 μl
Random hexamers	3 μl of 50 μM stock
RNA	2–4 μg
RT enzyme	1 μl
RNase-free water	Up to 30 μl

Mix contents with a quick vortex and pulse centrifuge then incubate in a water bath at 37 °C for 1 h.

2.6.3.3.2.3 Quantitative real-time PCR. Typical experiments use duplicate or replicate samples from each condition. We routinely include nontransfected control samples and samples from cells transfected with a nontargeting or control siRNA. The qPCR protocol compares the amplification of the target gene to an internal control, performed in parallel samples. We typically use RNA polymerase II as the control gene. Oligonucleotide primers are required for the amplification. These can be designed using the free online Primer 3 tool (http://primer3.wi.mit.edu/). For amplification of RNA polymerase II, we use a 5′-GCACCACGTCCAATGACAT-3′ forward and 5′-GTGCGGCTGCTTCCATAA-3′ reverse primers.

Per well, make up the following on ice:

SYBR green Mastermix	10 μl
Forward primer	0.5 μl of 50 μM stock
Reverse primer	0.5 μl of 50 μM stock
Nuclease-free water	7.5 μl

Mix with a quick vortex and pulse centrifuge, then put 18.5 μl of the above into a well and add 1.5 μl of cDNA. Centrifuge at 1000 rpm for 30 s at 4 °C, then place into the qPCR machine. A standard amplification is as follows: 35 cycles of 95 °C, 30 s; 60 °C, 30 s; 72 °C, 30 s.

We usually perform melting curve analysis after the PCR cycling has finished to check for correct product amplification. A good range for the melting curve is 55–100 °C. Expression is then calculated using the comparative Ct method of analysis, which is explained in detail on the ABI Web site (http://www.appliedbiosystems.com/).

Note about primer concentrations. It might be beneficial to perform a primer concentration curve with control samples before you do your actual experiment, as these can vary between cell lines and the targets to be measured. A starting point is to use primers at 50 μM, then do serial dilutions to check for the optimal concentration. You want to use the concentration that gives you one product peak, and no primer dimer peak, in the melting curve analysis. A peak in the melting curve at approximately 70 °C is likely to correspond to only primer dimers, so further troubleshooting would be required. After every experiment, the products (or a subset of the products) can be run on a 1% agarose gel to check for correct product size.

2.6.4 Rescue experiments

Often considered as the "gold standard" in relation to off-target effects is what is often called a rescue experiment. These experiments are conceptually the same as those often used to define the target of small molecule inhibitors of specific pathways. In these cases, mutagenesis of the active site of an enzyme can overcome the effects of an inhibitor. In the case of RNAi, "rescue" is typically achieved through expression of a cDNA encoding a form of the gene of interest that is resistant to the siRNA or shRNA employed. These experiments can be achieved in one of three ways. First, one can introduce a cDNA that has been mutated to be resistant to the targeting RNA molecule (only a few changes, readily introduced by site-directed mutagenesis are usually sufficient). Second, one can use an siRNA/shRNA that targets the 5′- or 3′-noncoding regions of specific mRNAs and introduce a cDNA that does not include this noncoding region. Third, one can sometimes use a cDNA encoding a species orthologue which differs in nucleotide sequence within the region targeted by the interfering RNA. This last option of course has the limitation of species differences where the recombinant form might act as a dominant negative. Indeed, in some cases we have experienced such dominant negative phenotypes when using the same species of cDNA as the cell type under investigation. Here we describe our most commonly used approach to a rescue experiment.

An alternative here is to use lentiviral delivery of the cDNA to stably express the gene product of interest. GFP-tagging of the rescue construct can also facilitate easy identification of transfected cells. This latter option suffers from the additional complexity of nonhomologous expression, often a rat cDNA expressed in human cells could act as a dominant negative, especially at high levels of expression. Indeed, many components of the trafficking machinery act in this way in general (even in the same species) making rescue experiments more difficult. Our typical approach is to use transient transfection of a cDNA and immunofluorescence readouts when possible. This facilitates quantitative analysis of transgene expression and scoring of the "rescue" phenotype only in those cells expressing an appropriate level. Stable expression from lentivirus can overcome this problem. A typical experiment would involve transient introduction of the transgene using lipid-based transfection 24–48 h after transient transfection with siRNA duplexes. Phenotypes are then scored after a further 24 h.

Where such a rescue experiment is not possible, for example when the transgene acts in dominant negative fashion, it is imperative to ensure that phenotypes are validated in experiments using multiple independent siRNAs or shRNAs. In terms of Golgi function, additional experimental validation might come in the form of morphological assays using EM or advanced light microscopy techniques, functional assays of trafficking, for example, using the RUSH system (see Chapter 11), or other experiments such as the repolarization of the Golgi in response to scratch wounding of cell monolayers.

2.7 DISCUSSION

In any experiment involving RNAi-mediated knockdown, one must bear in mind that the end point is a knockdown and not a knockout. The remaining protein in a cell could at least partially compensate for one or more functions. This is likely to be the case especially for enzymes that act catalytically and therefore might compensate for reduced expression by increasing activity. In these experiments, there is often high variability between the efficiency of suppression on a single dish of cells. Ideally, one should correlate a level of knockdown with phenotype on a cell-by-cell basis. There is also significant variability between cell types, transfection methods, and individual researchers, all of which must be borne in mind when considering RNAi projects, especially those that run across many years.

The biggest caveat to RNAi remains that of off-target effects. This has been discussed in detail elsewhere (Jackson et al., 2003). Our approach is typically to ensure that we use multiple siRNA or shRNAs against any target and if pooled initially, then we would also encourage deconvolution of that pool to identify a single duplex that acts efficiently, or better still two independent duplexes. Rescue experiments present a best-case for defining the specificity of a phenotype. Complexity is introduced if one is exploring anything that might act on a transcriptional or translational pathway modulating gene expression. Here, the consequences might not be due to off-target effects but changes in gene expression resulting from depletion of the specific target.

RNAi experiments are now frequently allied to animal models, including ultimately engineered gene knockouts. The technology is beginning to emerge to be able to perform similar experiments in cell culture. The use of zinc-finger nucleases, transcription activator-like effector nucleases, and most recently clustered regularly interspaced short palindromic repeats provide opportunities to permanently inhibit gene expression in cells and tissues. These tools will be widely adopted as in many cases could replace RNAi. Transient knockdown remains a simple, relatively cheap, and highly effective method. For essential genes, it is also likely to retain a place in the repertoire of organelle research.

SUMMARY

The ability to accurately and specifically deplete gene products from cells is likely to be a key tool in our work toward a full understanding of Golgi function for many years to come. Many genes involved in trafficking to, through, and from the Golgi are likely to be lethal if knocked out which suggests that depletion techniques such as RNAi will remain a method of choice for some time. Careful use of controls, validation, and interpretation of data means that RNAi will likely provide significant insight into Golgi function on a cell, tissue, and organism level over the coming years.

Acknowledgments

Work in the Stephens laboratory is funded by the Medical Research Council and Biotechnology and Biological Sciences Research Council. Work in the Mellor laboratory is funded by the British Heart Foundation.

References

Bard, F., Casano, L., Mallabiabarrena, A., Wallace, E., Saito, K., Kitayama, H., et al. (2006). Functional genomics reveals genes involved in protein secretion and Golgi organization. *Nature*, *439*, 604–607.

Bramsen, J. B., Laursen, M. B., Nielsen, A. F., Hansen, T. B., Bus, C., Langkjaer, N., et al. (2009). A large-scale chemical modification screen identifies design rules to generate siRNAs with high activity, high stability and low toxicity. *Nucleic Acids Research*, *37*, 2867–2881.

Chen, C. A., & Okayama, H. (1988). Calcium phosphate-mediated gene transfer: A highly efficient transfection system for stably transforming cells with plasmid DNA. *BioTechniques*, *6*, 632–638.

Chen, X. W., Wang, H., Bajaj, K., Zhang, P., Meng, Z. X., Ma, D., et al. (2013). SEC24A deficiency lowers plasma cholesterol through reduced PCSK9 secretion. *Elife*, *2*, e00444.

Chia, J., Goh, G., Racine, V., Ng, S., Kumar, P., & Bard, F. (2012). RNAi screening reveals a large signaling network controlling the Golgi apparatus in human cells. *Molecular Systems Biology*, *8*, 629.

Demmel, L., Melak, M., Kotisch, H., Fendos, J., Reipert, S., & Warren, G. (2011). Differential selection of Golgi proteins by COPII Sec24 isoforms in procyclic Trypanosoma brucei. *Traffic*, *12*, 1575–1591.

Donze, O., & Picard, D. (2002). RNA interference in mammalian cells using siRNAs synthesized with T7 RNA polymerase. *Nucleic Acids Research*, *30*, e46.

Elbashir, S. M., Harborth, J., Lendeckel, W., Yalcin, A., Weber, K., & Tuschl, T. (2001). Duplexes of 21-nucleotide RNAs mediate RNA interference in cultured mammalian cells. *Nature*, *411*, 494–498.

Erfle, H., Neumann, B., Liebel, U., Rogers, P., Held, M., Walter, T., et al. (2007). Reverse transfection on cell arrays for high content screening microscopy. *Nature Protocols*, *2*, 392–399.

Farhan, H., Wendeler, M. W., Mitrovic, S., Fava, E., Silberberg, Y., Sharan, R., et al. (2010). MAPK signaling to the early secretory pathway revealed by kinase/phosphatase functional screening. *The Journal of Cell Biology*, *189*, 997–1011.

Folker, E. S., Ostlund, C., Luxton, G. W., Worman, H. J., & Gundersen, G. G. (2011). Lamin A variants that cause striated muscle disease are defective in anchoring transmembrane actin-associated nuclear lines for nuclear movement. *Proceedings of the National Academy of Sciences of the United States of America*, *108*, 131–136.

Harborth, J., Elbashir, S. M., Bechert, K., Tuschl, T., & Weber, K. (2001). Identification of essential genes in cultured mammalian cells using small interfering RNAs. *Journal of Cell Science*, *114*, 4557–4565.

Jackson, A. L., Bartz, S. R., Schelter, J., Kobayashi, S. V., Burchard, J., Mao, M., et al. (2003). Expression profiling reveals off-target gene regulation by RNAi. *Nature Biotechnology*, *21*, 635–637.

Petri, S., & Meister, G. (2013). siRNA design principles and off-target effects. *Methods in Molecular Biology*, *986*, 59–71.

Root, D. E., Hacohen, N., Hahn, W. C., Lander, E. S., & Sabatini, D. M. (2006). Genome-scale loss-of-function screening with a lentiviral RNAi library. *Nature Methods*, *3*, 715–719.

Simpson, J. C., Joggerst, B., Laketa, V., Verissimo, F., Cetin, C., Erfle, H., et al. (2012). Genome-wide RNAi screening identifies human proteins with a regulatory function in the early secretory pathway. *Nature Cell Biology*, *14*, 764–774.

Tang, B. L., Kausalya, J., Low, D. Y., Lock, M. L., & Hong, W. (1999). A family of mammalian proteins homologous to yeast Sec24p. *Biochemical and Biophysical Research Communications*, *258*, 679–684.

Tisdale, E. J., Kelly, C., & Artalejo, C. R. (2004). Glyceraldehyde-3-phosphate dehydrogenase interacts with Rab2 and plays an essential role in endoplasmic reticulum to Golgi transport exclusive of its glycolytic activity. *The Journal of Biological Chemistry*, *279*, 54046–54052.

Townley, A. K., Feng, Y., Schmidt, K., Carter, D. A., Porter, R., Verkade, P., et al. (2008). Efficient coupling of Sec23-Sec24 to Sec13-Sec31 drives COPII-dependent collagen secretion and is essential for normal craniofacial development. *Journal of Cell Science*, *121*, 3025–3034.

Watson, P., & Stephens, D. J. (2006). Microtubule plus-end loading of p150(Glued) is mediated by EB1 and CLIP-170 but is not required for intracellular membrane traffic in mammalian cells. *Journal of Cell Science*, *119*, 2758–2767.

Zala, D., Hinckelmann, M. V., Yu, H., Lyra da Cunha, M. M., Liot, G., Cordelieres, F. P., et al. (2013). Vesicular glycolysis provides on-board energy for fast axonal transport. *Cell*, *152*, 479–491.

Zanetti, G., Pahuja, K. B., Studer, S., Shim, S., & Schekman, R. (2011). COPII and the regulation of protein sorting in mammals. *Nature Cell Biology*, *14*, 20–28.

Ziauddin, J., & Sabatini, D. M. (2001). Microarrays of cells expressing defined cDNAs. *Nature*, *411*, 107–110.

CHAPTER

Trafficking Along the Secretory Pathway in Drosophila Cell Line and Tissues: A Light and Electron Microscopy Approach

3

Margarita Zacharogianni* and Catherine Rabouille*,†

*Hubrecht Institute for Developmental Biology and Stem Cell Research, Utrecht, The Netherlands

†The Department of Cell Biology, UMC Utrecht, Utrecht, The Netherlands

CHAPTER OUTLINE

Introduction and Rationale **36**

3.1 Materials **37**

3.2 Methods **38**

3.2.1 Drosophila Cultured S2 Cells 38

3.2.1.1 Protein Localization in the Early Secretory Pathway (Endogenous or Overexpressed) 38

3.2.1.2 RNAi in S2 Cells 41

3.2.1.3 Transport/Secretion Assays in S2 Cells 42

3.2.1.4 Live Cell Imaging 43

3.2.2 Protein Localization in Drosophila Tissues 43

3.2.2.1 Immunofluorescence Localization of Proteins of the Early Secretory Pathway in Whole Mount Tissues 43

3.2.2.2 Immunofluorescence on Thick Sections 46

3.2.2.3 EM and IEM of Drosophila Tissues 46

Conclusions/Perspectives **47**

Acknowledgments **47**

References **47**

Abstract

In the past, Drosophila has been used for molecular and developmental biology studies that have led to many important conceptual advances. In the last decade, this model organism has also been utilized to address cell biology issues, in particular

Methods in Cell Biology, Volume 118

ISSN 0091-679X
http://dx.doi.org/10.1016/B978-0-12-417164-0.00003-3

those related to membrane traffic through the secretory pathway. This has confirmed that the functional organization of the secretory pathway is conserved and it allowed further integrating secretion to signaling and development. Furthermore, Drosophila tissue culture S2 cells have been the basis of many RNAi screens, some addressing aspects of the functional organization of the secretory pathway and others identifying proteins of the secretory pathway in seemingly unrelated processes.

Taken together, studying the protein trafficking and the organization of the secretory pathway both in S2 cells and in tissues has become important. Here, we review light and electron microscopy techniques applied to Drosophila that allow gaining insight into the secretory pathway, and can easily be extended to other cell biology-related fields.

INTRODUCTION AND RATIONALE

One-third of the genome encodes proteins, which rely on the secretory pathway to reach their final destination (Almen, Nordstrom, Fredriksson, & Schioth, 2009; Stevens & Arkin, 2000). This pathway comprises the endoplasmic reticulum (ER), the ER exit sites (ERES) or transitional ER (tER), the intermediate compartment (ERGIC), the Golgi complex, the trans-Golgi network, and post-Golgi carriers. The function of the secretory pathway is to allow the synthesis, folding, transport, modification, sorting, and dispatching of proteins and lipids to their final destinations, such as the plasma membrane, the extracellular medium, and the endosomal/lysosomal compartments as well as the compartments of the secretory pathway itself.

In the context of celebrating the 110 anniversary of the discovery of the Golgi complex, we claimed and demonstrated that Drosophila is a suitable organism for studying the organization of the early secretory pathway (Kondylis & Rabouille, 2009). Indeed, the secretory pathway in flies has the same basic features as in mammalian cells. It comprises a single lumen ER pervading the entire cell, ERES characterized by the accumulation of COPII-coated vesicles, and a Golgi complex organized in stacked polarized cisternae. Yet, this organization also displays characteristic differences as compared to mammalian cells: the ERES appear larger; the Golgi is not a single copy organelle as the stacks are not linked together and remain discreet but next to the ERES, thus forming tER-Golgi units (Kondylis & Rabouille, 2009). However, despite the differences, most of the key players identified in mammalian cells have similar or overlapping roles in Drosophila, thus making this organism an excellent alternative, not only to mammalian cells but also to yeast that, so far, has been the genetic model of choice.

Drosophila S2 cells also allow very effective protein knockdown by RNAi and this makes them a very popular system for genome-wide screens (Boutros et al., 2004; Goshima et al., 2007; Kondylis, Tang, Fuchs, Boutros, & Rabouille, 2011; Philips, Rubin, & Perrimon, 2005; Zacharogianni et al., 2011; Zhang et al., 2006). Furthermore, high transfection efficiencies can be achieved allowing the production of recombinant eukaryotic proteins (Chyb, Raghu, & Hardie, 1999; Park, Chang, Lee, Lee, & Chung, 2001).

There are at least two good reasons why it is important to study trafficking through the secretory pathway in fly; one is the link to development: Despite the widely spread view that proteins functioning in the secretory pathway are ubiquitously expressed, genetic experiments have revealed specific roles in development as well as alternative functions outside the protein complexes where they were studied in mammalian cells (Kondylis & Rabouille, 2003). Second, screens for one given process seemingly unrelated to membrane traffic, either forward genetic screens in Drosophila or genome-wide RNAi screens in Drosophila S2 cells, sometimes identify candidates that function in the secretory pathway, thus tools and methods need to be available to understand their biology. This has recently been the case of the role for COPI in lipid droplet biogenesis (Beller et al., 2008; Guo et al., 2008) as well as Rab10 in the shape of the Drosophila egg chamber (Lerner et al., 2013).

Here, we outline protocols to visualize the early compartments of the secretory pathway in Drosophila cell lines and tissues and we describe functional secretion assays to study trafficking.

3.1 MATERIALS

STOCK SOLUTIONS

- Schneider's medium (Sigma, Cat. No. S0146)
- Fetal bovine serum (FBS) heat inactivated, insect tested (Sigma, Cat. No. F4135)
- Phosphate buffer (PB) 0.2 M, pH 7.4
- Phosphate buffered saline (PBS)
- Paraformaldehyde (PFA), 16% in water: Fully dissolve 16 g of FPA in 70 ml of water preheated to 70 °C under constant stirring for a maximum of 10 min. Adjust the final volume to 100 ml, filter using Whatman paper, and store aliquotes at −20 °C
- Glutaraldehyde (GA) grade I, 8% in water, specially purified for use as an electron microscopy (EM) fixative or other sophisticated use (Sigma-Aldrich, G5882).
- Formaldehyde 37% in 10% methanol (Sigma, Cat. No. F8775)
- Methanol 100% (−20 °C)
- Triton X-100 1% in PBS
- $CuSO_4(5H_2O)$ 100 mM stock solution in water
- Ringer solution: 7.2 g NaCl, 0.17 g $CaCl_2$, 0.37 g KCl in 1 l water, pH 7.4 Sterilize by passing through a 0.22 μm filter

WORKING SOLUTIONS

- 4% PFA in PBS for light microscopy
- 4% PFA in 0.1 M PB (pH 7.4) for immunoelectron microscopy (IEM)
- 2% GA in 0.1 M PB for EM
- Mixture of 2% PFA and 0.2% GA in PB for IEM microscopy
- 8% formaldehyde–methanol in PBS
- Blocking solution: 1% fish skin gelatin in PBS (Sigma, Cat. No. G7765)

REAGENTS

- Primary antibodies. For tagged proteins we often use mouse monoclonal anti-V5 (Life technologies, Cat. No. R960) but GFP or HA (Covance, Cat. No. MMS-101P) is also an option. GFP can of course be visualized by fluorescence microscopy, but it is also possible to boost the signal after fixation by using an anti-GFP antibody (Cat. No. GFP-12E6, DSHB, University of Iowa).
- Secondary antibodies coupled to fluorophores (Life Technologies, Alexa series)
- T7 Megascript kit (Ambion, Cat. No. AM1334)
- Effectene transfection reagent (Qiagen, Cat. No. 301425)
- Prolong-DAPI (Life Technologies, Cat. No. P-36931) or vectashield (Vector Labs, Cat. No. H-1200)

3.2 METHODS

3.2.1 Drosophila cultured S2 cells

3.2.1.1 Protein localization in the early secretory pathway (endogenous or overexpressed)

S2 cells are the most popular Drosophila cell line but research (e.g., high-throughput screening) is also conducted on Kc cells and a S2 cell derivative S2R$^+$ (Boutros et al., 2004; Liu, Sims, & Baum, 2009). Here we focus on S2 cells but most of the techniques described below can be applied to other cell lines. S2 cells are cultured in Schneider's medium supplemented with FBS at final concentration 10%. They are split every 5 days, 1:10, when cell density is approximately 10^7 cells/ml. It is important to pass some of the medium that the cells have been growing in (conditioned medium) every time the cells are split.

To visualize the different compartments of the secretory pathway, specific antibodies to different endogenous markers along the secretory pathway are available, such as Sec16 and Sec23 for the ERES (Ivan et al., 2008; Kondylis & Rabouille, 2003), dGRASP and dGMAP for the Golgi (Friggi-Grelin, Rabouille, & Therond, 2006; Kondylis, Spoorendonk, & Rabouille, 2005), and GRIP-domain proteins for the Trans-Golgi network (Sinka, Gillingham, Kondylis, & Munro, 2008).

An excellent alternative is to moderately overexpress tagged proteins under the control of a metallothionein promoter. In our experience, V5 tagging gives very good results, but other epitopes, such as HA and GFP, can also be used. We recommend transient transfections for 48 h using Effectene (Qiagen) that can lead up to 50% transfection (although some constructs will only be expressed in 10–15% of the cells). Stable S2 cell lines can also be generated, but in our experience, the metallothionein promoter starts to leak after a number of passages, leading to a basal expression even in the absence of $CuSO_4$ (that is used to induce protein expression).

We also recommend "pulsing" the expression of the chosen tagged protein by adding 1 mM $CuSO_4$ for 2–4 h (or even shorter when possible) before removing it and chasing for at least one additional hour, so that the protein reaches its final destination. Cycloheximide can be added during this chase, leading to a tighter pulse of expression. This is particularly important for resident proteins of the Golgi (such as Fringe-GFP) that are synthesized in the ER. The lack of chase might lead to misinterpretation of their localization due to signal from the newly synthesized peptide, which has not reached its final destination yet (Kondylis, van Nispen tot Pannerden, Herpers, Friggi-Grelin, & Rabouille, 2007). This treatment has to be implemented when the localization of an unknown protein is assessed by this technique.

Last, primary antibodies fail to reveal a clear pattern sometimes due to the low expression of the antigen. Mild overexpression of the untagged protein can solve this issue as in the case of SMSr (Vacaru et al., 2009).

3.2.1.1.1 Transient transfection protocol

Transfection is done using Effectene transfection reagent (Qiagen) according to the following protocol. The kit comprises the solutions used here:

- Seed 1–2 × 10^6 S2 cells.
- Use a clear, plastic tube to make the transfection mix and add 1 μg of DNA to EC buffer up to a total volume of 150 μl.
- Add 8 μl enhancer, vortex for 1 s, incubate 5 min at RT.
- Add 10 μl effectene, vortex for 15 sec, incubate 5 min at RT.
- Add 1 ml of the medium the cells are growing in and transfer the mix back to the cells, drop by drop.
- After 48 h, induce the expression of the tagged protein by adding $CuSO_4$ (1 mM final concentration) in the culture for 2–4 h at 26 °C.
- Remove the $CuSO_4$ by aspirating the medium and replacing it with new and incubate the cells for 2 h (with or without cycloheximide to block further protein synthesis) before fixing for immunofluorescence or immunoelectron microscopy.

On average, 25% transfection efficiency was achieved but for some constructs, we do not get more than 10–15%.

Note that S2 cells only mildly adhere to the coverslips, so the washes and changes of medium have to be performed very gently.

3.2.1.1.2 Stable transfection protocol

Generation of stable cell lines is also possible with the effectene transfection system:

- Co-transfect the desired construct with the pCoHygro, which confers hygromycin B resistance, keeping a ratio 10 μg desired construct: 1 μg pCoHygro.
- Culture transfected cells in antibiotic-free medium for 2 days.
- After 2 days start the selection with hygromycin B 150 μg/ml. Split the culture every 4 days (1:4). Stable clones should start appearing approximately 1–2 weeks later.

3.2.1.1.3 Fixation for light microscopy

We use either PFA or cold methanol:

- Plate S2 cells (WT or overexpressing tagged proteins) on coverslips.
- Make 4% PFA in PBS as described above and warm it up to room temperature (cold fixative will lead to microtubule depolymerization).
- Remove culture medium.

- Gently add 2 ml (if the experiment is done in a 3.5 cm dish) 4% PFA for 15–20 min or cold 100% methanol for 6 min.
- Rinse 3 × with PBS.

3.2.1.1.4 Immunolabeling protocol for S2 cells (Fig. 3.1A and B)

- If the cells are fixed with PFA, place the coverslips in the wells of a 12-well plate containing 1 ml of 50 mM NH_4Cl in PBS and incubate for 5 min.
- Add 110 μl Triton X-100 1% in PBS, mix gently, and incubate for another 5 min to permeabilize the plasma membrane.
- Wash 3 × with PBS. These three steps can be omitted if the cells are fixed with methanol.
- Add 1 ml 1% Fish Skin Gelatin (blocking buffer). Aspirate and add another 1 ml 1% FSG. Incubate for 20 min.
- Dilute the primary antibodies in 1% FSG and place them in 20 μl drops on parafilm (recommended dilutions: anti-V5 1:500, anti-GFP 1:200).
- Wipe gently the coverslips from the blocking buffer, place them on the drops and incubate for 25 min.
- Transfer the coverslips back in the 12-well plate and wash 3 × with 1% FSG.
- Dilute the secondary antibodies in blocking buffer and place them in 20 μl drops on parafilm.
- Wipe gently the coverslips from the blocking buffer, load them on the drops, and incubate for 20 min.
- Add dH_2O in a Petri dish and wash the coverslips by dipping them 10 times.
- Wipe them dry and mount on Prolong-DAPI or vectashield.

3.2.1.1.5 GFP fluorescence (Fig. 3.1A and B)

Fixation described above does not quench the GFP fluorescence, so GFP-tagged proteins can be visualized in the context of labeling other antigens. The signal can also be boosted by using anti-GFP antibodies. Note that GFP folds slowly in the lumen of organelles and low level of expression as well as time to localize are critical to achieve proper localization.

GFP fluorescence also allows for live cell imaging (see Section 3.2.1.4).

3.2.1.1.6 Confocal microscopy

The details of the early secretory pathway in S2 cells can be best visualized by confocal microscopy with at least 63 × magnification. To visualize the whole cell, we normally project 15–18 confocal planes with an interval of 0.4 μm (Fig. 3.1).

3.2.1.1.7 EM and IEM (Fig. 3.1C and D)

S2 cells are grown in the same way as above. The fixation is then carried out by adding either 3 ml 4% PFA in 0.1 M PB (not PBS) at pH 7.4 for 3 h at RT followed by 16 h at 4 °C, or 3 ml of 2% PFA, 0.2% GA in 0.1 M PB for 3 h. Both materials can be stored in 1% PFA (in PBS) at 4 °C for months. Fixed cells are then pelleted and processed for EM and IEM as described in (Ivan et al., 2008; Kondylis & Rabouille, 2003; Slot & Geuze, 2007).

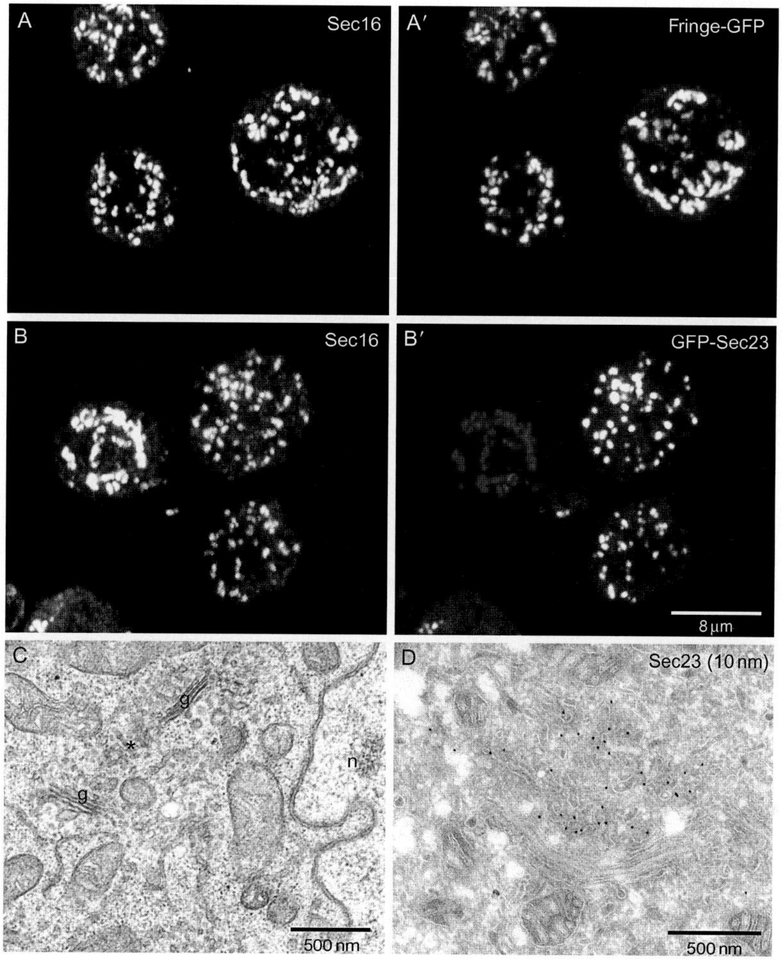

FIGURE 3.1

Visualization of the early secretory pathway of Drosophila S2 cells by light and electron microscopy. (A-A') Visualization of endogenous Sec16 (A) by immuno-fluorescence and Fringe-GFP (A') marking the ER exit sites and the Golgi, respectively. (B-B') Visualization of endogenous Sec16 (B) and the COPII subunit Sec23-GFP (B') at ERES. (C–D) Visualization of the early secretory pathway (ERES and Golgi stacks) on ultrathin plastic section by conventional electron microscopy (EM) (C) and immuno-EM on ultrathin frozen section labeled for Sec23 (10 nm gold) (D). n indicates the nucleus, * indicates the ERES characterized by accumulation of COPII vesicles, and g the Golgi. (See color plate.)

(C) and (D) are adapted from (Ivan et al., 2008) and (Kondylis et al., 2005).

3.2.1.2 RNAi in S2 cells

RNAi is very easy to perform in S2 cells and this has been exploited in multiple screens scoring the regulation of secretion and the organization of the secretory pathway (Bard et al., 2006; Kondylis et al., 2011; Zacharogianni et al., 2011):

– The dsRNA used for RNAi in S2 cells is usually 400–800 bp long.

- The primers for generating the targeting dsRNAs are designed using E-RNAi webservice (http://www.dkfz.de/signaling/e-rnai3/). A number of dsRNA libraries are available (http://e-rnai.dkfz.de, http://www.flyrnai.org).
- dsRNA is synthesized using the Megascript T7 kit from Ambion as described by the manufacturer.
- Centrifuge cells for 4 min at 1000 rpm.
- Remove the medium and resuspend the cells in serum-free Schneider's at the density of 7×10^5 cells/ml.
- Plate 1 ml (7×10^5 cells) in a 3.5 cm dish, add dsRNA at final concentration 37 μM, and incubate for 1 h at 26 °C.
- Add 2 ml Schneider's supplemented with FBS 10%.
- Incubate at 26 °C for 5 days while monitoring cell proliferation daily to assess the effect of the depletion. If the knockdown is lethal the RNAi period can be adjusted.

Note that depletion by RNAi is compatible with transient transfection, which is performed on day 3 of the RNAi. This can be useful to investigate the fate of an organelle (marked by a tagged marker) upon RNAi. For large RNAi screen, we recommend to use stable cell line expressing a given marker.

3.2.1.3 Transport/secretion assays in S2 cells

To monitor plasma membrane delivery, we use a stable cell line expressing the plasma membrane protein Delta under the control of the metallothionein promoter:

- Induce the expression of Delta with 1 mM $CuSO_4$ for 2 h at 26 °C.
- Change the medium into $CuSO_4$-free medium and chase in the presence of cycloheximide to allow the protein to reach its destination at the PM. This is important to prevent further Delta synthesis and detection in the ER.
- Fix in 4% PFA and process for IF using Delta specific antibodies C594.9B (DSHB), at dilution 1:600.

If the secretory pathway is functional, Delta is found at the plasma membrane (Fig. 3.2A) but if transport is impaired, Delta will be retained intracellularly (Fig. 3.2B). The efficiency of Delta delivery at the plasma membrane can be assessed either by imaging as described in (Kondylis & Rabouille, 2003) or by flow cytometry (Cherry et al., 2006).

Anterograde transport has also been assessed using the secreted proteins ss-HRP (Bard et al., 2006) and recombinant luciferase (Wendler et al., 2010). In these approaches the secreted protein is detected in the medium. Anterograde transport can also be assessed in Drosophila tissues after inducing the expression of a heterologous mammalian plasma membrane marker (CD8-GFP) using the UAS–Gal4 system (see Section 3.2.2.1.4) and examining its localization in different genetic backgrounds (Xu et al., 2002).

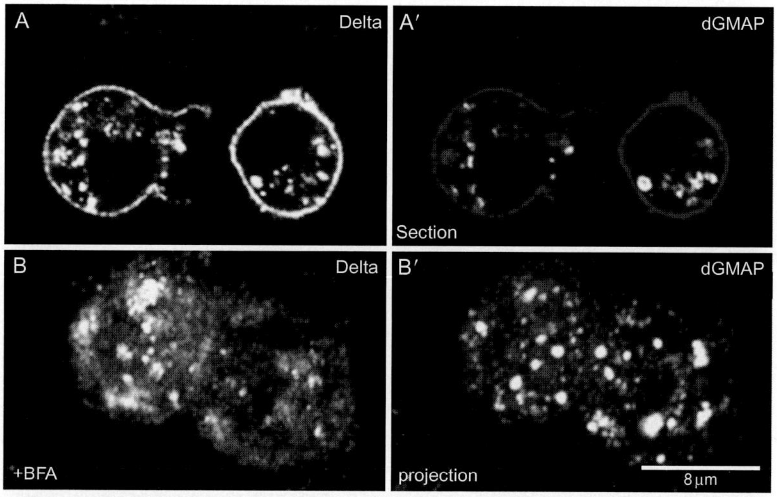

FUGURE 3.2

The Delta anterograde transport assay in the secretory pathway of S2 cells. (A-A') Visualization by immunofluorescence of Delta (A) and dGMAP (marking the Golgi, A') in S2 cells stably transfected with Delta-myc after induction of its expression for 2 h by $CuSO_4$ followed by a chase of 2 h in the presence of cycloheximide. Equatorial single confocal section. Note that Delta is decorating the plasma membrane and endosomes. (B-B') Same visualization as above in cells treated with Brefeldin A (BFA, inhibiting the ER to Golgi transport) prior to induction. Projection of 15 confocal sections. Note that Delta is trapped intracellularly. (See color plate.)

3.2.1.4 Live cell imaging

Live cell imaging is used to visualize changes in the morphology of the early secretory pathway under different conditions (Kondylis et al., 2007; Zacharogianni et al., 2011). Cells transiently or stably transfected with GFP-tagged components of the early secretory pathway (Sec23 for COPII, Fringe for the Golgi, Sec16-ΔNC1 for the ER exit sites) are plated on glass and induced to produce the fluorescent protein for 2 h. The protein is allowed to localize for 2 h by chasing in $CuSO_4$-free medium with or without cycloheximide. The cells are visualized using a confocal microscope equipped for live imaging or a spinning disc microscope. Fluorescent microscopes can also be used. To reduce the movement of the cells, they can be plated on concanavalin A-coated coverslips.

3.2.2 Protein localization in Drosophila tissues

3.2.2.1 Immunofluorescence localization of proteins of the early secretory pathway in whole mount tissues

3.2.2.1.1 Ovaries (Fig. 3.3C–F)

- Sort virgin females and place them on vials with yeast paste for 4 days at 18 °C (or 2 days at 25 °C) to allow the ovaries to develop.

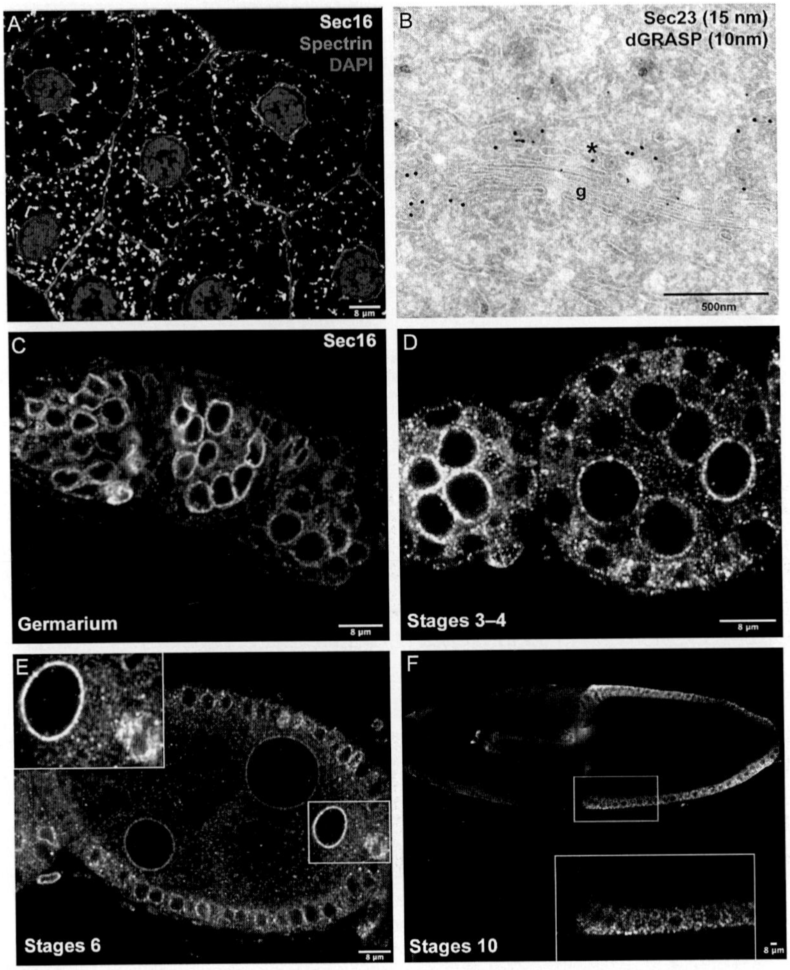

FIGURE 3.3

Visualization of Sec16 in Drosophila tissues. (A) Visualization of Sec16 (white) and α-spectrin (gray, outlining the plasma membrane) in salivary gland fixed in 8% formaldehyde in methanol by immuno-fluorescence. DAPI (gray) marks the nucleus. Note that the Sec16 pattern is consistent with its localization to ERES. (B) IEM of salivary gland fixed in 2% PFA-0.2% GA and labeled for the COPII subunit Sec23 (15 nm gold) and dGRASP (10 nm gold). Adapted from (Kondylis et al., 2005). (C–F) Visualization of Sec16 in the Drosophila egg chamber fixed in 8% formaldehyde and 2% methanol at different stages of development. Note that regarding the germ cells, till stage 4, Sec16 is localized around the nuclear envelope of nurse cells and oocyte (C–D). At stage 6, Sec16 is enriched in the oocyte nuclear envelope as well as in faint punctae pattern consistent with ERES (E and inset). At later stages, Sec16 is no longer visible in the germ cells (F). Regarding the follicle cells, Sec16 appears as a punctate pattern in the follicle cells at all stages consistent with ERES (see inset in F). It also seems enriched in the border cells (arrow in F) and the posterior-most cells of the follicular epithelium (inset in E) as well as on the nuclear envelope till stage 4. (See color plate.)

- Dissect the ovaries in Ringer's solution and fix them immediately according to either protocols:
  - In 8% formaldehyde–methanol in PBS for 15 min. This is followed by incubation of the ovaries 3 × for 5 min in dehydration methanol gradient (from 25% to 100% methanol in PBS). The ovaries can be stored at −20 °C in 100% methanol. Pass the ovaries through the reverse methanol gradient to bring them at 100% PBS before proceeding to the immunolabeling steps.
  - In 4% PFA in PBS for 15 min. If the antibody does not tolerate methanol, PFA can be used which is directly followed by immunostaining.

The choice of the fixative depends on the antibody, therefore pilot experiments might be required to determine the optimal fixative. For instance, phalloidin staining of actin does not tolerate methanol. Manipulating the tissue is a lot easier if you keep the ovaries as a whole and not separate ovarioles and egg chambers until the final step:

- Block with 2% BSA in PBS for 1 h.
- Incubate the ovaries with the primary antibodies in blocking buffer, overnight, rotating at 4 °C.
- Wash for 30 min in PBS (at least three washes).
- Incubate with the secondary antibodies for 1 h at RT.
- Wash for 30 min in PBS (at least three washes).
- After the final washing step, break the ovaries to separate ovarioles and single egg chambers by pipetting them up and down with a yellow tip with the tip chopped off.
- Allow the egg chambers to get to the bottom of the tube and remove as much PBS as possible.
- Add 20 μl Prolong-DAPI, transfer on glass slide, and cover with coverslip.

If there is high background staining you can perform the washing steps in PBS–Tween-20 0.1%.

3.2.2.1.2 Salivary glands (Friggi-Grelin et al., 2006) (Fig. 3.3A)

- Pick wandering 3rd instar larvae. The protocol is the same for earlier stages; however, the glands will be smaller.
- Dissect the glands in Ringer's solution by pulling the mouthparts and fix them in 4% PFA or 8% FA in methanol (see above Section 3.2.2.1.1).
- Keep the glands attached to the mouthparts, as it is easier to see them this way.
- Continue as with ovaries (no need for the methanol gradient unless you want to store the tissue).
- Place the glands on a drop of Prolong-DAPI on glass slide and cover with coverslip.

3.2.2.1.3 Imaginal discs (Dunne, Kondylis, & Rabouille, 2002)

- Use 3rd instar larvae as above.
- Cut the larvae into two in Ringer's solution and flip them so that the discs are still attached to the carcass.

- Fix as above (Section 3.2.2.1.1) and continue as with the ovaries.
- The discs are separated from the carcass before mounting them on prolong.

3.2.2.1.4 Visualization of overexpressed proteins

As for S2 cells expressing tagged markers of the secretory pathway using transfected cDNA under the control of a metallothionein promoter, flies can also express similar markers by using the UAS–GAL4 system (Brand & Perrimon, 1993). The cDNA of a protein of interest (such as dGRASP, for instance) is tagged with GFP in a pUAS vector compatible with transgenesis (Schotman, Karhinen, & Rabouille, 2008; Weil et al., 2012). The transgenic flies stably expressing this tagged marker can be crossed to GAL4 drivers (http://flystocks.bio.indiana.edu/Browse/misc-browse/gal4.php) either under the control of a tissue specific promoter (such as 3750 for follicle cells, or engrailed for wing imaginal disc) or a heat shock promoter. The tissue is then processed as above.

There are many GFP trapped (http://flystocks.bio.indiana.edu/Browse/in/protein-trap-GFP.htm) or transgenic lines available to mark the secretory pathway, such as Boca-GFP (ER), dGRASP-GFP (Golgi), and others.

3.2.2.2 Immunofluorescence on thick sections

It is often the case that the penetration of antibodies in the tissue is limited (e.g., in the oocyte, especially after stage 11 when the vitellin membrane makes it impermeable to fixation and incubation). Overexpression of GFP-tagged proteins may circumvent the issue. However, for endogenous proteins, labeling 250–500 nm frozen sections of aldehyde fixed tissues is the solution (Sommer, Oprins, Rabouille, & Munro, 2005):

- 250–500 nm thick sections are cut at −80 °C on an ultra-cryomicrotome, collected with a loop of methyl cellulose/sucrose and deposited on a Silan-coated glass slides. The areas of deposition are circled with wax pen. These slidescan be stored at 4 °C.
- The sections (adhering to the glass slide) are washed thoroughly with PBS at 37 °C followed by one at RT and incubated 5 min with freshly prepared sodium borohydride (1 mg/ml in PBS) followed by PBS washes.
- The rest of the procedure is identical to the gold labeling for IEM (see below) except that the primary antibody is incubated for 1 h and the secondary coupled to the fluorophore for 45 min in the dark.
- The following washes are also performed in the dark.
- The sections are finally rinsed with dH_2O. A drop of Prolong-DAPI is added and a coverslip is placed on top of the sample.

Sections can be viewed under a confocal microscope at 40× and 63×.

3.2.2.3 EM and IEM of Drosophila tissues

- The same tissues as for IF are fixed for IEM in either 2% PFA + 0.2% GA in PB 0.1 M (pH 7.4) or 4% PFA and can be stored in 1% PFA as for S2 cells. In the last step, a small drop of the stain Trypan blue is added that will accumulate around the tissues, making it much easier to visualize them in the subsequent manipulation steps (especially the embedding) (Fig. 3.3B).

- For the egg chambers, the stage of interest is selected and a single egg chamber is embedded in a small block of 10% gelatin (Delanoue, Herpers, Soetaert, Davis, & Rabouille, 2007; Sommer et al., 2005; Vanzo et al., 2007). These blocks are then infiltrated in 2.3 M sucrose and ultrathin frozen sections are cut on an ultra-cryomicrotome at −120 °C. The rest of the procedure is identical as for cells except that the entire tissue is cut.

CONCLUSIONS/PERSPECTIVES

The protocols described above provide a solid framework to study secretion and the functional organization of the compartments of the secretory pathway, which still have a lot to reveal. A rising field that has been gaining attention is the regulation of the secretory pathway by environmental signals. Recently, light and electron microscopy on fixed Drosophila cells as well as live cell imaging allowed us to observe a dramatic morphological change of the organization of the tER-Golgi units during nutrient stress (Zacharogianni et al., 2011). Combination of the light and electron microscopy techniques described above can help elucidate the details of the organization and behavior of the membrane compartments of the secretory route and add to our understanding of the regulation of secretion.

Acknowledgments

We thank the members of the lab of Catherine Rabouille, past and present, for establishing the techniques and sharing the know-how. We thank Angelica Aguilera for critically reading the manuscript and Vangelis Kondylis for the panels of Fig. 3.2.

References

Almen, M. S., Nordstrom, K. J., Fredriksson, R., & Schioth, H. B. (2009). Mapping the human membrane proteome: A majority of the human membrane proteins can be classified according to function and evolutionary origin. *BMC Biology*, *7*, 50.

Bard, F., Casano, L., Mallabiabarrena, A., Wallace, E., Saito, K., Kitayama, H., et al. (2006). Functional genomics reveals genes involved in protein secretion and Golgi organization. *Nature*, *439*, 604–607.

Beller, M., Sztalryd, C., Southall, N., Bell, M., Jackle, H., Auld, D. S., et al. (2008). COPI complex is a regulator of lipid homeostasis. *PLoS Biology*, *6*, e292.

Boutros, M., Kiger, A. A., Armknecht, S., Kerr, K., Hild, M., Koch, B., et al. (2004). Genome-wide RNAi analysis of growth and viability in Drosophila cells. *Science*, *303*, 832–835.

Brand, A. H., & Perrimon, N. (1993). Targeted gene expression as a means of altering cell fates and generating dominant phenotypes. *Development*, *118*, 401–415.

Cherry, S., Kunte, A., Wang, H., Coyne, C., Rawson, R. B., & Perrimon, N. (2006). COPI activity coupled with fatty acid biosynthesis is required for viral replication. *PLoS Pathogens*, *2*, e102.

Chyb, S., Raghu, P., & Hardie, R. C. (1999). Polyunsaturated fatty acids activate the Drosophila light-sensitive channels TRP and TRPL. *Nature*, *397*, 255–259.

Delanoue, R., Herpers, B., Soetaert, J., Davis, I., & Rabouille, C. (2007). Drosophila Squid/hnRNP helps Dynein switch from a gurken mRNA transport motor to an ultrastructural static anchor in sponge bodies. *Developmental Cell*, *13*, 523–538.

Dunne, J. C., Kondylis, V., & Rabouille, C. (2002). Ecdysone triggers the expression of Golgi genes in Drosophila imaginal discs via broad-complex. *Developmental biology*, *245*, 172–186.

Friggi-Grelin, F., Rabouille, C., & Therond, P. (2006). The cis-Golgi Drosophila GMAP has a role in anterograde transport and Golgi organization in vivo, similar to its mammalian ortholog in tissue culture cells. *European Journal of Cell Biology*, *85*, 1155–1166.

Goshima, G., Wollman, R., Goodwin, S. S., Zhang, N., Scholey, J. M., Vale, R. D., et al. (2007). Genes required for mitotic spindle assembly in Drosophila S2 cells. *Science*, *316*, 417–421.

Guo, Y., Walther, T. C., Rao, M., Stuurman, N., Goshima, G., Terayama, K., et al. (2008). Functional genomic screen reveals genes involved in lipid-droplet formation and utilization. *Nature*, *453*, 657–661.

Ivan, V., de Voer, G., Xanthakis, D., Spoorendonk, K. M., Kondylis, V., & Rabouille, C. (2008). Drosophila Sec16 mediates the biogenesis of tER sites upstream of Sar1 through an arginine-rich motif. *Molecular Biology of the Cell*, *19*, 4352–4365.

Kondylis, V., & Rabouille, C. (2003). A novel role for dp115 in the organization of tER sites in Drosophila. *The Journal of Cell Biology*, *162*, 185–198.

Kondylis, V., & Rabouille, C. (2009). The Golgi apparatus: Lessons from Drosophila. *FEBS Letters*, *583*, 3827–3838.

Kondylis, V., Spoorendonk, K. M., & Rabouille, C. (2005). dGRASP localization and function in the early exocytic pathway in Drosophila S2 cells. *Molecular Biology of the Cell*, *16*, 4061–4072.

Kondylis, V., Tang, Y., Fuchs, F., Boutros, M., & Rabouille, C. (2011). Identification of ER proteins involved in the functional organisation of the early secretory pathway in Drosophila cells by a targeted RNAi screen. *PloS One*, *6*, e17173.

Kondylis, V., van Nispen tot Pannerden, H. E., Herpers, B., Friggi-Grelin, F., & Rabouille, C. (2007). The golgi comprises a paired stack that is separated at G2 by modulation of the actin cytoskeleton through Abi and Scar/WAVE. *Developmental Cell*, *12*, 901–915.

Lerner, D. W., McCoy, D., Isabella, A. J., Mahowald, A. P., Gerlach, G. F., Chaudhry, T. A., et al. (2013). A Rab10-dependent mechanism for polarized basement membrane secretion during organ morphogenesis. *Developmental Cell*, *24*, 159–168.

Liu, T., Sims, D., & Baum, B. (2009). Parallel RNAi screens across different cell lines identify generic and cell type-specific regulators of actin organization and cell morphology. *Genome Biology*, *10*, R26.

Park, J. H., Chang, K. H., Lee, J. M., Lee, Y. H., & Chung, I. S. (2001). Optimal production and in vitro activity of recombinant endostatin from stably transformed Drosophila melanogaster S2 cells. *In Vitro Cellular & Developmental Biology. Animal*, *37*, 5–9.

Philips, J. A., Rubin, E. J., & Perrimon, N. (2005). Drosophila RNAi screen reveals CD36 family member required for mycobacterial infection. *Science*, *309*, 1251–1253.

Schotman, H., Karhinen, L., & Rabouille, C. (2008). dGRASP-mediated noncanonical integrin secretion is required for Drosophila epithelial remodeling. *Developmental Cell*, *14*, 171–182.

Sinka, R., Gillingham, A. K., Kondylis, V., & Munro, S. (2008). Golgi coiled-coil proteins contain multiple binding sites for Rab family G proteins. *The Journal of Cell Biology*, *183*, 607–615.

Slot, J. W., & Geuze, H. J. (2007). Cryosectioning and immunolabeling. *Nature Protocols*, *2*, 2480–2491.

Sommer, B., Oprins, A., Rabouille, C., & Munro, S. (2005). The exocyst component Sec5 is present on endocytic vesicles in the oocyte of Drosophila melanogaster. *The Journal of Cell Biology*, *169*, 953–963.

Stevens, T. J., & Arkin, I. T. (2000). Do more complex organisms have a greater proportion of membrane proteins in their genomes? *Proteins*, *39*, 417–420.

Vacaru, A. M., Tafesse, F. G., Ternes, P., Kondylis, V., Hermansson, M., Brouwers, J. F., et al. (2009). Sphingomyelin synthase-related protein SMSr controls ceramide homeostasis in the ER. *The Journal of Cell Biology*, *185*, 1013–1027.

Vanzo, N., Oprins, A., Xanthakis, D., Ephrussi, A., & Rabouille, C. (2007). Stimulation of endocytosis and actin dynamics by Oskar polarizes the Drosophila oocyte. *Developmental Cell*, *12*, 543–555.

Weil, T. T., Parton, R. M., Herpers, B., Soetaert, J., Veenendaal, T., Xanthakis, D., et al. (2012). Drosophila patterning is established by differential association of mRNAs with P bodies. *Nature Cell Biology*, *14*(12), 1305–1313.

Wendler, F., Gillingham, A. K., Sinka, R., Rosa-Ferreira, C., Gordon, D. E., Franch-Marro, X., et al. (2010). A genome-wide RNA interference screen identifies two novel components of the metazoan secretory pathway. *The EMBO Journal*, *29*, 304–314.

Xu, H., Brill, J. A., Hsien, J., McBride, R., Boulianne, G. L., & Trimble, W. S. (2002). Syntaxin 5 is required for cytokinesis and spermatid differentiation in Drosophila. *Developmental Biology*, *251*, 294–306.

Zacharogianni, M., Kondylis, V., Tang, Y., Farhan, H., Xanthakis, D., Fuchs, F., et al. (2011). ERK7 is a negative regulator of protein secretion in response to amino-acid starvation by modulating Sec16 membrane association. *The EMBO Journal*, *30*, 3684–3700.

Zhang, S. L., Yeromin, A. V., Zhang, X. H., Yu, Y., Safrina, O., Penna, A., et al. (2006). Genome-wide RNAi screen of Ca(2+) influx identifies genes that regulate Ca(2+) release-activated Ca(2+) channel activity. *Proceedings of the National Academy of Sciences of the United States of America*, *103*, 9357–9362.

CHAPTER

Studying Membrane Trafficking in the Worm *C. elegans* by RNA Interference

4

Zita Balklava[*] and Elizabeth Sztul[†]

[*]*School of Life and Health Sciences, Aston University, Aston Triangle, Birmingham, United Kingdom*

[†]*Department of Cell, Developmental and Integrative Biology, University of Alabama at Birmingham, Birmingham, Alabama, USA*

CHAPTER OUTLINE

Introduction and Rationale 52
4.1 Materials 55
4.2 Methods 56
4.2.1 Worm Maintenance 56
4.2.2 Generating RNAi 57
4.2.3 RNAi Delivery by Feeding 58
4.2.4 Live Imaging 60
4.2.5 Microscopy and Analysis 61
4.3 Discussion 64
Summary 65
Acknowledgment 65
References 65

Abstract

A powerful approach to gain understanding of molecular machinery responsible for membrane trafficking is through inactivation of gene function by RNA interference (RNAi). RNAi-mediated gene silencing occurs when a double-stranded RNA is introduced into cells and targets a complementary mRNA for degradation. The subsequent lack of mRNA prevents the synthesis of the corresponding protein and ultimately causes depletion of a particular gene product from the cell. The effects of such depletion can then by analyzed by functional, morphological, and biochemical assays. RNAi-mediated knockdowns of numerous gene products in cultured

Methods in Cell Biology, Volume 118

ISSN 0091-679X
http://dx.doi.org/10.1016/B978-0-12-417164-0.00004-5

cells of mammalian and other species origins have provided significant new insight into traffic regulation and represent standard approaches in current cell biology. However, RNAi in the multicellular nematode *Caenorhabditis elegans* model allows RNAi studies within the context of a whole organism, and thus provides an unprecedented opportunity to explore effects of specific trafficking regulators within the context of distinct developmental stages and diverse cell types. In addition, various transgenic *C. elegans* strains have been developed that express marker proteins tagged with fluorescent proteins to facilitate the analysis of trafficking within the secretory and endocytic pathways. This chapter provides a detailed description of a basic RNAi approach that can be used to analyze the function of any gene of interest in secretory and endosomal trafficking in *C. elegans*.

INTRODUCTION AND RATIONALE

One of the major advances in recent cell biology was the introduction by Sydney Brenner of the free-living nonparasitic soil nematode *C. elegans* as a model organism to study development and neuronal function (Brenner, 1974). A number of features make *C. elegans* a perfect model for analyzing complex intracellular and organismal processes: (1) it is easily and cheaply cultured in Petri dishes on lawns of bacteria *Escherichia coli* and in 1974 Brenner also developed a technique to freeze and thaw the worms (Brenner, 1974); (2) it grows rapidly and within 3 days develops from an egg to an adult worm of about 1 mm in length; (3) it produces approximately 300 progenies per self-fertilizing hermaphrodite, and this allows phenotypic analysis of many "identical" organisms; (4) it is transparent and allows easy imaging of fluorescently tagged marker proteins; (5) adult worms have distinct organs and tissues including intestine and reproductive system, and this allows monitoring of trafficking in tissue-specific manner; and (6) it is relatively easy to "knockdown" the function of any gene of interest by RNAi. In addition, wild-type *C. elegans* contains a constant 959 somatic cells and the complete cell lineage, depicting which cells are derived from which, was described in the 1980s by John Sulston (Sulston, Schierenberg, White, & Thomson, 1983), providing a necessary foundation for research into cell-fate determination. The genome of *C. elegans* is about a hundred million base pairs and encodes approximately 20,470 protein-coding genes. The genome of *C. elegans* was completely sequenced at the end of 1998 (Consortium CeS, 1998) and was the first completely sequenced genome of a multicellular animal.

Numerous studies over more than 40 years since the introduction of *C. elegans* firmly established that the pathways and molecular machineries that regulate numerous aspects of development and physiology are highly conserved between worms and mammals (reviewed in Markaki & Tavernarakis, 2010). Thus, *C. elegans* represents a valid invertebrate model for probing key questions in cell biology and is now indispensable for analyzing various cellular processes, including membrane trafficking.

A powerful approach to gain understanding of molecular machinery responsible for membrane trafficking is through inactivation of gene function through silencing of specific genes by RNA interference (RNAi). RNAi-based approaches were first described in plants and fungi, but are now routinely used in other organisms from various animal phyla. The discovery that RNAi-mediated silencing of gene products in *C. elegans* is relatively easy started a new era in research, and RNAi in worms has become one of the most powerful tools to study gene function in metazoans. Unlike the significantly more costly and lengthy approaches to knockout genes in a vertebrate model such as the zebrafish or in a mammalian model such as mice, RNAi-mediated knockdown of target genes in the worm allows evaluation of the function of the gene inexpensively and in a short amount of time. Moreover, with the generation and the general availability of multiple RNAi libraries targeting the vast majority of *C. elegans* genes, many labs can now probe the function of individual genes or perform genome-wide high-throughput screens to identify components operational within a particular function. Thus, RNAi approaches in the worm have become routine methods in many labs.

RNAi-mediated gene silencing occurs when a double-stranded (ds) RNA is introduced into cells and targets a complementary mRNA for degradation (Fire et al., 1998). The molecular mechanism of RNAi has been elucidated and it is now known that dsRNA introduced into cells is cleaved by Dicer, a type III endonuclease, into 21–23 nucleotide small-interfering (si)RNAs that associate with the RNA-induced silencing complex (RISC) containing an RNA-binding protein of the Argonaute/PIWI/PAZ family (reviewed in Grishok, 2005). RISC guides the siRNA to target mRNAs that are then cleaved (and degraded) by RISC enzymatic activity. *C. elegans*, like plants and fungi, can synthesize additional (secondary) siRNAs by RNA polymerase (RdRP) activity using the target mRNA as a template, thus amplifying the RNAi response (Grishok, Tabara, & Mello, 2000). The degradation of the targeted mRNA prevents synthesis of the corresponding protein and ultimately causes protein depletion (knockdown) that leads to functional defects and altered cellular phenotypes.

In *C. elegans*, siRNAs are long (minimal fragment length of 100 base pairs), double-stranded RNA duplexes that can be delivered by injection, soaking, or feeding (described in detail later). The response is usually systemic and can elicit mRNA degradation in most, if not all, the cells in the organism. When germ cells are affected, the RNAi effect can persist into the next generation (reviewed in Grishok, 2005). Phenotypes develop when the levels of the gene product are reduced below a functional threshold. The length of time necessary to observe a phenotype depends on the stability of the gene product and the robustness of the read-out. Proteins with short half-lives ($t_{1/2}$) will be rapidly eliminated from cells and phenotypes may be visible within hours after the addition of the RNAi, while proteins with long $t_{1/2}$ will persist in cells and phenotypes may not become apparent for days after addition of the RNAi. Importantly, RNAi can be used at any time during the animal's life, and thus processes specific to particular tissues or developmental stage can be probed.

RNAi has been used in *C. elegans* for: (1) large-scale functional genomics to identify genes required for early embryonic division (Sonnichsen et al., 2005), genes required for body fat storage (Ashrafi et al., 2003), genes that mediate increased life span (Samuelson, Carr, & Ruvkun, 2007), genes essential for heme homeostasis (Severance et al., 2010), or genes that regulate the formation of P bodies (Sun et al., 2011); (2) analyses of smaller gene sets to identify genes required for the function or development of neuromuscular synapses (Sieburth et al., 2005) or genes involved in neuronal dysfunction (Kuwahara et al., 2008); and (3) for probing the functions of individual genes (e.g., the function of *cdc-42* in control of embryonic and cellular polarity; Kay & Hunter, 2001) or the function of the GTPase dynamin in cytokinesis (Thompson, Skop, Euteneuer, Meyer, & McNiven, 2002).

Similarly, RNAi has been used to identify and characterize components regulating the secretory and endocytic trafficking pathways (Balklava, Pant, Fares, & Grant, 2007; Giuliani et al., 2009; Kobuna et al., 2010; Luo et al., 2011; Michaux et al., 2011; Poteryaev, Squirrell, Campbell, White, & Spang, 2005). A genome-wide RNAi screen performed by Balklava et al. used the yolk protein-170 YP170-GFP trafficking assay (described in detail below) to identify novel regulators of endocytosis and secretion (Balklava et al., 2007). Another genome-wide RNAi screen performed by Michaux et al. identified and characterized genetic interactors of AP-1 and characterized their functions in trafficking of von Willebrand factor (Michaux et al., 2011). A smaller RNAi screen carried out by Kobuna et al. looked at specific conditions and identified components of multivesicular body formation that are linked to correct cholesterol levels (Kobuna et al., 2010).

Numerous studies used RNAi to target multiple components of known trafficking complexes or machineries, such as depleting the eight known *C. elegans* syntaxins (Jantsch-Plunger & Glotzer, 1999), different components of the ESCRT complex (Kim et al., 2011), the exocyst complex (Jiu, Jin, Liu, Holmberg, & Jantti, 2012), or the BLOC-1 complex (Hermann et al., 2012), and assess the phenotypic consequences of such depletions. In addition, many studies used RNAi to explore the function of a single traffic regulator within the secretory and/or endosomal pathways, including the function of the intra-Golgi GS28 SNARE (Maekawa et al., 2009), the endosomal SNARE snap-29 (Kang, Bai, Zegarek, Grant, & Lee, 2011), the clathrin adaptor DAB-1, endosomal RAB-8 and RAB-5 (Kamikura & Cooper, 2006), and the Golgi tethering factor p115 (Grabski et al., 2012).

RNAi-based analysis of genes regulating the secretory and endocytic pathways has been greatly facilitated by the generation of transgenic worm strains that carry a tagged (usually with a fluorescent moiety) reporter protein. Two of the most often utilized models to study secretory and endocytic traffic in *C. elegans* focuses on intestinal cells and oocytes in transgenic strains expressing the major yolk protein YP170 (VIT-2) or the YP170-receptor RME-2 tagged with GFP (Grant & Hirsh, 1999). YP170 is a cholesterol-binding/transport protein related to human ApoB-100, the major protein component of serum low-density lipoprotein (LDL). YP170-GFP is produced by intestinal cells. RME-2 is the receptor for YP170 and is related to the human LDL-receptor. RME-2 is specifically expressed in oocytes (Grant & Hirsh, 1999).

Both YP170-GFP and RME-2-GFP transgenic worms can be used to study the integrity of the secretory and endocytic pathways. In wild-type animals,

YP170-GFP is continuously synthesized in the intestine and is efficiently secreted into the body cavity (pseudocoelom) so that wild-type animals show minimal accumulation of YP170-GFP in intestinal cells. Increased presence of YP170-GFP in the intestine is indicative of a secretory defect and can be used as a phenotypic read-out of gene function. In wild-type animals RME-2-GFP expressed in oocytes localizes to the plasma membrane and cortical endosomes. Defects in secretory trafficking cause RME-2-GFP to accumulate within the ER and the Golgi, and subsequently reduce its expression at the oocyte surface.

Defects in endocytic trafficking cause impaired uptake of YP170-GFP into intracellular compartments of the oocyte and its subsequent build-up in the body cavity. In addition, some defects in endocytic traffic lead to RME-2-GFP accumulation in endosomal compartments, but in the majority of cases RME-2-GFP localization is largely indistinguishable from its distribution in wild-type animals (Balklava et al., 2007). Usually, examining the phenotypic consequences of RNAi-mediated depletion of a particular gene product in these two strains is sufficient to assign a trafficking function for that gene within either the secretory or the endocytic pathway. As mentioned earlier, these strains have been used for RNAi-based large-scale functional genomics, analyses of smaller gene sets, and for probing the functions of individual genes. Although the overall complexity of these studies varies, all utilize the same basic set of approaches and technologies. In this chapter, we describe the general set of steps used to perform RNAi in *C. elegans* to study the role of a particular gene in secretion or endocytosis in the intestine and oocytes. Specifically, we describe a work flow composed of the following steps:

A. Growth and propagation of *C. elegans*
B. Preparation of RNAi-containing bacteria
C. Delivery of RNAi by feeding
D. Immobilization of worms and live imaging of YP170-GFP and RME-2-GFP
E. Analysis of YP170-GFP and RME-2-GFP localization relative to marker proteins.

4.1 MATERIALS

1. 35, 60, and 95-mm plastic Petri dishes (Greiner)
2. 500 ml glass Erlenmeyer flasks
3. NGM agar (for 1 l mix 17 g BactoAgar, 3 g NaCl, 2.5 g Peptone, and 1 ml cholesterol (5 mg/ml in ethanol)), autoclave and let it cool until 50 °C, then add following sterile solutions: 25 ml of 1 M KPO_4 (pH 6.0), 1 ml of 1 M $MgSO_4$, and 1 ml of 1 M $CaCl_2$. Dispense 4 ml for small plates (35 mm × 10 mm) and 10 ml for medium plates (60 mm × 15 mm)
4. NGM-lite agar (for 1 l mix 17 g BactoAgar, 1.5 g NaCl, 4 g Tryptone, 3 g KH_2PO_4, 0.5 g K_2HPO_4, and 1 ml cholesterol (5 mg/ml in EtOH)), autoclave and let it cool until 50 °C, then add 1 ml of 1 M $MgSO_4$, 1 ml of 1 M $CaCl_2$, 2 ml of 1 M IPTG, and 1 ml carbenicillin (50 mg/ml). Dispense 4 ml for small plates (35 mm × 10 mm) and 10 ml for medium plates (60 mm × 15 mm)

5. IPTG 1 M stock solution in distilled water, filter sterilized, store at −20 °C
6. Polypropylene culture tubes, 18 mm × 95 mm, (Greiner)
7. 1 M $CaCl_2$, filter sterilized
8. 1 M $MgSO_4$, filter sterilized
9. 1 M KPO_4, pH 6.0 (for 1000 ml mix 108.3 g KH_2PO_4 and 35.6 g K_2HPO_4, adjust the pH to 6 with phosphoric acid), autoclave
10. 5 mg/ml cholesterol in ethanol
11. 50 mg/ml carbenicillin, filter sterilized, store at −20 °C
12. 5 mg/ml tetracycline in ethanol, store at −20 °C
13. 100 mg/ml ampicillin, filter sterilized, store at −20 °C
14. Agarose
15. M9 buffer (for 1 l mix 5 g NaCl, 3 g KH_2PO_4, 6 g Na_2HPO_4), autoclave, when cool, add 1 ml of sterile 1 M $MgSO_4$
16. 0.1 M tetramisole in M9 buffer
17. Plain glass microscope slides, ground edges, 76 mm × 26 mm (1.0–1.2 mm thick) (Fisher)
18. Glass coverslips, 22 mm × 22 mm (Fisher)
19. LB media (for 1 l dissolve 25 g of LB high-salt powder (Melford) in water), autoclave. To grow bacteria for the RNAi experiment, LB media is supplemented with 100 μl/ml of ampicillin
20. LB media with 2% (w/v) agar (for 1 l add 25 g of LB high-salt powder and 20 g of agar), autoclave, cool until 55 °C, and dispense 20 ml into large (95 mm × 15 mm) dishes. For growth of RNAi clones add 1 ml of sterile 100 mg/ml ampicillin and 0.125 ml of 5 mg/ml tetracycline before pouring into the plates
21. *E. coli* OP50 and HT115 strains can be obtained from *C. elegans* Genetics Centre (http://www.cbs.umn.edu/CGC/strains/)
22. *pwIs23* (VIT-2::GFP) transgenic worms expressing YP170-GFP, *pwIs116* (RME-2::GFP) transgenic worms expressing GFP-tagged yolk receptor, and ojIs37 (pie1p::GFP::UGTP-1) expressing GFP-tagged Golgi marker can be obtained from *C. elegans* Genetics Centre (http://www.cbs.umn.edu/CGC/strains/)
23. pie-1p::GFP::SP12 transgenic worms expressing GFP-tagged ER marker and *pwIs42 (ppie1::GFP::rme-1)* transgenic worms expressing GFP-tagged recycling endosome marker have been described before by Poteryaev et al. (2005) and Balklava et al. (2007), respectively. These transgenic strains can be requested from the respective laboratories where they have been generated. L4440 feeding vector can be obtained from AddGene (http://www.addgene.org/).

4.2 METHODS

4.2.1 Worm maintenance

Worms are routinely grown on NGM agar plates seeded with *E. coli* OP50. OP50 is kept as a frozen bacterial stock at −80 °C and streaked on LB agar plate prior to use (a streaked plate can be kept at 4 °C for a month). To prepare bacterial lawns, a single

OP50 colony is picked from the streaked LB agar plate and inoculated into an Erlenmeyer flask containing 200 ml of LB media and grown overnight at 37 °C in a shaking incubator (200 rpm). Next day NGM agar plates are seeded with 100 or 200 μl of OP50 overnight cultures into small (3 cm) or medium (6 cm) plates, respectively, and left overnight at room temperature to dry and form a lawn. Plates seeded with OP50 can be kept refrigerated for a month.

Plates seeded with *E. coli* OP50 lawns are prewarmed at 20 °C and around 5–20 worms are placed per plate and kept there till they exhaust the bacterial lawn. When all the bacterial lawn has disappeared (depending on the amount of worms transferred, the strain and the growth temperature it can take from 3 days to 2 weeks), a few worms are picked and transferred as discussed earlier or alternatively chunks containing a few hundred worms are cut with a flame-sterilized spatula and transferred onto fresh OP50 seeded plate.

4.2.2 Generating RNAi

There are four general ways of delivering dsRNA into *C. elegans*: (1) injection of dsRNA into worms (Fire et al., 1998), (2) *in vivo* production of dsRNA from transgenic promoters (Tavernarakis, Wang, Dorovkov, Ryazanov, & Driscoll, 2000), (3) soaking worms in dsRNA (Tabara, Grishok, & Mello, 1998), and (4) feeding worms with bacteria expressing dsRNA (Timmons, Court, & Fire, 2001). We focus this chapter on delivery of RNAi by feeding, and thus describe libraries and protocols for generating bacterial strains expressing relevant RNAi for this mode of delivery. Generation of RNAi for other modes of delivery and the delivery methods are nicely summarized in a recent review (Wang & Barr, 2005).

There are two *C. elegans* RNAi feeding libraries that are readily available to researchers. One is from the Ahringer laboratory (Kamath & Ahringer, 2003) and the other from the Vidal laboratory (Rual et al., 2004). There is some overlap between the two libraries, and together they are predicted to cover about 94% of all *C. elegans* genes. Both libraries were made by inserting a fragment of each individual gene into the L4440 feeding vector. L4440 is a modified version of the Bluescript cloning vector with a T7 promoter on each side of the multiple cloning site driving transcription of each DNA strand (Timmons et al., 2001). While Kamath et al. used conventional PCR to amplify genomic DNA fragments as a template for "TA cloning" into the EcoRV site modified by addition of 3′-T overhangs (Kamath & Ahringer, 2003), Rual et al. utilized version 1.1 of the *C. elegans* ORFeome resource which contains 11,942 cloned ORFs (Reboul et al., 2001) and transferred them into L4440 vector by Gateway cloning. L4440 vectors containing the insert were then transformed into HT115 (DE3), an RNase III-deficient *E. coli* strain with IPTG-inducible T7 polymerase activity (Timmons et al., 2001). RNase III is an enzyme that normally degrades a majority of dsRNAs in the bacterial cell and bacteria deficient of RNase III were specifically engineered to produce high quantities of specific dsRNA segments; in addition, the deficiency is marked with tetracycline resistance (Timmons et al., 2001).

The RNAi libraries in the form of bacterial feeding clones are available from commercial sources. The Ahringer's RNAi library is provided as frozen glycerol stocks of bacterial strains arrayed in 384-well plates (or as individual clones) and is distributed by Source BioScience LifeSciences (http://www.lifesciences.sourcebioscience.com/clone-products/mirna–rnai-resources/c-elegans-rnai-library.aspx). Source Bioscience provides a wide range of online tools and support information regarding the RNAi clones and protocols. The Vidal lab RNAi library is also distributed as frozen glycerol stocks from either Source BioScience or ThermoFisher (http://www.thermoscientificbio.com/non-mammalian-cdna-and-orf/c.-elegans-rnai/) and a range of supplemental and support information is available from ThermoFisher Website.

If the gene of interest is not covered by any of the commercial libraries, it is possible to make the feeding construct in house using standard cloning techniques. A fragment of 200–2000 bp (dsRNAs shorter than 150 bp are less effective) of the gene of interest is amplified by PCR using either genomic or cDNA as a template and ligated between two opposite-facing T7 promoters in the feeding vector L4440. If genomic DNA is used as a template, maximum overlap with the coding sequence should be targeted. Standard molecular cloning methods are used for PCR and cloning and can be found in Maniatis (1982). The protocols used to generate the Ahringer library are described in Kamath and Ahringer (2003). The construct is then sequence verified and transformed into HT115 bacterial strain. Bacterial strains can be stored as frozen stocks at −80 °C.

4.2.3 RNAi delivery by feeding

For RNAi experiments, bacterial RNAi clones are first streaked from the frozen stock on LB agar plates containing 100 μg/ml ampicillin and 12.5 μg/ml tetracycline and grown overnight at 37 °C (Fig. 4.1). Next day, a single colony is inoculated into 5 ml of LB media containing 100 μg/ml ampicillin and grown overnight at 37 °C. 100 or 200 μl of the overnight culture is then seeded on either 30 or 60 mm NGM-lite plates, respectively, and the plates are left overnight at room temperature to induce expression of the dsRNA (NGM-lite contains IPTG to induce expression of the dsRNA).

Next day 20–50 L1 stage worms or eggs are transferred to the seeded NGM-lite plates and incubated at 20 °C for 3–4 days. At that time the young adults are scored for phenotypes, as described in detail below. It is important to set the experiment with synchronized or semi-synchronized worms, so that the phenotype is scored in age-matched animals. If studying more than one gene of interest and using many plates, it is beneficial to start with synchronized starved L1s which can be obtained by sodium hypochlorite treatment of gravid adults as described in Stiernagle (2006) and letting the eggs to hatch in sterile M9 buffer. If only studying one or two genes at the same time it is easier to work with semi-synchronized worms. There are several ways of obtaining semi-synchronized eggs or L1 stage worms: (1) place 7–10 gravid adults on RNAi plates for 4–6 h, let them lay eggs and then remove the adults; (2) transfer 20–50 freshly laid eggs to the RNAi plates; or (3) individually pick 20–50 L1 stage

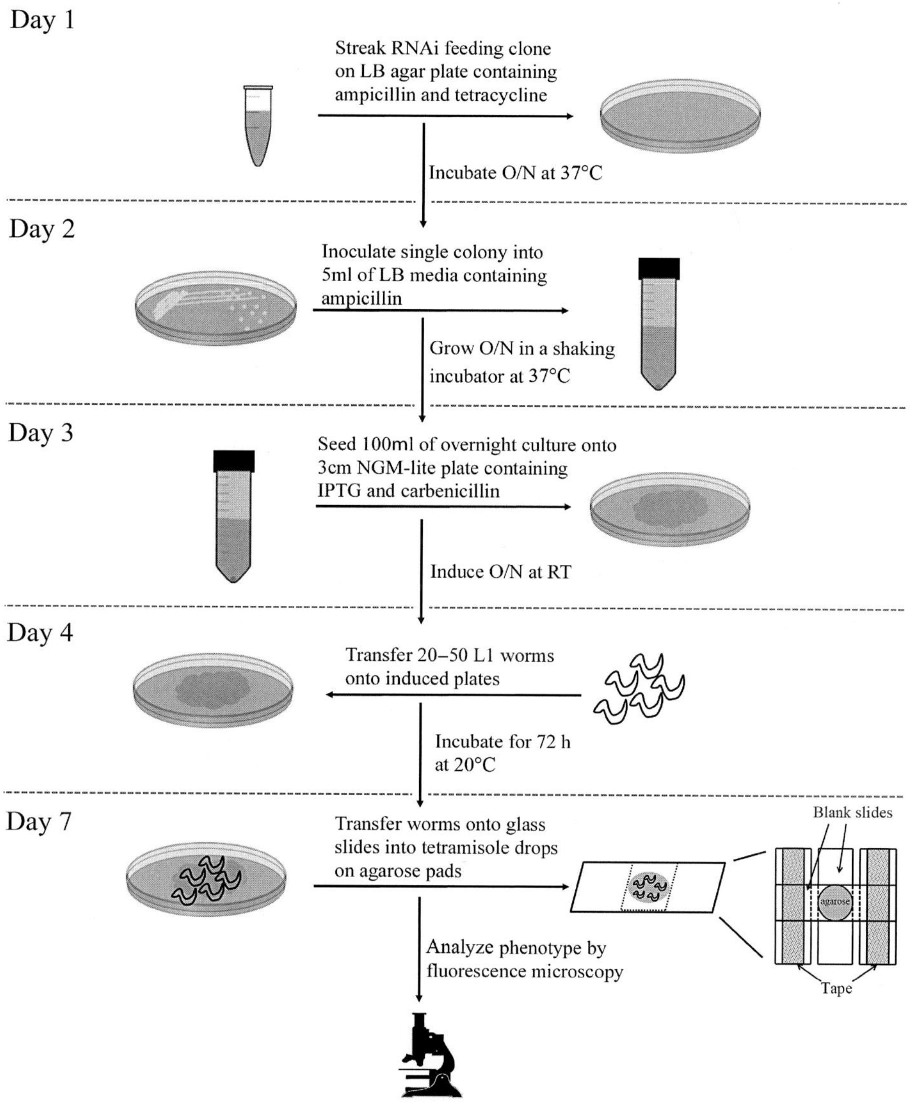

FIGURE 4.1

Diagram of the overall protocol for RNAi in *C. elegans*. See text for detailed description. (See color plate.)

worms and transfer them to the RNAi plates. Freshly starved plates often contain large numbers of L1 animals. In some cases, when an essential gene is targeted, starting with L1 stage worms will result in arrested larvae or sterile adults, and in that case it is advisable to reduce the exposure to RNAi and start with L3/L4 larval stage worms and score the phenotypes of young adults 24–36 h later. Routinely, the

RNAi experiment is repeated two to three times to ensure the reproducibility of the phenotypes.

All RNAi experiments must include a negative control, which involves feeding worms with the L4440 vector lacking the RNAi (empty vector). It is often possible to also include a positive control, especially when the process studied is well characterized and other regulators are already known. One possible positive control for secretion is RNAi for SNAP-29 (Sato, Saegusa, Sato, Hara, & Harada, 2011). SNAP-29 encodes a target-SNARE involved in the fusion of secretory vesicles with the plasma membrane and SNAP-29 RNAi inhibits yolk secretion from the intestine (Kang et al., 2011). A positive control for inhibition of endocytosis is RNAi for receptor-mediated-endocytosis-1 (RME-1) (Balklava et al., 2007). RME-1 encodes an Eps15-homology-domain protein that together with AMPH-1 regulates the formation of carriers mediating transport between the recycling endosome and the plasma membrane (Grant et al., 2001; Pant et al., 2009). RME-1 RNAi impairs yolk receptor RME-2 recycling to the plasma membrane and subsequently reduces YP170 uptake from the body cavity.

Detection of a phenotype is usually a reliable indication of a positive RNAi result, as the level of false positives is $<1\%$, (Kamath et al., 2003). Conversely, it is more difficult to draw conclusions from a negative RNAi experiment as the false negative rate is about 30% (Kamath et al., 2003). In this case, it is always possible to optimize the RNAi assay and as a starting point to increase the length of exposure to the RNAi and scoring the F1 adults. RNAi effectiveness can also be potentiated in *C. elegans* by transferring a transgene into a mutant strain with an enhanced RNAi response and using the resulting strain. Mutations in genes such as *rrf-3* (RNA-direct RNA polymerase family; Simmer et al., 2002) and *eri-1* (enhanced RNAi; Kennedy, Wang, & Ruvkun, 2004) produce an enhanced RNAi response in many tissues. In fact, the YP170-GFP assay to discover novel membrane-trafficking regulators was performed in *rrf-3* mutant background (Balklava et al., 2007). Despite efforts to optimize the RNAi procedure, there is extensive variability in effectiveness of particular dsRNA using different conditions, and the final design must consider different technical approaches as well as the biological question to be addressed.

4.2.4 Live imaging

Many phenotypic changes (e.g., body length, larval development, growth arrest) can be used to assess the possible function of a particular gene in membrane trafficking. However, one of the most direct and easiest approaches involves performing the RNAi in transgenic *C. elegans* strains expressing GFP-tagged YP170 and GFP-tagged RME-2 (developed by the Grant laboratory; http://www.barthgrant.com/), as described in detail and referenced in the Introduction. Such transgenic worms are subject to RNAi as described above. After 3–4 days, live young adult worms are mounted on slides containing 2% agarose pads in a 3–5 μl drop of 0.1 M tetramisole (a paralytic that eliminates movement), covered with a coverslip and imaged within 1 h.

Agarose pads are prepared as follows (Fig. 4.1): (1) put two layers of lab tape lengthwise on two glass slides; (2) place the two taped slides on a flat surface and place a clean slide between them; (3) melt 1 g of agarose in 50 ml of M9 buffer and using a Pasteur pipette, place a drop of agarose onto the clean slide; (4) quickly cover the agarose drop with another clean slide placed on top of the three slides in a perpendicular fashion; (5) gently press until the agarose drop is flattened to a circle about 0.5 mm thick (the thickness of the tape spacers); (6) when agarose has solidified, separate the two blank slides by sliding one relative to the other. The agarose pad usually sticks to one of the slides. Prepared slides with agarose pads should be used immediately or can be stored in humidifying chambers for a few hours before they dry out.

Imaging of the worms can be performed on any fluorescent microscope, but we routinely use a fully motorized Zeiss Axiovert 200M fluorescence microscope (Carl Zeiss Ltd., Welwyn Garden City, UK) and a Hamamatsu Orca camera driven by Volocity software (Improvision, Coventry, UK).

Images are obtained for control worms (fed empty L4440 vector) and RNAi worms, focusing on the localization of YP170-GFP within the entire worm and the localization of RME-2-GFP within the oocytes. In wild-type worms, YP170-GFP is detected in the intestine, oocytes, and in the developing embryos (Fig. 4.2A), while RME-2-GFP localizes to the cortical endosomes and plasma membrane of oocytes (Fig. 4.2B). The accumulation of YP170-GFP in the body cavity (as shown for RNAi-mediated depletion of the Golgi tethering factor *uso-1*; Grabski et al., 2012; Figure 4.2C) or the endosomal regulator *cdc-42* (Fig. 4.2E) indicates lack of efficient uptake into oocytes. This phenotype may be caused by defects within the endosomal pathway or may result from defects in the exocytic delivery of the RME-2 receptor to the plasma membrane of oocytes. To assign the lesion to either the endosomal or the secretory pathway, the localization of RME-2-GFP within oocytes is assessed. The observed retention of RME-2-GFP within intracellular compartments in worms fed *uso-1* RNAi (Fig. 4.2D) and the delivery of RME-2 to the cell surface in worms fed *cdc-42* RNAi (Fig. 4.2F) indicate that depletion of *uso-1* inhibits the secretory pathway while depletion of *cdc-42* inhibits endocytic trafficking.

4.2.5 Microscopy and analysis

The analysis described above is often sufficient to determine whether the gene product silenced by the tested RNAi functions within the secretory or endocytic pathways. However, it is often advisable to further refine the information content by assessing the particular stage of trafficking that is blocked by depletion of the gene product. For these analyses, the localization of RME-2-GFP in RNAi-depleted worms is analyzed relative to markers of specific secretory and endocytic compartments. Numerous strains expressing markers of the various compartments tagged with fluorescent proteins have been developed and characterized (Poteryaev et al., 2005; Sato et al., 2005, 2006, 2008). Each marker shows a characteristic distribution within each tissue (Fig. 4.3). Importantly, the distribution patterns are highly consistent between individual animals, and comparison of the distribution of RME-2-GFP

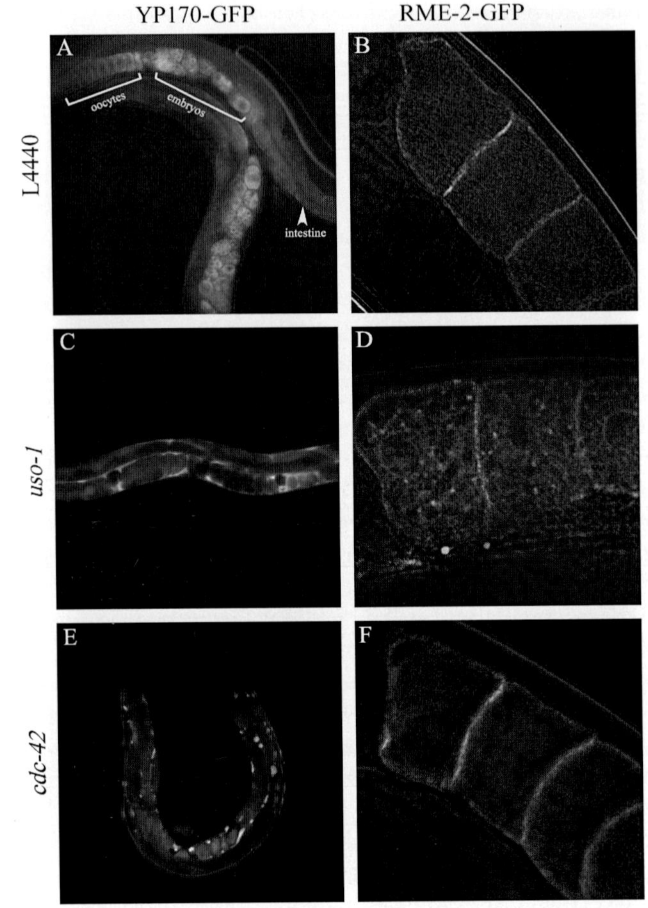

FIGURE 4.2

Assessing gene function in transgenic *C. elegans*. Transgenic L1 animals expressing YP170-GFP or RME-2-GFP were fed control RNAi (A and B), *uso-1* RNAi (C and D), or *cdc-42* RNAi (E and F) and imaged 3 days later as young adults. Depletion of *uso-1* causes secretory defect while depletion of *cdc-42* results in inhibition in endocytosis (details in the text). (See color plate.)

to the published pattern of each marker can provide significant insight into the stage of traffic that is blocked by the RNAi.

A block in secretory traffic at the level of the ER generally results in a diffuse cellular distribution of RME-2-GFP within oocytes, similar to that shown for the SP12-GFP marker, while an arrest at the Golgi causes the accumulation of RME-2-GFP in numerous puncta scattered throughout the oocyte, analogous to that shown by the GFP-UGTP-1 marker (Fig. 4.3A and B). A block at the level of recycling endosome causes RME-2-GFP accumulation within cortical vesicles immediately

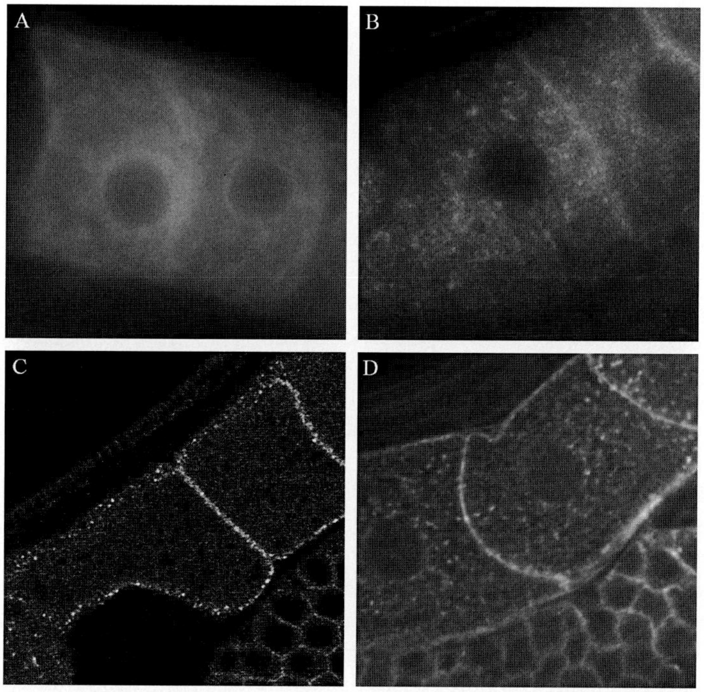

FIGURE 4.3

Marker distribution in *C. elegans* oocytes. Transgenic animals expressing the ER marker SP12-GFP (A), the Golgi marker GFP-UGTP-1 (B), the recycling endosome marker GFP-RME-1 (C), or the lysosome marker GFP-LMP-1 (D) were imaged. (See color plate.)

adjacent to the PM, a distribution shown by the GFP-RME-1 marker, while defects in trafficking at the lysosomes manifest as accumulation of RME-2-GFP in slightly larger, more centrally located vesicles as shown by the localization of the GFP-LMP-1 marker (Fig. 4.3C and D). Using the localization of RME-2-GFP relative to such compartment markers can inform on the function of each gene product within the secretory and endosomal pathway and form the basis for assigning individual genes into stage-specific groups. This in turn can inform on possible interactions (direct or indirect) between the genes.

To precisely delineate the compartment at which RME-2-GFP is arrested after RNAi, it is sometimes possible to generate a doubly transgenic worms that coexpress the relevant compartment marker and RME-2-GFP and then use such double transgenes for RNAi. Unfortunately, germline often does not tolerate expression of two transgenes and results in silencing of the expression of either one or both transgenes. Alternatively, it is possible to replace the expression of one of the transgenes with immunofluorescent labeling with the appropriate antibody of the endogenous protein. Double transgenic approach is more easily achievable in the intestine where it is possible to either express fluorescently tagged protein of interest or endocytic cargo protein such as human transferrin receptor or human TAC (alpha-chain of the IL-2 receptor; Chen et al., 2006) together with the marker of intracellular membrane compartment tagged with a different fluorescent moiety and look for the general colocalization as well as colocalization following exposure to RNAi.

4.3 DISCUSSION

Genetic analyses in model animals (including the worm *C. elegans*) over the last few decades have provided unprecedented knowledge of gene-specific functions. Studying gene function using a worm carrying a mutation in a particular gene is of great benefit, but unfortunately is not always possible for several reasons: there is no such mutant allele available; mutations in essential genes cause lethality and restrict use of homozygous mutant worms; and the mutation does not necessarily cause loss-of-function phenotype. Forward genetics is also widely used and is preferred over reverse genetics, but it is very time consuming and the outcomes can be unpredictable. There are currently two *C. elegans* knockout consortia that can generate a deletion allele for a gene of interest "on request," but the process requires a long wait and there is no guarantee that the resulting deletion will cause a scorable loss-of-function phenotype. It is possible to do a mutagenesis and obtain a mutant in a standard laboratory, bearing in mind that the generation of mutants as well as the characterization of the actual mutant allele is time consuming and does not always result in generation of desired mutant allele.

Because of the difficulties inherent in genetic approaches, the discovery of RNA-mediated interference in *C. elegans* instigated a new era in research, and RNAi in worms has become one of the most powerful tools to study gene function. The RNAi technique offers several advantages over the use of mutant worms. First, most genes can be silenced using RNAi clones from commercial libraries. Even when the RNAi clone is not represented in a commercially distributed library it is relatively easy to make your own RNAi construct from worm gDNA or cDNA using standard cloning techniques. Second, once the RNAi clone has been obtained, it is easy and fast to modify the experimental conditions to achieve optimal gene knockdown by varying parameters such as time and temperature of RNAi treatment, induction strength, worm stage, etc. Third, once the experimental conditions are optimized, the results can be obtained in a relatively short time on many animals, thus allowing statistical analysis of the phenotype (however, it needs to be mentioned that RNAi often produces greater phenotypic variability between individuals when compared to the population of mutant worms). Fourth, because RNAi leads to the removal of the gene product, the resulting phenotype is by definition due to loss of function as opposed to mutations, which may lead to aberrant function or even activation.

Analysis of genes regulating membrane trafficking in the worm has benefited from both forward and reverse genetics approaches and continues to do so. However, the advent of RNAi and the increasing availability of relevant reagents and improved experimental approaches has positioned the worm as a model organism that is seductive to many researchers. With a constant growth of transgenic worm strains carrying fluorescent markers for intracellular membrane compartments in different tissues, it is becoming easier to dissect the function of gene of interest and its role in regulation of membrane transport. It is clear that the utility of *C. elegans* is growing and that it represents a valid and informative model system that is accessible to many researchers. We believe that the combination of easy maintenance with the availability of relevant reagents and transgenic strains will increase the range of questions that can be answered using RNAi in worms.

SUMMARY

RNAi-mediated gene silencing is commonly used to analyze gene function in *C. elegans* and can be complemented with mutational, molecular, and biochemical approaches to define precise physiological relevance of gene products. The RNAi approach has been extensively used to provide information about key aspects of development and cell and biochemical processes. While the approach has limitations, it provides an excellent means to probe the physiological importance of genes within a multicellular organism and allows the characterization of tissue-specific and stage-specific cellular functions. Hence, numerous studies have used and continue to use RNAi to target components of known trafficking complexes or machineries and reveal their physiological functions.

Acknowledgment

This work was supported by the Royal Society Dorothy Hodgkin Fellowship to Z. B.

References

Ashrafi, K., Chang, F. Y., Watts, J. L., Fraser, A. G., Kamath, R. S., Ahringer, J., et al. (2003). Genome-wide RNAi analysis of Caenorhabditis elegans fat regulatory genes. *Nature*, *421*(6920), 268–272.

Balklava, Z., Pant, S., Fares, H., & Grant, B. D. (2007). Genome-wide analysis identifies a general requirement for polarity proteins in endocytic traffic. *Nature Cell Biology*, *9*(9), 1066–1073.

Brenner, S. (1974). The genetics of Caenorhabditis elegans. *Genetics*, *77*(1), 71–94.

Chen, C. C., Schweinsberg, P. J., Vashist, S., Mareiniss, D. P., Lambie, E. J., & Grant, B. D. (2006). RAB-10 is required for endocytic recycling in the Caenorhabditis elegans intestine. *Molecular Biology of the Cell*, *17*(3), 1286–1297.

Consortium CeS (1998). Genome sequence of the nematode C. elegans: A platform for investigating biology. *Science*, *282*(5396), 2012–2018.

Fire, A., Xu, S., Montgomery, M. K., Kostas, S. A., Driver, S. E., & Mello, C. C. (1998). Potent and specific genetic interference by double-stranded RNA in Caenorhabditis elegans. *Nature*, *391*(6669), 806–811.

Giuliani, C., Troglio, F., Bai, Z., Patel, F. B., Zucconi, A., Malabarba, M. G., et al. (2009). Requirements for F-BAR proteins TOCA-1 and TOCA-2 in actin dynamics and membrane trafficking during Caenorhabditis elegans oocyte growth and embryonic epidermal morphogenesis. *PLoS Genetics*, *5*(10), e1000675.

Grabski, R., Balklava, Z., Wyrozumska, P., Szul, T., Brandon, E., Alvarez, C., et al. (2012). Identification of a functional domain within the p115 tethering factor that is required for Golgi ribbon assembly and membrane trafficking. *Journal of Cell Science*, *125*(Pt. 8), 1896–1909.

Grant, B., & Hirsh, D. (1999). Receptor-mediated endocytosis in the Caenorhabditis elegans oocyte. *Molecular Biology of the Cell*, *10*(12), 4311–4326.

Grant, B., Zhang, Y., Paupard, M. C., Lin, S. X., Hall, D. H., & Hirsh, D. (2001). Evidence that RME-1, a conserved C. elegans EH-domain protein, functions in endocytic recycling. *Nature Cell Biology*, *3*(6), 573–579.

Grishok, A. (2005). RNAi mechanisms in Caenorhabditis elegans. *FEBS Letters*, *579*(26), 5932–5939.

Grishok, A., Tabara, H., & Mello, C. C. (2000). Genetic requirements for inheritance of RNAi in C. elegans. *Science*, *287*(5462), 2494–2497.

Hermann, G. J., Scavarda, E., Weis, A. M., Saxton, D. S., Thomas, L. L., Salesky, R., et al. (2012). C. elegans BLOC-1 functions in trafficking to lysosome-related gut granules. *PLoS One*, *7*(8), e43043.

Jantsch-Plunger, V., & Glotzer, M. (1999). Depletion of syntaxins in the early Caenorhabditis elegans embryo reveals a role for membrane fusion events in cytokinesis. *Current Biology*, *9*(14), 738–745.

Jiu, Y., Jin, C., Liu, Y., Holmberg, C. I., & Jantti, J. (2012). Exocyst subunits Exo70 and Exo84 cooperate with small GTPases to regulate behavior and endocytic trafficking in C. elegans. *PLoS One*, *7*(2), e32077.

Kamath, R. S., & Ahringer, J. (2003). Genome-wide RNAi screening in Caenorhabditis elegans. *Methods*, *30*(4), 313–321.

Kamath, R. S., Fraser, A. G., Dong, Y., Poulin, G., Durbin, R., Gotta, M., et al. (2003). Systematic functional analysis of the Caenorhabditis elegans genome using RNAi. *Nature*, *421*(6920), 231–237.

Kamikura, D. M., & Cooper, J. A. (2006). Clathrin interaction and subcellular localization of Ce-DAB-1, an adaptor for protein secretion in Caenorhabditis elegans. *Traffic*, *7*(3), 324–336.

Kang, J., Bai, Z., Zegarek, M. H., Grant, B. D., & Lee, J. (2011). Essential roles of snap-29 in C. elegans. *Developmental Biology*, *355*(1), 77–88.

Kay, A. J., & Hunter, C. P. (2001). CDC-42 regulates PAR protein localization and function to control cellular and embryonic polarity in C. elegans. *Current Biology*, *11*(7), 474–481.

Kennedy, S., Wang, D., & Ruvkun, G. (2004). A conserved siRNA-degrading RNase negatively regulates RNA interference in C. elegans. *Nature*, *427*(6975), 645–649.

Kim, D. W., Sung, H., Shin, D., Shen, H., Ahnn, J., Lee, S. K., et al. (2011). Differential physiological roles of ESCRT complexes in Caenorhabditis elegans. *Molecules and Cells*, *31*(6), 585–592.

Kobuna, H., Inoue, T., Shibata, M., Gengyo-Ando, K., Yamamoto, A., Mitani, S., et al. (2010). Multivesicular body formation requires OSBP-related proteins and cholesterol. *PLoS Genetics*, *6*(8), 1–16.

Kuwahara, T., Koyama, A., Koyama, S., Yoshina, S., Ren, C. H., Kato, T., et al. (2008). A systematic RNAi screen reveals involvement of endocytic pathway in neuronal dysfunction in alpha-synuclein transgenic C. elegans. *Human Molecular Genetics*, *17*(19), 2997–3009.

Luo, L., Hannemann, M., Koenig, S., Hegermann, J., Ailion, M., Cho, M. K., et al. (2011). The Caenorhabditis elegans GARP complex contains the conserved Vps51 subunit and is required to maintain lysosomal morphology. *Molecular Biology of the Cell*, *22*(14), 2564–2578.

Maekawa, M., Inoue, T., Kobuna, H., Nishimura, T., Gengyo-Ando, K., Mitani, S., et al. (2009). Functional analysis of GS28, an intra-Golgi SNARE, in Caenorhabditis elegans. *Genes to Cells*, *14*(8), 1003–1013.

Maniatis, T. (1982). *Molecular cloning : A laboratory manual*. In T. Maniatis, E. F. Fritsch, & J. Sambrook (Eds.), *Manual for genetic engineering*. Cold Spring Harbor, N.Y.: Cold Spring Harbor Laboratory. http://nla.gov.au/nla.cat-vn1699650.

Markaki, M., & Tavernarakis, N. (2010). Modeling human diseases in Caenorhabditis elegans. *Biotechnology Journal*, *5*(12), 1261–1276.

Michaux, G., Dyer, C. E., Nightingale, T. D., Gallaud, E., Nurrish, S., & Cutler, D. F. (2011). A role for Rab10 in von Willebrand factor release discovered by an AP-1 interactor screen in C. elegans. *Journal of Thrombosis and Haemostasis*, *9*(2), 392–401.

Pant, S., Sharma, M., Patel, K., Caplan, S., Carr, C. M., & Grant, B. D. (2009). AMPH-1/Amphiphysin/Bin1 functions with RME-1/Ehd1 in endocytic recycling. *Nature Cell Biology*, *11*(12), 1399–1410.

Poteryaev, D., Squirrell, J. M., Campbell, J. M., White, J. G., & Spang, A. (2005). Involvement of the actin cytoskeleton and homotypic membrane fusion in ER dynamics in Caenorhabditis elegans. *Molecular Biology of the Cell*, *16*(5), 2139–2153.

Reboul, J., Vaglio, P., Tzellas, N., Thierry-Mieg, N., Moore, T., Jackson, C., et al. (2001). Open-reading-frame sequence tags (OSTs) support the existence of at least 17,300 genes in C. elegans. *Nature Genetics*, *27*(3), 332–336.

Rual, J. F., Ceron, J., Koreth, J., Hao, T., Nicot, A. S., Hirozane-Kishikawa, T., et al. (2004). Toward improving Caenorhabditis elegans phenome mapping with an ORFeome-based RNAi library. *Genome Research*, *14*(10B), 2162–2168.

Samuelson, A. V., Carr, C. E., & Ruvkun, G. (2007). Gene activities that mediate increased life span of C. elegans insulin-like signaling mutants. *Genes and Development*, *21*(22), 2976–2994.

Sato, M., Saegusa, K., Sato, K., Hara, T., & Harada, A. (2011). Caenorhabditis elegans SNAP-29 is required for organellar integrity of the endomembrane system and general exocytosis in intestinal epithelial cells. *Molecular Biology of the Cell*, *22*(14), 2579–2587.

Sato, K., Sato, M., Audhya, A., Oegema, K., Schweinsberg, P., & Grant, B. D. (2006). Dynamic regulation of Caveolin-1 trafficking in the germ line and embryo of Caenorhabditis elegans. *Molecular Biology of the Cell*, *17*(7), 3085–3094.

Sato, M., Sato, K., Fonarev, P., Huang, C. J., Liou, W., & Grant, B. D. (2005). Caenorhabditis elegans RME-6 is a novel regulator of RAB-5 at the clathrin-coated pit. *Nature Cell Biology*, *7*(6), 559–569.

Sato, M., Sato, K., Liou, W., Pant, S., Harada, A., & Grant, B. D. (2008). Regulation of endocytic recycling by C. elegans Rab35 and its regulator RME-4, a coated-pit protein. *EMBO Journal*, *27*(8), 1183–1196.

Severance, S., Rajagopal, A., Rao, A. U., Cerqueira, G. C., Mitreva, M., El-Sayed, N. M., et al. (2010). Genome-wide analysis reveals novel genes essential for heme homeostasis in Caenorhabditis elegans. *PLoS Genetics*, *6*(7), e1001044.

Sieburth, D., Ch'ng, Q., Dybbs, M., Tavazoie, M., Kennedy, S., Wang, D., et al. (2005). Systematic analysis of genes required for synapse structure and function. *Nature*, *436*(7050), 510–517.

Simmer, F., Tijsterman, M., Parrish, S., Koushika, S. P., Nonet, M. L., Fire, A., et al. (2002). Loss of the putative RNA-directed RNA polymerase RRF-3 makes C. elegans hypersensitive to RNAi. *Current Biology*, *12*(15), 1317–1319.

Sonnichsen, B., Koski, L. B., Walsh, A., Marschall, P., Neumann, B., Brehm, M., et al. (2005). Full-genome RNAi profiling of early embryogenesis in Caenorhabditis elegans. *Nature*, *434*(7032), 462–469.

Stiernagle, T. (2006). Maintenance of C. elegans. In *The C. elegans research community*: WormBook. http://dx.doi.org/10.1895/wormbook.

Sulston, J. E., Schierenberg, E., White, J. G., & Thomson, J. N. (1983). The embryonic cell lineage of the nematode Caenorhabditis elegans. *Developmental Biology*, *100*(1), 64–119.

Sun, Y., Yang, P., Zhang, Y., Bao, X., Li, J., Hou, W., et al. (2011). A genome-wide RNAi screen identifies genes regulating the formation of P bodies in C. elegans and their functions in NMD and RNAi. *Protein and Cell*, *2*(11), 918–939.

Tabara, H., Grishok, A., & Mello, C. C. (1998). RNAi in C. elegans: Soaking in the genome sequence. *Science*, *282*(5388), 430–431.

Tavernarakis, N., Wang, S. L., Dorovkov, M., Ryazanov, A., & Driscoll, M. (2000). Heritable and inducible genetic interference by double-stranded RNA encoded by transgenes. *Nature Genetics*, *24*(2), 180–183.

Thompson, H. M., Skop, A. R., Euteneuer, U., Meyer, B. J., & McNiven, M. A. (2002). The large GTPase dynamin associates with the spindle midzone and is required for cytokinesis. *Current Biology*, *12*(24), 2111–2117.

Timmons, L., Court, D. L., & Fire, A. (2001). Ingestion of bacterially expressed dsRNAs can produce specific and potent genetic interference in Caenorhabditis elegans. *Gene*, *263*(1–2), 103–112.

Wang, J., & Barr, M. M. (2005). RNA interference in Caenorhabditis elegans. *Methods in Enzymology*, *392*, 36–55.

CHAPTER

Endomembrane and Golgi Traffic in Plant Cells

5

Giovanni Stefano*,†, Anne Osterrieder†, Chris Hawes‡ and Federica Brandizzi*,†

**Department of Energy Plant Research Laboratory, Michigan State University, East Lansing, Michigan, USA*

†Department of Plant Biology, Michigan State University, East Lansing, Michigan, USA

‡Department of Biological and Medical Sciences, Oxford Brookes University, Oxford, United Kingdom

CHAPTER OUTLINE

Introduction 70

5.1 Leaves as a Model System for Analyzing the Plant Golgi and Secretory Pathway ... 71

5.1.1 Methods 72

5.1.1.1 Transformation 72

5.2 Mutant Screens for Trafficking Proteins 73

5.2.1 Methods 74

5.2.1.1 EMS Treatment, Confocal Screening, and Mapping 74

5.3 Photobleaching for Transport Studies 75

5.3.1 Methods 75

5.3.1.1 Bleach Protocol 76

5.4 Förster Resonance Energy Transfer for Protein Interactions 77

5.4.1 Methods 80

Summary 81

Acknowledgments 81

References 81

Abstract

Leaf epidermal cells make ideal specimens for the investigation of the plant secretory pathway in that it is relatively easy to tag with fluorescent proteins and visualize *in vivo* the various organelles of the pathway. A number of techniques can be employed to identify and study proteins within the endomembrane organelles and to study their dynamics and interactions. Here, we discuss the most commonly used approaches to express proteins within arabidopsis and tobacco leaves, the use of

Methods in Cell Biology, Volume 118

ISSN 0091-679X
http://dx.doi.org/10.1016/B978-0-12-417164-0.00005-7

mutant screens to identify trafficking proteins, and the use of two *in vivo* techniques, Fluorescence recovery after photobleaching and Förster resonance energy transfer, to study protein dynamics in plant cells.

INTRODUCTION

At the heart of the cell lies the secretory pathway: an intricate system of organelles that work in concert to produce and modify one-third of all cellular proteins, as well as lipids, which make up the majority of the cell membranes, and carbohydrates, which are attached to most secretory proteins. In plant cells, these organelles also make important components of the cell wall. The first organelle of the secretory pathway is the ER. The synthesis of most of the plant secretory products is initiated in the reticulated network of the ER, from which they traffic to the Golgi to be distributed to other secretory compartments and, as recently discovered, also to chloroplasts (Matheson, Hanton, & Brandizzi, 2006). The Golgi apparatus, which in plants is made of dispersed, mobile, and polarized stacks in close proximity to the ER (Brandizzi, Snapp, Roberts, Lippincott-Schwartz, & Hawes, 2002), collects and processes secretory materials before sorting them to their destinations. The Golgi also receives materials from the vacuole and the plasma membrane (Foresti & Denecke, 2008; Matheson et al., 2006). In plants, the Golgi synthesizes materials of a cellulosic cell wall that is not generally found in nonplant organisms and sorts the machinery involved in wall deposition to the plasma membrane (Keegstra, 2010). The secretory pathway is upregulated in response to biotic stresses and serves as a reservoir of membrane-tethered transcription factors that help cells to cope with abiotic stress and developmental clues (Chen, Slabaugh, & Brandizzi, 2008). In addition, the secretory pathway has important roles in synthesizing and retaining important receptors for plant hormones (e.g., the ethylene receptor; Chen, Randlett, Findell, & Schaller, 2002) and mediating their subcellular traffic (e.g., BRI; Geldner, Hyman, Wang, Schumacher, & Chory, 2007). Implicit in the vital roles of the secretory pathway for the inner workings of the cell as well as plant growth and development is the existence of a tight regulatory network to ensure membrane homeostasis during the processes of membrane traffic which entails movement of lipids, membrane, and sugars among distinct cellular subcompartments.

Compared to other systems that are less amenable than plants to visual inspection of endomembranes, such as yeast, or that do not benefit from powerful genetic tools, such as mammalian culture cells, the use of model plant species enables modern scientists to address specific questions that are fundamental in plants and all other eukaryotes using a powerful system for live cell biology combined with genetics studies. In this chapter, we discuss a number of techniques and experimental setups that are enabling plant researchers to discover the identity of the players and their dynamics and interactions that influence endomembrane traffic in living cells. The applications of the approaches described in this chapter are instrumental in defining the mechanisms that ensure establishment and maintenance of the secretory organelles with relevance beyond the field of plant biology for morphological and functional characteristics of the secretory pathway that are conserved among eukaryotes.

5.1 LEAVES AS A MODEL SYSTEM FOR ANALYZING THE PLANT GOLGI AND SECRETORY PATHWAY

Leaf epidermal cells are perfect material for high-resolution fluorescence live cell imaging of the components of the plant secretory pathway and as such can be considered to represent self-contained miniature laboratories where besides observational imaging experiments can be carried out on protein transport and dynamics, protein interactions, and organelle interactions. They have the advantage over root cells of being relatively large with a massive central vacuole and thin layer of cortical cytoplasm with few chloroplasts, within which one can carry out imaging and other manipulations. Therefore, such cells constitute optimal material for microscopy analyses of the plant endomembranes because they allow capturing fine details of the subcellular structures.

Expression of fluorescent protein (FP) fusions targeted to various secretory organelles has for a number of years been successfully used to investigate protein localization and protein–protein interactions (Lippincott-Schwartz, Cole, & Presley, 1998; Sparkes & Brandizzi, 2012). Furthermore, thanks to advanced confocal techniques developed over the past few years, more detailed information about protein movement both in aqueous and membrane environments have become possible. A protein of interest tagged with a FP is usually introduced in the plant tissue either by transient or stable transformation. Transient transformation allows the observation of the protein tagged with FPs within 2–4 days from the transformation event, while it takes about a couple of months or longer to obtain transgenic plants through stable transformation (Sparkes, Runions, Kearns, & Hawes, 2006).

Both tobacco species and arabidopsis possess specific characteristics making them useful for the different forms of transformation. Transient transformation can be obtained through various means including infection with a virus carrier (Boevink et al., 1998) or biolistic bombardment of DNA into cells. However, one of the cheapest and safest methods that does not require specialized equipment is infiltration using Agrobacterium carrying a binary vector encoding the FP-fusion of interest (Sparkes et al., 2006). This technique is fast and reliable and especially suitable for the expression of multiple constructs tagged to different colored FPs. The reader is referred to a paper by Denecke et al. (2012) for a discussion on the merits of transient versus stable expression and native versus constitutive promoters for the study of the secretory pathway. To date, most work has been restricted to *Arabidopsis thaliana*, with the advantage of genetic manipulation, or various tobacco varieties (*Nicotiana tabacum, N. benthamiana, N. clevelandii*) which are easy to exploit for transient expression. Transient transformation in tobacco epidermal cells usually shows a higher frequency of transformed cells compared to Arabidopsis.

With transient expression, the protein of interest is only expressed for few days before the fluorescent signal disappears. Because of this, however, transient expression gives the advantage to study proteins or combination thereof that are potentially lethal and would result in no stable transformants. Transient expression generally provides a useful window to monitor localization and dynamics of proteins of interest before cell death processes start. To control the level of protein expression, the ubiquitin promoter instead of the most commonly used CaMV35S promoter can be adopted. The latter can on occasion cause the production of very high level of proteins that could alter or disturb their subcellular distribution (Grefen et al., 2010). The ubiquitin promoter may provide more homogeneous tissue transformation in stable transformation.

In transformation of Arabidopsis or tobacco plants, the T-DNA carried within the transforming DNA vector is inserted randomly into the genome; as consequence, one or more genes could be interrupted in the same time. In this case, in stable transformation, backcrosses can be made to ensure that the integration of the T-DNA is limited to one insertion. Upon obtaining stable lines, the ideal situation is to perform tail-PCR and identify where and how many genes are disrupted by the T-DNA insertion. It is also important to analyze multiple independent lines, at least five, to ensure that the FP-fusion shows a similar pattern of distribution among the transformants. In transient expression, this issue is solved by analyzing multiple cells within the transformed areas in the tissue.

5.1.1 Methods

Full protocols for the production and expression of FP constructs have been detailed in a previous volume (Runions, Hawes, & Kurup, 2006) and by Sparkes et al. (2006).

5.1.1.1 Transformation

5.1.1.1.1 Transient transformation in tobacco

Tobacco transient transformation can be performed as described in a published detailed protocol (Sparkes et al., 2006). We have found that higher transformation efficiency using plants that are 3–4-weeks old, grown in a controlled chamber with a cycle of 23 °C for 18 h light and 18 °C for 6 h night or grown in a greenhouse and transferred to a chamber for a few days before infiltration. In case of very low efficiency transformation, we recommend infiltrating agrobacteria carrying the gene of interest in plants at different stages (ranging from 2–3 weeks to 5–6 weeks).

5.1.1.1.2 Transient and stable transformation in Arabidopsis

Different protocols have been published for transient transformation of Arabidopsis plants; unfortunately, the percentage rate of fluorescent cells obtained after transformation is generally lower compared to tobacco plants (Campanoni, Sutter, Davis, Littlejohn, & Blatt, 2007; Li, Park, von Arnim, & Nebenführ, 2009; Marion et al., 2008). One of the most important factors that could increase the efficiency is the age of seedlings used, which we recommend to be not older than 4–5 day after germination.

Stable transformation has been performed in different ways using different media, chemicals, or methods, such as dipping, vacuum infiltration, or a combination of the two. However, the most used and successful method remain the protocol described by Clough and Bent (1998) which uses floral dipping. In all the methods used, the most important factor, which affects the percentage of stable transformants, is the health/age of the plants used.

5.1.1.1.3 Imaging leaf material

Arabidopsis cotyledons (7–10-days old) or 0.5 cm^2 square section of leaves (tobacco or Arabidopsis) are mounted on a microscope slide (1 in. × 3 in., 1.2 mm thick) in water with the abaxial side oriented toward the objective lens and enclosed with a coverslip. For confocal microscopy, the use of oil or water immersion lenses of the highest possible numerical aperture is recommended.

5.2 MUTANT SCREENS FOR TRAFFICKING PROTEINS

The introduction of genetically encoded FP reporters has permitted the adoption of new mutant screen opportunities, which are profoundly different from earlier screens set to identify plant phenotypic change (Sparkes & Brandizzi, 2012; Zwiewka & Friml, 2012). Microscopy-based screens allow the identification of mutants with misplaced subcellular FP-based reporters, which can be informative for specific defects in protein traffic routes and organelle integrity. This approach benefits from the availability of the sequenced model plant, *A. thaliana*. The screens also benefit from the availability of fluorescent markers targeted to all the endomembrane compartments that permit the analyses of specific trafficking pathways inside the cell and the morphology of specific organelles. Furthermore, FPs have been successfully engineered to improve maturation time as well as resistance to various degrees of environmental conditions including chloride, pH, and light irradiation. Extremely important for plant cells studies are the FP variants that are resistant to changes in pH and enable the monitoring of highly acidic microenvironments such as vacuole and apoplast where original FPs are not visible. On the other hand, however, FP quenching due to the acidic ambient can be exploited as system to identify mutants. Based on this approach, the secreted marker SEC-green fluorescent protein (GFP) in which the sporamin signal peptide is fused to the GFP allows the GFP to be secreted/transported to the apoplast. Seedlings mutagenized have been successfully screened for intracellular accumulation of SEC-GFP where the pH situation is favorable for fluorescence (Zheng, Kunst, Hawes, & Moore, 2004).

The possibility to visualize FP reporters in combination with the available strategies to produce mutagenized plants with ethyl methane sulfonate (EMS), gamma-ray, fast neutron, or T-DNA insertions can be used in forward genetic screens to identify proteins involved in protein trafficking and organelle integrity (Quesada, Ponce, & Micol, 2000; Stefano, Renna, & Brandizzi, 2012). EMS treatment produces point mutation, mainly G/C-to-A/T transitions with high frequency, while gamma-ray and fast neutron produce many types of mutations including base substitutions, deletions, and chromosomal alterations, but the mutation frequency is lower compared to EMS treatment. As EMS induces a high frequency of mutations, a relatively small number of plants is needed to reach saturation and the generation of a large number of gene knockouts as well as hypomorphic knockdowns is possible. Once the mutants have been identified, the mutation can be mapped using classical mapping or next generation sequencing which can be performed in a relatively short time (Stefano, Renna, & Brandizzi, 2012). With these approaches, it has proved possible to identify a number of proteins that are key to the functioning of the plant secretory pathway such as potential Golgi–ER tethers as well as proteins involved in Golgi trafficking and membrane integrity (Renna et al., 2013; Sparkes & Brandizzi, 2012).

5.2.1 Methods

5.2.1.1 EMS treatment, confocal screening, and mapping

Arabidopsis lines stably expressing a fluorescent marker are mutagenized using EMS (for mutagenesis procedures refer to detailed protocols (Stefano et al., 2012)). This provides the so-called M1 generation. Seeds are collected from individual M1 plants to generate M2 lines. M2 seeds from individual lines are plated and screened for aberrant localization of the fluorescent marker using either a laser scanning confocal microscope or a fluorescence microscope. The use of confocal versus conventional fluorescence microscopy has some advantages. Confocal gives us a good resolution in each focal plane eliminating out of focus light, and it is possible to perform 3D reconstruction of protein distribution along the Z-axis. Our screening method starts with plating 60 seeds from each M2 lines as well as 5–10 control seeds of the nonmutagenized background. Seedlings are left to grow for 7–10 days. It is important to note that in some case the seeds show aberrant phenotype early in the development and other later in development, so we suggest observing the same line both in early and late developmental stages. With seedlings, it is possible to mount them directly on slides. As observing material we use cotyledon cells from the abaxial region that are very easy to observe compared to other cell types. Of course, depending on the fluorescent marker used and the tissue of interest, any part of the seedlings can be used in the screening. Often, the CaMV 35S promoter provides a patchy expression pattern with some of the fluorescent marker expressing only in some tissues. An alternative could be to use protein tagged with fluorescent marker expressed under the ubiquitin promoter that gives a very even distribution of the fluorescence in the plants. At the microscope, the samples should be observed from the cortical to the medial region of the cell as any altered distribution of the fluorescent marker could be affecting just specific regions of the cytoplasm.

Identified mutants are transferred to soil and reobserved in the next generation (M3) to make sure that the subcellular phenotypes are maintained. Mutants are then allowed to flower for crossing with an alternate genotype or the same wild-type genotype. This permits assessment of whether the mutation is recessive or dominant through confocal analyses of the F2 progeny of the crosses. For confirmed recessive mutations, the mutation can be mapped using a combination of DNA hybridization and sequencing protocols recently described (Stefano et al., 2012), which is faster if compared to traditional mapping methods. This relatively new approach is based on the ability to identify in a single assay numerous single feature polymorphisms using high-density oligonucleotide arrays, like the Affymetrix Arabidopsis ATH1 GeneChip array that can analyze circa 24,000 genes. The advantages of this method compared to the classical approach is the fact that mapping can be performed using a relatively small number of F2 individuals to locate the mutation allowing the identification of the mutated gene in a shorter time.

5.3 PHOTOBLEACHING FOR TRANSPORT STUDIES

Fluorescence recovery after photobleaching (FRAP) is a powerful tool enabling the study of active protein movement or diffusion within or on and off membranes as well as interconnected soluble environments (Martinière et al., 2012), or for the study of transport of fluorescently tagged proteins between compartments of the secretory pathway in real time (Brandizzi et al., 2002; Da Silva et al., 2004; Matheson et al., 2007; Renna et al., 2013). As such we have used the technique to quantify membrane protein transport between the ER and the Golgi apparatus, showing that such movement is energy-dependent but independent of the cytoskeleton (Brandizzi et al., 2002), as well as reversible binding of small GTPases, such as ARFs and Sar1, on and off ER, and Golgi membranes. Essentially, with FRAP, a high-energy laser beam is used to irreversibly quench the fluorophore fluorescence in a predetermined area of the specimen and recovery of fluorescence is monitored through time-lapse imaging. As such, movement of fluorescent molecules into the bleached area can be followed and in most cases it must be assumed that this is balanced by an equal loss of quenched molecules out of the bleached area.

5.3.1 Methods

Most of the organelles of the plant secretory pathway demonstrate a high degree of motility which is almost exclusively actin-mediated (Boevink et al., 1998; Brandizzi et al., 2002). Therefore, although it is possible to bleach FPs targeted to motile organelles such as Golgi bodies if they show at least partial interruptions in movement (Da Silva et al., 2004), it is generally best to inhibit movement with an actin depolymerizing drug such as latrunculin B or cytochalasin D (Brandizzi et al., 2002). Depending on the confocal microscope being used, it is possible to bleach multiple regions of interest; however, it is advisable to make sure that there are unbleached organelles in the field of view that can serve as control. In other words, from such regions, a time series of fluorescence intensity can be established to monitor any photobleaching by the imaging laser during the recovery period being monitored. Such bleaching will have to be factored into calculations of fluorescence recovery on the bleached regions. Furthermore, a region in the field of view where the FP fusions do not localize is necessary to calculate the background fluorescence. This will be subsequently subtracted from the fluorescence values of the bleached areas as well as the selected unbleached areas that serve as control reference.

5.3.1.1 Bleach protocol

Below is a typical series of steps to be taken when photobleaching an region of interest (ROI). Most modern confocal microscopes come supplied with easily programmable bleach protocols.

- To initiate the experiment it is necessary to treat specimens with actin depolymerizing agents to stop movement of organelles. We find that for leaf material, treatment with 25 μM latrunculin B from a 10 mM stock solution in DMSO for at least 30 min is suitable. Alternatively, treatment with 20 μg/mL cytochalasin D from a 1 mg/mL stock solution in DMSO for at least 30 min can also be used. Generally, we submerge the leaf tissue in a tube containing the working solution and the tissue is then mounted on the slide with the same solution. We recommend leaving the tissue in the actin-depolymerizing solution for no longer than 90 min in order to reduce excessive cellular stress that may confound the results. We also find that the thickness of the coverslip can influence the results; therefore, we recommend using coverslips with consistent thickness (No. 1). It is also important to use consistent settings of magnification (i.e., same objective and optical zoom).
- The excitation and emission wavelengths for imaging will depend on the fluorochrome of choice and they should provide maximum signal yield with minimal laser intensity power. Avoid imaging oversaturation or undersaturation by monitoring pixel saturation levels with the microscope software. If the purpose of the experiment is quantitative (e.g., comparing recovery rates of FPs in the Golgi or in ER regions), the size (i.e., area and shape) of the ROIs should be identical.
- To initiate the FRAP experiment at the microscope, select 2–3 bleach regions using ROI settings on the confocal. As control ROIs, select two additional ROIs outside the bleach region: (a) an ROI in an area where the FP signal is visible (reference ROI), and (b) an ROI where no fluorescent signal is visible (background ROI). As explained above, the reference ROI will allow the levels of nonspecific bleaching that can be achieved during scanning to be assessed. This region should be therefore placed relatively far away from the bleach ROIs within the imaging field. The background ROI will define the levels of fluorescence that should be achieved upon the bleach scan in the FRAP analysis and will be subtracted from the bleach ROI value in the computational phase of the FRAP experiment.
- Record 10 full field scans with the lowest possible laser beam intensity (i.e., generally, 1–5% transmittance). Then bleach ROIs by scanning with full-power laser beam (i.e., 100% transmittance for a predetermined number of scans).

Generally, the number of scans should be sufficient to lower the fluorescence of the bleached area to near background levels, although this is not an absolute necessity. This can be established by comparing the levels of fluorescence in the ROI set on the organelle with the background ROI intensity.

- Record full-field scans with low laser settings during recovery period until a plateau of signal intensity values is reached. The number of scans during the recovery period is generally established empirically and will depend on the experiment in hand. As scanning causes intrinsic bleaching, it is generally advisable to keep the number of the fluorescence recovery scans to a minimum. On the other hand, it is also important to ensure that the number of scans is sufficient to record the fluorescence recovery in the bleached ROI over time. Therefore, as an initial step to get familiar with the FRAP protocol, it may be useful to set up imaging conditions for the FRAP experiments with cells expressing the FP targeted to the organelle of interest that have been fixed with paraformaldehyde. The material should be used to record ROI fluorescence with low laser intensity in order to define the conditions to obtain no significant bleaching during the time course of the experiment.
- It is customary to express fluorescence recovery in terms of the half time ($t½$) which is the time required for fluorescence in the bleached region to recover 50% of its initial fluorescence. This can be computed by scaling the postbleach fluorescence recovery to a 0–100% scale with the immediate postbleach intensity being 0% and the asymptote of the recovery being 100%. The bleached ROI intensity values should be subtracted from the background ROI values as well as normalized to the reference ROI values at equivalent imaging time points. The number of ROIs to be analyzed will depend of the FRAP experiment but they should be set to achieve statistical significance.

5.4 FÖRSTER RESONANCE ENERGY TRANSFER FOR PROTEIN INTERACTIONS

Whilst coexpression of proteins tagged to different FPs is a powerful technique to assess colocalization of proteins within specific organelles or regions of organelles (French, Mills, Swarup, Bennett, & Pridmore, 2008), there is no information in such images on any interactions between the proteins. Several techniques have been developed that permit the *in vivo* assessment of protein interactions. FRET (Förster or fluorescence resonance energy transfer), being the most powerful, permits the *in vivo* analysis of protein–protein interaction through measurement of nonradiative energy transfer from a donor fluorophore to an acceptor fluorophore of high excitation wavelength (Majoul, Jia, & Duden, 2006). Thus, FP constructs, where donor and acceptor are fused to putative interaction partners, are particularly suitable for this form of analysis. Suitable pairings that display a sufficient spectral overlap between excitation and emission spectra may be used, such as CFP/YFP or GFP/mRFP. As FRET efficiency decreases with the universal sixth power of the separation distance between the fluorophores, FRET can only be measured when the fluorophores are very close to each other, between 1 and 10 nm (Sparkes et al., 2011). This proximity only occurs when the two fusion proteins interact, and not when they simply colocate to the same cell structure (Majoul et al., 2006). FRET can be easily measured on a confocal microscope by a decrease in donor emission (quenching of the fluorescence) and with a corresponding increase in acceptor fluorophore fluorescence. A popular method is to photobleach the acceptor fluorophore and measure the accompanying increase in fluorescence of the donor fluorophore (Fricker, Runions, & Moore, 2006; Harholt et al., 2012).

A considerable improvement over the classic FRET measurement techniques is to utilize changes in the fluorescence excited-state lifetime of the donor fluorophore as it loses energy to the acceptor (Gerritsen, Draaijer, van den Heuvel, & Agronskaia, 2006; Sun, Day, & Perisamy, 2011). This has several advantages over conventional FRET in that fluorophore lifetime is unaffected by fluorophore concentration of excitation intensity and spectral bleed-through is not a problem as only the donor lifetime is measured. The lifetime of a fluorescent molecule is the average time that it stays in its excited state before decaying to its ground state, although it can change under different microenvironmental and physiological conditions in the cell (Becker, 2012). For FP constructs lifetimes can vary depending on the specific microenvironment of a protein spliced to the FP (Osterrieder et al., 2009). Lifetimes can be measured using photon counting detectors linked to time-correlated single-photon counting (SPC) modules available from specialist suppliers such as Becker and Hickl, and Picoquant, time gating modes as intensified charge coupled devices and frequency modulated methods. All of these techniques need pulsed light sources. These can be interfaced with confocal microscopes (Stahl et al., 2013) or two-photon systems, the latter giving better signal to noise ratio, Z resolution, and less background fluorescence in thick samples such as whole leaves, resulting in cleaner lifetime signals (Schoberer, Liebminger, Botchway, Strasser, & Hawes, 2013). Interactions between pairs of fluorophores can be measured through a reduction in the lifetime of the donor molecule (quenching) as a result of energy transfer to the acceptor molecule. This can simply be calculated by measuring the fluorescence lifetime of the donor alone or in combination with the acceptor fluorophore. This transfer or FRET percentage efficiency ($E\%$) is calculated as follows:

$$E\% = [1 - (\tau_{DA}/\tau_{D})] \times 100$$

τ_{DA} and τ_{D} are the mean pixel-by-pixel excited-state lifetimes of the donor fluorophore in the presence and absence, respectively, of the acceptor fluorophore determined for each pixel. For GFP the decrease in average donor lifetimes by a minimum of 0.2 ns or 8% (where donor lifetime is determined to be 2.5 ns) in the presence of the acceptor can be considered relevant to indicate FRET and thus indicate potential protein–protein interaction (Schoberer et al., 2013; Sun et al., 2011).

There are not many reports of the use of FRET to measure interactions within the plant secretory pathway (Harholt et al., 2012). We have concentrated on using a two-photon FLIM system for assessing the interactions of Golgi matrix proteins with small GTPases, such as ARL1 (Osterrieder et al., 2009; Fig. 5.1) and interactions between *N*-glycan processing enzymes in the Golgi stack (Schoberer et al., 2013). Likewise, homotypic interactions between the ER membrane proteins, the reticulons which are responsible for inducing tubule curvature, have been assessed by FRET-FLIM (Sparkes et al., 2010).

A typical two-photon microscope set-up connected to a confocal microscope would produce 700–1000 nm laser beam from a mode-locked titanium sapphire laser producing 180-fs pulses at 75 MHz pumped by a solid state, continuous wave, 532-nm laser. Emission signals are collected without descanning, filtered to remove near infra-red light, detected with a fast microchannel plate photomultiplier tube. Line, frame, and pixel clock signals generated by the scanning system are linked to a time-correlated SPC module to generate the raw FLIM data. FLIM images

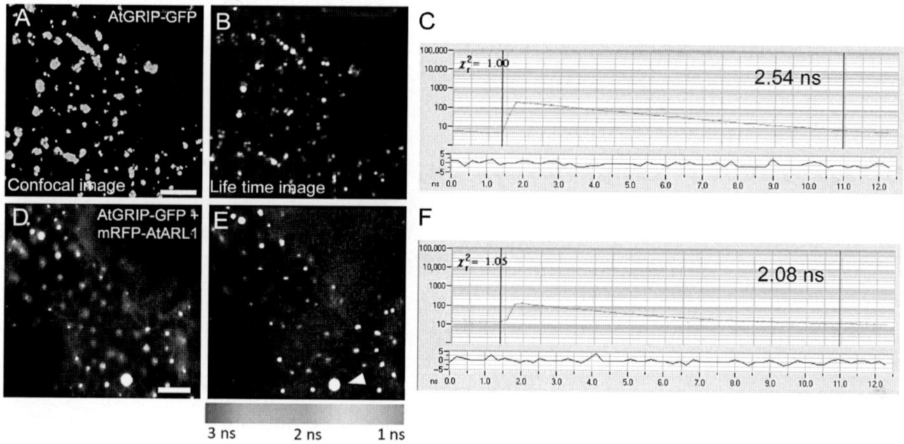

FIGURE 5.1

Example of FLIM data—interactions between coiled-coil proteins and small regulatory GTPases. The subcellular localization of interactions between coiled-coil proteins fused to GFP and small regulatory GTPases fused to mRFP were visualized with pseudo-colored lifetime maps generated in SPCImage (Becker & Hickl, Germany). (A) Confocal image of a tobacco leaf epidermal cell expressing the *trans*-Golgi matrix protein AtGRIP-GFP as control. (B) Pseudo-colored lifetime map of the same cell. Golgi bodies appear in blue, indicating an unquenched lifetime around 2.5 ns. (C) Representative decay curve for a single point analysis of AtGRIP-GFP with a lifetime of 2.54 ns and a χ^2 value of 1, indicating an optimal single exponential fit. (D) Confocal image of a cell expressing AtGRIP-GFP and the small regulatory GTPase mRFP-AtARL1. (E) Lifetime map showing quenching of AtGRIP-GFP in punctate structures colocating with mRFP–AtARL1. The lifetime reduction is reflected in the green color. Occasionally, large bright aggregates labeled by both proteins were observed, which did not show reduced lifetime values (arrowhead). (F) Representative decay curve for a single point analysis of AtGRIP-GFP+mRFP-ARL1 with a reduced lifetime of 2.08 ns. The χ^2 of 1.05 indicates a single exponential optimal fit.

From Osterrieder et al. (2009).

are analyzed using proprietary SPC image analysis software, although other analysis software such as TIMP are available.

5.4.1 Methods

Exact procedures for FRET measurements will depend on the microscope and systems being used. Our standard procedure for two-photon FRET-FLIM of leaf material expressing GFP and mRFP constructs is given below.

In order to achieve adequate photon counts for fluorescence lifetime acquisition (>100 in the peak channel of 256 bins), samples should be scanned several times to accumulate sufficient signal. This means that the area of interest needs to be immobilized in order to collect cumulative data. Depending on the cell structure of interest, this could mean drug-mediated cytoskeleton depolymerization (Osterrieder et al., 2009) or even fixation with paraformaldehyde, or by restricting analysis to an area of the structure that has less remodeling (e.g., measuring interaction of endoplasmic reticulum proteins near the nuclear envelope, Sparkes et al., 2010).

Mount $5 \times 5\ \text{mm}^2$ leaf samples or if using arabidopsis whole young leaves, pretreated as required on a glass slide with coverslip thickness no. 1. To prevent sample drift it is advisable to secure coverslip edges with sticky tape. Find a suitable region of the sample to measure using confocal microscopy and save a reference confocal image. Ideally, data should be collected from cells showing double expression of both constructs and also from cells expressing just the GFP construct for acquisition of control FLIM data.

Switch to two-photon laser to perform lifetime reading. A signal intensity of at least 100 photons is required in the peak channel. For Golgi bodies labeled with GFP we usually perform between 3–5 scans, each of 30 s. For brighter structures one scan with 30 s may be sufficient. The laser power should be adjusted for every construct depending on fluorochrome brightness, and care should be taken as exposing samples to too high laser powers can cause photodamage and dielectric breakdown of water in the cells.

Save images and import into FLIM analysis software. Refer to manufacturer instructions for specific procedures, for example, adjusting the decay model in the software or binning (the number of pixels the program takes into account for analysis). Becker & Hickl software translates lifetime image into a false-colored map in which different colors indicate different lifetimes. The first step is to perform curve-fitting analysis. Chi-square is used as an indicator for best exponential curve fit, where a $\chi^2 = 1.0$ in the structure is best fit of the data points to the lifetime value generated. Values of Chi-square that deviate from 1 indicates poor fitting to the data points when a single exponential condition is applied. Since we expect the lifetime value of GFP to be a single exponential decay, we discard data under $\chi^2 = 0.8$ and over $\chi^2 = 1.5$. (These indicate multiple decay components.)

Data can be extracted from lifetime images either on pixel basis and plotted into distribution histograms (see Osterrieder et al., 2009) or by drawing regions of interest and collecting average (Sparkes et al., 2010).

SUMMARY

Here, we have detailed just a few of the techniques that form just part a much larger toolbox available to the plant biologist for the study of the Golgi and the secretory pathway, but which can just as equally be applied to other cytoplasmic organelles. These include the use of photoactivatable and photoswitchable FPs, laser trapping and ablation techniques, and the whole range of recently developed high-resolution fluorescence microscope technologies. In conclusion, FP expression in leaf epidermal cells offers an extremely tractable experimental environment in which to apply such laser technologies to explore *in vivo* the dynamics and interactions of components and proteins of the secretory pathway.

Acknowledgments

Some of this work was supported by grants from the National Institutes of Health (R01 GM101038-01), Chemical Sciences, Geosciences and Biosciences Division, Office of Basic Energy Sciences, Office of Science, U.S. DOE (DE-FG02-91ER20021), NASA (NNX12AN71G), and the National Science Foundation (MCB 0948584 and MCB1243792) (FB) and the BBSRC (BB/F008147/1) (CH). We also thank Stan Botchway (STFC, Central Laser Facility, Harwell, UK) for reading the chapter.

References

Becker, W. (2012). Fluorescence lifetime imaging—Techniques and applications. *Journal of Microscopy*, *247*, 119–136.

Boevink, P., Oparka, K., Santa Cruz, S., Martin, B., Betteridge, A., & Hawes, C. R. (1998). Stacks on tracks: The plant Golgi apparatus traffics on an actin/ER network. *Plant Journal*, *15*, 441–447.

Brandizzi, F., Snapp, E., Roberts, A., Lippincott-Schwartz, J., & Hawes, C. R. (2002). Membrane protein transport between the ER and Golgi in tobacco leaves is energy dependent but cytoskeleton independent: Evidence from selective photobleaching. *Plant Cell*, *14*, 1293–1309.

Campanoni, P., Sutter, J. U., Davis, C. S., Littlejohn, G. R., & Blatt, M. R. (2007). A generalized method for transfecting root epidermis uncovers endosomal dynamics in Arabidopsis root hairs. *Plant Journal*, *51*, 322–330.

Chen, Y. F., Randlett, M. D., Findell, J. L., & Schaller, G. E. (2002). Localization of the ethylene receptor ETR1 to the endoplasmic reticulum of Arabidopsis. *Journal of Biological Chemistry*, *277*, 19861–19866.

Chen, Y. N., Slabaugh, E., & Brandizzi, F. (2008). Membrane-tethered transcription factors in *Arabidopsis thaliana*: Novel regulators in stress response and development. *Current Opinion in Plant Biology*, *11*, 695–701.

Clough, S. J., & Bent, A. F. (1998). Floral dip: A simplified method for Agrobacterium-mediated transformation of *Arabidopsis thaliana*. *The Plant Journal*, *16*, 735–743.

Da Silva, L. L. P., Snapp, E. L., Denecke, J., Lippincott-Schwartz, L., Hawes, C., & Brandizzi, F. (2004). Membrane cargo-mediated recruitment of GTPase Sar1p to specific export sites of the ER membrane. *Plant Cell*, *16*, 1753–1771.

Denecke, J., Aniento, F., Frigerio, L., Hawes, C., Hwang, I., Mathur, J., et al. (2012). Secretory pathway research: The more experimental systems the better. *The Plant Cell*, *24*, 1316–1326.

Foresti, O., & Denecke, J. (2008). Intermediate organelles of the plant secretory pathway: Identity and function. *Traffic*, *9*, 1599–1612.

French, A. P., Mills, S., Swarup, R., Bennett, M. J., & Pridmore, T. P. (2008). Colocalization of fluorescent markers in confocal microscope images of plant cells. *Nature Protocols*, *3*, 619–628.

Fricker, M., Runions, J., & Moore, I. (2006). Quantitative fluorescence microscopy: From art to science. *Annual Review of Plant Biology*, *57*, 79–107.

Geldner, N., Hyman, D. L., Wang, X., Schumacher, K., & Chory, J. (2007). Endosomal signaling of plant steroid receptor kinase BRI1. *Genes & Development*, *21*, 1598–1602.

Gerritsen, H. C., Draaijer, D. A., van den Heuvel, D. J., & Agronskaia, A. V. (2006). Fluorescence lifetime imaging in scanning microscopy. In J. B. Pawley (Ed.), *Handbook of biological confocal microscopy* (pp. 516–534). New York: Springer.

Grefen, C., Donald, N., Hashimoto, K., Kudla, J., Schumacher, K., & Blatt, M. R. (2010). A ubiquitin-10 promoter-based vector set for fluorescent protein tagging facilitates temporal stability and native protein distribution in transient and stable expression studies. *The Plant Journal*, *64*, 355–365.

Harholt, J., Jensen, J. K., Verhertbruggen, Y., Søgaard, C., Bernard, S., Nafisi, M., et al. (2012). ARAD proteins associated with pectic arabinan biosynthesis form complexes when transiently overexpressed in planta. *Planta*, *236*, 115–128.

Keegstra, K. (2010). Plant cell walls. *Plant Physiology*, *154*, 483–486.

Li, J. F., Park, E., von Arnim, A. G., & Nebenführ, A. (2009). The FAST technique: A simplified Agrobacterium-based transformation method for transient gene expression analysis in seedlings of Arabidopsis and other plant species. *Plant Methods*, *5*, 6.

Lippincott-Schwartz, J., Cole, N., & Presley, J. (1998). Unravelling Golgi membrane traffic with green fluorescent protein chimeras. *Trends in Cell Biology*, *8*, 16–20.

Majoul, I., Jia, Y., & Duden, R. (2006). Practical fluorescence resonance energy transfer or molecular nanobioscopy of living cells. In J. B. Pawley (Ed.), *Handbook of biological confocal microscopy* (pp. 788–807). New York: Springer.

Marion, J., Bach, L., Bellec, Y., Meyer, C., Gissot, L., & Faure, J. D. (2008). Systematic analysis of protein subcellular localization and interaction using high-throughput transient transformation of Arabidopsis seedlings. *The Plant Journal*, *56*, 169–179.

Martinière, A., Lavagi, I., Nageswaran, G., Rolfe, D. L., Maneta-Peyret, L., Luu, D.-T., et al. (2012). The cell wall constrains lateral diffusion of plant plasma-membrane proteins. *Proceedings of the National Academy of Sciences of the United States of America*, *109*, 12805–12810.

Matheson, L. A., Hanton, S. L., & Brandizzi, F. (2006). Traffic between the plant endoplasmic reticulum and Golgi apparatus: To the Golgi and beyond. *Current Opinion in Plant Biology*, *9*, 601–609.

Matheson, L. A., Hanton, S. L., Rossi, M., Latijnhouwers, M., Stefano, G., Renna, L., et al. (2007). Multiple roles of ADP-ribosylation factor 1 in plant cells include spatially regulated recruitment of coatomer and elements of the Golgi matrix. *Plant Physiology*, *143*, 1615–1627.

Osterrieder, A., Carvalho, C. M., Latijnhouwers, M., Johansen, J. M., Stubbs, C., Botchway, S., et al. (2009). Fluorescence lifetime imaging of interactions between Golgi tethering factors and small GTPases in plants. *Traffic*, *10*, 1–13.

Quesada, V., Ponce, M. R., & Micol, J. L. (2000). Genetic analysis of salt-tolerant mutants in *Arabidopsis thaliana*. *Genetics*, *154*, 421–436.

Renna, L., Stefano, G., Majeran, W., Micalella, C., Meinnel, T., Giglione, C., et al. (2013). Golgi traffic and integrity depend on N-myristoyl transferase-1 in Arabidopsis. *Plant Cell*, *25*, 1756–1773.

Runions, J., Hawes, C., & Kurupp, S. (2006). Fluorescent protein fusions for protein localization in plants. *In Methods in Cell Biology*, *390*, 239–255.

Schoberer, J., Liebminger, E., Botchway, S. W., Strasser, R., & Hawes, C. (2013). Time-resolved fluorescence imaging reveals differential interactions of *N*-glycan processing enzymes across the Golgi stack *in planta*. *Plant Physiology*, *161*, 1737–1754.

Sparkes, I., & Brandizzi, F. (2012). Fluorescent protein-based technologies: Shedding new light on the plant endomembrane system. *The Plant Journal*, *70*, 96–107.

Sparkes, I. A., Graumann, K., Martiniére, A., Schoberer, J., Wang, P., & Osterrieder, A. (2011). Bleach it, switch it, bounce it, pull it: Using lasers to reveal plant cell dynamics. *Journal of Experimental Botany*, *62*, 1–7.

Sparkes, I., Runions, J., Kearns, A., & Hawes, C. (2006). Rapid, transient expression of fluorescent protein fusions in tobacco plants and generation of stably transformed plants. *Nature Protocols*, *1*, 2019–2025.

Sparkes, I., Tolley, N., Aller, I., Svozil, J., Osterrieder, A., Botchway, S., et al. (2010). Five plant reticulon isoforms share ER localisation, topology, ER membrane shaping properties. *Plant Cell*, *22*, 1333–1343.

Stahl, Y., Grabowski, S., Bleckmann, A., Kühnemuth, R., Weidtkamp-Peters, S., Pinto, K. G., et al. (2013). Moderation of *Arabidopsis* root stemless by CLAVATA1 and ARABIDOPSIS CRINKLY4 receptor kinase complexes. *Current Biology*, *23*, 362–371.

Stefano, G., Renna, L., & Brandizzi, F. (2012). Fluorescence-microscopy screening and next-generation sequencing: Useful tools for the identification of genes involved in organelle integrity. *Journal of Visualized Experiments*, *62*, e3809.

Sun, Y., Day, R. N., & Perisamy, A. (2011). Investigating protein-protein interactions in living cells using fluorescence lifetime imaging microscopy. *Nature Protocols*, *6*, 1324–1340.

Zheng, H., Kunst, L., Hawes, C., & Moore, I. (2004). A GFP-based assay reveals a role for RHD3 in transport between the endoplasmic reticulum and Golgi apparatus. *Plant Journal*, *37*, 398–414.

Zwiewka, M., & Friml, J. (2012). Fluorescence imaging-based forward genetic screens to identify trafficking regulators in plants. *Frontiers in Plant Science*, *3*, 97.

CHAPTER

Biochemical Analysis of Secretory Trafficking in Mammalian Cells

6

Peristera Roboti[1], Tomasz M. Witkos[1] and Martin Lowe
Faculty of Life Sciences, University of Manchester, Manchester, United Kingdom

CHAPTER OUTLINE

Introduction **86**
6.1 Materials **89**
6.1.1 Cell Culture 89
6.1.2 Sample Preparation and Analysis 91
6.2 Methods **91**
6.2.1 Infection of Cells with Recombinant Adenovirus-encoding ts045-VSVG-GFP 91
6.2.2 Cycloheximide Chase, Endo H Digestion, and Cell-surface Biotinylation 92
6.2.3 Pulse-chase and Immunoprecipitation 94
6.2.4 Pulse-chase Analysis of Protein Secretion 97
6.3 Discussion **99**
Summary **101**
Acknowledgments **101**
References **102**

Abstract

Protein trafficking within the secretory pathway of mammalian cells is amenable to analysis by biochemical methods. This can be achieved by monitoring posttranslational modifications that occur naturally within the secretory pathway, or by measuring the delivery of cargo to the cell surface or extracellular medium. These approaches can be combined with additional manipulations such as specific temperature blocks that permit analysis of distinct trafficking steps. Biochemical analysis is advantageous in that it permits both a sensitive and quantitative measure of trafficking along the pathway. The methods discussed in this chapter permit the analysis of trafficking of both

[1]These authors contributed equally to the manuscript.

Methods in Cell Biology, Volume 118

ISSN 0091-679X
http://dx.doi.org/10.1016/B978-0-12-417164-0.00006-9

endogenous cargo proteins and ectopically expressed model cargos, which can be followed using either Western blotting or metabolic pulse-chase approaches. These methods are relatively straightforward and suitable for use in most modern cell biology laboratories. In addition to the well-established methods that we describe here in detail, we also refer to the development of more recent tailored approaches that add further to the arsenal of tools that can be used to assess trafficking in the secretory pathway.

INTRODUCTION

The canonical secretory pathway comprises the endoplasmic reticulum (ER), Golgi apparatus, and secretory granules that are present in specialized secretory cells (Fig. 6.1). Approximately 30% of all cellular proteins are synthesized at the ER, which is the entry portal to the secretory pathway. Within the ER proteins undergo folding, and in many cases posttranslational modifications such as N-linked glycosylation and disulphide bond formation also occur. Proteins that are not ER residents are transported onward to the Golgi apparatus, which in mammalian cells occurs via an intermediate compartment sometimes referred to as the ERGIC (ER to Golgi intermediate compartment). Within the Golgi apparatus, cargo proteins undergo extensive glycan processing as well as additional modifications including phosphorylation, sulfation, and proteolytic processing. At the *trans*-Golgi network (TGN), cargo is sorted into carriers for delivery either to the plasma membrane, secretory granules, or the endocytic pathway. In polarized cells, cargos are sorted at the TGN into carriers destined for either the apical or basolateral plasma membrane domains. In addition to the extensive forward trafficking of cargo proteins, there is also retrograde trafficking (not shown in Fig. 6.1) to retrieve escaped organelle resident proteins and recycle components of the trafficking machinery.

Various biochemical approaches can be used to monitor the trafficking of cargo proteins along the secretory pathway in intact mammalian cells. These approaches permit analysis of trafficking under different experimental conditions, for example, depletion or loss of proteins of interest by siRNA or the use of mutant cell lines respectively, stable expression of dominant negative constructs, or treatment of cells with small molecule inhibitors. Biochemical analysis can be used to study entire populations of cargo proteins, endogenous or ectopically expressed model cargo proteins, or particular cargo proteins of interest. The use of specific temperature blocks and analysis of posttranslational modifications that occur at different compartments within the secretory pathway permits the analysis of distinct trafficking steps. Analysis of cargo protein trafficking can be performed using a pulse-chase approach, with metabolic labeling of newly synthesized cargo at the ER during the pulse followed by a chase in which the cargo is transported along the secretory pathway. An alternative approach is to monitor trafficking using Western blotting of the cargo protein of interest, but in this case a method for synchronizing production or transport is required.

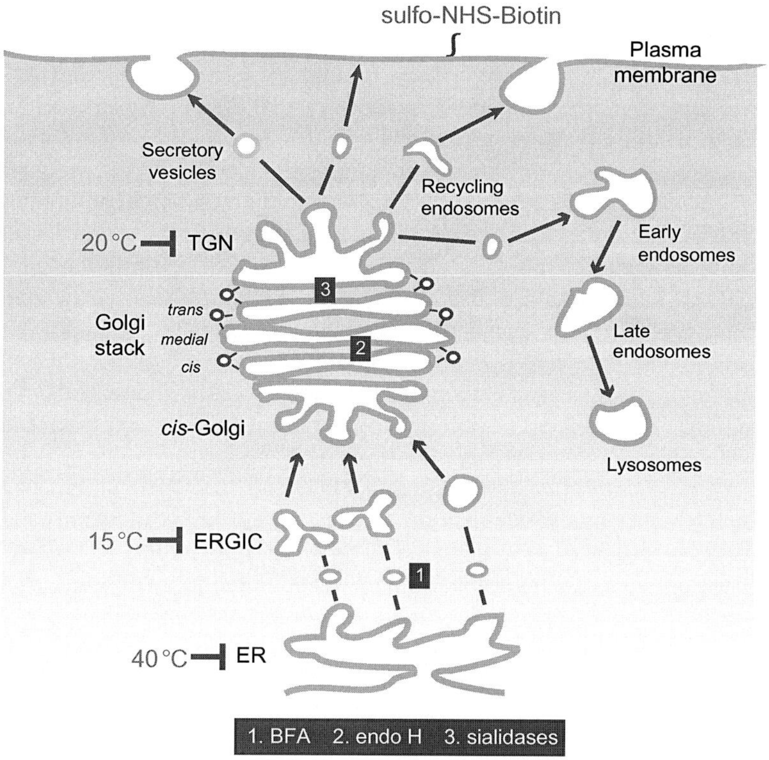

FIGURE 6.1

Major routes of protein trafficking in the mammalian secretory pathway. Membrane and soluble proteins destined for different compartments of the endomembrane system or the plasma membrane are translocated into the endoplasmic reticulum (ER) and then sorted into transport vesicles that merge together to form the ER-Golgi intermediate compartment (ERGIC). Subsequently, anterograde cargo is targeted to the *cis*-Golgi and then transported through the Golgi stack to the *trans*-Golgi network (TGN), where it is sorted to the appropriate destination (e.g., the endosomal system, secretory granules, or the plasma membrane). Technical approaches that are used to study protein trafficking, such as temperature blocks and cell surface biotinylation, are shown. Note that the 40 °C block applies only to thermosensitive mutants of cargo proteins. Inhibitory compounds and glycosidases are commonly used to monitor subcellular protein transport. Some examples are (1) brefeldin A (BFA), which can be used in living cells to inhibit ER to Golgi transport; (2) Endoglycosidase H (Endo H), which removes high-mannose, but not complex, oligosaccharides from *N*-glycoproteins, permitting measurement of transport to the *medial*-Golgi; (3) sialidases (also called neuraminidases), which remove terminal sialic acid residues from N- and O-linked oligosaccharides, tracking protein movement to the *trans*-Golgi/TGN. (See color plate.)

Pulse-chase labeling of newly synthesized proteins is usually performed with [^{35}S]methionine, which is often supplied mixed with [^{35}S]cysteine. A brief labeling pulse (10–20 min) is followed by a chase of appropriate length to allow delivery of the labeled proteins to the Golgi apparatus, endosomal system, plasma membrane, or extracellular medium if the protein(s) are secreted. All labeled proteins can be analyzed by autoradiography or phosphorimaging. If a specific cargo protein is being analyzed, then it must first be isolated by immunoprecipitation prior to analysis. Delivery to the Golgi apparatus can be detected in various ways, but most commonly is determined by the analysis of posttranslational modifications that are Golgi-specific. Thus, delivery to the *medial*-Golgi can be determined by the acquisition of complex glycan chains that confer resistance to cleavage by Endoglycosidase H (Endo H), while delivery to the *trans*-Golgi/TGN can be assessed by acquisition of terminal sialic residues, as determined by cleavage with sialidases (also known as neuroaminidases). Transport of proteins to the plasma membrane can be monitored in two ways, depending on whether the proteins of interest are present at the membrane itself or secreted into the medium. Cell surface proteins can be specifically labeled using biotinylation, which is done by incubating cells with membrane impermeant reagents that add a biotin group to those proteins (Elia, 2008; Orr, 1981). Biotinylated proteins can then be isolated on streptavidin-coupled beads prior to further analysis. Proteins secreted into the medium can be enriched by precipitation prior to subsequent analysis (Bensadoun & Weinstein, 1976), or can be analyzed directly if sufficiently abundant.

Allied to the use of the above methods for assessing secretory trafficking, generic temperature blocks can also be employed. In mammalian cells, trafficking from the ERGIC compartment is arrested at 15 °C, allowing the accumulation of cargo at this compartment prior to a subsequent shift to 37 °C to allow a synchronized wave of trafficking to the Golgi apparatus and later compartments (Saraste & Kuismanen, 1984). Similarly, incubating cells at 20 °C blocks the exit of cargo from the TGN, allowing the analysis of TGN to plasma membrane transport (Matlin & Simons, 1983).

Trafficking of endogenous or ectopically expressed proteins of interest can be analyzed in mammalian cells. For example, transferrin receptor has frequently been used as an endogenous model cargo to study trafficking along the secretory pathway (Omary & Trowbridge, 1981). This glycoprotein is expressed in most cultured cell types and undergoes defined posttranslational processing that can be used to assess its delivery to various stages of the secretory pathway. The trafficking of extracellular matrix proteins has also been extensively studied. In the case of procollagen, exit from the ER can be synchronized through the addition of ascorbic acid to the medium, which promotes synthesis, folding, assembly, and ER exit (Murad et al., 1981). One of the most extensively studied cargo proteins is VSVG, a transmembrane glycoprotein of vesicular stomatitis virus (VSV). A temperature-sensitive mutant version of this protein (ts045-VSVG) is misfolded and accumulates in the ER at the nonpermissive temperature (40 °C) (Zilberstein, Snider, Porter, & Lodish, 1980). Upon shift to permissive temperature (32 °C), the mutant glycoprotein undergoes folding and synchronous exit from ER and trafficking along the secretory pathway.

ts045-VSVG can be expressed in the cell line of choice using either the mutant virus (ts045-VSV), recombinant viral vectors such as the adenoviral vector described in this chapter, or using DNA transfection. Another viral glycoprotein that can be used to study secretory trafficking is influenza hemagglutinin (HA) (Gething & Sambrook, 1981). In polarized cells, HA is delivered to the apical membrane while VSVG is delivered to the basolateral membrane, permitting the biochemical analysis of secretory trafficking to these distinct membrane domains.

In this chapter, we document protocols for monitoring secretory trafficking of ts045-VSVG, using both Western blotting and metabolic labeling. A protocol for measuring trafficking of an endogenous model cargo, transferrin receptor, is also described. Finally, we describe a protocol for the measurement of the trafficking of total secreted and cell surface cargo proteins. An overview of the experimental procedures presented in this chapter is shown in Fig. 6.2.

6.1 MATERIALS

6.1.1 Cell culture

1. Cell line of choice (HeLa M or HepG2 cells in this example, grown at 37 °C and 5% CO_2)
2. Cell culture medium: in this example, DMEM supplemented with 10% fetal bovine serum (FBS) and 2 mM L-glutamine
3. Phosphate-buffered saline (PBS), pH 7.4 containing 135 mM NaCl, 2.7 mM KCl, 10 mM Na_2HPO_4, and 1.8 mM KH_2PO_4
4. Concentrated recombinant adenovirus expressing ts045-VSVG-GFP was prepared using an Adenovirus Expression Vector Kit (Takara Bio) (Kanegae et al., 1995; Yoshimura et al., 2004); store 5–20 μl aliquots at −80 °C and avoid more than two freeze/thaw cycles. Note that for biochemical experiments, adenovirus expressing untagged ts045-VSVG can also be used (see Diao, Frost, Morohashi, & Lowe, 2008).
5. Infection medium: DMEM, or the appropriate culture medium, supplemented with 5% FBS and 2 mM L-glutamine
6. CO_2-independent medium (Invitrogen, cat. no. 18045-054) supplemented with 10% FBS and 2 mM L-glutamine
7. Labeling medium, methionine/cysteine-free (Invitrogen, cat. no. 21013-024) supplemented with 2 mM L-glutamine
8. EasyTagT EXPRESS [^{35}S] protein labeling mix (>1000 Ci/mmol; PerkinElmer, cat. no. NEG772014M)—a mixture of [^{35}S]methionine and [^{35}S] cysteine utilized for radiolabeling of cellular proteins
9. 2.5 mg/ml of brefeldin A (BFA; Sigma, cat. no. B7651) in ethanol; store at −20 °C
10. 10 mg/ml of cycloheximide (CHX; Sigma, cat. no. C7698) in water; store 1 ml aliquots at −20 °C
11. Mixture of 0.25 M L-methionine (Sigma, cat. no. M5308) and 0.25 M L-cysteine (Sigma, cat. no. C7352) in water; store 1 ml aliquots at −20 °C.

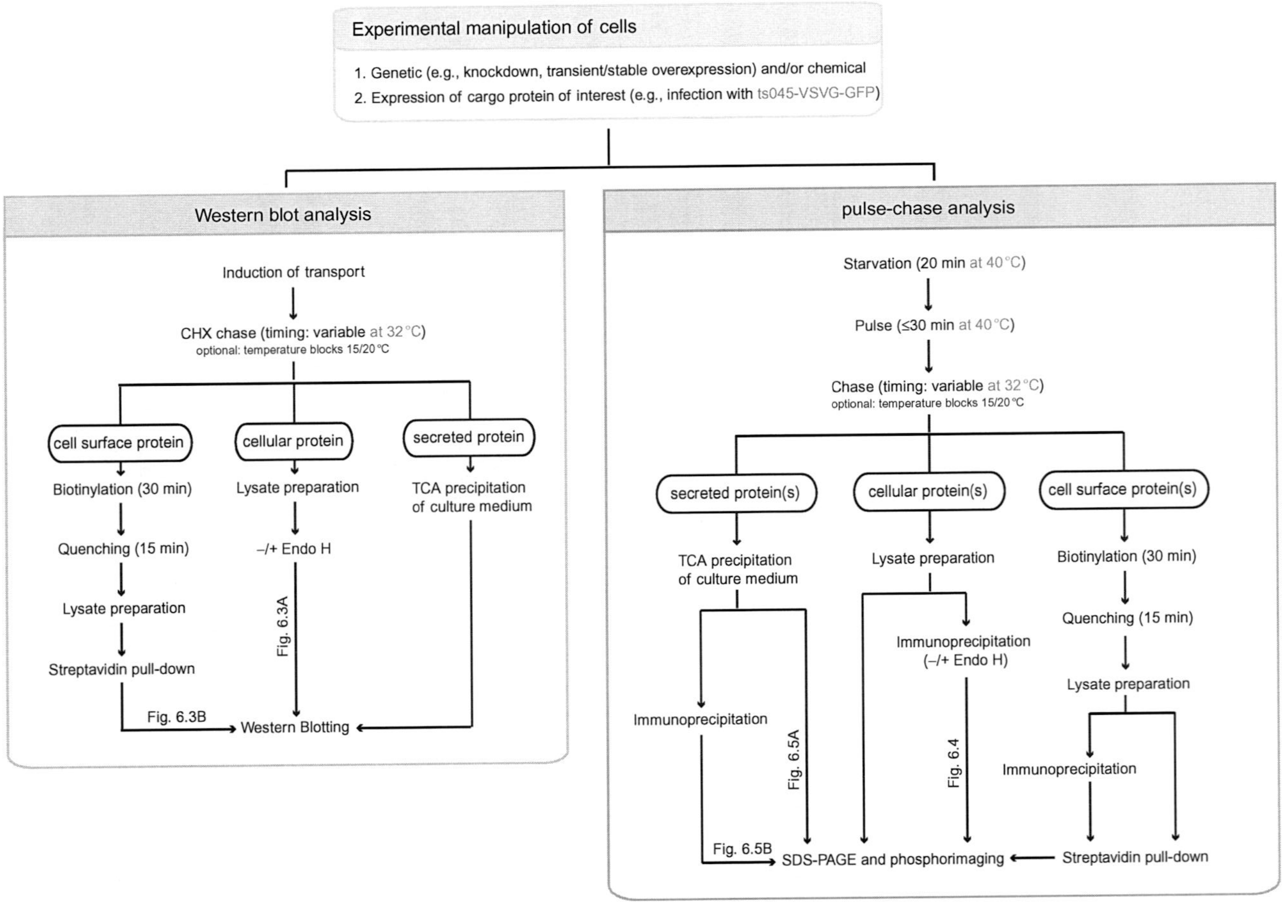

FIGURE 6.2

Biochemical analysis of protein trafficking in the mammalian secretory pathway. Flowchart describing the outline of analyses presented in this chapter, with the associated figures indicated. Experimental steps relevant to VSVG trafficking are shown in red. Note that these assays can be applied to various experimental setups. (See color plate.)

6.1.2 Sample preparation and analysis

1. EZ-link Sulfo-NHS-LC-Biotin (Thermo Scientific, cat. no. 21335)
2. Ammonium chloride (NH_4Cl): 500 mM solution in PBS
3. High-Performance Streptavidin Sepharose (GE Healthcare); wash beads twice with the appropriate lysis buffer and resuspend them in the same buffer at 1:1
4. Protease inhibitor cocktail (Sigma, cat. no. P8340)
5. 0.5% TX-100 lysis buffer: 20 mM Hepes (pH 7.4), 100 mM KCl, 5 mM $MgCl_2$, and 0.5% (v/v) Triton X-100
6. RIPA buffer: 10 mM Tris–HCl (pH 7.6), 150 mM NaCl, 1% (v/v) NP-40, 0.2% (w/v) sodium deoxycholate, and 0.1% (w/v) SDS
7. Bicinchonic acid (BCA) protein assay kit (Pierce, cat. no. 23227)
8. Endoglycosidase H (Endo H; Calbiochem, cat. no. 324717)
9. Phenylmethylsulfonyl fluoride (PMSF): 100 mM in ethanol; store at −20 °C
10. Pansorbin cells (Calbiochem, cat. no. 507861)
11. Antibodies of choice: monoclonal anti-VSVG antibody clone P5D4 (Sigma, cat. no. V5507), monoclonal anti-TfR clone H68.4 (Invitrogen, cat. no. 136800), sheep anti-GFP (Diao, Rahman, Pappin, Lucocq, & Lowe, 2003), and goat anti-saposin D (Konrad Sandhoff, University of Bonn, Germany)
12. Protein G-agarose (GenScript, cat. no. L002009); wash beads twice with the appropriate lysis buffer and resuspend them in the same buffer to prepare a 50% protein G-agarose bead slurry
13. Trichloroacetic acid (TCA): dissolve 22 g of TCA in 10 ml water to obtain a 100% solution; maintain in dark bottle at 4 °C
14. 2 × SDS gel-loading buffer: 100 mM Tris–HCl (pH 6.8), 10 mM EDTA, 1% (w/v) L-methionine, 4% (w/v) SDS, 20% (v/v) glycerol, 200 mM DTT, and 0.04% (w/v) bromophenol blue
15. Fixation mix: 20% methanol and 10% acetic acid in water.

6.2 METHODS

6.2.1 Infection of cells with recombinant adenovirus-encoding ts045-VSVG-GFP

For infection with recombinant adenovirus-encoding ts045-VSVG-GFP, cells are plated into 35 mm dishes one or several days before infection (depending upon the treatment conditions; see also Fig. 6.2, steps 1 and 2) so that they are 70–80% confluent on the day of infection. Before infecting cells, rinse them with prewarmed PBS. Dilute the appropriate amount of virus in prewarmed infection medium to obtain a multiplicity of infection of 200–500 PFUs/cell. We generally use 0.5 μl high-titer adenovirus stock per 35 mm dish to achieve sufficient expression levels of VSVG for biochemical studies. It is important to use reduced serum in the infection medium (see Section 6.1.1) to improve virus adherence to the cell surface, and to keep the volume of infection medium containing the virus as low as possible (0.4 ml per 3.5 cm dish is recommended) to maximize infection efficiency. Infection is performed at 37 °C for 1 h, gently tilting the plates every 10–15 min to prevent the cells from drying out. Following the initial infection, add 1.8 ml complete DMEM per dish and shift cells to 40 °C for 14–18 h before further analysis (see Fig. 6.2). Expression of adenovirus-encoded ts045-VSVG-GFP should be detectable by 4–6 h and will rise gradually over the next 18 h.

6.2.2 Cycloheximide chase, Endo H digestion, and cell-surface biotinylation

In mammalian cells, protein trafficking from the site of synthesis on the rough ER to the Golgi apparatus or beyond can be monitored by studying sensitivity to digestion by Endo H, an enzyme that cleaves high-mannose (ER/*cis*-Golgi forms) but not complex (*medial*-Golgi and post-Golgi forms) N-linked oligosaccharides. Since a trimmed glycosylated protein produced by Endo H digestion migrates faster by SDS-PAGE than the corresponding untrimmed protein, they can be readily identified. Protein traffic to the cell surface can be studied by biotinylation of cell surface proteins and streptavidin pull-down. These assays can be used to track movement of temperature-sensitive mutant VSVG under different treatment conditions. When ts045-VSVG-GFP-infected cells are grown at the restrictive temperature of 40 °C, the protein misfolds and is retained in the ER by quality control. When the cells are subsequently shifted to the permissive temperature of 32 °C, the protein folds correctly and is transported through the Golgi apparatus to the cell surface.

We have monitored the trafficking of ts045-VSVG-GFP to and through the Golgi to the cell surface using Endo H digestion or biotinylation analysis, respectively, with detection by Western blotting (see Figs. 6.2 and 6.3). Approximately 18 h postinfection at 40 °C, ts045-VSVG-GFP-expressing cells are shifted to a 32 °C water bath by incubation in prewarmed CO_2-independent medium containing 100 μg/ml CHX to inhibit protein synthesis. It is important to note that in temperature-shift experiments, an acute temperature change is achieved by replacing the medium with medium warmed at the required temperature and direct transfer of the cells in a water bath adjusted to the appropriate temperature. Furthermore, in the absence of a 5% CO_2 atmosphere, the additional buffering compounds present in the CO_2-independent medium are required to maintain physiological pH.

Directly after the 40 °C block or after incubation for an indicated time period at 32 °C, cells are washed twice with ice-cold PBS and lysed in 100 μl ice-cold 0.5% TX-100 lysis buffer supplemented with protease inhibitor cocktail. Following solubilization of cells by incubation on ice for at least 15 min, the cell lysate is clarified by centrifugation at 13,200 rpm for 15 min at 4 °C. At this point, total protein concentration is measured using the BCA assay. For treatment with Endo H, 30 μl cell lysate corresponding to approximately 60 μg of total protein is denatured by the addition of an equal volume of 2 × SDS gel-loading buffer and boiling at 95 °C for 5 min. Denatured samples (60 μl) are treated with 0.5 μl of Endo H, vortexed, and incubated at 37 °C overnight. Samples are then spun briefly and loaded onto an

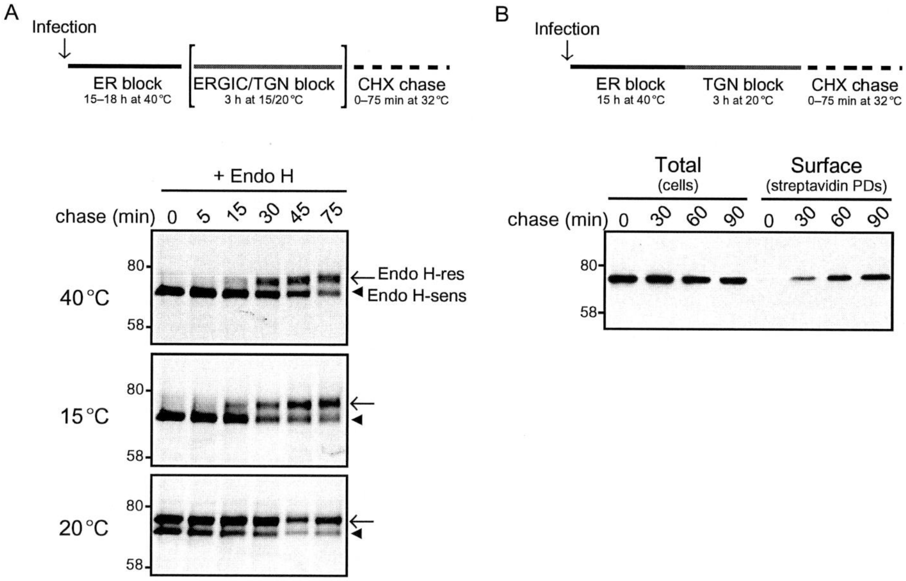

FIGURE 6.3

Trafficking of ts045-VSVG to the Golgi apparatus and plasma membrane monitored by Western blotting. HeLa M cells infected with adenovirus-encoding ts045-VSVG-GFP were incubated at 40 °C for 15–18 h to retain misfolded VSVG in the ER. (A) Anterograde trafficking was induced by shifting the cells to 32 °C either directly after the 40 °C block or after a 3 h CHX block at 15 °C or 20 °C that results in accumulation of VSVG in the ERGIC or TGN, respectively. Cell lysates collected after the indicated block or at different points of the 32 °C release were treated with Endo H and analyzed by immunoblotting using anti-GFP antibody. Analysis of Endo H-treated samples resolves a faster migrating band representing the sensitive, trimmed form (arrowheads) of ts045-VSVG-GFP and a slower migrating band corresponding to the resistant, untrimmed form (arrows), indicative of delivery to the *medial*-Golgi. Note the slightly faster rate of delivery to the Golgi when shifting from 15 °C compared to 40 °C. At 20 °C the majority of ts045-VSVG-GFP is Endo H-resistant consistent with it already having travelled through the *medial*-Golgi. (B) Surface biotinylation of ts045-VSVG-GFP after a TGN block (20 °C, 3 h) and subsequent release at 32 °C. Plasma membrane proteins were pulled down using streptavidin beads (streptavidin PDs) and surface VSVG-GFP was detected by immunoblotting with an anti-GFP antibody. (See color plate.)

8% SDS-PAGE gel, and ts045-VSVG-GFP is detected by Western blotting with an anti-GFP antibody (Fig. 6.3A).

Low temperature-induced arrest of trafficking at 15 or 20 °C results in protein accumulation in the ERGIC (Saraste & Kuismanen, 1984) or TGN (Matlin & Simons, 1983), respectively. CHX-chase assays based on these reversible temperature blocks can be used in combination with Endo H digestion (Fig. 6.3A,

15 and 20 °C blocks) or surface biotinylation (see below) for biochemical dissection of individual steps of the secretory pathway (see also Fig. 6.2). Briefly, ts045-VSVG-GFP-infected cells are shifted from 40 °C to a 15 or 20 °C water bath in CO_2-independent medium containing CHX and incubated for 3 h, prior to release at 32 °C for the desired time and subsequent analysis.

Selective biotinylation of cell surface proteins at the end of the indicated chase period is performed by washing the cells twice with ice-cold PBS and incubating them in 0.5 ml 0.5 mg/ml solution of sulfo-NHS-LC-Biotin in PBS for 30 min on ice. The specificity of this reagent for cell surface labeling is based on the fact that it is membrane-impermeable, and therefore as long as the cells remain intact, only cell surface proteins are biotinylated. It is important that the biotinylation reagent is prepared immediately before addition to cells because its NHS-ester moiety readily hydrolyzes and becomes nonreactive. Next, cells are washed twice with ice-cold PBS and quenching of the biotinylation reaction is performed by incubating the cells in 0.5 ml 50 mM NH_4Cl for 15 min on ice. Cells are washed twice with ice-cold PBS and lysed by the addition of 250 μl ice-cold 0.5% TX-100-containing PBS supplemented with protease inhibitor cocktail. For recovery of the cell surface biotinylated proteins, 200 μl clarified lysates corresponding to approximately 150 μg of total protein are incubated with prewashed 25 μl streptavidin-sepharose for 2 h at 4 °C with continuous mixing. Beads are washed three times with 0.5% TX-100-containing PBS and biotinylated proteins are eluted from the beads by boiling in 2 × SDS gel-loading buffer for 5 min at 95 °C. Total and surface (biotinylated) VSVG are detected by Western blotting (Fig. 6.3B). Quantification of the bands can be performed using densitometry with ImageJ software, and surface delivery is the ratio of biotinylated versus total VSVG.

6.2.3 Pulse-chase and immunoprecipitation

Pulse-chase assays integrated with Endo H digestion or surface biotinylation can be used to quantitatively measure rates of protein trafficking in the cell. Pulse-chase involves labeling of cellular proteins with a radioactive amino acid(s) for a brief period of time (≤30 min; pulse), followed by incubation in the absence of radioactivity over desired periods of time (chase). At the end of the chase, the protein of interest is recovered by immunoprecipitation, digested with Endo H, and the processing of N-glycans in newly synthesized radiolabeled protein visualized by autoradiography or phosphorimaging (see Fig. 6.2, pulse-chase analysis, cellular protein). It is also possible to carry out surface biotinylation at the end of the chase, followed by immunoprecipitation and subsequent binding of immunoprecipitated proteins to streptavidin beads to determine levels of biotin incorporation and thus surface delivery (Fig. 6.2, pulse-chase analysis, cell surface protein). These quantitative methods are able to discern small differences in the rate of trafficking after different experimental manipulations. We describe pulse-chase assays to study the kinetics of trafficking of ts045-VSVG-GFP or endogenous transferrin receptor using Endo H digestion (Fig. 6.4) in conditions where cells have been pretreated with siRNAs (siLuc control or siGMAP210) or with an inhibitory molecule (BFA).

Because chase experiments involve harvesting a number of samples at different time points, plate the cells into individual 35-mm culture dishes. If necessary, infect cells with ts045-VSVG-GFP adenoviral particles as described in Section 6.2.1. After an 18 h postinfection period at 40 °C, the medium is removed and the cells are washed twice with 1 ml prewarmed PBS. In order to deplete intracellular pools of methionine and cysteine, cells are incubated with 1 ml prewarmed (40 °C) methionine/cysteine-free labeling medium for 20 min at 40 °C. Subsequently, cells are labeled for 20 min at 40 °C in 1 ml fresh labeling medium containing 25 μCi/ml [^{35}S]Met/Cys protein labeling mix. The medium is carefully removed and the cells are washed twice with PBS. To conduct the chase, add 2 ml of prewarmed (32 °C) CO_2-independent medium supplemented with 1 mM nonradioactive methionine and cysteine and incubate the radiolabeled cells for appropriate time points in a 32 °C water bath. It should be noted that when the effect of a small molecule compound is studied by pulse-chase, the cells often have to be pretreated with that compound for 1 h prior to starvation. Furthermore, the compound should be included throughout the starvation, pulse, and chase (see treatment of cells with BFA in Fig. 6.4A).

At the end of the chase period, cells are washed twice with ice-cold PBS and lysed on ice by incubation with 100 μl of ice-cold RIPA buffer supplemented with protease inhibitor cocktail. Clarified lysates are denatured in the presence of 1% SDS at 37 °C for 30 min. It should be noted that although most antibodies (including anti-VSVG P5D4 used in this example) recognize epitopes in native proteins, immunoprecipitation under native conditions is not recommended because it may result in high background of nonspecifically bound proteins. The samples are diluted with 0.5 ml (i.e., 5 volumes) of RIPA buffer containing 8 mM unlabeled methionine and cysteine and 1 mM PMSF. The excess 1% NP-40 in RIPA buffer sequesters SDS into NP-40 micelles, protecting the antigen–antibody interaction from a potential interference by SDS. The lysates are then incubated with antibodies specific for the reporter protein overnight on a rotator at 4 °C. This can be preceded by an optional preclearing step in which the lysate is incubated with 10 μl pansorbin at 4 °C for 1 h in order to preclear any nonspecific binding of the lysate to beads. Next day, add 10 μl of Protein G-agarose to each sample and incubate for 1 h at a rotator at 4 °C. The beads are recovered by centrifugation for 1 min at maximum speed and washed three times with 1 ml of ice-cold RIPA buffer. Forty microliters of 2× sample buffer that had prewarmed to 95 °C is added, and samples are immediately incubated at 95 °C for 5 min. Half of the denatured samples are treated with 0.5 μl of Endo H, vortexed, and incubated at 37 °C overnight.

Samples are analyzed by SDS-PAGE on an appropriate percentage gel (see Fig. 6.4A). Gels are carefully removed from the glass plates, and after incubation in the fixation mix for at least 15 min, they are dried on a gel dryer for 2 h at

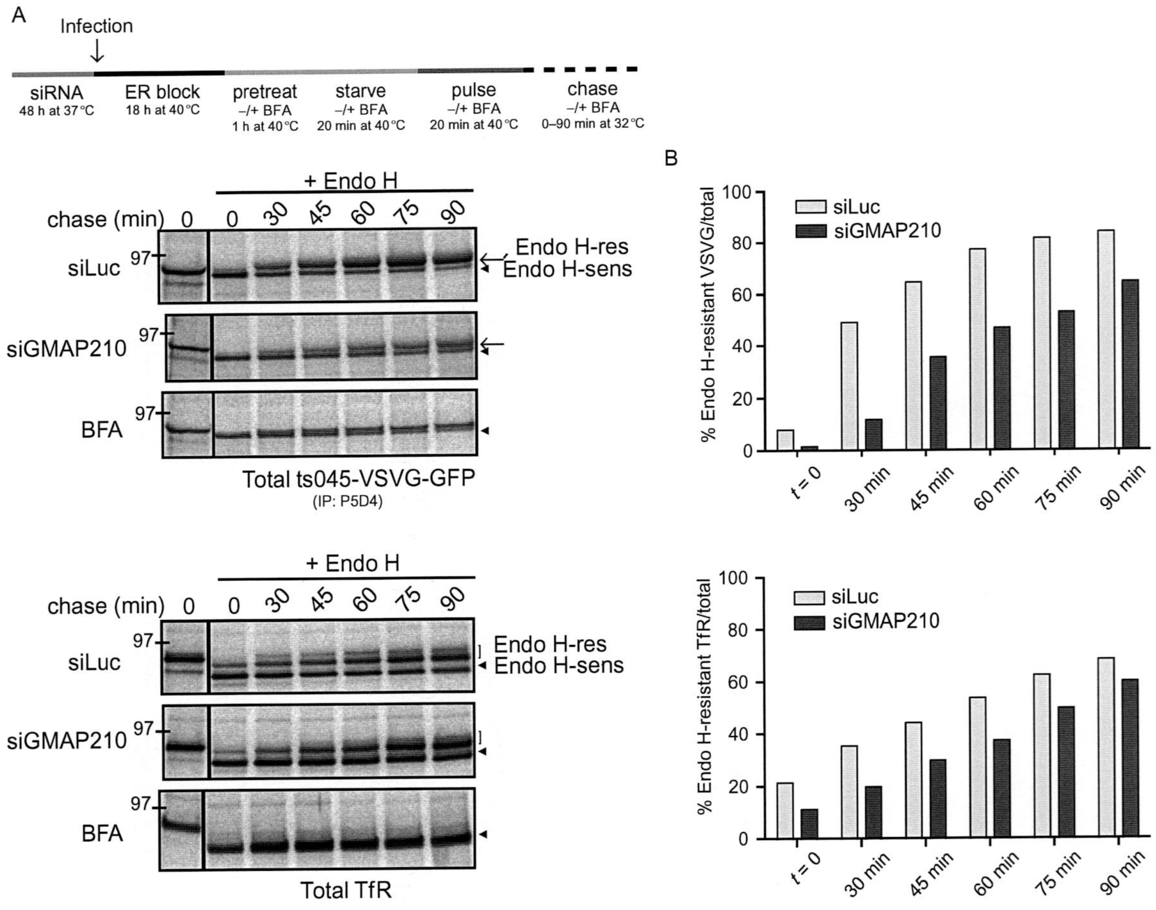

FGIURE 6.4 For legend see next page.

65 °C. Expose the dried gels to clean phosphorimaging plates and develop the exposures after 24–48 h using an image analyzer equipped with lasers for detection of radioactive isotopes (e.g., FLA-3000; Fuji). Signals from radiolabeled proteins are quantified using compatible software (e.g., AIDA) (Fig. 6.4B).

6.2.4 Pulse-chase analysis of protein secretion

The general transport of secretory proteins under different conditions can be studied by analysis of the amount and/or kinetics of release of newly synthesized proteins into the culture medium using a pulse-chase approach (see Fig. 6.2, pulse-chase analysis, secreted proteins and Fig. 6.5A). Surface biotinylation can also be used to monitor total protein delivery to the cell surface (Fig. 6.2, pulse-chase analysis, cell surface protein). HeLa M or HepG2 cells grown in 35 mm dishes at a confluency of 80–90% are washed twice with PBS, methionine/cysteine-starved for 20 min and labeled with 25 μCi/ml [^{35}S]Met/Cys, as described in Section 6.2.3, for 10 min at 37 °C. At the end of the pulse, cells are washed twice with PBS and chased

FIGURE 6.4

Trafficking of membrane glycoproteins from the ER to the Golgi apparatus monitored using pulse-chase. (A) HeLa M cells were transfected with a control siRNA (siLuc) or an siRNA targeting the *cis*-Golgi protein GMAP210 (siGMAP210) for 48 h. Alternatively, they were left untransfected but were treated with a drug inhibiting secretion (Brefeldin A, BFA). After infection with ts045-VSVG-GFP, cells were incubated at 40 °C for an additional 17–18 h to accumulate the viral glycoprotein in the ER. BFA (5 μg/ml) was added to some dishes for 1 h at 40 °C. Newly synthesized proteins were radiolabeled for 20 min at 40 °C, and cells were incubated at 32 °C in the presence of an excess of unlabeled Met/Cys. BFA was included throughout the pulse and chase in BFA-treated cells. At the indicated times, cells were lysed and cell extracts were subjected to immunoprecipitation (IP) with antibodies to VSVG (P5D4) or endogenous transferrin receptor (TfR). Half of the immunoprecipitated material was digested with Endo H, and proteins were separated by SDS-PAGE and visualized by phosphorimaging. Bands corresponding to Endo H-sensitive and Endo H-resistant forms of VSVG and TfR are indicated. Note that trafficking of both VSVG and TfR was slowed down in GMAP210-knockdown cells compared with control siLuc-treated cells, as demonstrated by the kinetics of appearance of Endo H-resistant protein. BFA treatment blocked the exit of both VSVG and TfR from the ER based on the sensitivity of these two *N*-glycoproteins to Endo H during the entire chase period. (B) The levels of Endo H-sensitive and Endo H-resistant forms were quantified and the percentage of Endo H-resistant form with respect to total protein was calculated. Note that the difference between GMAP210-knockdown cells and control siLuc-treated cells was most obvious at earlier time points of chase, suggesting that GMAP210 is important for the initial steps in the secretory pathway of exogenous (VSVG) and endogenous (TfR) model cargo proteins. (See color plate.)

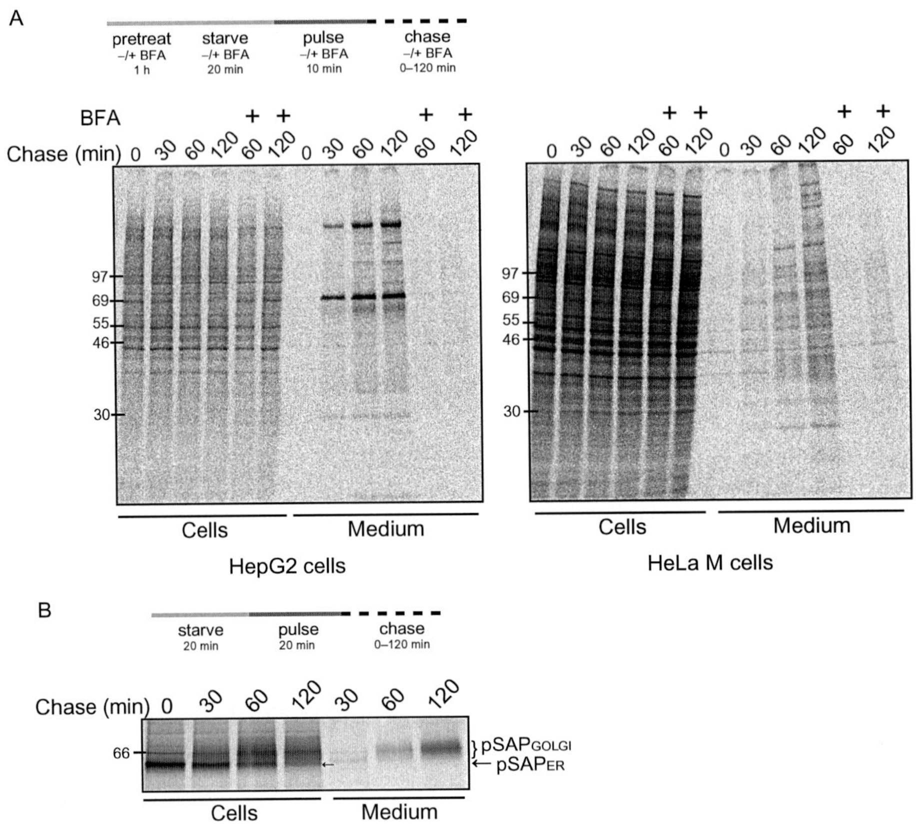

FIGURE 6.5

Protein secretion monitored by pulse-chase. (A) HeLa M or HepG2 cells were left untreated or treated with BFA at 5 μg/ml for 1 h (+BFA). Newly synthesized proteins were radiolabeled for 10 min and then chased in medium containing unlabeled Met/Cys for the indicated times. BFA was included throughout the pulse and chase in BFA-treated cells. At each time, the medium was collected and secreted proteins were precipitated with TCA, whereas cells were lysed. Radiolabeled proteins in cell lysates and TCA precipitates were resolved using a 12% SDS-PAGE gel and visualized by phosphorimaging. Note the different profile of secreted proteins from the two cell types. (B) Pulse-chase labeling of prosaposin (pSAP) in HeLa M cells. Intracellular (cells) and secreted (medium) pSAP were recovered by denaturing immunoprecipitation with anti-saposin D antibody. The ER and Golgi forms of the protein are indicated. Note that pSAP species bearing complex oligosaccharides were evident at chase $t=0$, indicating some pSAP species were transported from the ER to the Golgi within 20 min (pulse step). pSAP was detected in cell culture medium at the 60 min chase point and was as abundant as intracellular pSAP after 2 h of chase. (See color plate.)

in 1 ml prewarmed (37 °C) DMEM (or other medium formulation depending on the cell line used) supplemented with 1 mM nonradioactive methionine and cysteine for the indicated time periods. For secreted proteins, medium is collected at the end of the chase and placed in an Eppendorf tube on ice and supplemented with protease inhibitor cocktail. The cellular proteins are collected by washing twice with ice-cold PBS, and lysis in 100 μl RIPA buffer supplemented with protease inhibitor cocktail. Solubilization and denaturation of clarified lysates is carried out as previously described.

The culture medium is centrifuged at 2000 rpm for 5 min at 4 °C to pellet detached cells and cellular debris, and secreted proteins are precipitated by addition of 100% TCA to a final concentration of 10% and incubation at 4 °C overnight. The following day, the samples are left at room temperature to thaw completely and protein precipitates are collected by centrifugation at 13,200 rpm for 15 min at 4 °C. Pellets are resuspended in 0.4 ml 100% chilled acetone by vortexing, and protein precipitates are recovered by centrifugation at 13,200 rpm for 10 min at 4 °C and air-dried at room temperature for approximately 30 min. Finally, pellets are dissolved in 2 × SDS gel-loading buffer and the samples are denatured by heating at 37 °C for 1 h with shaking. For studying the secretion of a particular model protein (see Fig. 6.2, pulse-chase analysis, secreted proteins and Fig. 6.5B), cell lysates and secreted material in the medium are subjected to immunoprecipitation with the relevant antibody, and samples are analyzed by SDS-PAGE and phosphorimaging as described in Section 6.2.3.

6.3 DISCUSSION

We have presented some of the most commonly used biochemical assays for studying the secretory pathway in mammalian cells. Biochemical assays can be appropriately tailored according to the aim of experiment and type of cells used. For instance, some cells constitutively secrete more proteins than others and therefore are more amenable to the analysis of global protein secretion. It is worth noting that although most components of the secretory pathway are common in all mammalian cells, specific cell lines differ in the relative abundance of some proteins, which may cause variability in trafficking rates in different cell lines. It is important to first validate the levels of endogenous cargo proteins before studying their trafficking. Exogenous cargoes require introduction of the nucleic acid sequence encoding the cargo protein into cells. However, it is technically challenging to transfect certain cell types with DNA vector constructs, for example, primary cell lines. This difficulty can be solved through the use of lentiviral or adenoviral vectors, such as the one described here for ts045-VSVG, that are able to infect practically every type of mammalian cell.

Biochemical assays can be used for studying all secretory proteins or specific cargo proteins, which can be endogenous or ectopically expressed. Each of these strategies has its advantages and disadvantages. Analysis of all proteins, for instance those delivered to the cell surface or secreted into the extracellular medium, permits a global view of trafficking efficiency and also allows changes in the trafficking of specific proteins to be detected, which can subsequently be identified by mass spectroscopy techniques. It also does not require use of specific antibodies since all proteins can be visualized by autoradiography/phosphorimaging or Coomassie blue/silver staining. Endogenous cargo proteins typically provide more physiological data than ectopically expressed proteins; however, examination of their trafficking is often more technically challenging. It requires specific antibodies for immunoprecipitation or immunoblotting that may not be commercially available, and low abundance or synthesis rates may result in a low detectable signal.

An alternative approach to analyze secretory trafficking is the use of model cargos that are enzymes. In this case, secretion is assessed by measuring the enzymatic activity present in the medium, negating the use of antibodies. One of the commonly used enzymatic cargoes is a fusion protein of horseradish peroxidase with the N-terminal signal sequence of human growth hormone (ssHRP) (Connolly, Futter, Gibson, Hopkins, & Cutler, 1994). It is worth mentioning that ssHRP is also suitable for electron microscopy studies. Other ectopically expressed enzymatic cargoes frequently used for studying secretion are *Gaussia* luciferase (Gluc) (Badr, Hewett, Breakefield, & Tannous, 2007) and a truncated version of human placental alkaline phosphatase (SEAP) which lacks a glycosyl-phosphatidylinosytol (GPI) anchor and is therefore secreted from cells (Berger, Hauber, Hauber, Geiger, & Cullen, 1988). The abundance of these enzymatic cargos in the medium and cell lysates is easily measured using specific assays.

The posttranslational modification of cargo proteins within the secretory pathway can be analyzed in various ways. We describe the use of Endo H to monitor delivery to the *medial*-Golgi apparatus, while cleavage by sialidases or the measurement of terminal sulfation of glycans using radioactive sulfate can be used to assess delivery to the *trans*-Golgi/TGN. Alternatively, delivery of certain proteins to the TGN can be assessed by defined proteolytic cleavage events that are mediated by endogenous TGN-localized proteases such as furin. All these techniques can serve as valuable tools to detect even small differences in the rate of protein trafficking. However, the results should be interpreted with caution. Since these methods rely on endogenous Golgi-localized enzymes, any experimental manipulations that alter the localization or activity of the enzymes will alter the results obtained, irrespective of the rate of cargo trafficking.

More tailored approaches have recently been developed to study secretory trafficking in mammalian cells. One method is to analyze trafficking of engineered proteins comprising a fusion with the F_M tag, a mutant version of the FKBP12 protein. This tag results in formation of reversible aggregates that disassociate upon addition of the AP21998 ligand to cells (Rivera et al., 2000). Fusing secretory cargo proteins with the F_M tag allows their accumulation in the ER in an aggregate form and synchronous release from this compartment upon addition of the ligand. This method has been successfully used in the examination of protein trafficking at physiological temperature both by biochemical and imaging-based assays. A noteworthy modification of this technique is the addition of GFP to the F_M tag which allows the analysis of secretion into the medium using flow cytometry or Western blotting (Gordon, Bond, Sahlender, & Peden, 2010).

A recently developed alternative method for examination of secretion in mammalian cells is the RUSH system (Boncompain et al., 2012, see Chapter 11). This approach is based on expressing two fusion proteins in a cell line of interest. The first one, referred to as the hook, is a resident protein of a compartment within the secretory pathway, such as the ER or Golgi cisternae, that is tagged with streptavidin. The second one, called the reporter, is the cargo protein of interest, which is fused with streptavidin-binding peptide (SBP) and a fluorescent tag. Cells expressing both proteins accumulate the reporter in the donor compartment because of the streptavidin–SBP interaction. Addition of biotin to the medium triggers dissociation of the reporter from the hook, resulting in a synchronized wave of trafficking of the reporter along the secretory pathway. A major advantage of the RUSH system is that it enables examination of trafficking from different compartments within the secretory pathway without the need for temperature blocks, and allows greater versatility in terms of the number of trafficking steps that can be analyzed. Although the RUSH system was primarily established to assess protein trafficking by imaging, it can also be applied to biochemical analysis. Indeed, it was shown that wild-type VSVG tagged with SBP was released from the ER and transported through Golgi-acquired Endo H resistance with similar kinetics to ts045-VSVG.

SUMMARY

Biochemical analysis of trafficking within the secretory pathway remains an important approach to assess the functioning of this pathway. With the advent of technologies that can ablate the function of specific gene products, such as RNA interference or mutagenesis with zinc-finger nucleases, together with the development of increasing numbers of small molecule inhibitors, it is more important than ever that we can sensitively and quantitatively assess rates of protein trafficking within cells. The use of established biochemical approaches such as those described here affords this possibility for the secretory pathway. Therefore, we strongly believe that the continued use of such assays will be important for fully dissecting the mechanisms that underpin secretory trafficking in mammalian cells.

Acknowledgments

Work in the Lowe lab is funded by the BBSRC (BB/1007717/1), MRC (MR/K000810/1), and Lowe Syndrome Trust (MU/ML/1010 and 92589). T. M. W. is supported by a Wellcome Trust PhD studentship (096601/Z/11/Z).

References

Badr, C. E., Hewett, J. W., Breakefield, X. O., & Tannous, B. A. (2007). A highly sensitive assay for monitoring the secretory pathway and ER stress. *PLoS One*, *2*(6), e571.

Bensadoun, A., & Weinstein, D. (1976). Assay of proteins in the presence of interfering materials. *Analytical Biochemistry*, *70*(1), 241–250.

Berger, J., Hauber, J., Hauber, R., Geiger, R., & Cullen, B. R. (1988). Secreted placental alkaline phosphatase: A powerful new quantitative indicator of gene expression in eukaryotic cells. *Gene*, *66*(1), 1–10.

Boncompain, G., Divoux, S., Gareil, N., de Forges, H., Lescure, A., Latreche, L., et al. (2012). Synchronization of secretory protein traffic in populations of cells. *Nature Methods*, *9*(5), 493–498.

Connolly, C. N., Futter, C. E., Gibson, A., Hopkins, C. R., & Cutler, D. F. (1994). Transport into and out of the Golgi complex studied by transfecting cells with cDNAs encoding horseradish peroxidase. *Journal of Cell Biology*, *127*(3), 641–652.

Diao, A., Frost, L., Morohashi, Y., & Lowe, M. (2008). Coordination of golgin tethering and SNARE assembly: GM130 binds syntaxin 5 in a p115-regulated manner. *Journal of Biological Chemistry*, *283*(11), 6957–6967.

Diao, A., Rahman, D., Pappin, D. J., Lucocq, J., & Lowe, M. (2003). The coiled-coil membrane protein golgin-84 is a novel rab effector required for Golgi ribbon formation. *Journal of Cell Biology*, *160*(2), 201–212.

Elia, G. (2008). Biotinylation reagents for the study of cell surface proteins. *Proteomics*, *8*(19), 4012–4024.

Gething, M. J., & Sambrook, J. (1981). Cell-surface expression of influenza haemagglutinin from a cloned DNA copy of the RNA gene. *Nature*, *293*(5834), 620–625.

Gordon, D. E., Bond, L. M., Sahlender, D. A., & Peden, A. A. (2010). A targeted siRNA screen to identify SNAREs required for constitutive secretion in mammalian cells. *Traffic*, *11*(9), 1191–1204.

Kanegae, Y., Lee, G., Sato, Y., Tanaka, M., Nakai, M., Sakaki, T., et al. (1995). Efficient gene activation in mammalian cells by using recombinant adenovirus expressing site-specific Cre recombinase. *Nucleic Acids Research*, *23*(19), 3816–3821.

Matlin, K. S., & Simons, K. (1983). Reduced temperature prevents transfer of a membrane glycoprotein to the cell surface but does not prevent terminal glycosylation. *Cell*, *34*(1), 233–243.

Murad, S., Grove, D., Lindberg, K. A., Reynolds, G., Sivarajah, A., & Pinnell, S. R. (1981). Regulation of collagen synthesis by ascorbic acid. *Proceedings of the National Academy of Sciences of the United States of America*, *78*(5), 2879–2882.

Omary, M. B., & Trowbridge, I. S. (1981). Biosynthesis of the human transferrin receptor in cultured cells. *Journal of Biological Chemistry*, *256*(24), 12888–12892.

Orr, G. A. (1981). The use of the 2-iminobiotin-avidin interaction for the selective retrieval of labeled plasma membrane components. *Journal of Biological Chemistry*, *256*(2), 761–766.

Rivera, V. M., Wang, X., Wardwell, S., Courage, N. L., Volchuk, A., Keenan, T., et al. (2000). Regulation of protein secretion through controlled aggregation in the endoplasmic reticulum. *Science*, *287*(5454), 826–830.

Saraste, J., & Kuismanen, E. (1984). Pre- and post-Golgi vacuoles operate in the transport of Semliki Forest virus membrane glycoproteins to the cell surface. *Cell*, *38*(2), 535–549.

Yoshimura, S., Yamamoto, A., Misumi, Y., Sohda, M., Barr, F. A., Fujii, G., et al. (2004). Dynamics of Golgi matrix proteins after the blockage of ER to Golgi transport. *Journal of Biochemistry*, *135*(2), 201–216.

Zilberstein, A., Snider, M. D., Porter, M., & Lodish, H. F. (1980). Mutants of vesicular stomatitis virus blocked at different stages in maturation of the viral glycoprotein. *Cell*, *21*(2), 417–427.

CHAPTER

Methods for Analysis of Apical Lumen Trafficking Using Micropatterned 3D Systems

7

Alejo E. Rodríguez-Fraticelli and Fernando Martín-Belmonte
Centro de Biología Molecular "Severo Ochoa", Madrid, Spain

CHAPTER OUTLINE

Introduction **106**
7.1 3D Culture of MDCK Cysts on Micropatterns **107**
7.1.1 Protocol Overview 108
7.1.2 Working with Matrigel 108
7.1.3 Working with Micropatterns 109
7.1.4 MDCK Cell Strains 110
7.1.5 Trypsinization, Resuspension, and Seeding Cells on Micropatterns (Basic Protocol) 110
7.1.6 Matrix Overlay Method 111
7.1.7 Fixation and Analysis 112
7.1.7.1 PFA/formalin Fixation 112
7.1.7.2 Methanol Fixation 112
7.1.7.3 Acetone Fixation 113
7.1.7.4 Permeabilization 113
7.1.7.5 Blocking and Antibody Incubations 113
7.1.7.6 Secondary Antibodies and Other Staining Reagents 114
7.1.7.7 Mounting and Imaging 114
7.1.8 Imaging and Quantification 114
7.1.8.1 Number of Lumens 115
7.1.8.2 Cell Death 117
7.1.8.3 Spindle Orientation 117
7.1.9 Preparation of Preactivated Micropatterns Coated with Other Proteins. 118
7.1.10 *In Vivo* Experiments 119
7.2 Discussion **120**
Acknowledgments **120**
References **121**

Methods in Cell Biology, Volume 118

ISSN 0091-679X
http://dx.doi.org/10.1016/B978-0-12-417164-0.00007-0

Abstract

Epithelial organs are made of interconnected branched networks of tubules, with a central lumen lined by a monolayer of epithelial cells. Certain epithelial cell lines can be converted into organotypic cultures by the addition of extracellular matrix components. When cultured in these conditions, epithelial cells reorient the axis of polarity, reorganize the membrane surfaces, and transport apical proteins to form the lumen in a process that recapitulates essential aspects of *de novo* apical membrane formation during epithelial organ morphogenesis. Micropatterns are a simple technique that allows cell culture in a controlled adhesive environment with extremely high precision, close to the nanometer scale. We have recently developed a method to culture MDCK cysts on micropatterns of different sizes and composition. Using this method we found that changes in micropattern shape and size can be used to modify cell contractility to understand its contribution to apical membrane formation. When imaging cysts on micropatterns the main advantage is that apical-directed vesicle trafficking is visualized in the *x–y* plane, which presents higher resolution on confocal microscopes. Thus, the use of micropatterns is an efficient setup to analyze polarized secretion with unprecedented higher resolution in both time and space.

INTRODUCTION

Epithelial organs are made of interconnected networks of tubules, constituted by a monolayer of epithelial cells surrounding a central cavity or lumen. Epithelial cells are constitutively polarized cells that present different plasma membrane compartments, which are maintained through the organization of different molecular fences and a specialized sorting machinery that directs protein trafficking specifically to each compartment (Datta, Bryant, & Mostov, 2011). Most studies on polarized trafficking have been performed using traditional bidimensional (2D) cell cultures using (Transwells®) filters. These studies have been crucial for the elucidation of protein-trafficking pathways controlling epithelial polarity and morphogenesis, because both the apical and basolateral membranes were accessible to biochemical preparation and analysis (Rodriguez-Boulan, Kreitzer, & Musch, 2005; Rodriguez-Boulan & Nelson, 1989). However, epithelial cells can also be converted into organotypic cultures by the addition of extracellular matrix (ECM) components. In these conditions, the cell cultures closely resemble the three-dimensional (3D) architecture of the organs from which they are derived (Zegers, O'Brien, Yu, Datta, & Mostov, 2003). When cultured in 3D, epithelial cells orient the axis of polarity through the cues originating at cell-to-cell and cell-to-matrix contacting regions, reorganize the membrane surfaces, and then transport apical proteins to the available free surfaces to form the lumen, in a process that recapitulates essential aspects of *de novo* apical membrane formation during epithelial organ morphogenesis (Apodaca, Gallo, & Bryant, 2012). Thus, 3D culture methods provide a powerful framework for the study of apical trafficking in a functionally relevant process such as lumen formation during organ development.

Epithelial cyst formation *in vitro* was discovered early in the 1980s using thyroid follicles (Chambard, Gabrion, & Mauchamp, 1981). Shortly thereafter, epithelial cysts were developed using Madin–Darby canine kidney (MDCK) cells cultured on collagen-I gels (Mangoo-Karim, Uchic, Lechene, & Grantham, 1989; Wang, Ojakian, & Nelson, 1990a, 1990b). As imaging methods improved, especially with the extensive use of confocal microscopes, the analysis of organotypic 3D cultures became easier to approach, which allowed the elucidation of the mechanisms associated with essential pathways of epithelial morphogenesis. Interestingly, most of these mechanisms, including single lumen formation, tubule branching, and mitotic spindle orientation, were found to contribute specifically to the 3D epithelial phenotype, as their disruption had little-to-no phenotype in conventional 2D cultures. Thus, these pioneering studies proved that 3D culture of epithelial cells was fundamental to unravel the complex process of epithelial organ morphogenesis, and provided a bridge model between traditional *in vitro* cell culture methods and animal studies, where these mechanisms could not be approached by these easy but sophisticated means of genetic and biochemical manipulation.

7.1 3D CULTURE OF MDCK CYSTS ON MICROPATTERNS

In physiological conditions, cells are able to sense a variety of stimuli that dictates their behavior during tissue morphogenesis. Apart from biochemical stimuli, such as growth factors, morphogens, or ion concentrations, cells are able to respond to physical stimuli, such as pressure and stretching (Nelson & Bissell, 2006). Moreover, cells sense these biochemical and biophysical cues through the physical interaction with the ECM (mainly by integrin receptors), and thus, physical properties of the ECM can control important aspects of cell physiology (DuFort, Paszek, & Weaver, 2011). For instance, cells respond to changes in ECM stiffness through modifications in their cytoskeletal contractility.

Micropatterns are a simple technique that allows cell culture in a controlled adhesive environment with extremely high precision, close to the nanometer scale (Thery, 2010). Micropatterned coverslips are coated with a hydrophobic gel layer that repels cell adhesion, except in specific regions where the hydrophobic coating has been eliminated (micropatterns) causing cells to adhere according to the shape of the micropatterns. The current precision in the micropatterning techniques used by certain manufacturers (i.e., CYTOO) produces highly reproducible chips where cells can be seeded in arrays that are easily traceable for high-throughput image quantifications. We have recently developed a method to culture MDCK cysts on micropatterns of different sizes and composition (Rodriguez-Fraticelli, Auzan, Alonso, Bornens, & Martin-Belmonte, 2012). Using this method, we found that changes in micropattern shape and size can be used to modify cell contractility to understand its contribution to epithelial morphogenesis. MDCK cells can bind collagen-I through interaction with the β1-integrin receptor. Collagen-I elicits an adhesive response that triggers extensive formation of focal adhesions and cell spreading up to around 1600 μm^2. Taking advantage of these features, we plated MDCK cells in collagen-I-coated micropatterns, where cell contractility could be modulated by the size of the micropattern. Cells on large micropatterns presented increased cell contractility, which prevented lumen formation, while cells on small micropatterns produced fewer stress fibers and lumen formation proceeded normally (Rodriguez-Fraticelli et al., 2012).

7.1.1 Protocol overview

In brief, MDCK cells are trypsinized, centrifuged, and resuspended in fetal bovine serum-supplemented Modified Eagle's Medium (MEM). Cells are then seeded on a micropatterned coverslip previously coated with laminin or collagen-I on a 35-mm plastic dish. Cells are allowed to sediment and attach to the adhesive micropatterns for 1 h. The coverslip is then washed briefly and attached cells are incubated at 37 °C for 3–4 h until they spread on the micropattern surface. Then, Matrigel-supplemented MEM is added, and MDCK cysts form over the course of 3 days. Lumens can be observed from 24 h after Matrigel addition. Cells are fixed using different protocols and then analyzed by immunofluorescence using confocal microscopy. Alternatively, cells expressing fluorescent tags can be also analyzed *in vivo* in micropatterns using live cell microscopy.

7.1.2 Working with Matrigel

Early studies provided evidence that laminin is perhaps the most important signaling ECM molecule required for epithelial morphogenesis (Bello-DeOcampo, Kleinman, Deocampo, & Webber, 2001; Ekblom, 1996; O'Brien et al., 2001; Santos & Nigam, 1993; Yu et al., 2005). More recently, our own work showed that cell confinement, which is regulated *in vivo* by the stiffness of the matrix, is also involved in epithelial morphogenesis and in particular is required to prevent cell spreading and contractility, which affect the capacity of the cell to reorganize the surrounding matrix (Rodriguez-Fraticelli et al., 2012). The result of these processes is the correct vectorial orientation of cell polarity, with the Golgi apparatus and the apical endosomal components localized toward the cell–cell junctions, and the nuclei separated from the cyst center, and closer to the ECM. Remarkably, purified laminin seems to be sufficient to prevent cell spreading and acts as a dominant cue (over other molecules such as collagen-I or IV) to organize epithelial polarity. The fastest method to provide laminin to cells is to culture them in the presence of basement membrane extracts (BMEs) such as Matrigel (an ECM extract from an Engelbreth–Holm–Swarm mouse sarcoma that is enriched in basement membrane components such as laminin and entactin). Therefore, culture of cysts overlaid with Matrigel results in faster and more consistent cell polarization and lumen formation rates, which is optimal for analyzing the process of epithelial morphogenesis using MDCK cells.

Matrigel (BD #356234) is a very unstable solution that may rapidly precipitate even when kept refrigerated. In fact, Matrigel starts to precipitate, forming gels over 6–8 °C, making it impossible to restore it to its original condition. In this situation, cells are unable to acquire the appropriate cell polarization since they lose the spatial cue provided by the laminin in suspension. Thus, it is important to keep in mind certain guidelines for working with Matrigel in optimal conditions for cyst formation.

PROCEDURE:

1. Matrigel (or other BMEs) will be shipped and received frozen in a glass vial on dry ice. Keep at −80 °C until aliquoting.
2. The day before aliquoting, thaw the Matrigel vial on ice overnight (in a cold room or inside a refrigerator).
3. The next morning, the Matrigel solution should be liquid. Slowly vortex the Matrigel vial to produce a homogenous solution. Keep at 4 °C.
4. Under a cell culture hood, place sterile 1.5 ml tubes on ice and allow them to chill. At the same time, prepare a box with dry ice.
5. Open the Matrigel vial cap and aliquot 400 μl of the solution into each tube using prechilled 1 ml pipette tips. Immediately close each tube after pipeting in the Matrigel, and place them on dry ice for fast freezing.
6. Matrigel aliquots can be stored at −80 °C up to 1 year maximum.
7. To thaw an aliquot, place them on ice overnight (in a cold room or inside a refrigerator). Always keep it on ice with the tube cap closed. Thawed aliquots can be kept refrigerated (≤5 °C) for up to 1 week.
8. When preparing a Matrigel solution with MEM, dilute the appropriate amount of Matrigel on cold MEM medium, invert the tube until Matrigel is dissolved, and then keep the solution on ice.
9. Dissolved Matrigel can be kept refrigerated for 1 week maximum.

7.1.3 Working with micropatterns

Although different methods exist for producing micropatterns for cell culture, we recommend CYTOOchips (available worldwide from www.cytoo.com). CYTOOchips present highly reproducible conditions (shape, size, coating) for each micropattern, which is an advantage for high-throughput experiments. Self-manufactured micropatterns can be useful as well, especially when assaying multiple pattern shapes and sizes that are not commercially available, which usually requires the manufacture of custom chips that is very time consuming.

When considering different pattern shapes and sizes for cyst formation, we have found that the optimal shape is a round-shaped micropattern (such as a disk or circle). Round shapes probably allow easier cyst rotation, which is required for proper organization of the basal lamina and lumen formation (Wang, Lacoche, Huang, Xue, & Muthuswamy, 2013). The optimal coating condition is laminin-coated micropatterns, which inhibit cell contractility and thus induce lumen initiation after only 24 h. We also observe that for MDCK cysts 500–700 μm^2 disk-shaped micropatterns present a slight advantage when performing lumen formation experiments, although the difference was more notable when culturing cells without laminin (Rodriguez-Fraticelli et al., 2012). Therefore, 500–700 μm^2 disk-shaped laminin-coated micropatterns will provide optimal conditions for cyst formation experiments (close to 80% of single lumen formation efficiency 3 days after culture; Rodriguez-Fraticelli et al., 2012).

When manipulating micropatterned chips, it is important to keep in mind that the majority of the coverslip surface is coated with a highly hydrophobic material that repels water and thus will dry cells faster than in normal coverslips (usually within 10–20 s outside of liquid medium). For this reason, we always keep the micropatterns inside liquid medium until cells are properly fixed.

7.1.4 MDCK cell strains

Different strains of MDCK cells are available to researchers from the ATCC and labs worldwide. Although MDCK strain 2 (MDCK-II) is the most frequently used, researchers have also used MDCK-clone 7, clone 8, MDCK-I, and NBL-2 cells. These cell strains present important variations in the passaging method and cyst formation efficiencies. In our hands, MDCK-II and NBL-2 cells are the best strains for 3D cyst culture (having over 80% single lumen formation efficiencies after 3 days of culture in Matrigel), and thus we recommend working with these cell strains when analyzing lumen formation using micropatterns. The cells should be passaged every 2 days at 70–80% confluence in Glutamine and 5–10% serum-supplemented MEM. High passage number MDCK cells have been recently described to present frequent chromosomal abnormalities and should be avoided for lumen formation experiments (Cassio, 2013). Cells should be split 2 days before starting with the micropattern-culture protocol.

7.1.5 Trypsinization, resuspension, and seeding cells on micropatterns (basic protocol)

MDCK cells are highly adherent cells that require several washes with Ca/Mg-free PBS and long trypsin incubation times (>15 min) in order to produce a single cell suspension that generates the best results on micropatterns. We also recommend using a cell strainer to prevent clump formation (BD, catalog #352340). Ideally, the strainer will produce a solution where cells will be as separated as possible.

The following protocol is for a single 20 × 20 mm micropatterned coverslip cultured on a 35-mm dish.

1. Remove culture medium and wash MDCK cells twice with 10 ml PBS.
2. The second wash is usually 10–20 min long, until cell–cell junctions appear brilliant (highly contrasted) using a light microscope.

3. Remove PBS and add 1 ml of prewarmed trypsin solution (PBS, 0.05 g/l trypsin, 25 mM EDTA).
4. Incubate the cells with trypsin solution at 37 °C until cells detach from the plate.
5. Add 10 ml of FBS-supplemented MEM, resuspend cells, and place in a 15-ml tube.
6. Centrifuge the tube at $90 \times g$ for 5 min and resuspend the pellet containing the cells in 5 ml of MEM medium.
7. Filter the cell solution using a cell strainer into a 50-ml tube.
8. Dilute cells into prewarmed MEM at 10,000 ml^{-1} cells. We use 4 ml (40,000 cells) per 35 mm dish (containing a single 20×20 mm CYTOOchip). It is extremely important that MEM is prewarmed to create uniform cell seeding.
9. Add the 4 ml of cell solution into the dish containing the micropatterns.
10. Without rocking the dish, place inside a 37 °C incubator and allow cells to sediment for 1 h.
11. Wash out the cells that have not adhered to the patterns using MEM. Rapidly pipette 5 ml of MEM over the center of the chip, and then remove the same volume with an aspirator. Repeat two to three times or until floating cells have disappeared.
12. Place the dish back inside the incubator and allow cells to spread for 3–4 h.

7.1.6 Matrix overlay method

As discussed before, the basic technique for polarizing epithelial cells into cysts is to add BMEs such as Matrigel into the medium (Debnath, Muthuswamy, & Brugge, 2003). The high content of laminin present in Matrigel provides sufficient signaling to adhering cells to induce cystogenesis. When properly prepared, Matrigel in the medium homogenously precipitates over the cell culture overnight and covers the cells (overlay method). Purified laminin may in theory produce the same results, although it needs to be added in such high concentration in the medium that it is extremely expensive considering current methods for purification from BME. The amount of Matrigel that is added to the medium is very important for proper induction of lumen formation. MDCK cells require at least 20 μl of Matrigel per milliliter of medium (at 10,000 cells/ml). For a 35-mm dish we use 80 μl of Matrigel diluted in 4 ml of medium. In order to prevent cell drying, instead of removing all the medium from the 35 mm dish containing the micropatterns, it is highly recommended to prepare 2 ml of 4% Matrigel and then add it to a 35-mm dish containing 2 ml of normal MEM medium (from the previous cell seeding and adhesion step) to achieve a final concentration of 2% Matrigel.

1. Thaw a Matrigel aliquot overnight on ice.
2. Pipette 80 μl of Matrigel in 2 ml of chilled MEM (4 °C) to create a 4% Matrigel solution. Mix thoroughly by pipetting up and down.
3. Warm the 4% Matrigel solution to room temperature.
4. Add 2 ml of Matrigel solution dropwise into the 35 mm dish containing the micropatterned cells. Note that it is very important that cells are correctly spread

on the micropatterns before the addition of Matrigel to prevent its precipitation and binding to the free micropattern surfaces, which would inhibit normal cell spreading.

5. Incubate cells with Matrigel for 24–72 h. Change medium every 48 h.
6. Longer incubation times are possible, but cells detach from the micropatterns after incubation for more than 5 days.

7.1.7 Fixation and analysis

Micropatterned cells on chips typically contain a small number of cells (about 100–200,000 per chip after 72 h). While these numbers are sufficient for RNA extraction and analysis, they are typically not adequate for most biochemical studies. Thus, light microscopy techniques are usually the preferred method of choice to analyze micropatterned cells. Additionally, 4% paraformaldehyde solution (prepared in phosphate buffer) or formalin is highly recommended to fix cells. However, methanol, acetone, ethanol, or other fixatives are also suitable for cell fixation on micropatterns.

7.1.7.1 PFA/formalin fixation

Prepare fresh 4% PFA solution. Remove culture medium from the culture dish until only 1 or 2 ml remain (to prevent micropatterns from drying). Add 4 ml of 4% PFA solution dropwise, without rocking the dish. Incubate the fixative for 10 min. Remove all the media from the dish and add 2 ml of fresh fixative. Incubate for an additional 20 min. Remove fixative and wash twice with large volume of PBS. Aldehyde quenching can be performed using 0.1 M glycine solution in phosphate buffer. Fixation with PFA can result in heavy ECM crosslinking around the basal membrane of the cysts, which may prevent certain antibodies or other reagents from diffusing into the entire structure (especially the lumen). In this case, fixation with cold methanol or acetone is recommended. Alternatively, reduced incubation times, or using 3% PFA, can reduce crosslinking and enhance detection with certain antibodies. If using formalin, just add 2 ml of formalin solution into the medium, without rocking the dish. Incubate for 10 min, remove all media, and then add 2 ml of fresh formalin. Incubate the formalin for an additional 10 min and proceed to the washing step.

7.1.7.2 Methanol fixation

Prepare cold pure methanol (−20 °C) and cold PBS in two separate dishes. With a pair of tweezers or forceps, pick up the micropatterned coverslip and rapidly dip it into the methanol solution, making sure it is fully covered. Place the dish inside the −20 °C freezer and let the cells dehydrate for 5 min. Allow the cells to rehydrate for 20 min in cold PBS before proceeding to next steps (blocking and staining). Dehydration with methanol will prevent phalloidin staining of F-actin. In some cases, methanol fixation can be too harsh for certain antibodies. In this case, acetone fixation can result in a milder dehydration that keeps most epitopes available for binding.

7.1.7.3 Acetone fixation

Prepare cold pure acetone (−20 °C) in a separate glass dish. With a pair of tweezers or forceps, pick up the micropatterned coverslip and rapidly dip it into the acetone solution. Place the dish inside the −20 °C freezer and let the cells dehydrate for 5 min. Rehydrate the cells in cold PBS for 20 min before proceeding to blocking and staining. Dehydration with acetone will prevent phalloidin staining of F-actin.

Fixed samples may be stored at 4 °C up to several weeks in PBS. Wrap up the plastic dish containing the coverslip in Parafilm® to prevent sample dehydration. Addition of 0.05% sodium azide is recommended for sample preservation. Remove azide before proceeding with the rest of the protocol.

7.1.7.4 Permeabilization

If cells are fixed using formaldehyde, detergent- or methanol-based permeabilization is required for antibody penetration. Typically, the permeabilization method will depend on the fixative reagent used. However, permeabilization is especially important when dealing with Matrigel-embedded cysts, since Matrigel can block entry of many antibodies into the cyst structure. For most preparations, incubate the micropatterns with 0.2% Triton X-100 in PBS for 15 min at 4 °C. For a harsher solubilization of membranes and lipids, add 0.1% sodium dodecyl sulfate (SDS) to the buffer.

7.1.7.5 Blocking and antibody incubations

All blocking and staining steps are performed at room temperature on a single day. Blocking solution may be supplemented with fish skin gelatin, or with 0.02% Triton X-100 or 0.3% saponin, to enhance antibody permeability and eliminate potential background. Volumes are measured for staining a single micropatterned coverslip in a 35-mm dish.

1. Prepare blocking solution (PBS, 3% Bovine Serum Albumin) and warm to room temperature.
2. Wash coverslips twice with PBS.
3. Replace the PBS with blocking solution (at least 2 ml) and incubate for 1 h at room temperature.
4. Dilute the appropriate primary antibodies in 2 ml of blocking solution.
5. Replace the blocking solution with antibody solution and incubate for 2 h at room temperature with gentle rocking.

Optional: Incubation of primary antibody solutions overnight at 4 °C may enhance signal detection, but it could also result in higher background.

When dealing with expensive or limited reagents, small volumes of antibody solutions may be prepared and incubations can be performed by inverting and placing the micropatterned coverslip on a 100 μl drop of antibody solution using a Parafilm®-coated incubation chamber. In this case, it is very important to create a humid environment by adding a wet sponge or other wet material in order to prevent the micropattern from drying.

7.1.7.6 Secondary antibodies and other staining reagents

1. Dilute fluorochrome-conjugated secondary antibodies (or fluorochrome-conjugated streptavidin) in 2 ml of blocking solution to create the staining solution. Other staining reagents such as fluorochrome-conjugated phalloidin or DNA-stains can be added to the staining solution.
2. Remove the primary antibody solution and rapidly add PBS (avoid drying the coverslip). Wash the micropattern coverslip twice with PBS, and then perform three washes with PBS with 10-min intervals. Perform a final 10-min incubation with 2 ml of blocking solution.
3. Replace the blocking solution with 2 ml of staining solution and incubate for 1 h at room temperature with gentle rocking.

7.1.7.7 Mounting and imaging

1. Wash the stained micropattern coverslip five times with PBS (10-min intervals).
2. Add a 100 μl drop of ProLong-GOLD anti-fading mounting medium to the center of a microscopy-quality glass slide. Use a fine tip to eliminate air bubbles. Allow the drop to rigidify slightly over 15 min.
3. Dry the coverslip using a tissue paper wipe.
4. Invert the coverslip (cells side down) and place over the drop of mounting medium. Remove excess mounting medium with a tissue paper.
5. Allow drying for 2 days at room temperature in a light-protected dry environment.
6. *Optional*: Seal the edges of the coverslip by using transparent nail polish. This will help preventing the coverslip from shifting around the glass slide, and will also prevent excessive drying of the specimen.

7.1.8 Imaging and quantification

After days of culturing MDCK cells on micropatterns, the resulting cyst structures normally have a *z*-depth of tens of microns, which makes imaging with wide-field microscopy very difficult. Confocal scanning technique eliminates most light from unfocused planes, resulting in a clean image of the central slice of a large structure.

Tubular epithelia are characterized by the presence of a single lumen, lined by the interconnected apical plasma membranes of all the cells in the tissue. When epithelial morphogenesis is disrupted, the loss of a single lumen is the most usual and conspicuous phenotype. When single lumen formation is affected, cell death, lumen filling, and the orientation of cell divisions are also quantified in order to establish the cause of lumen disruption.

7.1.8.1 Number of lumens

Staining with apical membrane markers, such as Podocalyxin-like (gp135 clone 3F2/D8, developed by George Ojakian at State University of New York) or Ezrin (BD cat. #610602), is the easiest way of specifically visualizing the apical membrane. Alternatively, apical accumulation of F-actin in microvilli and the terminal actin network can be stained with fluorescent phalloidin (available from various suppliers). Other reports of valid commercial apical membrane staining include the use of antibodies against aPKC (PKC-ζ, Santa Cruz, C-20, cat. #sc-216), Stx3 (rabbit serum, Synaptic systems cat. #110-032), or prominin-1 (Chemicon, Millipore #MAB4310).

Once the lumen is clearly visualized, we sample 4–5 fields of 25–30 cysts each. The total cyst number quantified per condition is about 200. Normally, we exclude counting aggregates that sometimes form in the borders of the wells or chambers. The main feature that requires quantification is the amount of lumens per cyst (Fig. 7.1). Normal cysts will have one (and sometimes two) large central lumens, with a round shape. We consider cysts to be abnormal if:

A. Cysts contain more than two lumens, or if (in the case of having less than three lumens).
B. Lumens are not centered in the cyst structure.
C. Apical proteins are in contact with the surrounding ECM after 48 h (inverted polarity).
D. At least a single cell is in contact with more than one lumen, and thus presents more than one apical plasma membrane at the same time (apical bipolarity).
E. Internal apical membranes are visualized in the cytosol after 48 h.
F. Lumen is filled with cells (which may be undergoing apoptosis).
G. Luminal apical membranes are enlarged ("star-shaped" lumens).

Correctly fused apical membranes are observed as localized between two cells, separating two lateral membranes and surrounded by tight junctions. In contrast, internalized apical membranes are localized inside the cell volume (separated from basolateral membranes). For this purpose, costaining with a basolateral marker such as β-catenin (Santa Cruz, H-102, cat. #sc-7199) or E-cadherin (rr1 mouse monoclonal antibody, DSHB, or BD cat. # 610405) and tight-junction markers such as ZO-1 (R26.4C rat monoclonal, DSHB) or Occludin (Life Technologies, OC-3F10, cat. #33-1500) is required to establish if apical membranes are being correctly transported and fused in the central region of the cyst. When MDCK cells accumulate apical markers in internal membranes, which typically colocalize with apical-transport machinery (such as Rab11, Life Technologies #71-5300 or Rab8, BD #610845), this phenotype usually reflects a defect in intracellular trafficking. For instance, disruption of the exocyst complex, Cdc42 or Rab8, causes accumulation of apical proteins in endomembranes that are retained intracellularly and do not fuse to form a lumen (Bryant et al., 2010).

When cysts contain more than two lumens (without accumulating apical markers in internal membranes), the phenotype can be explained by a disruption in

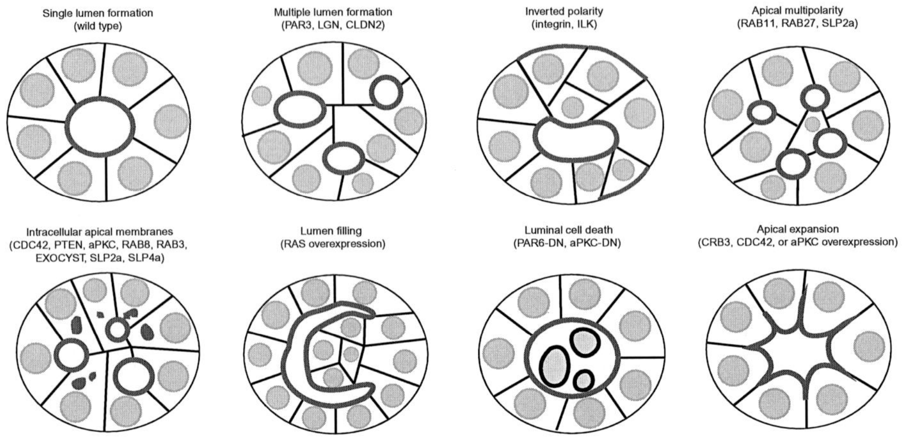

FIGURE 7.1

Phenotypes of lumen formation defects. During analysis of epithelial morphogenesis, MDCK cysts may present a wide array of abnormalities depending on the affected mechanisms of study. Multiple lumen formation is caused by disruption of the spindle orientation machinery, which is controlled by apical polarity proteins (such as Cdc42, aPKC, and Par3), and microtubule-binding protein complexes (such as Pins/LGN-NuMA-Gαi). Cell division orientation controls the positioning of the cytokinetic midbodies around the center of the cyst, which serve as landmarks for apical membrane assembly. When integrins and their downstream signaling molecules (such as Integrin-linked kinase, ILK) are affected, microtubule organization is disrupted and may present as cysts with inverted apical membrane markers. Disruption of vesicle tethering or targeting proteins results in apical multipolarity (more than one apical membrane per cell in the cyst aggregate) or retention of apical proteins in intracellular membranes (i.e., upon disruption of Rab3, Rab8, Rab11, Rab27, exocyst proteins, aPKC, Slp2, Slp4, and others). Lumen filling with cells is another frequent phenotype that may result from disrupted spindle orientation or abnormal cell extrusion (i.e., expression of dominant negative Par6 or aPKC inhibition) or from defects in the mechanisms of cell competition, which may be accompanied by apoptotic resistance (i.e., Ras overexpression). Finally, overexpression of apical polarity proteins can induce apical membrane expansion and formation of star-shaped lumens, where apical proteins leak into the lateral domains (i.e., Crumbs, aPKC, or constitutively active Cdc42 overexpression). (See color plate.)

microtubule polarity or a defect in the machinery that control spindle orientation during cell division. Indeed, spindle orientation controls the position of the midbody during cytokinesis, which in turn, determines the position of initial lumen since the machinery that initiates the apical membrane localizes to the midbody during cytokinesis close to the site of abscission (Jaffe, Kaji, Durgan, & Hall, 2008; Schluter et al., 2009). The process of cyst maturation would then require the coordination of apical trafficking and cell division, and consequently disruption of either of these two processes would result in defects in single lumen formation. For instance, when cells in the tubular epithelium divide in a nonplanar orientation (i.e., perpendicular to the planar orientation of the monolayer), this could result in cell extrusion from the epithelium (toward the ECM or toward the lumen). However, a thick layer of basal lamina surrounds the cyst and instead of getting extruded toward the ECM the cells remain attached to the aggregate and form a secondary lumen away from the center of the cyst. Thus, multiple lumen formation (phenotypes A or B) may result from alterations in the spindle orientation machinery. For instance, disruption of LGN or Par3 affects spindle orientation and causes multiple lumen formation in MDCK cysts (Hao et al., 2010; Zheng et al., 2010). Meanwhile, inverted polarity or apical bipolarity (phenotype C or D) usually results from disrupted ECM signaling, or from abnormal localization/polarization of the endovesicular trafficking machinery. For example, inhibition of integrin-linked kinase (ILK), which is required for ECM signaling-mediated regulation of microtubule polarity, causes inverted apical polarity, while disruption of synaptotagmin-like protein 2, which tethers apical vesicles to $PtdIns(4,5)P_2$-enriched apical membranes, causes apical bipolarity (Akhtar & Streuli, 2013; Galvez-Santisteban et al., 2012).

7.1.8.2 Cell death

When cells are extruded toward the lumen, they usually undergo apoptosis through a mechanism-denominated anoikis. Apoptotic cells in the lumen can be stained using activated Caspase-3 antibody (Cell Signaling #9661). Thus, observation of caspase-3 positive cells in the lumen usually suggests defects in spindle orientation. However, several other mechanisms, including defects in cell–cell junctions, cell cycle, etc., may also result in apical extrusion in the lumen, and cell death. However, when cells are extruded in the lumen, but no cell death is observed causing lumen filling, the phenotype suggests that these cells have developed resistance to apoptosis (Sakurai, Matsuda, & Kiyokawa, 2012).

7.1.8.3 Spindle orientation

Tubular epithelial cells divide with a planar orientation of the mitotic spindle to ensure that both daughter cells remain in the same plane of the monolayer, which is essential to maintain the physiological functions of the epithelium. The mechanism that controls planar cell division in epithelia is complex and robust, and it is implicated in the maintenance of a single lumen during epithelial morphogenesis (Durgan, Kaji, Jin, & Hall, 2011; Fujiwara et al., 2008; Guilgur, Prudencio, Ferreira, Pimenta-Marques, & Martinho, 2012; Hao et al., 2010; Jaffe et al., 2008; Qin, Meisen, Hao, & Macara, 2010; Rodriguez-Fraticelli et al., 2010; Wei, Bhattaram, Igwe, Fleming, & Tirnauer, 2012; Zheng et al., 2010). Thus, when multiple lumen phenotypes are observed, it is necessary to measure if cell division orientation is affected. It is advisable to quantify cell division orientation after the initial lumen is already formed, since it will help quantify the cell division angle more easily. Thus, the use of inducible interfering RNA or inhibitors is strongly recommended. To measure the angle of spindle orientation, tubulin antibodies (to stain metaphase and anaphase microtubules, Sigma-Aldrich, DM1α) and phospho-Histone-3 (which stains chromatin at

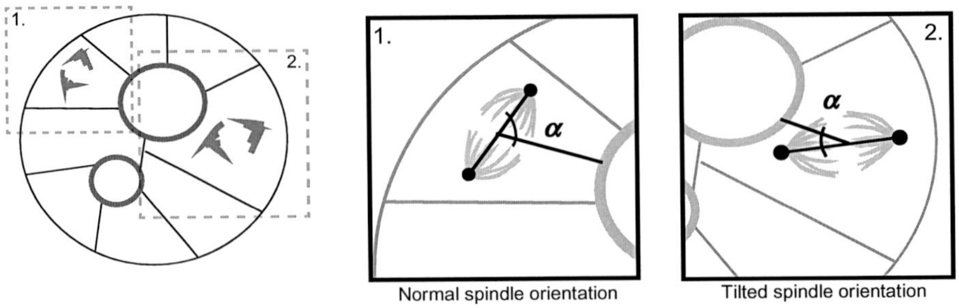

FIGURE 7.2

Spindle orientation quantifications. MDCK cells divide in the plane of the monolayer to maintain the architecture of the tube, including the formation of a single lumen. To quantify the orientation of the mitotic spindle, we measure the angle formed between the center of the spindle axis (the imaginary line crossing both spindle poles) and the center of the apical plasma membrane. When this angle is close to 90°, the orientation is considered normal, resulting in daughter cells that maintain the architecture of the monolayer (1). When the angle is closer to 0°, the orientation is considered abnormal, resulting in mispositioning of the midbody and ectopic lumen initiation (2). (See color plate.)

the end of metaphase, Cell Signaling #9701) are suggested. The angle between the line formed by the spindle poles and the apicobasal axis (from the center of the spindle to the center of the apical membrane) is measured in degrees, with 90° considered a normal division perpendicular to the AB axis, and 0° being an abnormal division parallel to the AB axis (Fig. 7.2). To quantify cell division more easily, cells can be treated to synchronize their cell cycles. MDCK cysts are hard to synchronize using cell division-blocking protocols, probably because junctions and ECM proteins play a major role in controlling cell cycle progression. However, the double thymidine block protocol does not affect cell division orientation, compared to controls, and can achieve 5–10% synchronization of cell cycles, which facilitates quantifications. As a starting-point protocol, perform an 18-h block with 2 mM thymidine-supplemented MEM, wash twice with MEM to "release" the block, then incubate for 5 h in normal MEM, and perform a second 18-h block with thymidine overnight. After the second thymidine release, start fixing cysts immediately every 30 min.

7.1.9 Preparation of preactivated micropatterns coated with other proteins

While commercial micropatterns can be ordered custom-made with different protein coatings, preactivated micropatterns can also be purchased and coated with different proteins produced in the lab. For instance, purified ECM-protein domains, cadherin-activating antibodies, inhibitory peptides, or growth factors can be added to coat micropatterns and analyze their effect on epithelial morphogenesis. Thus, we include a short protocol for coating preactivated micropatterns.

1. Dilute the desired purified proteins in 4 ml PBS (coating solution). A minimum protein concentration of 20 μg/ml is recommended.
2. Add a 100-μl drop of sterile water on a 35-mm dish.
3. Place the preactivated micropatterned coverslip on top of the drop in the dish. The drop of water will be used to prevent the coverslip from moving about the dish.
4. Pour the coating solution into the dish and incubate at room temperature for 2 h (or 4 h-to-overnight at 4 °C).
5. Add 5 ml of PBS and wash extensively by adding and removing PBS without letting the coverslip dry.
6. Wash twice with 10 ml of sterile water and then remove all media and dry the dish for 2 min (remove the lid).
7. The dried coated micropatterns can be kept at 4 °C in a parafilm-sealed dish, for up to 2 days.

7.1.10 *In vivo* experiments

When imaging cyst formation using traditional 3D cultures, we observed that cyst structures tend to rotate and shift in all three dimensions, and frequently end up tens of microns away from the initial position (Martin-Belmonte et al., 2008). One of the greatest advantages of culturing cysts on micropatterns is the fact that the imaged structures cannot displace in the 2D imaging plane (neither in the *z*-axis). This results in far easier tracking of cyst structures for the long recording times required to image epithelial morphogenesis, without the need of complex tracking equipment. Moreover, the arrayed disposition facilitates video recording in multiple positions to record several videos at the same time, and without the need to design complex multiple image acquisition macros. Finally, different shapes and sizes of micropatterns can be analyzed at the same time in the exact same conditions, just by changing the acquired position a few microns in the same well. Since micropatterned coverslips are still not offered in chambered coverslip-glass slides, a reusable chamber device is required for sealing the micropatterned chip for inverted microscopy.

1. After cells are seeded and spread on the micropatterns (after Basic Protocol step 12), prepare the CYTOOchamber (CYTOO, #30-010) by washing with SDS detergent, then extensively with water and drying with an air-nozzle.
2. Place the rubber seal into the indentation of the top chamber lid.
3. Place the precoated micropatterned coverslip in the bottom plate.
4. While holding the bottom plate, place the magnetic lid on top of the plate, until sealed.
5. If the rubber seal moves out of the indentation, use a thin tip pipette to place it back into position.
6. Add 2 ml of Matrigel solution dropwise into the CYTOOchamber containing the micropatterned cells. Steps 3–6 have to be performed rapidly to prevent cells from drying.

7. Image acquisition requires an inverted microscope and a 36-mm dish adapter. We use confocal laser scanning microscopy for short recording times (up to 4 h of recording), and epifluorescence wide field microscopy for longer recording times (from 4 to 72 h), both equipped with incubation chambers to control humidity (>90% saturation), temperature (37 °C), and CO_2 concentration (5%).

7.2 DISCUSSION

We have introduced a novel method for 3D-MDCK cyst culture using micropatterns. Micropatterning cyst formation presents several advantages over other methods. First, it provides a more reproducible niche (the micropattern), where cells are seeded in a single-cell manner. Second, it enables the control of this environment, since micropatterns can be of different shapes, sizes, and coatings, to study the role of the microenvironment in epithelial morphogenesis. Third, the interspersed distribution of micropatterns allows faster quantifications, and facilitates faster high-throughput image acquisition. And finally, live imaging experiments benefit from reduced 3D shift of the observed specimen, minimizing the need for tracking software and hardware. Additionally, it is notable that lumen initiation takes place closer to the coverslip on micropatterns, which results in higher resolution imaging, and should enable super-resolution confocal imaging. This is especially relevant for the visualization and analysis of trafficking pathways and vesicle transport events of epithelial cells. Since these cells are polarized in the apicobasal axis, the transport from the Golgi to the apical plasma membrane proceeds in the apicobasal plane. When imaging cysts, Golgi-to-apical vesicle is visualized in the x–y plane, which presents higher resolution on confocal microscopes. Thus, the use of micropatterns is an efficient setup to analyze Golgi-dependent polarized secretion with unprecedented higher resolution in both time and space. We also expect that further developments in the micropatterning techniques will make this method suitable for the culture and analysis of other epithelial cell types as well as the analysis of alternative pattern shapes in order to study other important morphogenetic processes such as tube elongation and branching.

Acknowledgments

We thank Carmen M. Ruiz-Jarabo for comments on the manuscript, and members of the Martin-Belmonte lab for discussions. This work was supported by grants from the Human Frontiers Science Program (HFSP-CDA 00011/2009), MICINN (BFU2011-22622), and CONSOLIDER (CSD2009-00016) to F. M.-B. A. E. R.-F. is a recipient of a JAE fellowship from CSIC. An institutional grant from Fundación Ramón Areces to CBMSO is also acknowledged.

References

Akhtar, N., & Streuli, C. H. (2013). An integrin-ILK-microtubule network orients cell polarity and lumen formation in glandular epithelium. *Nature Cell Biology*, *15*(1), 17–27. http://dx.doi.org/10.1038/ncb2646.

Apodaca, G., Gallo, L. I., & Bryant, D. M. (2012). Role of membrane traffic in the generation of epithelial cell asymmetry. *Nature Cell Biology*, *14*(12), 1235–1243. http://dx.doi.org/10.1038/ncb2635.

Bello-DeOcampo, D., Kleinman, H. K., Deocampo, N. D., & Webber, M. M. (2001). Laminin-1 and alpha6beta1 integrin regulate acinar morphogenesis of normal and malignant human prostate epithelial cells. *Prostate*, *46*(2), 142–153.

Bryant, D. M., Datta, A., Rodriguez-Fraticelli, A. E., Peranen, J., Martin-Belmonte, F., & Mostov, K. E. (2010). A molecular network for de novo generation of the apical surface and lumen. *Nature Cell Biology*, *12*(11), 1035–1045. http://dx.doi.org/10.1038/ncb2106.

Cassio, D. (2013). Long term culture of MDCK strains alters chromosome content. *BMC Research Notes*, *6*, 162. http://dx.doi.org/10.1186/1756-0500-6-162.

Chambard, M., Gabrion, J., & Mauchamp, J. (1981). Influence of collagen gel on the orientation of epithelial cell polarity: Follicle formation from isolated thyroid cells and from preformed monolayers. *Journal of Cell Biology*, *91*(1), 157–166.

Datta, A., Bryant, D. M., & Mostov, K. E. (2011). Molecular regulation of lumen morphogenesis. *Current Biology*, *21*(3), R126–R136. http://dx.doi.org/10.1016/j.cub.2010.12.003.

Debnath, J., Muthuswamy, S. K., & Brugge, J. S. (2003). Morphogenesis and oncogenesis of MCF-10A mammary epithelial acini grown in three-dimensional basement membrane cultures. *Methods*, *30*(3), 256–268.

DuFort, C. C., Paszek, M. J., & Weaver, V. M. (2011). Balancing forces: Architectural control of mechanotransduction. *Nature Reviews. Molecular Cell Biology*, *12*(5), 308–319. http://dx.doi.org/10.1038/nrm3112.

Durgan, J., Kaji, N., Jin, D., & Hall, A. (2011). Par6B and atypical PKC regulate mitotic spindle orientation during epithelial morphogenesis. *Journal of Biological Chemistry*, *286*(14), 12461–12474. http://dx.doi.org/10.1074/jbc.M110.174235.

Ekblom, P. (1996). Receptors for laminins during epithelial morphogenesis. *Current Opinion in Cell Biology*, *8*(5), 700–706.

Fujiwara, T., Kawakatsu, T., Tayama, S., Kobayashi, Y., Sugiura, N., & Kimata, K. (2008). Hyaluronan-CD44 pathway regulates orientation of mitotic spindle in normal epithelial cells. *Genes to Cells*, *13*(7), 759–770. http://dx.doi.org/10.1111/j.1365-2443.2008.01203.x.

Galvez-Santisteban, M., Rodriguez-Fraticelli, A. E., Bryant, D. M., Vergarajauregui, S., Yasuda, T., Banon-Rodriguez, I., et al. (2012). Synaptotagmin-like proteins control the formation of a single apical membrane domain in epithelial cells. *Nature Cell Biology*, *14*(8), 838–849. http://dx.doi.org/10.1038/ncb2541.

Guilgur, L. G., Prudencio, P., Ferreira, T., Pimenta-Marques, A. R., & Martinho, R. G. (2012). Drosophila aPKC is required for mitotic spindle orientation during symmetric division of epithelial cells. *Development*, *139*(3), 503–513. http://dx.doi.org/10.1242/dev.071027.

Hao, Y., Du, Q., Chen, X., Zheng, Z., Balsbaugh, J. L., Maitra, S., et al. (2010). Par3 controls epithelial spindle orientation by aPKC-mediated phosphorylation of apical Pins. *Current Biology*, *20*(20), 1809–1818. http://dx.doi.org/10.1016/j.cub.2010.09.032.

Jaffe, A. B., Kaji, N., Durgan, J., & Hall, A. (2008). Cdc42 controls spindle orientation to position the apical surface during epithelial morphogenesis. *Journal of Cell Biology*, *183*(4), 625–633. http://dx.doi.org/10.1083/jcb.200807121.

Mangoo-Karim, R., Uchic, M., Lechene, C., & Grantham, J. J. (1989). Renal epithelial cyst formation and enlargement in vitro: Dependence on cAMP. *Proceedings of the National Academy of Sciences of the United States of America*, *86*(15), 6007–6011.

Martin-Belmonte, F., Yu, W., Rodriguez-Fraticelli, A. E., Ewald, A. J., Werb, Z., Alonso, M. A., et al. (2008). Cell-polarity dynamics controls the mechanism of lumen formation in epithelial morphogenesis. *Current Biology*, *18*(7), 507–513. http://dx.doi.org/10.1016/j.cub.2008.02.076.

Nelson, C. M., & Bissell, M. J. (2006). Of extracellular matrix, scaffolds, and signaling: Tissue architecture regulates development, homeostasis, and cancer. *Annual Review of Cell and Developmental Biology*, *22*, 287–309. http://dx.doi.org/10.1146/annurev.cellbio.22.010305.104315.

O'Brien, L. E., Jou, T. S., Pollack, A. L., Zhang, Q., Hansen, S. H., Yurchenco, P., et al. (2001). Rac1 orientates epithelial apical polarity through effects on basolateral laminin assembly. *Nature Cell Biology*, *3*(9), 831–838. http://dx.doi.org/10.1038/ncb0901-831.

Qin, Y., Meisen, W. H., Hao, Y., & Macara, I. G. (2010). Tuba, a Cdc42 GEF, is required for polarized spindle orientation during epithelial cyst formation. *Journal of Cell Biology*, *189*(4), 661–669. http://dx.doi.org/10.1083/jcb.201002097.

Rodriguez-Boulan, E., Kreitzer, G., & Musch, A. (2005). Organization of vesicular trafficking in epithelia. *Nature Reviews. Molecular Cell Biology*, *6*(3), 233–247. http://dx.doi.org/10.1038/nrm1593.

Rodriguez-Boulan, E., & Nelson, W. J. (1989). Morphogenesis of the polarized epithelial cell phenotype. *Science*, *245*(4919), 718–725.

Rodriguez-Fraticelli, A. E., Auzan, M., Alonso, M. A., Bornens, M., & Martin-Belmonte, F. (2012). Cell confinement controls centrosome positioning and lumen initiation during epithelial morphogenesis. *Journal of Cell Biology*, *198*(6), 1011–1023. http://dx.doi.org/10.1083/jcb.201203075.

Rodriguez-Fraticelli, A. E., Vergarajauregui, S., Eastburn, D. J., Datta, A., Alonso, M. A., Mostov, K., et al. (2010). The Cdc42 GEF Intersectin 2 controls mitotic spindle orientation to form the lumen during epithelial morphogenesis. *Journal of Cell Biology*, *189*(4), 725–738. http://dx.doi.org/10.1083/jcb.201002047.

Sakurai, A., Matsuda, M., & Kiyokawa, E. (2012). Activated Ras protein accelerates cell cycle progression to perturb Madin-Darby canine kidney cystogenesis. *Journal of Biological Chemistry*, *287*(38), 31703–31711. http://dx.doi.org/10.1074/jbc.M112.377804.

Santos, O. F., & Nigam, S. K. (1993). HGF-induced tubulogenesis and branching of epithelial cells is modulated by extracellular matrix and TGF-beta. *Developmental Biology*, *160*(2), 293–302. http://dx.doi.org/10.1006/dbio.1993.1308.

Schluter, M. A., Pfarr, C. S., Pieczynski, J., Whiteman, E. L., Hurd, T. W., Fan, S., et al. (2009). Trafficking of Crumbs3 during cytokinesis is crucial for lumen formation. *Molecular Biology of the Cell*, *20*(22), 4652–4663. http://dx.doi.org/10.1091/mbc.E09-02-0137.

Thery, M. (2010). Micropatterning as a tool to decipher cell morphogenesis and functions. *Journal of Cell Science*, *123*(Pt. 24), 4201–4213. http://dx.doi.org/10.1242/jcs.075150.

Wang, H., Lacoche, S., Huang, L., Xue, B., & Muthuswamy, S. K. (2013). Rotational motion during three-dimensional morphogenesis of mammary epithelial acini relates to laminin matrix assembly. *Proceedings of the National Academy of Sciences of the United States of America*, *110*(1), 163–168. http://dx.doi.org/10.1073/pnas.1201141110.

Wang, A. Z., Ojakian, G. K., & Nelson, W. J. (1990a). Steps in the morphogenesis of a polarized epithelium. I. Uncoupling the roles of cell-cell and cell-substratum contact in establishing plasma membrane polarity in multicellular epithelial (MDCK) cysts. *Journal of Cell Science*, *95*(Pt. 1), 137–151.

Wang, A. Z., Ojakian, G. K., & Nelson, W. J. (1990b). Steps in the morphogenesis of a polarized epithelium. II. Disassembly and assembly of plasma membrane domains during reversal of epithelial cell polarity in multicellular epithelial (MDCK) cysts. *Journal of Cell Science*, *95*(Pt. 1), 153–165.

Wei, C., Bhattaram, V. K., Igwe, J. C., Fleming, E., & Tirnauer, J. S. (2012). The LKB1 tumor suppressor controls spindle orientation and localization of activated AMPK in mitotic epithelial cells. *PLoS One*, *7*(7), e41118. http://dx.doi.org/10.1371/journal.pone.0041118.

Yu, W., Datta, A., Leroy, P., O'Brien, L. E., Mak, G., Jou, T. S., et al. (2005). Beta1-integrin orients epithelial polarity via Rac1 and laminin. *Molecular Biology of the Cell*, *16*(2), 433–445. http://dx.doi.org/10.1091/mbc.E04-05-0435.

Zegers, M. M., O'Brien, L. E., Yu, W., Datta, A., & Mostov, K. E. (2003). Epithelial polarity and tubulogenesis in vitro. *Trends in Cell Biology*, *13*(4), 169–176.

Zheng, Z., Zhu, H., Wan, Q., Liu, J., Xiao, Z., Siderovski, D. P., et al. (2010). LGN regulates mitotic spindle orientation during epithelial morphogenesis. *Journal of Cell Biology*, *189*(2), 275–288. http://dx.doi.org/10.1083/jcb.200910021.

CHAPTER

Using Replication Defective Viruses to Analyze Membrane Trafficking in Polarized Epithelial Cells

8

Lucy Pigati[1], Richard S. Kang[1,2] and Heike Fölsch

Department of Cell and Molecular Biology, Northwestern University Feinberg School of Medicine, Chicago, Illinois, USA

CHAPTER OUTLINE

Introduction and Rationale **126**
8.1 Materials **129**
8.2 Methods **130**
8.2.1 Infection of Epithelial Cells with Replication Defective Lentiviruses for Gene Silencing 131
8.2.2 Infection of Epithelial Cells with Replication Defective Adenoviruses to Express Reporter Genes 131
8.2.3 Immunofluorescence Staining: Total and Surface Staining 133
8.2.4 Confocal Analysis 134
8.3 Discussion **135**
Summary **136**
Acknowledgments **136**
References **136**

Abstract

Epithelial cells in culture, especially once they are polarized, are extremely hard to manipulate by transient transfection methods. The use of replication defective adenoviruses for gene expression or replication defective retroviruses or lentiviruses to express shRNA for gene knockdown provides efficient tools to manipulate gene expression patterns even in hard-to-transfect cell lines. One of the advantages of

[1]These authors contributed equally.
[2]Current address: AuraSense, 8045 Lamon Avenue, Suite 410, Skokie, Illinois 60077, USA.

Methods in Cell Biology, Volume 118

ISSN 0091-679X
http://dx.doi.org/10.1016/B978-0-12-417164-0.00008-2

using defective adenoviruses for gene expression is that once the virus has been generated, it can easily be applied to a wide variety of cells. In addition, replication defective retro- and lentiviruses are used to stably deplete proteins from cell lines, which subsequently may be used for analyzing the polarized surface delivery of receptors that may be expressed using defective adenoviruses. The latter approach is especially useful if the expressed shRNA also encodes GFP for easy assessment of shRNA-expressing cells. Thus the use of defective viruses in epithelial cell research is convenient. This makes a detailed infection protocol a research tool that would be valuable to many laboratories. Here we describe in detail how cells are infected with defective retro- or lentiviruses and subsequently selected for stable gene knockdown. We then describe how these cells may be used for infection with defective adenoviruses and the subsequent analyses.

INTRODUCTION AND RATIONALE

Epithelial cell function is crucially dependent on the correct delivery of transmembrane receptors and ion channels to the apical or basolateral membrane domains. To understand the intracellular sorting machineries, researchers rely on epithelial cells in culture that faithfully recapitulate polarized sorting of proteins when seeded onto Transwell filter units at confluency and cultured for 3–4 days. For example, using stably transfected Madin–Darby canine kidney (MDCK) cells, the Mellman laboratory revealed the basolateral sorting signals in low-density lipoprotein receptor (LDLR) (Hunziker, Harter, Matter, & Mellman, 1991; Matter, Hunziker, & Mellman, 1992; Matter, Whitney, Yamamoto, & Mellman, 1993). This work was very labor intensive. For each LDLR mutant protein to be tested, stable MDCK clones were generated, and the generation of a stable epithelial cell line takes about 4–6 weeks. Furthermore, the LDLR mutant proteins could not easily be tested in other epithelial cell lines without generating another stable cell line. Stably expressing clones were generated because transient transfection of proteins into polarized epithelial cells is inefficient, which made biochemical analyses impossible to carry out using transient transfections.

In 1998, the Vogelstein laboratory published an advanced method to generate replication defective adenoviruses through homologous recombination in *Escherichia coli* and subsequent virus amplification in HEK293 cells (Kuehn, Herrmann, & Schekman, 1998). Adenoviruses have a broad host spectrum and can be found in at least 40 vertebrate species ranging from fish to humans (Li et al., 2010). Thus they will infect almost any cell line in the laboratory. Adenoviruses are especially useful for exogenous coexpression of multiple proteins (Keller, Toomre, Diaz, White, & Simons, 2001; Nokes, Fields, Collins, & Fölsch, 2008; Thompson et al., 2007). Figure 8.1 shows an example in which EGFP-tagged Rab13 (Rab13-EGFP) is virally coexpressed with TGN38 in MDCK cells also costained for endogenous GM130 to show TGN localization of Rab13 (see also Nokes et al., 2008). Note the relative low expression levels of Rab13 in the shown image as it manifests

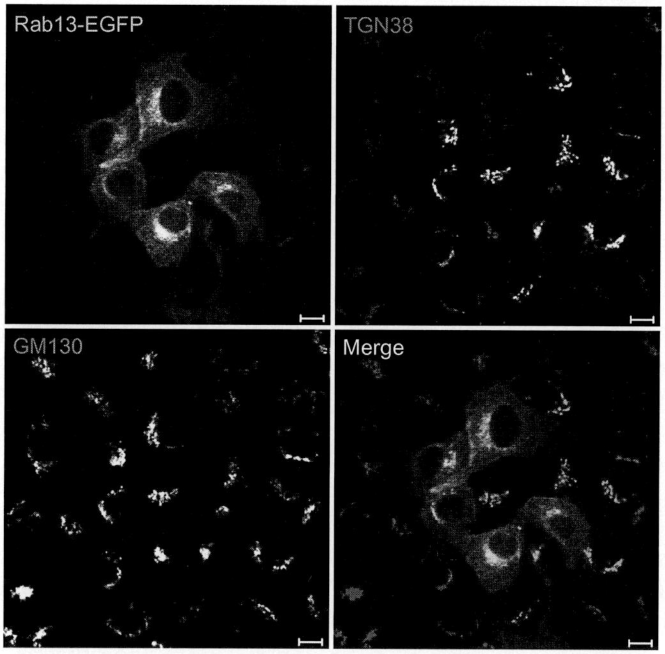

FIGURE 8.1

Rab13-EGFP localizes at the TGN. MDCK cells were grown on coverslips. Two days after seeding, cells were infected with defective adenoviruses encoding Rab13-EGFP and TGN38. One day after infection, cells were fixed and stained for TGN38 (in red) and endogenous GM130 (in blue). Specimens were analyzed by confocal microscopy. A representative image is shown. Scale bar is 10 μm. (See color plate.)

another advantage of defective adenoviruses. Through titering the amount of virus added to the cells, it is possible to gauge expression levels and infection efficiency, and to infect as little as about 10% or up to 80–100% of the cells. Thus adenoviruses may be used for the infection of a few cells for immunofluorescence analysis or most of the cells for biochemical assays.

Likewise, siRNA-mediated knockdown of genes became more efficient and cheaper with the use of replication defective retroviruses (Schuck, Manninen, Honsho, Fullekrug, & Simons, 2004). Instead of using transient transfection methods that necessitate the purchase of ever more RNA duplexes, replication defective retroviruses encoding shRNA for gene knockdown can be made rather easily at a low cost. Furthermore, retroviruses integrate their DNA into the host genome and thus it is possible to create and maintain cell lines stably depleted of the genes of interest (Anderson et al., 2005; Fields, King, Shteyn, Kang, & Fölsch, 2010). Lentiviruses belong to the family of retroviruses (Connolly, 2002). Companies like Thermo Scientific created shRNA libraries in lentiviral vectors such as pGIPZ. An added benefit of the pGIPZ system is that the green-fluorescent protein (GFP) is translated from the same mRNA that gives rise to the shRNA (Kang & Fölsch, 2011; Shteyn, Pigati, & Fölsch, 2011). Thus all cells that express the shRNA fluoresce in green. Frequently, replication defective retro- and lentiviruses are pseudotyped with the vesicular stomatitis virus glycoprotein (VSV-G) to guide cell infection. This is convenient because VSV can infect a wide variety of different animals including pigs, dogs, and humans (http://www.cfsph.iastate.edu/Factsheets/pdfs/vesicular_stomatitis.pdf). Thus VSV-G pseudotyped viruses are good for infection of MDCK cells, Lilly Laboratories cell porcine kidney (LLC-PK1) cells, and Human bronchial epithelial (HBE) cells (Anderson et al., 2005; Fields et al., 2010; Kang & Fölsch, 2011; Shteyn et al., 2011). Figure 8.2 shows an example of HBE cells stably depleted of ARH or GAPDH as control which were later infected with defective adenoviruses encoding the apically localized FcLR protein. FcLR is a fusion product of the ectodomain of Fc receptor and the first 22 amino acids of the cytosolic tail of LDLR (Matter et al., 1993; Thompson et al., 2007). Note that depletion of ARH does not interfere with the apical localization of the protein, which constitutes an important control for the specificity of the system (see also Kang & Fölsch, 2011).

Here we describe how to generate cell clones that stably deplete the gene of interest. Subsequently, we guide the readers through detailed protocols describing the infection of epithelial cells with defective adenoviruses followed by immunofluorescent labeling and confocal microscopy analysis.

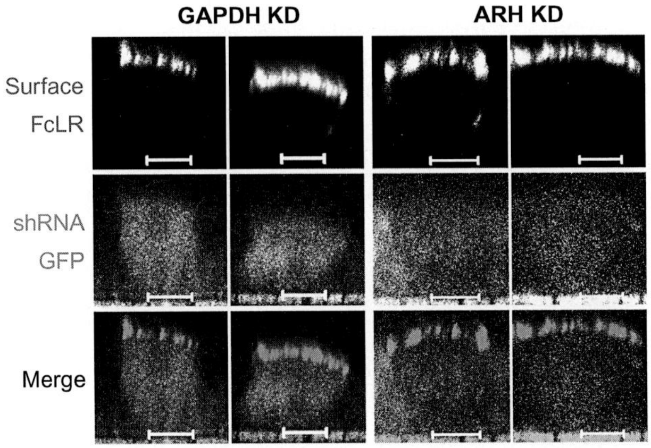

FIGURE 8.2

Depletion of ARH has no effect on apical sorting of FcLR. HBE cells stably expressing shRNA (labeled in green) to deplete ARH (right hand panels) or GAPDH as control (left hand panels) were grown on filter supports. Three days after seeding, cells were infected with defective adenovirus encoding FcLR. Twenty-four hour after infection, FcLR at the cell surface was stained with anti-ectodomain antibodies followed by Cy5-labeled secondary antibodies (shown in red). Specimens were analyzed by confocal microscopy and representative *XZ*-sections are shown. Scale bars are 5 μm. (See color plate.)

8.1 MATERIALS

1. Tissue culture plates.
2. 12-mm polycarbonate Transwell filter units (0.4 μm pore size) (Corning).
3. Alcian blue-treated round coverslips (12 mm diameter No. 1): Alcian blue is a polyvalent basic dye whose multiple, localized positive charges provide a great surface for cells to adhere to via their negatively charged glycocalyx. Add coverslips to a glass beaker containing ~200 ml 70% ethanol. Swirl a couple of times and discard ethanol. Aspirate remaining ethanol. Add a 0.2% Alcian blue solution (made in water and filtered to remove particles) to coverslips and microwave for 30 s. Do not boil. Swirl and microwave 2 × for 10 s each. Make sure solution does not boil. Let it stand for 10 min at the bench. Pour Alcian blue solution back into its container for reuse. In a tissue culture hood, rinse coverslips several times with sterile distilled water until water runs clear. Store coverslips in 50-ml Falcon tubes covered with sterile water. Coverslips may be used immediately after coating.
4. Fibronectin/collagen-coating solution: For 100 ml coating solution add 88 ml LHC basal medium, 10 ml bovine serum albumin (BSA) (1 mg/ml stock solution), 1 ml bovine collagen I (3 mg/ml stock solution), and 1 ml fibronectin (1 mg/ml stock solution). Do not vortex or swirl vigorously. Filter-sterilize the coating solution and store at 4 °C for up to 3 months.
5. Appropriate cell Media (Corning Cellgro or Invitrogen).
6. MEM without Ca^{2+} and Mg^{2+} (SMEM) (Corning Cellgro).
7. Puromycin solution (4 mg/ml): Solubilize 100 mg of puromycin powder (Calbiochem) in 25 ml PBS^{++}, filter-sterilize the stock solution, aliquot, and store at −20 °C.
8. Defective adenoviruses encoding FcLR, TGN38, and Rab13-EGFP (Nokes et al., 2008; Thompson et al., 2007).
9. Defective lentiviruses encoding shRNA directed against human ARH (Kang & Fölsch, 2011).
10. Phosphate-buffered saline (PBS): 2.67 mM KCl, 1.47 mM KH_2PO_4, 137.93 mM NaCl, 8.06 mM Na_2HPO_4 (=DPBS from Corning Cellgro).
11. PBS^{++}: PBS plus 0.901 mM Ca^{2+} and 0.493 mM Mg^{2+} (Corning Cellgro).
12. 3% Paraformaldehyde (PFA) solution in PBS: Measure 3.5 g PFA and place it into a 125-ml glass bottle. Measure 40 ml PBS in a graduated cylinder and add to PFA. Add 10 μl of 10 M NaOH to the solution and shake the bottle. Place the bottle into a 60 °C water bath until PFA dissolves. Avoid leaving the PFA solution in the water bath for too long. After complete solubilization check the pH with pH paper. The pH should be around 7. Pour the solution into a graduated cylinder and fill up to 50 ml with PBS. In the fume hood, aliquot PFA solution into 15-ml Falcon tubes, 5 ml each. Store at −20 °C for up to 3 months. Note this solution contains 7% PFA. On the day of your experiment thaw one 7% PFA aliquot and adjust to 3% with PBS.

13. Blocking/permeabilization buffer (BPB): 2% (wt/vol) BSA, 0.2–0.4% (wt/vol) saponin with or without 10% goat serum in PBS^{++}. Weigh 800 mg of BSA and 80–160 mg of saponin, place into 50-ml Falcon tube and add PBS^{++} to the 40-ml mark of the tube. Completely solubilize BSA and saponin by rotating end-over-end at room temperature. Make fresh on the day of your immunofluorescence experiment. Add 10% goat serum as necessary. Note, if your primary or secondary antibodies were raised in goats, you cannot use goat serum for blocking; use fetal bovine serum instead.
14. Primary antibodies: rat anti-FcLR antibodies were produced by the hybridoma cell line 2.4G2 (from ATCC); mouse anti-GM130 (IgG1 isotype) were purchased from BD Transduction Laboratories™ (clone 35/GM130); rabbit anti-TGN38 antibodies were a gift from Dr. Graham Warren (Max F. Perutz laboratories, Vienna, Austria).
15. Secondary antibodies: goat anti-rat-Cy5 (Jackson ImmunoResearch), goat anti-rabbit-Alexa594 (Molecular Probes), and goat anti-mouse IgG1-Cy5 (Southern Biotech).
16. Mount: 10% (wt/vol) 1,4-diazabicyclo[2.2.2]octane (Dabco), 50% (wt/vol) glycerol. Weigh 5 g glycerol into a 15-ml Falcon tube. Next weigh 1 g DABCO and add to the glycerol. Fill the Falcon tube to the 10-ml mark with distilled water using a 5 ml pipet. Cap the tube, seal with parafilm, and wrap entire tube with aluminum foil to protect the DABCO from light. Rotate solution end-over-end at room temperature for 1 h or until DABCO is completely dissolved. Make 1 ml aliquots in 1.5-ml centrifuge tubes. Store aliquots at −20 °C for future use. Store one aliquot at 4 °C for immediate use.
17. Tweezers, facial tissues, slides, clear nail polish, 18 × 18-mm square micro cover glass No. 1, and surgical blade (Feather Surgical Blade, stainless steel, No. 11).

8.2 METHODS

Handling of cell cultures that were infected with defective retro/lentiviruses, defective adenoviruses, or both requires strict adherence to safety guidelines. For example, always perform experiments with live viruses in tissue culture hoods and collect waste materials in biohazard waste containers for immediate inactivation after usage according to the safety guidelines of your institution. Once your samples are fixed and transferred into clean plates, you may proceed with working outside of the tissue culture hood. For the generation of defective lentiviruses, defective retroviruses or defective adenoviruses, we follow well-established methods as published in the literature (Barton & Medzhitov, 2002; Dull et al., 1998; Kuehn et al., 1998; Luo et al., 2007; Salmon & Trono, 2006).

All tissue culture plates, Transwell filters, or coverslips that are used in the following protocols for the growth of HBE cells are coated with a fibronectin and collagen matrix. To this end, add enough fibronection/collagen-coating solution to the tissue culture device to cover its surface and incubate in a tissue culture incubator at 37 °C for 2–24 h. Remove the coating solution before seeding the cells.

8.2.1 Infection of epithelial cells with replication defective lentiviruses for gene silencing

To infect epithelial cells with defective lentiviruses or retroviruses that encode shRNAs to deplete the genes of interest seed cells into six-well plates 1 day before infection. Make sure the cells are only ~40–50% confluent on the day of infection, because incorporation of the retroviral DNA into the cell genome occurs during cell division (Roe, Reynolds, Yu, & Brown, 1993). Note, however, that lentiviruses will also insert their genome into nonreplicating cells (Connolly, 2002). For infection, quickly thaw the tubes containing viruses by rolling the tubes in your (gloved) hands. We usually have about 1 ml of virus in the tubes at a concentration of about 10^7–10^8 transduction units (TU)/ml. Mix 500 μl viral stock with 500 μl of growth medium and add polybrene to a final concentration of 6 μg/ml. We add 1 μl of a 6 mg/ml polybrene stock solution to 1 ml of virus. Mix virus stock gently without vortexing to avoid bubble formation. Allow for the formation of polybrene–virus interaction by incubating 5 min at room temperature. Next, remove growth medium from your cells and add 1 ml of virus/polybrene solution per well of the six-well plate. Spread the virus solution to cover the entire plate by gentle rocking. Seal the plates with parafilm and spin at $580 \times g$ for 1 h at 4 °C. After the spin, slowly add 2 ml fresh growth medium (do not remove the virus-containing medium), be careful not to disturb the viruses attached onto the cells after the spin. One day after infection, start selection for cells that incorporated the viral genome. Our defective lenti/retroviruses typically carry a puromycin-resistance gene. We add puromycin to a final concentration of 16 μg/ml. Change the medium daily including freshly added puromycin until the cells are confluent, which is typically about 2–3 days after infection. Once confluent, split cells into 10-cm plates and continue with puromycin selection (12 μg/ml). Your cells are now ready to be seeded for your choice of experiments. We successfully used this protocol to stably deplete gene expression in MDCK, LLC-PK1, and HBE cells (Anderson et al., 2005; Fields et al., 2010; Kang & Fölsch, 2011; Shteyn et al., 2011). Note that your cell line of interest may have different sensitivities to the defective lenti/retroviruses. Thus dependent on your infection and knockdown efficiency, you may have to use more or less virus. Knockdown efficiency should be determined by quantitative Western blotting, quantitative RT-PCR, or quantitative immunofluorescence (Anderson et al., 2005; Gravotta et al., 2007; Kang & Fölsch, 2011; Nokes et al., 2008; Shteyn et al., 2011).

8.2.2 Infection of epithelial cells with replication defective adenoviruses to express reporter genes

Defective adenoviruses can be used to express low to high amounts of your gene of interest for immunofluorescence or biochemical experiments in, for example, coimmunoprecipitations. We use defective adenoviruses for gene expression in MDCK cells, LLC-PK1 cells stably expressing μ1A or μ1B, and epithelial cell lines that stably deplete genes of interest after lenti/retrovirus infection. For immunofluorescence experiments with coverslip-grown cells, we seed about 2 to 5×10^4 cells per coverslip, and infect with defective adenoviruses at 0.1–1 plaque forming units (pfu)/cell typically 2 days after seeding and 1 day before immunofluorescence analysis. To infect cells on filters, we seed 4×10^5 cells per 12-mm Transwell filter, infect the cells 3 days after seeding with about 1 pfu/cell, and perform the immunofluorescence analysis 1 day later. In order to have enough gene expression for biochemical analysis, we infect cells with about 10–15 pfu/cell 1 day before preparing the cells for immunoprecipitation, Western blot analysis, or both. We typically use epithelial cells grown in six-well plates for biochemistry. We seed about 2.5×10^6 cells per well of a six-well plate 1 day before virus infection and 2 days before the biochemical analysis. Alternatively, we seed 8×10^5 cells per 24-mm Transwell filter. Three days after seeding, cells are infected with 10–15 pfu/cell. Cells are analyzed 24 h after infection.

To infect the cells with defective adenoviruses, quickly thaw out the tubes containing viruses by rolling the tubes in your (gloved) hands and vortex the thawed virus stocks to separate potential viral aggregates. We typically have 1 ml aliquots of high titer virus stocks in the range of 10^7–10^9 pfu/ml. Calculate how much virus you need for infection. For example, to infect 4×10^5 cells with 15 pfu/cell using a virus with a titer of 2×10^8 pfu/ml, you will need 30 μl of the virus stock (4×10^5 cells times 15 pfu/cell $= 6 \times 10^6$ pfu total). Dilute the viruses in infection media and vortex. For MDCK cells we use Ca^{2+}- and Mg^{2+}-free MEM (SMEM) without serum. We use serum-free media because adenovirus infection is inhibited by serum proteins (Shortridge, 1970). The absence of Ca^{2+} and Mg^{2+} will result in the opening of tight junctions so that the virus can access the basolaterally localized coxsackie and adenovirus receptor. Note that in cells that do not express the epithelial cell-specific clathrin adaptor AP-1B, this receptor is localized to the apical membrane domain (Carvajal-Gonzalez et al., 2012). LLC-PK1 and HBE cells form less robust monolayers and typically do not survive treatment in SMEM. Thus we perform the infection in respective growth medium without serum. The total amounts of infection media and virus are as follows: 100 μl for cells grown on 12-mm Transwell filters, 150 μl for cells grown on coverslips, 250 μl for cells grown on 24-mm Transwell filters, and 1 ml for cells grown in six-well plates.

Now you are ready to infect your cells. Wash cells 1 × with infection media, and in the case of cells grown on 12- or 24-mm Transwell filters add 1 ml or 2.5 ml of infection media to the basolateral chamber, respectively. Now add the virus-containing media directly onto cells (apical chamber of filter-grown cells). Gently rock the cells in a tissue culture incubator at 37 °C, 5% CO_2 for 1 h in case of cells grown on coverslips or in six-well plates and 2 h for cells grown on Transwell filters. Because filter-grown cells are typically polarized they are harder to infect and need more incubation time. Subsequently, remove the virus-containing solutions (and infection media from basolateral chambers) and add normal growth medium to the cells. Further incubate for 24 h in a tissue culture incubator at 37 °C, 5% CO_2 to allow for gene expression. Proceed with your choice of biochemical experiment or immunofluorescence analysis.

8.2.3 Immunofluorescence staining: Total and surface staining

Remember to perform all procedures in the tissue culture hood until the specimens are fixed and placed into uncontaminated plates. It is important to protect the specimens from light with aluminum foil (or by placing them into a dark environment) if any of your expressed proteins carries a fluorescent tag to prevent bleaching of the signals. This is especially important after fixation.

For surface staining, start with placing the specimens, which are typically cells grown on 12-mm Transwell filters, onto a metal plate on ice to chill the cells, and wash the cells 3 × with ice-cold PBS^{++}. Subsequently, place a piece of parafilm onto the metal plate on ice and pipette 30 μl antibody recognizing the ectodomain of your protein of interest onto this surface. Next add 50 μl of the same antibody to the apical chamber of the Transwell filter for staining of proteins at the apical surface and place the whole filter unit onto the 30 μl drop of antibody to stain the basolateral surface. In our example shown in Fig. 8.2 we used the rat monoclonal antibody 2.4G2 that recognizes the ectodomain of the FcLR fusion protein. After 1 h incubation on ice, specimens are washed 3 × with ice-cold PBS^{++} and fixed with 3% PFA on ice for 20 min. This is followed by 5 min incubation in 1 ml PBS^{++} at room temperature. During this time cut out the filters from the filter holders using a surgical blade. Place the filters with the cells facing up into clean wells of a 12-well plate containing 1 ml PBS^{++}. At your bench, incubate specimens in 1 ml of BPB containing 10% goat serum for 1 h at room temperature. If you need to stain for intracellular epitopes, prepare additional primary antibodies in BPB. Spin the antibody dilutions at 15,700 × g for 10 min at room temperature in a microcentrifuge and place 30 μl antibody drops onto clean parafilm placed in a wet chamber. Place your specimens with the cells facing down onto the antibody drops and incubate for 1 h at room temperature. Subsequently, place the specimens, cells facing up, back into the BPB solution for 6 min, and continue to wash 4 × with BPB for 6 min each. If you do not need additional stainings, directly proceed to the next step and prepare your fluorescently labeled secondary antibodies. In our example as shown in Fig. 8.2, we diluted 1 μl of Cy5-labeled goat anti-rat antibodies in 100 μl BPB, followed by a 10 min spin at 15,700 × g at room temperature in a microcentrifuge. Pipette 30 μl antibody drops onto clean parafilm in a wet chamber and place specimens cells facing down onto the antibody drops. Incubate for 1 h at room temperature. Place specimens back into BPB in a 12-well plate for 6 min and wash four more times in BPB for 6 min each. Dip specimens 3 × into clean water, collect excessive water with facial tissue, and place onto slides with the cells facing up. Add ~15 μl mounting solution on top of each specimen and cover with 18 × 18-mm micro glass. Gently pat dry using facial tissues and seal with nail polish. Analyze specimens with an inverted microscope.

If we want to localize proteins intracellularly, we often express our proteins of interest in coverslip-grown cells. In our example in Fig. 8.1, we virally expressed Rab13-EGFP together with TGN38. To begin the staining, wash the specimens 3 × with PBS^{++} at room temperature in a tissue culture hood. Next the cells are fixed with 3% PFA for 15 min at room temperature followed by a 5 min incubation in 1 ml PBS^{++}

at room temperature. During the last 5 min incubation time, transfer the coverslips into clean wells of a 12-well plate containing 1 ml PBS^{++} and continue working at your bench. Block and permeabilize the specimens in 1 ml BPB containing 10% goat serum for 1 h at room temperature. Prepare your primary antibody solution. For our example we diluted 1 μl rabbit anti-TGN38 antibodies and 2 μl mouse IgG1 anti-GM130 antibodies in 200 μl BPB, followed by a 10 min spin at 15,700 × *g* at room temperature in a microcentrifuge. Pipette 30 μl drops of antibody solution onto clean parafilm in a wet chamber and place your specimens cells facing down onto these drops. Incubate for 1 h at room temperature. To wash the specimens, place them back into the BPB solution in your 12-well plate, cells facing up, and incubate for 6 min. Wash four more times in BPB for 6 min each. Prepare your secondary antibody solution. For our example we diluted 1 μl of Alexa594-labeled goat anti-rabbit and 2 μl of Cy5-labeled goat anti-mouse IgG1 antibodies in 200 μl BPB and spun at 15,700 × *g* for 10 min at room temperature in a microcentrifuge. Pipette 30 μl antibody drops onto clean parafilm in a wet chamber and place specimens onto these drops cells facing down. Incubate for 1 h at room temperature. Place the specimens back into the 12-well plate containing BPB, cells facing up. Incubate for 6 min and wash for four more times for 6 min each in BPB. Prepare your slides by placing ~10 μl drops of mount onto their clean surfaces. Dip coverslips 3 × into clean water and collect excessive water with facial tissues. Place coverslips with cells facing down onto the drops of mount on the slide; gently press down using facial tissues to remove excessive mount, and seal with nail polish.

8.2.4 Confocal analysis

We analyze all our specimens with confocal microscopy, because even when grown on coverslips, epithelial cells are never flat; and confocal microscopy is the easiest way to ensure imaging of single focal planes without disturbance from out of focus fluorescence. When we image coverslip-grown cells, we typically image only one focal plane, which ideally is the plane in which all markers show maximum staining. This may be followed by an analysis of pixel overlap using the confocal software or the determination of colocalization coefficients using Volocity software (Kang & Fölsch, 2011; Shteyn et al., 2011).

There are basically two ways to image filter-grown cells. (1) You may collect galleries and image multiple individual focal planes from the apical to the basal surface of your cell in 0.3 μm steps. You may then pick individual focal planes to show in a publication or you may assemble galleries into 3D images using programs such as the Volocity software (Kang & Fölsch, 2011; Shteyn et al., 2011). Note that collecting a single gallery may take up to 30 min depending on how many fluorophores you are imaging. (2) If you are only interested in the localization of surface receptors to the apical or basolateral surfaces, using the fast *z*-line function of your microscope software is a faster option. Here, the microscope images pixels along a chosen line and assembles them directly into *XZ* images such as shown in Fig. 8.2. This scan typically takes less than 1 min and allows the user to obtain data quickly. If desired, it is possible to measure pixel counts at the apical membrane, basolateral membrane, or both using, for example, Volocity software. This allows you to determine the percentage of apical versus basolateral localization of your surface protein of interest under the chosen experimental conditions (Shteyn et al., 2011).

To generate the images shown in Figs. 8.1 and 8.2, we used a Microsytem LSM510, Carl Zeiss Microimaging, Inc. inverted microscope, and a 63 × objective.

8.3 DISCUSSION

The described methods for manipulating gene expression are a great way to minimize costs and to maximize flexibility when analyzing protein sorting from the TGN or recycling endosomes in epithelial cells that are hard to transfect using transient transfection methods. Once the viruses are generated, there is no further need for transfection reagents or RNA duplexes for gene depletion. Moreover, the replication defective retroviruses, lentiviruses, and adenoviruses show a broad host range meaning they can be used to infect virtually any cell line in the laboratory (Li et al., 2010) (http://www.cfsph.iastate.edu/Factsheets/pdfs/vesicular_stomatitis.pdf). In addition, replication defective lentiviruses or retroviruses for expression of shRNAs are suitable for the generation of cell lines stably depleted of the target gene. It should be noted, however, that it is possible that cells stably depleted of a certain gene will activate suppressor pathways over time. If it is noted that only newly generated cell lines show the desired knockdown phenotype, but that this phenotype is lost after several generations in culture, it is better to generate the knockdown cells freshly for each round of experiments (Anderson et al., 2005; Fields et al., 2010; Shteyn et al., 2011).

Adenoviruses enter the cells through clathrin-mediated endocytosis. Adenovirus infection leads to the rupture of endosomes within 30 min after virus addition to the cells (Maier, Marvin, Wodrich, Campbell, & Wiethoff, 2012). These endosomes typically underlie the plasma membrane and are mostly EEA1-negative, constituting perhaps pre-EEA1 endosomes (Maier et al., 2012). Thus by removing the viruses 1–2 h after infection, we ensure intact endosomes by the time we analyze the cells (Anderson et al., 2005). Defective adenoviruses may be used for biochemical analysis such as immunoprecipitations or radioactive pulse-chase experiments to analyze sorting during biosynthetic delivery (Fields et al., 2007). However, if analyzed by immunofluorescence, it is only possible to ascertain the steady-state sorting of proteins. Furthermore, in assays that rely on overexpression of dominant negative proteins such as the activated Rab13 mutant protein Rab13Q67L, prolonged overexpression results in the disintegration of the Golgi apparatus and opening of the tight junctions (Marzesco et al., 2002; Nokes et al., 2008). Thus, to analyze biosynthetic delivery of surface proteins in the presence of dominant negative protein mutants, we revert to a microinjection protocol (Cook, Ang, Kang, & Fölsch, 2011; Kang & Fölsch, 2011; Nokes et al., 2008; Shteyn et al., 2011). Microinjection is not covered here, but was the topic of a recent publication in the *Journal of Visualized Experiments* (Cook et al., 2011).

Although we used the described methods exclusively on epithelial cells, they can be easily adapted to nonepithelial cells. Thus we hope that the described methods will be useful for many researchers who wish to add more flexibility at a low cost to their research methods.

SUMMARY

The use of replication defective viruses is a viable alternative to transient transfection methods. Among its many advantages is the flexibility of the system, its high infection rates, and low cost. Especially when working with hard-to-transfect cell lines, defective viruses allow for nearly 100% infectivity, which is crucial for biochemical experiments. Viruses are easily titrated giving the researcher great control over infection efficiencies and expression levels. The protocol outlined here provides a detailed guide through a variety of different infection conditions, which should allow the interested reader to readily adopt these methods to her or his laboratory.

Acknowledgments

We would like to thank Dr. Graham Warren for the generous gift of rabbit antibodies against TGN38. This work was funded by a grant from the National Institutes of Health (GM070736) to H. Fölsch. R.S. Kang was supported by the Molecular and Cellular Basis of Disease Training Program (GM8061).

References

Anderson, E., Maday, S., Sfakianos, J., Hull, M., Winckler, B., Sheff, D., et al. (2005). Transcytosis of NgCAM in epithelial cells reflects differential signal recognition on the endocytic and secretory pathways. *The Journal of Cell Biology*, *170*, 595–605.

Barton, G. M., & Medzhitov, R. (2002). Retroviral delivery of small interfering RNA into primary cells. *Proceedings of the National Academy of Sciences of the United States of America*, *99*, 14943–14945.

Carvajal-Gonzalez, J. M., Gravotta, D., Mattera, R., Diaz, F., Perez Bay, A., Roman, A. C., et al. (2012). Basolateral sorting of the coxsackie and adenovirus receptor through interaction of a canonical YXXPhi motif with the clathrin adaptors AP-1A and AP-1B. *Proceedings of the National Academy of Sciences of the United States of America*, *109*, 3820–3825.

Connolly, J. B. (2002). Lentiviruses in gene therapy clinical research. *Gene Therapy*, *9*, 1730–1734.

Cook, R. N., Ang, S. F., Kang, R. S., & Fölsch, H. (2011). Analyzing the function of small GTPases by microinjection of plasmids into polarized epithelial cells. *Journal of Visualized Experiments*, *51*, 2645.

Dull, T., Zufferey, R., Kelly, M., Mandel, R. J., Nguyen, M., Trono, D., et al. (1998). A third-generation lentivirus vector with a conditional packaging system. *Journal of Virology*, *72*, 8463–8471.

Fields, I. C., King, S. M., Shteyn, E., Kang, R. S., & Fölsch, H. (2010). Phosphatidylinositol 3,4,5-trisphosphate localization in recycling endosomes is necessary for AP-1B-dependent sorting in polarized epithelial cells. *Molecular Biology of the Cell*, *21*, 95–105.

Fields, I. C., Shteyn, E., Pypaert, M., Proux-Gillardeaux, V., Kang, R. S., Galli, T., et al. (2007). v-SNARE cellubrevin is required for basolateral sorting of AP-1B-dependent cargo in polarized epithelial cells. *The Journal of Cell Biology*, *177*, 477–488.

Gravotta, D., Deora, A., Perret, E., Oyanadel, C., Soza, A., Schreiner, R., et al. (2007). AP1B sorts basolateral proteins in recycling and biosynthetic routes of MDCK cells. *Proceedings of the National Academy of Sciences of the United States of America, 104*, 1564–1569.

Hunziker, W., Harter, C., Matter, K., & Mellman, I. (1991). Basolateral sorting in MDCK cells requires a distinct cytoplasmic domain determinant. *Cell, 66*, 907–920.

Kang, R. S., & Fölsch, H. (2011). ARH cooperates with AP-1B in the exocytosis of LDLR in polarized epithelial cells. *The Journal of Cell Biology, 193*, 51–60.

Keller, P., Toomre, D., Diaz, E., White, J., & Simons, K. (2001). Multicolour imaging of post-Golgi sorting and trafficking in live cells. *Nature Cell Biology, 3*, 140–149.

Kuehn, M. J., Herrmann, J. M., & Schekman, R. (1998). COPII-cargo interactions direct protein sorting into ER-derived transport vesicles. *Nature, 391*, 187–190.

Li, Y., Ge, X., Zhang, H., Zhou, P., Zhu, Y., Zhang, Y., et al. (2010). Host range, prevalence, and genetic diversity of adenoviruses in bats. *Journal of Virology, 84*, 3889–3897.

Luo, J., Deng, Z. L., Luo, X., Tang, N., Song, W. X., Chen, J., et al. (2007). A protocol for rapid generation of recombinant adenoviruses using the AdEasy system. *Nature Protocols, 2*, 1236–1247.

Maier, O., Marvin, S. A., Wodrich, H., Campbell, E. M., & Wiethoff, C. M. (2012). Spatiotemporal dynamics of adenovirus membrane rupture and endosomal escape. *Journal of Virology, 86*, 10821–10828.

Marzesco, A. M., Dunia, I., Pandjaitan, R., Recouvreur, M., Dauzonne, D., Benedetti, E. L., et al. (2002). The small GTPase Rab13 regulates assembly of functional tight junctions in epithelial cells. *Molecular Biology of the Cell, 13*, 1819–1831.

Matter, K., Hunziker, W., & Mellman, I. (1992). Basolateral sorting of LDL receptor in MDCK cells: The cytoplasmic domain contains two tyrosine-dependent targeting determinants. *Cell, 71*, 741–753.

Matter, K., Whitney, J. A., Yamamoto, E. M., & Mellman, I. (1993). Common signals control low density lipoprotein receptor sorting in endosomes and the Golgi complex of MDCK cells. *Cell, 74*, 1053–1064.

Nokes, R. L., Fields, I. C., Collins, R. N., & Fölsch, H. (2008). Rab13 regulates membrane trafficking between TGN and recycling endosomes in polarized epithelial cells. *The Journal of Cell Biology, 182*, 845–853.

Roe, T., Reynolds, T. C., Yu, G., & Brown, P. O. (1993). Integration of murine leukemia virus DNA depends on mitosis. *The EMBO Journal, 12*, 2099–2108.

Salmon, P., & Trono, D. (2006). Production and titration of lentiviral vectors. *Current Protocols in Neuroscience, Chapter 4*, Unit 421.

Schuck, S., Manninen, A., Honsho, M., Fullekrug, J., & Simons, K. (2004). Generation of single and double knockdowns in polarized epithelial cells by retrovirus-mediated RNA interference. *Proceedings of the National Academy of Sciences of the United States of America, 101*, 4912–4917.

Shortridge, K. F. (1970). The specificity of adenovirus-human serum inhibitor interaction. *Archiv für die Gesamte Virusforschung, 30*, 238–244.

Shteyn, E., Pigati, L., & Fölsch, H. (2011). Arf6 regulates AP-1B-dependent sorting in polarized epithelial cells. *The Journal of Cell Biology, 194*, 873–887.

Thompson, A., Nessler, R., Wisco, D., Anderson, E., Winckler, B., & Sheff, D. (2007). Recycling endosomes of polarized epithelial cells actively sort apical and basolateral cargos into separate subdomains. *Molecular Biology of the Cell, 18*, 2687–2697.

CHAPTER

SNAP-Tagging the Retrograde Route

9

Ludger Johannes* and Massiullah Shafaq-Zadah†

*Institut Curie—Centre de Recherche, Trafic, Signaling and Delivery group, Paris Cedex 05, France

†UMR144 CNRS, Paris, France

CHAPTER OUTLINE

Introduction 140
9.1 Materials 141
9.2 Methods 142
9.2.1 Cell Surface Modification Reagent 142
9.2.2 Capture Reagent 144
9.2.3 Validation of the BG/SNAP-tag Strategy 144
9.2.3.1 SNAP-tag Localization to Golgi Membranes by Immunofluorescence 144
9.2.3.2 SNAP-tag Validation Using STxB 145
9.2.4 BG/SNAP-tag Strategy in Proteomics Format 147
9.2.4.1 BG-PEG$_9$-NHS is a Cell Impermeable Amine Reactive Linker ... 147
9.2.4.2 Proteomics Format 148
9.2.5 Validation of Retrograde Cargo Candidates—Antibody Uptake 150
9.2.5.1 BG Tagging of Anti-GFP Antibody 150
9.2.5.2 BG-tagged Antibody Uptake Assay 150
9.3 Conclusion/Discussion 152
Summary 153
Acknowledgments 153
References 153

Abstract

We have developed a chemical biology strategy to identify proteins that follow the retrograde transport route from the plasma membrane to the Golgi apparatus, via endosomes. The general principle is the following: plasma membrane proteins are covalently tagged with a first probe. Only the ones that are then transported to

Methods in Cell Biology, Volume 118

ISSN 0091-679X
http://dx.doi.org/10.1016/B978-0-12-417164-0.00009-4

trans-Golgi/TGN membranes are covalently bound to a capture reagent that has been engineered into this compartment. Specifically, the first probe is benzylguanine (BG) that is conjugated onto primary amino groups of plasma-membrane proteins. The capture reagent includes an O^6-alkylguanine-DNA alkyltransferase-derived fragment, the SNAP-tag, which forms a covalent linkage with BG. The SNAP-tag is fused to the GFP-tagged Golgi membrane anchor from galactosyl transferase for proper targeting to *trans*-Golgi/TGN membranes. Cell-surface BG-tagged proteins that are transported to *trans*-Golgi/TGN membranes (i.e., that are retrograde cargoes) are thereby covalently captured by the SNAP-tag fusion protein. For identification, the latter is immunopurified using GFP-Trap, and associated retrograde cargo proteins are identified by mass spectrometry. We here provide a step-by-step protocol of this method.

INTRODUCTION

The retrograde transport route ensures the trafficking of proteins and lipids from the plasma membrane to the Golgi apparatus, via endosomes, and in some cases on to the endoplasmic reticulum (Bonifacino & Rojas, 2006; Johannes & Popoff, 2008). Retrograde transport was described as an entry gate for protein toxins such as the bacterial Shiga (Johannes & Romer, 2010) and cholera toxins (Lencer & Tsai, 2003), and the plant toxin ricin (Sandvig et al., 2004). Some viral proteins including the HIV-1 envelop protein and the herpes virus glycoprotein M also traffic via this pathway, which thereby contributes to viral infection (Blot, Janvier, Le Panse, Benarous, & Berlioz-Torrent, 2003; Crump et al., 2004; Duncan & Kornfeld, 1988; Snider & Rogers, 1985).

Several endogenous proteins use the retrograde transport route. Historically, the first to be identified were the transferrin and mannose 6-phosphate receptors (Duncan & Kornfeld, 1988; Snider & Rogers, 1985). More recently, the list of retrograde cargoes has greatly increased (for a review, see Bonifacino & Rojas, 2006; Johannes & Popoff, 2008). Striking examples are glucose and copper transporters (Shewan et al., 2003; Voskoboinik & Camakaris, 2002), Wnt morphogen and glutamate receptors (Belenkaya et al., 2008; Coudreuse, Roel, Betist, Destree, & Korswagen, 2006; Yang et al., 2008; Zhang et al., 2012). The discovery of these proteins as retrograde cargoes highlights the importance of retrograde transport for essential cellular functions such as morphogen gradient formation, compartmentalization of ion transporter activity, and postsynaptic activity in neurons. Yet, no systematic study has ever been performed to identify retrograde cargoes.

We have taken advantage of the SNAP-tag system to develop a vectorial proteomics approach that should enable such systematic analysis with the ultimate goal of determining the retrograde proteome in different cell types and physiological or pathological situations (Shi et al., 2012) (Fig. 9.1). The SNAP-tag is derived from the human DNA repair protein O^6-alkylguanine-DNA alkyltransferase. It mediates the covalent linkage of O^6-benzylguanine (BG) to a specific residue of the protein, even when O^6-BG is itself conjugated to other chemical entities (Keppler et al., 2003).

Our vectorial proteomics approach involves the following steps (Fig. 9.1): (i) Cell surface proteins are modified with BG. (ii) BG surface-modified cells are incubated at

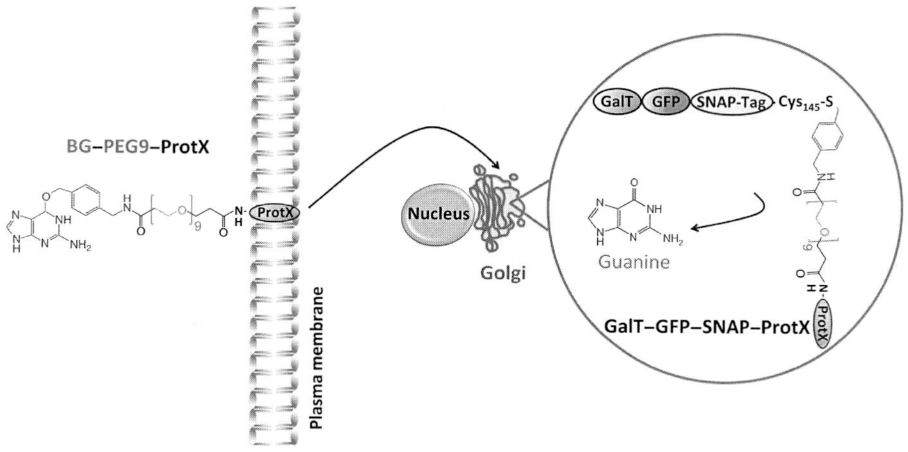

FIGURE 9.1

Schematic representation of the vectorial proteomics approach. Plasma membrane proteins are BG-tagged (BG-PEG_9-ProtX) on ice. Upon subsequent incubation at 37 °C, cargo proteins of the retrograde route are transferred to *trans*-Golgi/TGN membranes where they react with the GalT-GFP-SNAP capture reagent. After release of guanine, a GalT-GFP-SNAP-ProtX conjugate is obtained. (See color plate.)

37 °C to allow for retrograde transport to occur. (iii) At the TGN, BG-labeled plasma membrane proteins that have undergone retrograde transport react with a SNAP-GFP construct that has been positioned in this compartment by fusion with a cytosolic and transmembrane domain fragment of galactosyl transferase (GalT). (iv) The conjugates between BG-tagged retrograde cargoes and GalT-GFP-SNAP fusion protein are collected by immunopurification and identified by mass spectrometry.

Here, we expose the experimental details of this approach. Furthermore, we discuss its limitations, and describe how hits (i.e., retrograde cargo candidates) can be validated.

9.1 MATERIALS

PLASMIDS

- pGalT-EGFP vector (Frank Perez, UMR144 CNRS, Curie Institute, Paris, France)
- pSNAPm (New England Biolabs)

ANTIBODIES

- Rabbit antigiantin from Curie Institute Antibody Platform, reference A-R-R#05
- Secondary Cy3-labeled goat antirabbit (Reference 111-166-045, Jackson ImmunoResearch)
- Mouse anti-GFP antibody (Reference 11814460001, Roche) supplied in 50 mM sodium phosphate buffer, pH 7.5, 1 mM ethylenediaminetetraacetic acid (EDTA)

- Rabbit anti-SNAP-tag antibody (Reference P9310S, NEB), supplied in 10 mM HEPES, pH 7.5, 150 mM NaCl, 100 μg/ml bovine serum albumine (BSA), and 50% glycerol
- Mouse anti-STxB antibody (clone 13C4, purified from hybridoma, ATCC cat. no. CRL 1794)
- Rabbit anti-STxB antibody (homemade polyclonal)

BUFFERS

- PBS (phosphate-buffered saline): in 1 l distilled water dissolve 8 g NaCl, 0.2 g KCl, 1.44 g Na_2HPO_4, 0.24 g KH_2PO_4, adjust pH to 7.4 with HCl
- PBS^{++}: PBS containing 130 mM Ca^{2+} and 200 mM Mg^{2+}
- TNE buffer: 5 mM Tris, 150 mM NaCl, 5 mM EDTA, pH 8.0
- TNE lysis buffer: TNE buffer containing 1% NP-40 (Sigma-Aldrich) and protease inhibitor cocktail (1 mg/ml of each aprotinin, leupeptin, antipain, pepstatin, 1 M benzamidine, and 40 mg/ml phenylmethanesulfonyl fluoride in DMSO)

IMMUNOPRECIPITATION

- Protein G-Sepharose® Fast Flow beads supplied in 20% ethanol (Reference P3296, Sigma-Aldrich)
- GFP-Trap®_A beads supplied in 20% ethanol (Reference GTA-20, Chromotek)

SNAP-TAG REAGENTS

- SNAP-Cell® TMR-Star (Reference S9105S, NEB), stock solution at 0.6 mM in DMSO (Reference D2650, Sigma-Aldrich)
- SNAP-Cell™ Block stock solution at 2 mM in DMSO (Reference S9106S, NEB)
- BG-NHS stock solution at 4 mM in DMSO (Reference S9151S, NEB)
- BG-PEG_9-NHS obtained as described in Shi et al. (2012), stock solution at 100 mM in DMSO

IMMUNOFLUORESCENCE

- Fixation: 4% paraformaldehyde (PFA), in PBS
- Quenching: 50 mM NH_4Cl in PBS
- Permeabilization: 0.2% saponin (Reference 102855, MP Biomedicals), 2% BSA (Reference 04-100-811-C, Euromedex) in PBS
- Mounting: 0.13 g/ml mowiol, 0.33 g/ml glycerol, 0.2 M Tris/HCl, pH 8.5

9.2 METHODS

9.2.1 Cell surface modification reagent

The cell surface modification reagent is composed of the following elements (Fig. 9.2A): BG moiety for TGN capture (blue), polyethylene glycol (PEG) linker arm to render the reagent membrane impermeable (red), and primary amino-reactive *N*-hydroxysuccinimidyl (NHS) moiety for modification of cell surface proteins

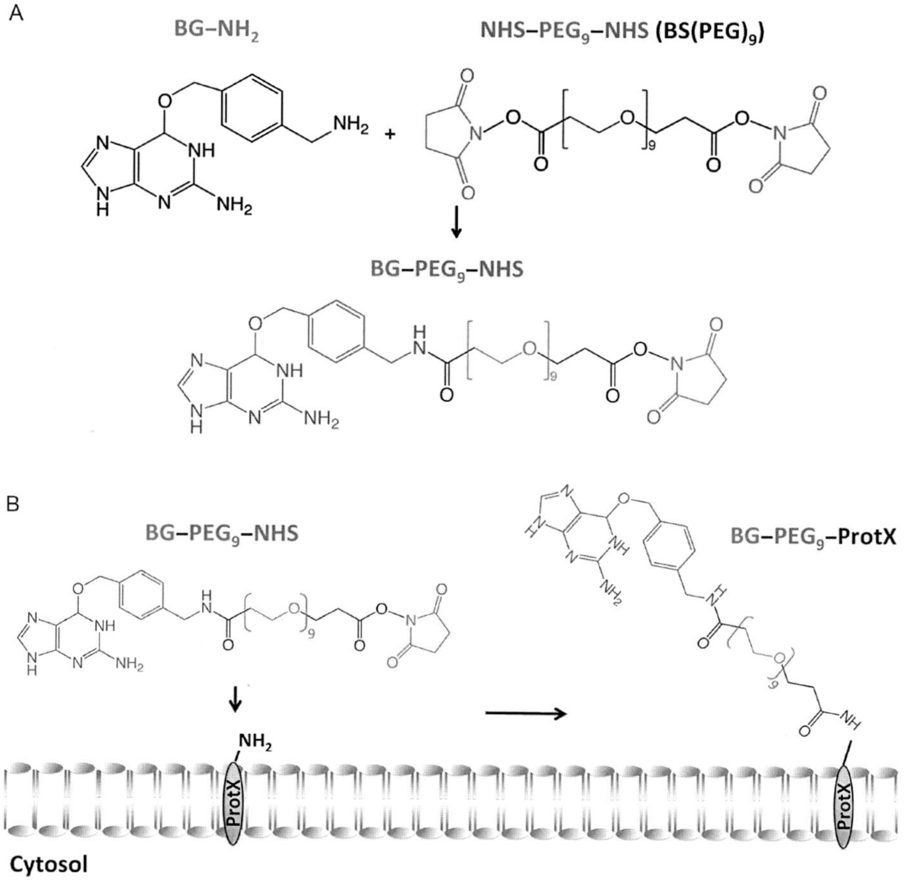

FIGURE 9.2

BG-PEG$_9$-NHS compound synthesis/cell surface modification. (A) BG-NH$_2$ is dissolved in anhydrous dimethylformamide and added to bis *N*-hydroxysuccinimide BS(PEG)$_9$. After further reactions as described in *Cell Surface Modification Reagent* section, the membrane impermeable BG-PEG$_9$-NHS cell surface modification reagent is obtained. This reagent is chemically composed of benzylguanine (blue), PEG$_9$ (red), and NHS (green). (B) Chemistry of cell surface modification leading to BG tagging of plasma membrane proteins (here: ProtX). (See color plate.)

(green). This reagent is termed BG-PEG$_9$-NHS. The chemical synthesis of this reagent is achieved as follows: BG-NH$_2$ (78.2 mg, 0.29 mmol; Fig. 9.2A) is obtained according to published procedure (Keppler et al., 2004), dissolved under argon in anhydrous dimethylformamide (10 ml), and successively added to commercially available bis-succinimide ester-activated PEG compound BS(PEG)$_9$ (266.9 mg, 0.38 mmol), 1-ethyl-3-(3-dimethylaminopropyl) carbodiimide (78.3 mg, 0.41 mmol), and hydroxybenzotriazole (55.4 mg, 0.41 mmol) solution at −5 °C. The solution is stirred at room temperature for 6 h. The solvent is removed under reduced pressure and the product is purified by preparative high-performance liquid chromatography (HPLC) (Shi et al., 2012).

NHS is an amino-specific functionality that covalently reacts with cell surface proteins, preferentially via primary amines on the side chains of lysine residues (Fig. 9.2B). This reaction can be achieved on living cells on ice, conditions under which endocytosis is inhibited. The NHS moiety transfers other chemical labels onto amino acceptors (here: plasma membrane proteins) if these are part of the same molecule. In the context of the current vectorial proteomics system, NHS is linked to BG for the reaction with the *trans*-Golgi/TGN-localized capture reagent (see below). Both NHS and BG can diffuse across membranes, and would thereby spontaneously (i.e., independent of retrograde transport) react with the capture reagent. We have therefore introduced nine PEG repeats into the molecule to render it hydrophilic, yielding BG-PEG_9-NHS (Fig. 9.2A). The lack of membrane permeability of this reagent is described in Section 9.3.4.

9.2.2 Capture reagent

The capture reagent is molecularly designed to be localized on *trans*-Golgi/TGN membranes. This is achieved by fusing the SNAP-tag to a fragment from GalT. GalT is a single pass membrane protein that is localized at the *trans*-cisternae of the Golgi apparatus and on TGN membranes (Roth & Berger, 1982). We use an existing pEGFP-GalT-GFP vector to obtain the GalT-GFP fragment, which is then inserted into the pSNAPm vector (New England Biolabs, NEB), yielding pGalT-GFP-SNAP. The GFP-tag allows for localization of the fusion protein by immunofluorescence microscopy, and its enrichment using GFP-Trap for proteomics.

As mentioned above, BG-tagged plasma membrane proteins that are shuttled to *trans*-Golgi/TGN membranes (putative retrograde cargoes, termed here ProtX) covalently react with the GalT-GFP-SNAP fusion protein, through the transfer of the benzyl group onto an internal cysteine residue within the SNAP-tag, yielding after the release of the guanine base the conjugate GalT-GFP-SNAP-ProtX (Fig. 9.1).

9.2.3 Validation of the BG/SNAP-tag strategy

In this section, we describe the step-by-step validation of the BG/SNAP-tag strategy on a stable HeLa cell line expressing the GalT-GFP-SNAP fusion protein. This cell line is termed HeLa-GGS in the following sections.

9.2.3.1 SNAP-tag localization to Golgi membranes by immunofluorescence

4×10^5 HeLa-GGS cells are seeded on glass coverslips in four-well plates 1 day prior to the experiment. The cells are cultured at 37 °C in complete DMEM medium (containing 10% fetal calf serum) in a 5% CO_2 incubator, washed three times with PBS^{++} (PBS containing 130 mM Ca^{2+} and 200 mM Mg^{2+}), and then fixed for 10 min at room temperature with PFA (4% in PBS). PFA is quenched by incubation of cells for 10 min at room temperature with 50 mM NH_4Cl prepared in PBS. Cells are permeabilized for 30 min at room temperature with 0.2% saponin in a solution of 2% BSA in PBS. Primary (here: rabbit antigiantin antibody from Curie antibodies platform, Reference A-R-R#05) and secondary antibodies (here: Cy3-labeled goat antirabbit from Jackson ImmunoResearch) are appropriately diluted in saponin–BSA–PBS solution and incubated with cells for 30–45 min. After three washes with saponin–BSA–PBS (5 min each), coverslips are dipped into water and mounted on Mowiol.

Confocal microscopy analysis reveals a consistent codistribution of GalT-GFP-SNAP (green)- and giantin (red)-specific labeling, strongly indicating that the GalT-GFP-SNAP fusion protein is indeed correctly localized in Golgi membranes (Fig. 9.3A).

To validate the enzymatic activity of the SNAP-tag in the context of the GalT-GFP-SNAP fusion protein, 4×10^5 HeLa-GGS cells are incubated for 15 min at 37 °C in complete DMEM medium with 3 μM TMR-Star (0.6 mM stock solution in DMSO, NEB), a rhodamine-labeled membrane permeable BG derivative. TMR-Star labeling (red) colocalizes with green GFP fluorescence in a perinuclear location (Fig. 9.3B), indicating that BG-SNAP reaction occurs on living cells within the membranes of the Golgi apparatus.

9.2.3.2 SNAP-tag validation using STxB

A model cargo of the retrograde route, the B-subunit of Shiga toxin (STxB) (Johannes & Romer, 2010), is chosen to demonstrate that the BG-SNAP reaction can also occur when BG is chemically linked onto the surface of a protein. STxB binds to its cellular receptor, the glycosphingolipid Gb3, is internalized by clathrin-dependent and -independent endocytosis, transported from early endosomes to the TGN, Golgi cisternae, and the endoplasmic reticulum (Johannes & Romer, 2010; Sandvig, Torgersen, Engedal, Skotland, & Iversen, 2010).

STxB is purified according to established procedures (Mallard & Johannes, 2003). Purified STxB (1 mg/ml final concentration in PBS) is incubated for 2 h at room temperature with BG-NHS (4 mM stock solution in DMSO, NEB) at a molar ratio of 1:3. Excess BG-NHS is removed by overnight dialysis at 4 °C against PBS, using 10 kDa molecular weight cut-off point dialysis cassettes (Thermo Scientific). HeLa-GGS cells are then incubated with 1 μM STxB-BG (or STxB as a negative control) for 30 min on ice in complete DMEM medium (membrane-binding step), washed, and shifted for 4 h to 37 °C (intracellular transport).

For immunofluorescence analysis, cells are washed with PBS^{++}, fixed with PFA, permeabilized with saponin, and labeled with anti-STxB antibody (clone 13C4), as described above. Confocal microscopy is used to analyze the localization pattern of wild-type STxB or STxB-BG in comparison to the GalT-GFP-SNAP fusion protein.

STxB is in part localized after 4 h in a perinuclear site and overlaps with the GalT-GFP-SNAP fusion protein (likely in Golgi membranes), and in part found in a peripheral network, likely the endoplasmic reticulum (Fig. 9.3C). Such distribution pattern between Golgi and endoplasmic reticulum has already been described in earlier studies

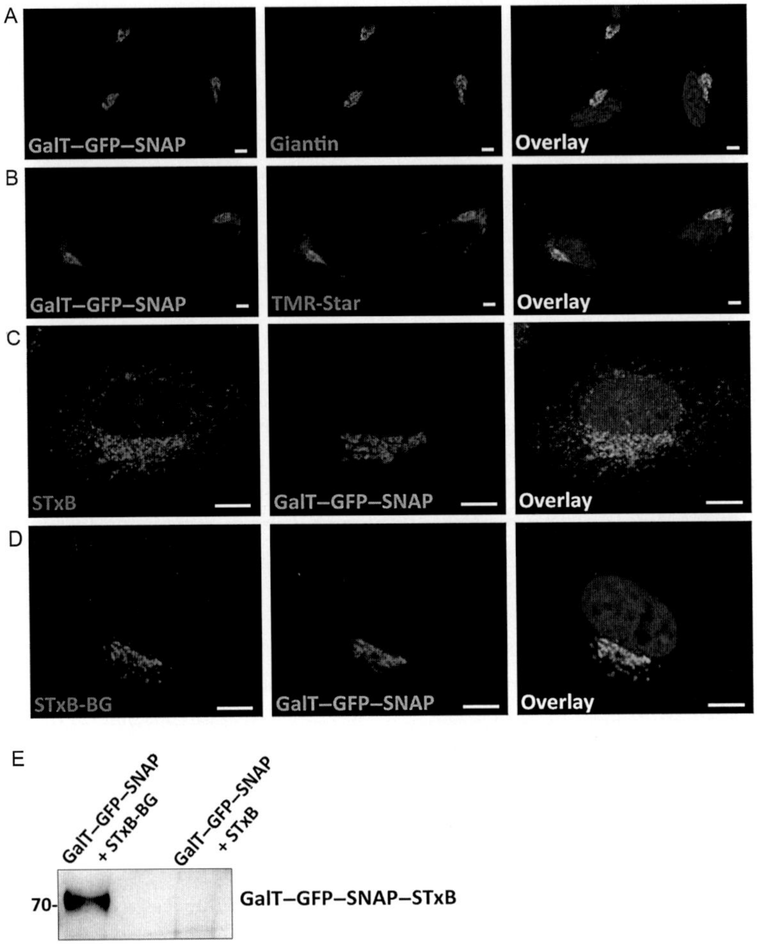

FIGURE 9.3

Strategy validations using STxB. (A) GalT-GFP-SNAP (green) localizes to Golgi membranes of HeLa-GGS cells, as deduced from its codistribution with giantin (red). (B) The TMR-Star reagent (red) reacts with GalT-GFP-SNAP (green), thereby validating the functionality of the SNAP-tag in this molecular context. (C–D) Without the BG tag, STxB (red) is found in Golgi membranes and in the cell periphery (likely ER) of HeLa-GGS cells (C), while BG-tagged STxB (red) is retained exclusively in the Golgi area (D). (E) A band of the expected size of GalT-GFP-SNAP-STxB reaction product is detected only on HeLa-GGS cells that are incubated with BG-tagged STxB, and not upon incubation with nontagged STxB. All images are acquired with Nikon A1R confocal microscope. Scale bars = 10 μm. (See color plate.)

(Johannes, Tenza, Antony, & Goud, 1997). In contrast, STxB-BG exclusively localizes in a perinuclear site with GalT-GFP-SNAP fusion protein (Fig. 9.3D). This differential behavior clearly demonstrates that BG-tagged STxB is retained in the perinuclear Golgi area, likely because of a reaction with the GalT-GFP-SNAP fusion protein.

The formation of such a GalT-GFP-SNAP-STxB reaction product is tested by immunoprecipitation. HeLa-GGS cells are continuously incubated for 4 h at 37 °C in complete DMEM medium with 1 μM of either unmodified STxB, or BG-STxB. The cells are then washed three times with DMEM to eliminate free ligands. Prior to cell lysis, unreacted GalT-GFP-SNAP fusion protein is quenched to prevent its reaction with STxB-BG in the cell lysate. The membrane permeable quenching reagent is termed SNAP-Cell-Block (2 mM stock solution in DMSO, NEB). Quenching is performed by incubation of cells for 20 min at 37 °C with 10 μM SNAP-Cell-Block in complete DMEM medium.

Cells are then lysed at 4 °C by a 30-min incubation under agitation on a rotating mixer with TNE buffer (50 mM Tris/HCl, 150 mM NaCl, 5 mM EDTA) containing 1% nonyl phenoxypolyethoxylethanol (NP-40). Cells are scrapped off the culture dish using a 1-ml end-cut tip, and the cell lysate is cleared of debris and DNA by a 13,200-rpm centrifugation for 10 min at 4 °C in an Eppendorf table top centrifuge. For immunoprecipitation, GFP-Trap beads (Chromotek), which recognize GFP with high affinity, are washed three times at 4 °C with TNE buffer, including bead collection by centrifugation for 3 min at 4600 rpm at the same temperature. 30 μl of beads are used for immunoprecipitation from 4×10^5 HeLa-GGS cells. The washed beads are suspended in cell lysate from above, and incubated overnight at 4 °C by end-over-end rotation. After three rapid washes at 4 °C in TNE buffer, excess washing buffer is removed using a Hamilton syringe, and beads are suspended in 3 × sample buffer (2 M Tris/HCl, pH 6.8, 20% SDS, 30% glycerol, 0.03% phenol red), and heated for 5 min at 95 °C. The protein eluate from the beads is separated by electrophoresis on 10% SDS-PAGE gels, and immunoblotted using rabbit anti-STxB antibodies (homemade rabbit polyclonal).

A protein species of the expected size of the GalT-GFP-SNAP-STxB reaction product can be detected only when cells are incubated with BG-tagged STxB, and not with untagged STxB (Fig. 9.3E) The inhibition of retrograde transport using specific small molecule tools (i.e., Retro-2 from Stechmann et al., 2010) or the disassembly of the Golgi apparatus using brefeldin-A prevent the formation of this reaction product (Shi et al., 2012), strongly indicating that the encounter of STxB-BG and GalT-GFP-SNAP indeed occurs in *trans*-Golgi/TGN membranes.

9.2.4 BG/SNAP-tag strategy in proteomics format

In this section, we describe the application of the SNAP-tag/BG reaction scheme to the proteomics analysis of the retrograde route.

9.2.4.1 BG-PEG$_9$-NHS is a cell impermeable amine reactive linker

To covalently label cell surface proteins with BG, we use a modified version of BG-NHS, termed NHS-PEG$_9$-BG (Fig. 9.2A). The PEG$_9$ unit is intended to increase the hydrophilicity of the compound, and thereby prevents its spontaneous diffusion across membranes, which would lead to the quenching of the TGN-localized SNAP-tag of the GalT-GFP-SNAP fusion protein. The membrane impermeability of NHS-PEG$_9$-BG is documented as follows.

HeLa-GGS cells are incubated for 1 h at 4 °C with 4 mM BG-NHS, 4 mM NHS-PEG$_9$-BG, or 10 μM SNAP-Cell-Block, all in PBS^{++} buffer, washed three times with the same buffer, and shifted for 15 min to 37 °C in complete DMEM medium. After this primary incubation, cells are again incubated in DMEM for 30 min at 37 °C with TMR-Star (NEB), a membrane permeable compound in which BG is linked to a red fluorophore, 6-carboxytetramethylrhodamine. This molecule reacts with the *trans*-Golgi/TGN-localized GalT-GFP-SNAP only under conditions under which the BG derivatives of the prior incubation have not reached this compartment (i.e., compounds that have been rendered sufficiently hydrophilic such that they cannot diffuse passively across membranes). After washing (three times, as described above), cells are lysed in TNE buffer with 1% NP-40, and postnuclear supernatents from cell lysates are loaded onto 10% SDS-PAGE gels, followed by direct visualization using a fluorescence scanner (Typhoon 9400, GE).

In control cells (i.e., cells that are not treated with BG-modified compounds during the primary incubation), a fluorescent band is detected at the expected size of GalT-GFP-SNAP (Fig. 9.4, lane 2; see arrow). For cells that are incubated during the primary phase with the SNAP-Cell-Block reagent, this band cannot be detected (lane 3), confirming that this compound indeed diffuses passively across membranes and quenches GalT-GFP-SNAP, such as to prevent the secondary modification by TMR-star. This condition thereby defines the maximal level of inhibition. A similar level of inhibition is observed with BG-NHS (lane 4), showing that this compound is also fully membrane permeable. In contrast, NHS-PEG$_9$-BG does not quench SNAP-tag reactivity on GalT-GFP-SNAP during the primary incubation (lane 5), demonstrating that this compound is not membrane permeable.

9.2.4.2 Proteomics format

4×10^6 HeLa-GGS cells are incubated for 30 min at 4 °C with 1 μM STxB in complete DMEM medium, which leads to STxB binding to cells via its glycolipid receptor Gb3 with high apparent affinity in the nanomolar range (Johannes et al., 1997). Cells are then washed in the same medium, and incubated for 1 h at 4 °C with 1 mM NHS-PEG$_9$-BG in PBS^{++} buffer, to BG-modify plasma membrane proteins. Cells are washed and incubated in complete DMEM medium for 10 h at 37 °C to allow for internalization of BG-tagged cell surface proteins.

Further treatments are as described above in Section 9.3.3.2: Quenching using SNAP-Cell Block reagent, cell lysis in TNE/NP-40 buffer, GFP-Trap, and SDS-PAGE. Proteins are then submitted to in-gel digestion.

For this, the SDS-PAGE gel is stained using colloidal blue; migration lanes are excised and subjected to 10 mM DTT in 100 mM ammonium bicarbonate (1 h at 60 °C) for reduction prior to alkylation with 55 mM iodoacetamide in 100 mM ammonium bicarbonate (45 min at room temperature). Gel pieces are washed with 100 mM ammonium bicarbonate and shrinked by incubation with 100% acetonitrile

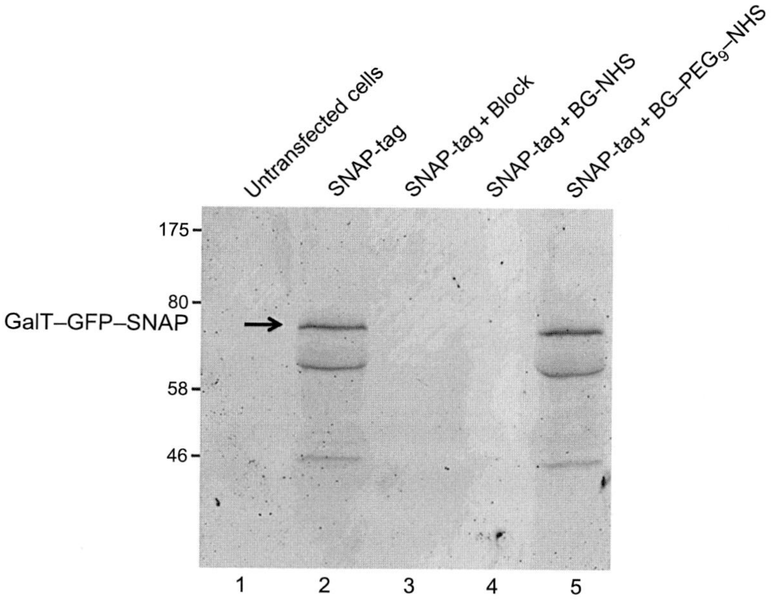

FIGURE 9.4

Membrane permeability test of BG reagents. Untransfected HeLa cells (lane 1) or HeLa-GGS cells (SNAP-tag, lanes 2–5) are incubated for 1 h with DMSO (lane 2), SNAP-Cell-Block reagent (lane 3), BG-NHS (lane 4), or NHS-PEG$_9$-BG (lane 5). After washing, cells are incubated at 37 °C with TMR-Star, lysed, and after SDS-PAGE gels are viewed with a fluorescence scanner. Note that a fluorescent band at the expected size of the GalT-GFP-SNAP capture reagent function proteins (arrow) is seen only in the DMSO (lane 2) and NHS-PEG$_9$-BG (lane 5) conditions, indicating that the NHS-PEG$_9$-BG cell surface modification reagent is indeed membrane impermeable.

From Shi et al. (2012).

(10 min at room temperature), before being subjected to an overnight trypsin digestion at 30 °C in 25 mM ammonium bicarbonate. Peptide concentration and separation are achieved using an actively split capillary HPLC system (Ultimate 3000 system) connected to the LTQ Orbitrap XL™ mass spectrometer (Thermo Scientific). The resulting spectra are analyzed via the Mascot™ Software created with Proteome Discoverer (version: 1.2.0.92, Thermo Scientific) using the NCBInr human Protein Database.

STxB is clearly identified among the retrograde cargo candidate proteins (Table 9.1). Other proteins of this list (data not shown) may represent plasma membrane proteins for which trafficking via the retrograde route has previously not been shown. We describe below a method by which such retrograde cargo candidates can be further validated.

Table 9.1 Peptide identification by LC–MS/MS

Identifier	Size (aa)	MW (Da)	Matching peptides	Coverage (%)
GalTGFP	366	39608	11	28.1
SNAP-tag	193	20239.3	4	38.3
STxB	70	7793.8	1	14.3

STxB-bound HeLa-GGS cells are incubated for 1 h at 4 °C with NHS-PEG$_9$-BG cell surface modification reagent, and then after washing for 10 h at 37 °C. Immunoprecipitated GFP-tag material is analyzed by mass spectrometry, as described in the proteomic format *section. Note that STxB peptides are clearly identified among other retrograde cargo candidates.*
From Shi et al. (2012).

9.2.5 Validation of retrograde cargo candidates—Antibody uptake

To confirm the trafficking via the retrograde route of putative retrograde cargo candidates from the proteomics list, we have developed an antibody uptake assay (Fig. 9.5). Here, this protocol is illustrated for a well-characterized cargo of the retrograde route, the cation-independent mannose 6-phosphate receptor (CI-MPR) (Duncan & Kornfeld, 1988). CI-MPR shuttles newly synthesized mannose 6-phosphate-modified lysosomal enzymes from the TGN to endosomes (Brown, Goodhouse, & Farquhar, 1986). After cargo delivery, the empty receptor is retrieved from endosomes to the TGN. A small fraction of CI-MPR is also found at the plasma membrane where it binds to extracellular ligands for subsequent internalization (Lin, Mallet, Huang, & Maxfield, 2004). For our study, we use a stable cell line expressing a GFP-tagged CI-MPR fusion protein (Waguri et al., 2003). An anti-GFP antibody (Roche) specifically recognizes this fusion protein whose GFP part is exposed to the extracellular medium (Fig. 9.5).

The experimental principle is the following: Upon transport of the GFP-CI-MPR fusion protein from the cell surface to the Golgi compartment, the anti-GFP-BG antibody (that will bind to GFP-CI-MPR at the cell surface upon addition to the cell culture medium) is dragged along and reacts with the *trans*-Golgi/TGN-localized GalT-GFP-SNAP capture reagent once it has reached this compartment (Fig. 9.5).

9.2.5.1 BG tagging of anti-GFP antibody

Anti-GFP antibody (Roche) is prepared at a final concentration of 5 mg/ml in 50 mM sodium phosphate buffer pH 7.5, containing 1 mM EDTA. BG-NHS reagent is added in 10-fold molar excess over antibody. This mixture is incubated 6 h at 4 °C prior to dialysis against 50 mM sodium phosphate pH 7.5, 1 mM EDTA, using 10 kDa molecular weight cut-off (MWCO, Thermo Scientific) dialysis cassettes. Additional quenching of the BG-NHS linker can be performed by adding 20 mM Tris (containing primary amine) for 20 min at room temperature prior to dialysis.

9.2.5.2 BG-tagged antibody uptake assay

4×10^5 MPR-GFP-expressing HeLa cells transiently transfected with pGalT-GFP-SNAP in 6-well culture plates, are incubated for 4 h at 37 °C with 15 μg/ml of unmodified or BG-tagged anti-GFP antibodies in complete DMEM medium. Cells

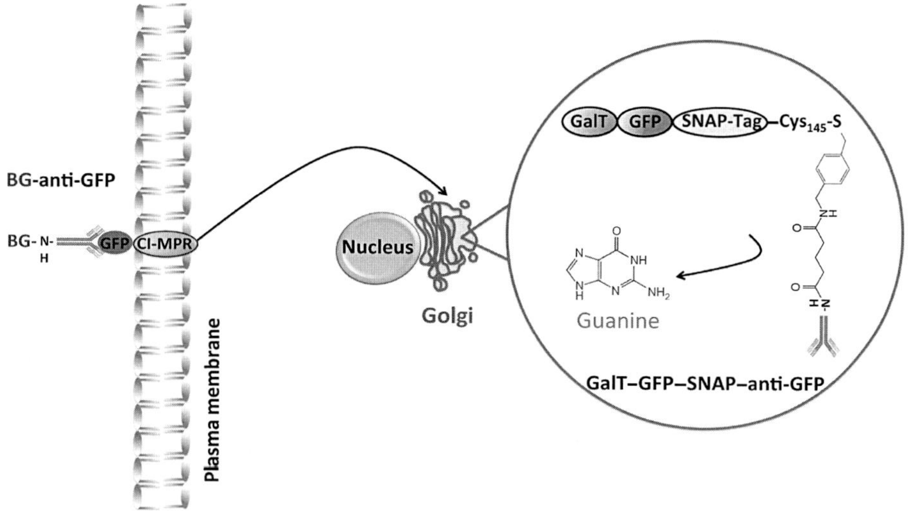

FIGURE 9.5

Schematic representation of the antibody uptake protocol. This protocol is illustrated with the well-characterized GFP-tagged retrograde cargo CI-MPR. If a given protein (here GFP-CI-MPR) indeed follows the retrograde transport route, the BG-tagged antibody that specifically recognizes this protein (here BG-anti-GFP) is dragged along to *trans*-Golgi/TGN membranes, where it is conjugated to the GalT-GFP-SNAP capture reagent (here: yielding the GalT-GFP-SNAP-anti-GFP conjugate). (See color plate.)

are washed, *trans*-Golgi/TGN-localized GalT-GFP-SNAP that has not reacted with BG-tagged cargo proteins is quenched with SNAP-Cell-Block reagent, and cells are lysed as described in Section 9.3.3.2. Cell lysate is incubated overnight at 4 °C by end-over-end mixing with 50 μl of protein G-Sepharose beads (Sigma-Aldrich). Immunoprecipitated proteins are separated on 8% SDS-PAGE gels in nonreducing condition to avoid the dissociation of antibody chains, and finally blotted with anti-GFP (lanes 1–4) or anti-SNAP (lanes 5–8) antibodies (Fig. 9.6)

Positive bands just below 250 kDa are observed only in conditions in which the anti-GFP antibody is BG-tagged and cells are expressing the GalT-GFP-SNAP capture reagent (Fig. 9.6, lanes 1 and 5). This size corresponds to the GalT-GFP-SNAP fusion protein (70 kDa) linked to the nonreduced antibody (150 kDa). Under conditions in which either non BG-tagged anti-GFP antibody (lanes 2 and 6) or the GalT-GFP-SNAP capture reagent (lanes 3 and 7) are missing, or cells are used that do not express GFP-tagged CI-MPR (lanes 4 and 8), no band is detected at the expected size.

This approach can in principle be transposed to any candidate protein from the retrograde proteomics hit list, provided that either antibody is available against the extracellular domain of the protein, or the protein can be GFP tagged.

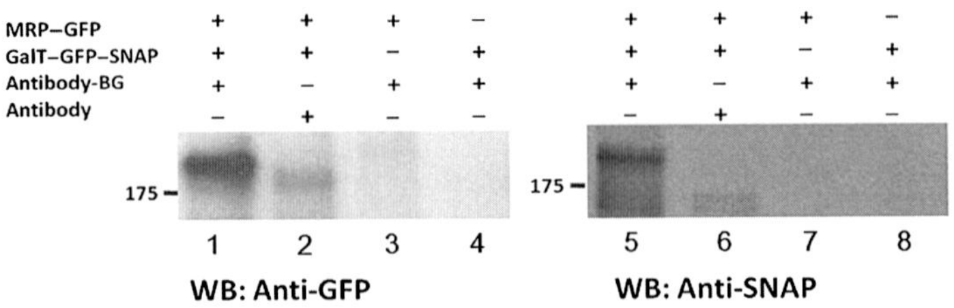

FIGURE 9.6

Antibody uptake assay on GFP-tagged CI-MPR. Cells stably expressing GFP-tagged CI-MPR (MPR-GFP +) or not (MPR-GFP −) are transiently transfected or not with GalT-GFP-SNAP, and incubated with BG-tagged anti-GFP antibody (Antibody-BG), or nontagged anti-GFP antibody (Antibody), as indicated. After 4 h incubation at 37 °C, cells are lysed, incubated with protein-G-sepharose, and immunoprecipitated material is analyzed by SDS-PAGE and Western blotting against GFP (lanes 1–4) or SNAP-tag (lanes 5–8). Note that a band corresponding to the GalT-GFP-SNAP-antibody conjugate (above 175 kDa) is detected only in conditions where BG-tagged anti-GFP antibody is incubated with cells expressing GFP-tagged CI-MPR and the GalT-GFP-SNAP capture reagent (lanes 1 and 5).

From Shi et al. (2012).

9.3 CONCLUSION/DISCUSSION

At the basis of the approach that is detailed in this methods review, the study by Shi et al. (2012) was the first to demonstrate that the BG-SNAP-tag ligation can occur on living cells between full size proteins within the lumen of an intracellular compartment. We apply this finding to the proteomics analysis of the retrograde transport route, for which plasma membrane proteins are BG-tagged, and the SNAP-tag capture reagent is localized to *trans*-Golgi/TGN membranes. In principle, this experimental design can be transposed more generally to other endocytic pathways, by placing the SNAP-tag into other intracellular compartments (early or late endosomes, endoplasmic reticulum, cytosol, etc.). SNAP-tag constructs for the endoplasmic reticulum and the cytosol are indeed already available (Geiger, Luisoni, Johnsson, Greber, & Helenius, 2013).

The current approach relies on the labeling of primary amino groups of plasma membrane proteins with BG-NHS. Only cargo proteins can thereby be identified whose intracellular trafficking characteristics are not abolished by this chemical modification. Weak hits in the proteomics format might also represent proteins that are partially affected by BG tagging. In such cases, the antibody uptake protocol (which does not require BG tagging of the cargo candidate itself) is a valid means to obtain complementary confirmation on the authenticity of apparently weak hits.

Another reason for obtaining apparently weak hits (i.e., proteins on the retrograde proteomics hit list that are present in low peptide numbers) that might in reality be efficient cargoes of the retrograde route resides in the experimental design of the current approach. Our method relies on the modification of the plasma membrane proteome on cells that are kept on ice, thereby introducing a bias toward proteins that are abundant at the cell surface at steady state. Some retrograde cargo proteins may only transiently cycle through the plasma membrane, and only small numbers of these molecules would be BG-tagged under our experimental conditions.

SUMMARY

Retrograde transport from the plasma membrane to the TGN is increasingly attracting attention of scientists in diverse fields ranging from cellular neurobiology, Alzheimer's disease research and infection biology to development. The vectorial proteomics approach that is described in this method's review enables an integrated analysis of the retrograde proteome. We describe step-by-step how to move from tagging plasma membrane proteins with an amino-reactive BG probe to the TGN-localized and SNAP-tag-based capture, enrichment, and identification of retrograde cargo candidates, including their validation using an antibody uptake protocol. In its current format, the approach is limited to cell lines that can be stably transfected with the capture reagent. The possible development of virus-based expression systems should in the future also allow working on primary nondividing cells.

Acknowledgments

The work in the laboratory of the authors was supported by Agence Nationale de la Recherche (ANR-09-BLAN-283, ANR-11 BSV2 018 03, ANR-11 BSV2 014 03) and Institut National du Cancer (PLBIO11-022-IDF-JOHANNES).

References

Belenkaya, T. Y., Wu, Y., Tang, X., Zhou, B., Cheng, L., Sharma, Y. V., et al. (2008). The retromer complex influences Wnt secretion by recycling wntless from endosomes to the trans-Golgi network. *Developmental Cell*, *14*(1), 120–131. http://dx.doi.org/10.1016/j.devcel.2007.12.003.

Blot, G., Janvier, K., Le Panse, S., Benarous, R., & Berlioz-Torrent, C. (2003). Targeting of the human immunodeficiency virus type 1 envelope to the trans-Golgi network through binding to TIP47 is required for env incorporation into virions and infectivity. *Journal of Virology*, *77*(12), 6931–6945.

Bonifacino, J. S., & Rojas, R. (2006). Retrograde transport from endosomes to the trans-Golgi network. *Nature Reviews. Molecular Cell Biology*, *7*(8), 568–579. http://dx.doi.org/10.1038/nrm1985.

Brown, W. J., Goodhouse, J., & Farquhar, M. G. (1986). Mannose-6-phosphate receptors for lysosomal enzymes cycle between the Golgi complex and endosomes. *Journal of Cell Biology*, *103*(4), 1235–1247.

Coudreuse, D. Y., Roel, G., Betist, M. C., Destree, O., & Korswagen, H. C. (2006). Wnt gradient formation requires retromer function in Wnt-producing cells. *Science*, *312*(5775), 921–924. http://dx.doi.org/10.1126/science.1124856.

Crump, C. M., Bruun, B., Bell, S., Pomeranz, L. E., Minson, T., & Browne, H. M. (2004). Alphaherpesvirus glycoprotein M causes the relocalization of plasma membrane proteins. *Journal of General Virology*, *85*(Pt 12), 3517–3527. http://dx.doi.org/10.1099/vir.0.80361-0.

Duncan, J. R., & Kornfeld, S. (1988). Intracellular movement of two mannose 6-phosphate receptors: Return to the Golgi apparatus. *Journal of Cell Biology*, *106*(3), 617–628.

Geiger, R., Luisoni, S., Johnsson, K., Greber, U. F., & Helenius, A. (2013). Investigating endocytic pathways to the endoplasmic reticulum and to the cytosol using SNAP-trap. *Traffic*, *14*(1), 36–46. http://dx.doi.org/10.1111/tra.12018.

Johannes, L., & Popoff, V. (2008). Tracing the retrograde route in protein trafficking. *Cell*, *135*(7), 1175–1187. http://dx.doi.org/10.1016/j.cell.2008.12.009.

Johannes, L., & Romer, W. (2010). Shiga toxins—From cell biology to biomedical applications. *Nature Reviews. Microbiology*, *8*(2), 105–116. http://dx.doi.org/10.1038/nrmicro2279.

Johannes, L., Tenza, D., Antony, C., & Goud, B. (1997). Retrograde transport of KDEL-bearing B-fragment of Shiga toxin. *Journal of Biological Chemistry*, *272*(31), 19554–19561.

Keppler, A., Gendreizig, S., Gronemeyer, T., Pick, H., Vogel, H., & Johnsson, K. (2003). A general method for the covalent labeling of fusion proteins with small molecules in vivo. *Nature Biotechnology*, *21*(1), 86–89. http://dx.doi.org/10.1038/nbt765.

Keppler, A., Kindermann, M., Gendreizig, S., Pick, H., Vogel, H., & Johnsson, K. (2004). Labeling of fusion proteins of O6-alkylguanine-DNA alkyltransferase with small molecules in vivo and in vitro. *Methods*, *32*(4), 437–444. http://dx.doi.org/10.1016/j.ymeth.2003.10.007.

Lencer, W. I., & Tsai, B. (2003). The intracellular voyage of cholera toxin: Going retro. *Trends in Biochemical Sciences*, *28*(12), 639–645. http://dx.doi.org/10.1016/j.tibs.2003.10.002.

Lin, S. X., Mallet, W. G., Huang, A. Y., & Maxfield, F. R. (2004). Endocytosed cation-independent mannose 6-phosphate receptor traffics via the endocytic recycling compartment en route to the trans-Golgi network and a subpopulation of late endosomes. *Molecular Biology of the Cell*, *15*(2), 721–733. http://dx.doi.org/10.1091/mbc.E03-07-0497.

Mallard, F., & Johannes, L. (2003). Shiga toxin B-subunit as a tool to study retrograde transport. *Methods in Molecular Medicine*, *73*, 209–220.

Roth, J., & Berger, E. G. (1982). Immunocytochemical localization of galactosyltransferase in HeLa cells: Codistribution with thiamine pyrophosphatase in trans-Golgi cisternae. *Journal of Cell Biology*, *93*(1), 223–229.

Sandvig, K., Spilsberg, B., Lauvrak, S. U., Torgersen, M. L., Iversen, T. G., & van Deurs, B. (2004). Pathways followed by protein toxins into cells. *International Journal of Medical Microbiology*, *293*(7–8), 483–490.

Sandvig, K., Torgersen, M. L., Engedal, N., Skotland, T., & Iversen, T. G. (2010). Protein toxins from plants and bacteria: Probes for intracellular transport and tools in medicine. *FEBS Letters*, *584*(12), 2626–2634. http://dx.doi.org/10.1016/j.febslet.2010.04.008.

Shewan, A. M., van Dam, E. M., Martin, S., Luen, T. B., Hong, W., Bryant, N. J., et al. (2003). GLUT4 recycles via a trans-Golgi network (TGN) subdomain enriched in Syntaxins 6 and 16 but not TGN38: Involvement of an acidic targeting motif. *Molecular Biology of the Cell*, *14*(3), 973–986. http://dx.doi.org/10.1091/mbc.E02-06-0315.

Shi, G., Azoulay, M., Dingli, F., Lamaze, C., Loew, D., Florent, J. C., et al. (2012). SNAP-tag based proteomics approach for the study of the retrograde route. *Traffic*, *13*(7), 914–925. http://dx.doi.org/10.1111/j.1600-0854.2012.01357.x.

Snider, M. D., & Rogers, O. C. (1985). Intracellular movement of cell surface receptors after endocytosis: Resialylation of asialo-transferrin receptor in human erythroleukemia cells. *Journal of Cell Biology*, *100*(3), 826–834.

Stechmann, B., Bai, S. K., Gobbo, E., Lopez, R., Merer, G., Pinchard, S., et al. (2010). Inhibition of retrograde transport protects mice from lethal ricin challenge. *Cell*, *141*(2), 231–242. http://dx.doi.org/10.1016/j.cell.2010.01.043.

Voskoboinik, I., & Camakaris, J. (2002). Menkes copper-translocating P-type ATPase (ATP7A): Biochemical and cell biology properties, and role in Menkes disease. *Journal of Bioenergetics and Biomembranes*, *34*(5), 363–371.

Waguri, S., Dewitte, F., Le Borgne, R., Rouille, Y., Uchiyama, Y., Dubremetz, J. F., et al. (2003). Visualization of TGN to endosome trafficking through fluorescently labeled MPR and AP-1 in living cells. *Molecular Biology of the Cell*, *14*(1), 142–155. http://dx.doi.org/10.1091/mbc.E02-06-0338.

Yang, P. T., Lorenowicz, M. J., Silhankova, M., Coudreuse, D. Y., Betist, M. C., & Korswagen, H. C. (2008). Wnt signaling requires retromer-dependent recycling of MIG-14/Wntless in Wnt-producing cells. *Developmental Cell*, *14*(1), 140–147. http://dx.doi.org/10.1016/j.devcel.2007.12.004.

Zhang, D., Isack, N. R., Glodowski, D. R., Liu, J., Chen, C. C., Xu, X. Z., et al. (2012). RAB-6.2 and the retromer regulate glutamate receptor recycling through a retrograde pathway. *Journal of Cell Biology*, *196*(1), 85–101. http://dx.doi.org/10.1083/jcb.201104141.

CHAPTER

Assessing ER and Golgi N-Glycosylation Process Using Metabolic Labeling in Mammalian Cultured Cells

10

Romain Péanne*, Jorick Vanbeselaere†, Dorothée Vicogne†, Anne-Marie Mir†, Christophe Biot†, Gert Matthijs*, Yann Guérardel† and François Foulquier†

**Center for Human Genetics, KU Leuven University, Herestraat 49, Leuven, Belgium*

†Structural and Functional Glycobiology Unit, UMR CNRS/USTL 8576, IFR 147, Avenue Mendeleïev, University of Lille 1, Villeneuve d'Ascq, France

CHAPTER OUTLINE

Introduction and Rationale **158**
10.1 Materials **162**
10.1.1 To Assess ER N-glycosylation Efficiency 162
10.1.2 To Assess Golgi N-glycosylation Efficiency 164
10.2 Methods **166**
10.2.1 To Assess ER N-glycosylation Efficiency 166
10.2.1.1 Metabolic Labeling 166
10.2.1.2 Sequential Extraction Procedure 167
10.2.1.3 Purification and Quantification of LLO, fOSs, and Glycoproteins 168
10.2.1.4 HPLC Separation of the Purified Oligosaccharides 170
10.2.2 To Assess Golgi N-glycosylation Efficiency 171
10.2.2.1 Cells Metabolic Labeling with Alkyne-tagged Sialic Acid and Fixation 171
10.2.2.2 "Click Chemistry": Probe Labeling Reactions 171
10.2.2.3 Immunolabeling 171
10.2.2.4 Microscopic Analysis of Fluorescent Labeling in Cells 172
10.3 Discussion **172**
Summary **175**
Acknowledgments **175**
References **175**

ISSN 0091-679X
http://dx.doi.org/10.1016/B978-0-12-417164-0.00010-0

Abstract

Modifications of *N*-glycosylation in disease states are common and illustrate the crucial requirement of glycosylation in human biology. Mainly based on glycan permethylation and the use of mass spectrometry analysis, we can easily understand that many different methods to analyze the *N*-glycome have seen the day. While extremely powerful, these methods are mainly used to analyze qualitative variations of *N*-glycosylation of human serum proteins and do not necessarily reflect the glycosylation status of derived mammalian cultured cells. This chapter summarizes two methods that we are routinely using in our laboratory to assess the ER and Golgi *N*-glycosylation process. The proposed methodology allows pinpointing ER as well as Golgi glycosylation deficiencies in mammalian cultured cells. The first approach is based on direct metabolic labeling of cultured mammalian cells with [2-^{3}H] mannose followed by sequential extraction and HPLC analysis of the purified oligosaccharides. The second one is based on the copper-catalyzed azide alkyne cycloaddition (CuAAC) strategy. We propose the use of alkyne-tagged sialic acid (SialNAl) to visualize the Golgi glycosylation efficiency. Their metabolic incorporation into newly synthesized glycoproteins can then be chemoselectively coupled to complementary azide-functionalized fluorophores, and visualized by using confocal laser scanning microscopy. To summarize, we present here a detailed description of our know-how in the field of ER and Golgi *N*-glycosylation.

INTRODUCTION AND RATIONALE

During the past decades, many co- or posttranslational modifications were identified, including phosphorylation, sulfatation, methylation, *N*-acetylation, palmitoylation, and glycosylation. Glycosylation (from the greek *glykys*: sweet) stands for an enzymatic reaction defined by the covalent attachment of monosaccharides or polysaccharides (glycans) onto proteins or lipids. This essential posttranslational modification produces an abundant, diverse, and highly regulated repertoire of glycan structures, frequently attached to proteins or lipids. In addition, glycosylation is found in every cell type of an organism and is distributed throughout the entire phylogenetic spectrum, ranging from archea and eubacteria to higher eukaryotes.

In mammals, protein glycosylation encompasses different glycosylation types and can be divided in two major forms, non-compartmentalized glycosylation and compartmentalized glycosylation, occurring in specific cellular organelles. However, if the structural diversity of the glycan structures in mammals is produced predominantly in the secretory pathway of the cells, some forms of glycosylation occur outside of the secretory pathway. *N*-Acetylglucosamine (GlcNAc) was reported to be linked to serine and threonine residues on many cytoplasmic and nuclear proteins (Hart, Holt, & Haltiwanger, 1988). All other types of glycosylation occur at the interface of two organelles: the endoplasmic reticulum (ER) and/or the Golgi compartment. Three major types can be distinguished and classified, according to the linkage of the monosaccharides to amino acid residues. *N*-Glycans are linked to the amide group of asparagine residues of proteins, *O*-glycosylation characterizes the attachment of glycans to the hydroxyl group of serine and threonine side chains, and *C*-glycosylation is defined by the attachment of glycans to the C2 atom of tryptophan residues.

N-Glycosylation occurs at the interface of the ER and the Golgi apparatus. This glycosylation is the most widespread form of glycosylation responsible for the modification of secreted and transmembrane proteins. This process is initiated in the ER by the assembly of an oligosaccharide precursor consisting of three glucoses (Glc), nine mannoses (Man), and two *N*-acetylglucosamine (GlcNAc) residues. The assembly of this Glc3Man9GlcNAc2 structure, linked to dolichol pyrophosphate (Dol-PP) carrier, is a stepwise process strongly conserved in the evolution, and known as the dolichol cycle. The dolichol cycle starts at the cytoplasmic face of the ER with the formation of the dolichol-linked oligosaccharide (lipid-linked oligosaccharide, LLO) intermediate Man5GlcNAc2-PP-Dol at the luminal face of the ER membrane. This intermediate is then flipped at the cytosolic face of the ER membrane by the RFT1 protein before its elongation to form the full Glc3Man9GlcNAc2-PP-Dol oligosaccharide precursor. Once synthesized, the oligosaccharide precursor is cotranslationaly linked in β to asparagine residues of proteins by the oligosaccharyltransferase complex. Once exported in the Golgi lumen, *N*-linked glycans are subjected to an extensive maturation, by the sequential action of glycosidases and glycosyltranferases, to generate a broad range of structures. The Golgi trimming and maturation of *N*-glycans chains start in the *cis*-Golgi compartment by catalytic removal of all α(1,2)-linked mannose residues by the Golgi mannosidase I. Removal of these three mannose residues gives rise to Man5GlcNAc2 branched structures. The *N*-acetylglucosaminyltransferase I (GlcNAc-TI) is a key enzyme in *N*-glycan processing. GlcNAC-TI subsequently catalyzes the conversion of high mannose type to hybrid- and complex-type *N*-glycans by adding a GlcNAC residue to the newly uncovered mannose residue. The α(1,3)- and α(1,6)-linked mannose residues are catalytically removed by the Golgi mannosidase II in the medial Golgi compartment. The addition of a second GlcNAc residue on the α(1,6) mannose branch of the glycan chain is mediated by the GlcNAc transferase II (GlcNAC-TII). The addition of this GlcNAc residue governs the conversion of hybrid glycan-type glycan structures to complex-type ones. Many more additional branches can be generated by the respective action of GlcNAc-TIV, GlcNAC-TV, and/or GlcNAc-TVI. Branching reactions can result in biantennary, triantennary, tetra-antennary, and penta-antennary oligosaccharides. The introduction of a bisecting GlcNAc residue attached to the β-mannose of the core by GlcNAc-TIII can prevent further extension of *N*-glycans. GlcNAc bisecting residues may be carried by complex and hybrid *N*-glycans and consequently inhibit the activity of GlcNAc-TII, GlcNAc-TIV, and GlcNAc-TV, as well as the function of Golgi mannosidase II (Schachter, 2000). Still in the medial Golgi compartment, a fucose residue can be added to the first GlcNAc residue, covalently linked to the Asn residue in the polypeptide, by the core fucolsyltransferase. Terminal Golgi glycosylation mostly occurs in the *trans*-Golgi cisternae. Galactose (Gal) and sialic acid residues are added symmetrically and sequentially to each of the nonreducing ends of the *N*-glycan antennae. Galactosylation of *N*-glycans is mostly catalyzed by the β(1,4)- and β(1,3)-galactosyltransferases. In humans, the most common terminal sialic acid is NeuAc, which is added to the Gal residue in a α(2,3), α(2,6), or α(2,8) linkage.

Even if 20 years ago the role of glycosylation was underestimated, the identification of a new group of diseases related to defects of glycosylation and called congenital disorders of glycosylation (CDG) dramatically illustrated the importance of this post-translational modification. CDG form a family of autosomal recessive genetic disorders, leading to defects in the biosynthesis of *N*- and *O*-glycoproteins, and glycolipids protein-linked glycans (Jaeken & Matthijs, 2007). To date, many different genetic defects have been described, comprising a very broad range of phenotypes (Theodore & Morava, 2011). According to the localization of the deficiency, *N*-glycosylation defects can be divided into two groups, CDG-I and CDG-II. CDG-I disorders relate to defects in the synthesis of the *N*-glycans in the ER, while CDG-II disorders arise from defects in the processing and elongation of the glycan chains, mainly in the Golgi apparatus.

Originally, mutations were found in ER/Golgi glycosyltransferases, remodeling glycosidases, and sugar nucleotide transporters that are known to have a direct role in the glycosylation process. Recently, however, forms of CDG caused by vesicular Golgi trafficking and Golgi pH homeostasis deficiencies were also identified (Rosnoblet, Peanne, Legrand, & Foulquier, 2013). One challenge resides in analyzing the ER and/or Golgi glycosylation. Most of the available methods are based on reporter *N*-glycoproteins such as transferrin or the use of high-performance liquid chromatography (HPLC) and mass spectrometry (MS) to establish the *N*-glycome. While powerful, these methods are almost exclusively performed on plasma sample and despite the constant evolution of glycan analysis, it is difficult to assess the ER/Golgi glycosylation efficiency in cultured mammalian cells. In this chapter, we propose two reliable optimized methods to analyze the ER and Golgi glycosylation. The first method is based on tritiated mannose metabolic labeling and the second method on alkyne-tagged sialic acid (SialNAl). Contrary to expectations, [2-^{3}H]-mannose and not [1,5,6-^{3}H]-glucose has been shown to be the preferred source of mannose for glycoprotein biosynthesis in human fibroblasts (Panneerselvam, Etchison, & Freeze, 1997). Considering the balance between their hydrophobic moiety (dolichol) and their hydrophilic part (glycan), the different types of oligosaccharides can be easily isolated using a sequential extraction procedure based on their solubility in organic solvent (Fig. 10.1). This sequential extraction procedure relies on a biphasic extraction with a mixture of $CHCl_3$/MeOH/H_2O (3/2/1 by volume), allowing the successive isolation of soluble oligosaccharides (free oligosaccharides, fOSs) and dolichol pyrophosphate-linked oligosaccharides (LLO), leaving a delipidated protein (and glycoprotein) pellet. The power of azide- or alkyne-modified sugars resides on their metabolic incorporation into newly synthesized glycoproteins that can then be chemoselectively coupled to complementary alkyne- or azide-functionalized fluorophores and biotinylation reagents for detection or affinity

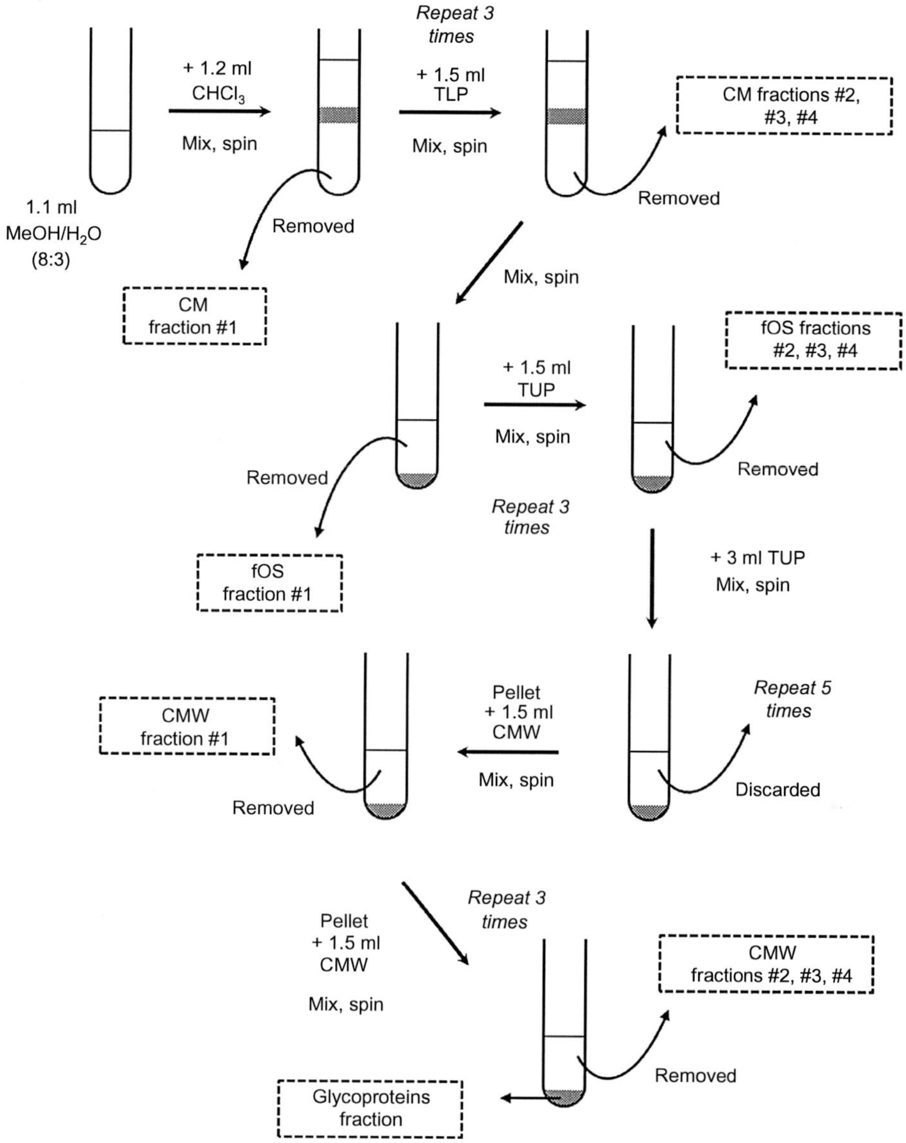

FIGURE 10.1

Sequential extraction procedure. The sequential extraction procedure relies on a biphasic extraction with a mixture of $CHCl_3/MeOH/H_2O$ (3/2/1 by volume), allowing the successive isolation of soluble oligosaccharides (free oligosaccharides: fOSs) and dolichol pyrophosphate-linked oligosaccharides (lipid-linked oligosaccharides: LLO), leaving a delipidated glycoprotein pellet. TLP, theoretical lower phase; CM, chloroform/methanol; TUP, theoretical upper phase; fOS, free oligosaccharides; CMW, chloroform/methanol/water (10/10/3 by volume).

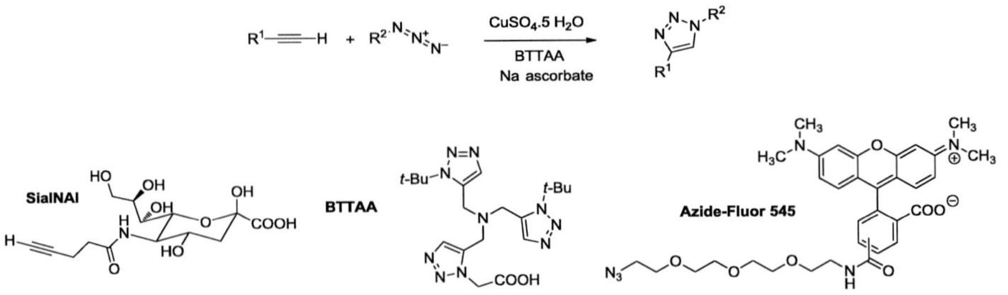

FIGURE 10.2

Huisgen azide–alkyne cycloaddition reaction. Chemical structures of the alkyne-tagged sialic acid analogue, *N*-5-pentynoyl-β-neuraminic acid (SiaNAl), BTTAA, and azide fluor 545.

capture. The small size of these alkyne-modified sugars not only allow them to be properly processed by the glycosylation enzymes but also prevent them from disturbing physiological processes, as encountered with the use of fluorescent tags. The choice of the *N*-acetylneuraminic acid (Neu5Ac) (Fig. 10.2) as the appropriate monosaccharide, instead of its metabolic precursor mannosamine, was made on account of its probable uptake by pinocytosis (Bardor, Nguyen, Diaz, & Varki, 2005). The alkyne-tagged sialic acid (SialNAl) can be incorporated metabolically onto glycoproteins and visualized by the use of azido-functionalized fluorophores (Fig. 10.3; Prescher & Bertozzi, 2005; Vanbeselaere et al., (2013)).

10.1 MATERIALS

10.1.1 To assess ER N-glycosylation efficiency

1. 175 cm^2 Cell Culture Flask (BD Falcon, Cat No. 353112)
2. Dulbecco's Modified Eagle's Medium (DMEM), no glucose (Sigma-Aldrich, Cat. No. D5030-10L)
3. L-Cysteine (Sigma-Aldrich, Cat. No. C7352-25G)
4. L-Methionine (Sigma-Aldrich, Cat. No. M5308-25G)
5. L-Glutamine (Sigma-Aldrich, Cat. No. G8540-25G)
6. D-(+)-Glucose (Sigma-Aldrich, Cat. No. G7021-100G)
7. Pre-incubation medium: DMEM 0% glucose + cysteine (1:1000 of stock solution 100 mg/ml) + methionine (1:1000 of stock solution 15 mg/ml) + glutamine (1:1000 of stock solution 6.5 mg/ml) + glucose at final concentration 0.5 mM
8. D-2-[^{3}H]-mannose (specific activity 15–30 Ci/mmol, Perkin Elmer, Cat. No. NET570A001MC)
9. Dulbecco's phosphate-buffered saline with Ca^{2+} and Mg^{2+} (DPBS) containing NaCl, KCl, Na_2HPO_4, KH_2PO_4 (Lonza, Cat. No. BE17-513F)
10. Methanol (CH_3OH) HPLC grade: 100%
11. Chloroform ($CHCl_3$) HPLC grade: 100%

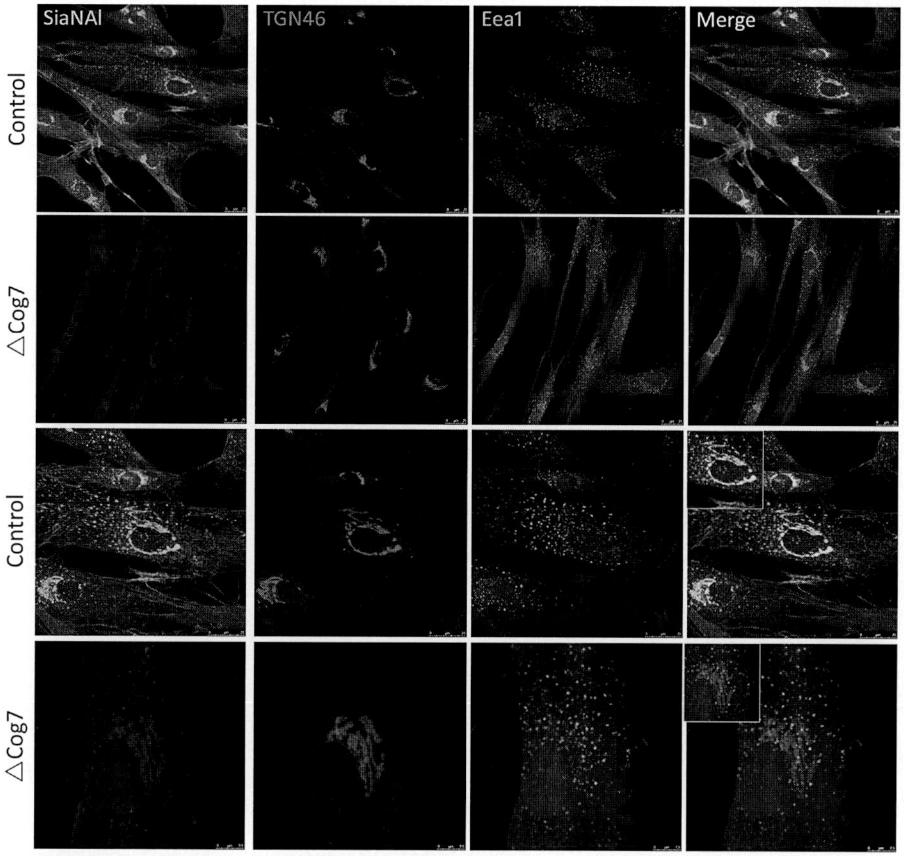

FIGURE 10.3

Example of alkyne-tagged sialic acid incorporation into control fibroblasts versus COG7-deficient CDG patient fibroblasts. Fibroblasts from healthy individuals and COG7-deficient CDG patient were metabolically labeled with 500 μM of SiaNAl for 24 h, fixed, and stained with antibodies against TGN46 (late Golgi marker). Azido fluor 545 as well as Alexa 700-conjugated secondary antibodies allowed visualizing TGN46 (blue), and incorporated alkyne-tagged sialic acid into glycoconjugates in red. Cells were imaged using confocal fluorescence microscopy. (See color plate.)

12. MeOH/H_2O (8/3 by volume) mixture: 8 volumes of methanol and 3 volumes of an aqueous solution containing 0.1 mM Tris/HCl, 4 mM $MgCl_2$, and 3 mg/ml immunoglobulin G in H_2O buffer (pH 7.4)
13. Glass disposable 15-ml centrifuge tubes (Sigma-Aldrich, Cat. No. CLS9950215-125EA)
14. Vortex mixer

15. Pasteur glass pipettes, unplugged, length 230 mm (Fischer Scientific, Cat. No. FB50253)
16. Theoretical lower phase: $CHCl_3$/MeOH/4 mM $MgCl_2$ in H_2O (86/14/1 by volume)
17. Theoretical upper phase: $CHCl_3$/MeOH/4 mM $MgCl_2$ in H_2O (3/48/47 by volume)
18. CMW mixture: $CHCl_3$/MeOH/H_2O (10/10/3 by volume)
19. ULTIMA Gold liquid scintillation cocktail (Perkin Elmer, Cat. No. 6013321)
20. 4 ml Plastic Scintillation Vials (Perkin Elmer, Cat. No. 6000253)
21. Bio-Gel P2 gel (BioRad, Cat. No. 150-4118)
22. 0.1 M acetic acid
23. Rotary evaporator
24. Acetonitrile HPLC grade
25. D-(+)-Mannose (Sigma-Aldrich, Cat. No. M6020-25G): 100 mM
26. QAE-Sephadex A-25 (GE Healthcare, Life Sciences, Cat. No. 17-0190-01)
27. 10 mM Tris/HCl buffer, pH 7.4
28. 20 mM NaCl, 10 mM Tris/HCl buffer, pH 7.4
29. 70 mM NaCl, 10 mM Tris/HCl buffer, pH 7.4
30. 100 mM NaCl, 10 mM Tris/HCl buffer, pH 7.4
31. Alkaline phosphatase from calf intestine (Roche Diagnostics, Cat. No. 10713023001)
32. Milli-Q H_2O
33. Carbograph Ultra-Clean Columns 150 mg (Grace, Cat. No. 510142)
34. 4 mM $ZnCl_2$ in PBS, pH 7.4
35. Inorganic Pyrophosphatase (Roche Diagnostics, Cat. No. 10108987001)
36. Aluminum thin layer chromatography (TLC) plates (Sigma-Aldrich, Cat. No. Z193275)
37. *n*-Butanol/ethanol/acetic acid/pyridine/water (10/100/3/10/30 by volume)
38. Tetrahydrofuran (THF), HPLC grade
39. 9-ml glass tubes 13 × 100 mm with Teflon-lined screw caps (Corning, Cat. No. 9826-13)
40. 0.1 M ammonium bicarbonate buffer, pH 7.9
41. Trypsin from bovine pancreas (Sigma-Aldrich, Cat. No. T1426-100MG)
42. 50 mM phosphate buffer, pH 7.2
43. Recombinant *N*-glycosidase F (Roche Applied Science, Cat. No. 11365169001)
44. ProFSA Plus software (Perkin Elmer)
45. HPLC
46. Asahipak NH2P-50 4E column (Shodex, Cat. No. F7630001)
47. Flo-one Beta detector (Packard)

10.1.2 To assess Golgi N-glycosylation efficiency

1. 12-well plates (ThermoScientific, "Nunc," Cat. No. 150628)
2. Coverslip Circles, Ø 12 mm, 0.13–0.16 mm thickness (ThermoScientific, "Menzel-Gläser," Cat. No. DXD-10151175)

3. DMEM (with Ultraglutamine 1, with 4.5 g/l glucose) (Lonza, Cat. No. BE12-604F)
4. Fetal bovine serum (FBS) (Lonza, Cat. No. DE14-801F)
5. *N*-5-Pentynoyl-β-neuraminic acid (SiaNAl): alkyne-tagged sialic acid: 500 μM (our laboratory but can be purchased from Sigma or Fischer Scientific)
6. Syringe Filters, PES, 33 mm, 0.22 μm (Fisher Scientific, Cat. No. 11977495)
7. Pasteur glass pipettes, unplugged, length 230 mm (Fisher Scientific, Cat. No. FB50253)
8. DPBS with Ca^{2+} and Mg^{2+} containing NaCl, KCl, Na_2HPO_4, KH_2PO_4 (Lonza, Cat. No. BE17-513F)
9. "Cooper" injectable preparation water (Cooper Laboratories: Pharmaxie, Cat. No. 3409528)
10. Paraformaldehyde (PFA): 4% solution in PBS 1 ×, pH 7.3. Add 110 ml of Cooper injectable preparation water into a glass beaker with a magnetic stirring bar, and warm to about 70 °C. Weigh out 8.0 g of PFA and add this to the beaker (under a ventilated hood). Mix for a few minutes and slowly raise the pH by adding 1 N NaOH dropwise to have a clear/transparent solution. Add 8.0 g of sucrose and adjust the final volume to 200 ml with 20 ml of PBS 10 × and 70 ml of Cooper injectable preparation water. Recheck the pH, and adjust it with small amounts of dilute HCl to approximately 7.3. Aliquot the solution and store at −20 °C.
11. Methanol (CH_3OH): 100%
12. Phosphate-buffered saline without Ca^{2+} and Mg^{2+} (PBS 10 ×) containing NaCl, KCl, Na_2HPO_4, KH_2PO_4 (Lonza, Cat. No. BE17-517Q)
13. Plastic box
14. Whatman paper (GE Healthcare, Cat. No. 3030-335)
15. Parafilm (Dutscher, Cat. No. 090260)
16. Permeabilizing solution: 0.5% Triton X-100 in PBS 1 ×. Add 50 μl of Triton X-100 to 10 ml of PBS 1 ×. Stir to mix.
17. Click reaction solution: 50 μM $CuSO_4 \cdot 5H_2O$, 300 μM BTTAA (bis[(*tert*-butyltriazoyl)methyl]-[(2-carboxymethyltriazoyl)methyl]-amine), 2.5 mM ascorbic acid, 100 mM HK_2PO_4, 100 μM azido-545 fluorescent probe. Weigh out 174.2 mg of HK_2PO_4 and dissolve this in 8 ml of H_2O. Adjust the pH to 7.4 with 0.2 M chlorydric acid (HCl). Then add 4.95 mg of ascorbic acid, 100 μl of a 100 × $CuSO_4 \cdot 5H_2O$ solution (100 × = 12.485 mg in 10 ml of H_2O), 1.2906 mg of BTTAA, 0.63 mg of azido-fluor-545. Adjust the final volume to 10 ml with H_2O.
18. Blocking solution: 1.2% gelatin in PBS 1 ×. Weigh out 0.6 g of gelatin. Dissolve in PBS 1 × up to a final volume of 50 ml.
19. Blocking buffer: 2% FBS, 2% bovine serum albumin (BSA), 0.2% gelatin in PBS 1 ×. Weigh out 10.0 g of BSA, 1.0 g of gelatin. Dissolve these in PBS 1 ×. Stir and add 10 ml of FBS and make up to a total volume of 500 ml with PBS 1 ×. Aliquot the solution and store at − 20 °C.
20. Primary rabbit polyclonal antibody against TGN46 (AbDseroTec, Cat No. AHP500G)

21. Primary mouse monoclonal antibody against EEA1 (BD Biosciences, Cat. No. 610456)
22. Primary mouse monoclonal antibody against Lamp 2 (Santa Cruz, Cat. No. sc-18822)
23. Anti-rabbit Alexa Fluor 700 antibody (Invitrogen Molecular Probes, Cat. No. A21038)
24. Anti-mouse Alexa Fluor 488 antibody (Invitrogen Molecular Probes, Cat. No. A11001)
25. Mowiol Mounting Medium: Add 2.4 g of Mowiol 4–88 to 6 g of glycerol. Mix well and add 6 ml of H_2O and leave for several hours at room temperature. Add 12 ml of 0.2 M Tris–Cl (pH 8.5) and heat to 50 °C for 10 min with occasional mixing. After the Mowiol dissolves, clarify by centrifugation at 5000 × *g* for 15 min. Aliquot and store at −20 °C.
26. Microscope Superfrost Slides, 76 × 26 mm (ThermoScientific, Cat. No. 10143560WCUT)
27. Tweezers

10.2 METHODS

10.2.1 To assess ER N-glycosylation efficiency

10.2.1.1 Metabolic labeling

Metabolic labeling of fibroblasts from healthy individuals or isolated from CDG patients starts by growing the cells (8×10^6) overnight in a 175-cm^2 tissue culture flask. Cells should reach about 90% of confluence for optimal labeling.

The day after, remove the medium from the tissue culture flask and gently wash cells with 5 ml pre-incubation medium containing 0.5 mM glucose. Then, pre-incubate cells for 45 min at 37 °C with 5 ml of this medium.

To pulse radiolabel cells, first remove the pre-incubation medium and incubate the cells for 1 h at 37 °C with 5 ml labeling medium containing 150 μCi of tritiated mannose.

Once the labeling period is over, remove and properly discard the labeling medium. Terminate the reaction by immediately washing the cell monolayer three times with 10 ml of ice-cold PBS. Then, add 1.1 ml of a mixture made of 8 volumes of methanol and 3 volumes of 0.1 mM Tris/HCl, pH 7.4, 4 mM $MgCl_2$ in H_2O containing 3 mg/ml of immunoglobulin G as carrier protein (MeOH/H_2O: 8/3 by volume) to scrape cells. Transfer the scraped cells to a 15-ml glass centrifuge tube and add 1.2 ml of chloroform to obtain a final mixture of chloroform/methanol/water (3/2/1 by volume, the volume of the cells being estimated as 0.1 ml).

Thoroughly mix (15 s) with a Vortex mixer before to proceed to the sequential extraction procedure described hereafter. If necessary, the obtained lysate can be stored at −20 °C, by sealing the tubes with Parafilm to prevent solvent evaporation.

NB: For chase experiments, the radioactive culture medium should be replaced by a medium containing 5 mM glucose and 5 mM mannose.

10.2.1.2 Sequential extraction procedure

This sequential extraction procedure relies on a biphasic extraction with a mixture of $CHCl_3$/MeOH/H_2O (3/2/1 by volume). Our method was applied to fibroblasts from healthy individuals and to a panel of fibroblast isolated from known CDG-I patients. The sequential extraction procedure is as follows and summarized in the flowchart in Fig. 10.1.

After thoroughly mixing (15 s) with a Vortex mixer, centrifuge the cell lysate for 7 min at 3900 rpm. Carefully remove the lower phase with a Pasteur pipette and transfer it to a 15-ml glass centrifuge tube. This phase, called CM fraction, contains the labeled dolichol phosphate mannose (Man-P-Dol). Wash then both the upper phase and the protein interphase by adding 1.5 ml of theoretical lower phase ($CHCl_3$/MeOH/4 mM $MgCl_2$ in H_2O; 86/14/1 by volume). Thoroughly mix (15 s) with a Vortex mixer, and centrifuge the mixture for 3 min at 3900 rpm. Remove the newly obtained CM extract with a Pasteur pipette and pool it with the initial CM fraction. Repeat this washing step twice and gather all CM fractions.

Following collection of the last CM extract, thoroughly mix (15 s) the remaining upper phase and protein interface with a Vortex mixer, and centrifuge the mixture for 3 min at 3900 rpm to pellet the protein fraction. Carefully remove the upper phase with a Pasteur pipette and transfer it to a new 15-ml glass centrifuge tube. This phase, called fOS fraction, contains the free soluble oligosaccharides. To collect all fOSs from the mixture, add 1.5 ml of theoretical upper phase ($CHCl_3$/MeOH/4 mM $MgCl_2$ in H_2O; 3/48/47 by volume). Thoroughly mix (15 s) with a Vortex mixer, and centrifuge the mixture for 3 min at 3900 rpm. Remove the newly obtained fOS extract with a Pasteur pipette and pool it with the initial fOS fraction. Repeat this washing step twice and gather all fOS fractions.

Finally, wash the protein pellet by adding 3 ml of theoretical upper phase ($CHCl_3$/MeOH/4 mM $MgCl_2$ in H_2O; 3/48/47 by volume). Thoroughly mix (15 s) with a Vortex mixer, and centrifuge the mixture for 3 min at 3900 rpm. Carefully remove the upper phase with a Pasteur pipette and discard it. Repeat this washing step five times, discarding the upper phase each time. Be careful to not disturb the protein pellet. To eliminate any solvent from the protein pellet, centrifuge the mixture for 3 min at 3900 rpm and carefully remove all solvents in the tube with a Pasteur pipette.

The last step of this sequential extraction procedure consists in the solubilization of the pyrophosphate dolichol-linked oligosaccharides (LLO) by the addition of $CHCl_3$/MeOH/H_2O (CMW, 10/10/3 by volume) from the protein pellet. Consequently, add 1.5 ml of CMW to the protein pellet. Thoroughly mix (30 s) with a Vortex mixer, and centrifuge for 3 min at 3900 rpm. Carefully remove the supernatant, called CMW fraction, with a Pasteur pipette and transfer it to a 15-ml glass centrifuge tube. Extract the pellet two additional times by adding 1.5 ml of CMW. Thoroughly mix (30 s) with a Vortex mixer, centrifuge for 3 min at 3900 rpm, and gather all CMW fractions. Importantly, all solvents should then be removed from the protein pellet, consisting in the glycoprotein fraction.

The fOS soluble extract, CMW extract, and protein pellet will be further treated for radioactivity counting and characterization (see below).

10.2.1.3 Purification and quantification of LLO, fOSs, and glycoproteins

The complete solubilization of samples into the scintillation liquid, as well as the complete elimination of quenching agents such as $CHCl_3$, is necessary for accurate determination of dpm.

10.2.1.3.1 Purification of neutral oligosaccharides

Remove solvent from the fOS fraction by placing the tube sample under a stream of nitrogen, and dry the fraction until about 1 ml is left in the tube.

The next step aims to purify and desalt the oligosaccharides. Load the fOS fraction on a Bio-Gel P2 column (1 cm diameter, 50 cm in length) using 0.1 M acetic acid at a flow rate of 6 ml/h for elution. In order to identify fractions containing the purified oligosaccharides, dilute 50 μl of each eluted fraction (1 ml in total) in 500 μl of the scintillation liquid. Free oligosaccharides are usually eluted within fractions 20–45. Then, measure the radioactivity for these fractions and pool all the fOS fractions.

Finally, use a rotary evaporator to dry the pooled fractions before to dissolve the fOSs in acetonitril/H_2O mixture (70/30 by volume) to obtain a final concentration of 50,000 dpm in 25 μl. To avoid excessive loss of the fOSs content, we recommend to first coat the glass of the evaporation flask with 15 μl of 100 mM mannose.

Structural analysis of fOSs can then be performed by HPLC as described further.

10.2.1.3.1.1 Ion-exchange chromatography of fOSs. Ion-exchange chromatography of fOSs allows to evidence the formation of free, phosphorylated oligosaccharide structures. Following the desalting step on Bio-Gel P2 columns, dissolve the dried fOSs in 1 ml of 10 mM Tris/HCl, pH 7.4. Next, equilibrate a 3-ml QAE-Sephadex A25 column (prepared according to the manufacturer's instructions) with 10 mM Tris/HCl, pH 7.4. Load the dissolved fOSs on the column, which is now ready for use. Then, wash the column with 1 ml of 10 mM Tris/HCl, pH 7.4. Proceed to four additional washing steps by adding 2 ml of 10 mM Tris/HCl, pH 7.4.

Elution is then achieved using an increasing salt gradient. Elute the phosphorylated oligosaccharides stepwise with increasing concentrations of NaCl, ranging from 20 to 100 mM, in 10 mM Tris/HCl. Initiate the elution with 2 ml of 10 mM Tris/HCl, pH 7.4 and 20 mM NaCl. Repeat this elution step four times, and pool the eluted phosphorylated oligosaccharides. Further proceed to the elution by increasing the NaCl gradient to 70 mM. Add 2 ml of 10 mM Tris/HCl, pH 7.4 and 70 mM NaCl to the column for further elution. Repeat this step four more times and pool the eluted phosphorylated oligosaccharides. Finally, elute the remaining phosphorylated oligosaccharides by adding five successive times 2 ml of 10 mM Tris/HCl, pH 7.4 and 100 mM NaCl to the column. Collect together the 10 last eluted fractions before to proceed to the dephosphorylation of the oligosaccharides and HPLC structural analysis.

10.2.1.3.1.2 Dephosphorylation of phosphorylated oligosaccharides. To confirm the charged nature of these fOSs, an *in vitro* alkaline phosphatase treatment can be performed, before to subsequently reanalyze the products by ion-exchange chromatography.

First, combine the phosphorylated oligosaccharides obtained at the end of the ion-exchange chromatography (70 and 100 mM NaCl elution steps) and completely dry the pooled fractions using a rotary evaporator. Dissolve then the phosphorylated oligosaccharides in 200 μl Milli-Q H_2O. Verify that the pH has a value of 8.0 and adjust with HCl if necessary.

To digest the phosphorylated oligosaccharides, add 20 units of calf alkaline phosphatase and incubate the reaction overnight at 37 °C in a water bath. The following day, purify the reaction on a carbograph column according to the manufacturer's instructions.

After alkaline phosphatase digestion, structural analysis of phosphorylated oligosaccharides can be performed by HPLC as described further.

10.2.1.3.1.3 TLC of monophosphorylated and pyrophosphorylated species. To discriminate between monophosphorylated and pyrophosphorylated species, the collected fOSs fractions (neutral and phosphorylated) need to be subjected to a treatment with an inorganic pyrophosphatase.

After desalting on Bio-Gel P2 columns, collect the fOS fractions, evaporate and dissolve the dried oligosaccharides in 200 μl PBS, pH 7.4, containing 4 mM of $ZnCl_2$. To perform the inorganic pyrophosphatase treatment, add 5 units inorganic pyrophosphatase and incubate the reaction overnight at 37 °C in a water bath. For alkaline phosphatase treatment, proceed as described above.

Load 10 μl of the reaction on a TLC plate and develop the plate in *n*-butanol/ethanol/acetic acid/pyridine/water (10/100/3/10/30 by volume) for 2 h. Heat-dry the plate before redeveloping it with the same solvent mixture for a further 2 h. To identify the oligosaccharide species, separate the TLC plate into equal pieces and scrape them for direct scintillation counting.

10.2.1.3.2 Release of the oligosaccharide moiety from oligosaccharide-PP-Dol

Mild acid treatment cleaves the oligosaccharide moiety from the dolichol pyrophosphate carrier. We recommend using 0.1 M HCl in THF.

First, completely evaporate the CMW fraction to dryness under a stream of nitrogen. After drying, dissolve the lipid extract in 1 ml of 0.1 M HCl in THF in sealed-capped tubes. Maintain the tubes for 2 h at 50 °C in a water bath.

Next, purify the released labeled oligosaccharides by depositing the total hydrolyzate (1 ml) on a Bio-Gel P2 column (1 cm diameter, 50 cm in length), using 0.1 M acetic acid at a flow rate of 6 ml/h for elution. As described above (see Section 10.2.1.3.1), measure the radioactivity for the eluted fractions and pool all the fractions containing the released oligosaccharides.

Finally, dry the pooled fractions and dissolve the released oligosaccharides in acetonitril/H_2O mixture (70/30 by volume) to obtain a final concentration of 50,000 dpm in 25 µl.

Structural analysis of the released labeled oligosaccharides can then be performed by HPLC as described further.

10.2.1.3.3 Release of the oligosaccharide moiety from newly synthesized glycoproteins

Even if chemical cleavage can be used to release the newly formed oligommanoside-type glycans, we recommend releasing the oligosaccharide moiety from newly synthesized glycoproteins with the endoenzyme peptide-*N*-glycanase (PNGase).

First, proceed to the proteolytic digestion of the protein pellet. Begin by completely drying the pellet under a stream of nitrogen. Then, resuspend the dried pellet in 300 µl of 0.1 M ammonium bicarbonate buffer, pH 7.9, containing 1 mg/ml of trypsin. Digest overnight at room temperature. The following day, inactivate trypsin activity by incubation for 10 min at 100 °C before to put the sample under a stream of nitrogen to eliminate the ammonium bicarbonate.

Once the proteins digested, dissolve the dried pellet in 100 µl of 50 mM phosphate buffer, pH 7.2 and 0.5 units of PNGAse F. Incubate the glycoproteins for 4 h at 37 °C, to release the oligosaccharides from the peptides.

To purify the released labeled oligosaccharides, add 1 ml of 50 mM phosphate buffer, pH 7.2 to the reaction and deposit the total volume (1.1 ml) on a Bio-Gel P2 column. Proceed then as described in Section 10.2.1.3.1.

Structural analysis of the released labeled oligosaccharides can then be performed by HPLC as described further.

10.2.1.4 HPLC separation of the purified oligosaccharides

To separate the oligosaccharides by HPLC, we use an amino derived Asahipak NH_2-P50 column (250 × 4.6 mm) and apply a gradient of acetonitrile/H_2O ranging from 70:30 to 50:50 over 90 min at a rate flow of 1 ml/min. Oligosaccharides are identified on the basis of their retention time, compared to standard glycans. Elution of the labeled oligosaccharides is then monitored by continuous β-counting with a flo-one β detector.

To analyze HPLC chromatograms we use the ProFSA software. To compare the percentage of specific oligomannoside structures, a fixed amount (50,000 dpm) of radioactivity is always injected into HPLC. The counts in each peak are calculated on the basis of the peak area and normalized against the total number of counts in the injected sample. Moreover, the differences in mannose residues between the different oligomannoside structures always have to be taken into account. For instance, the radioactivity associated to Man5 species hence needs to be multiplied by 9/5 to be comparable to the radioactivity associated to Man9 species.

10.2.2 To assess Golgi N-glycosylation efficiency

10.2.2.1 Cells metabolic labeling with alkyne-tagged sialic acid and fixation

Our method can be applied to fibroblasts from healthy individuals to a panel of fibroblasts isolated from known CDG-I and II patients or any other cell lines.

To start, seed the cells into 12-well plates containing sterile glass coverslips and cultivate them in 10% FBS/DMEM at 37 °C.

The day after, incubate all cell lines with 500 μl/well of growth medium supplemented with alkyne-tagged sialic acid (500 μM SiaNAl) and clarified through a 0.22-μm syringe filter for 16 h at 37 °C. Then, wash the cells three times with 2 ml/well of DPBS and fix them with either 500 μl/well of PAF for 30 min at room temperature or 500 μl/well of methanol (−20 °C) for 10 min at room temperature. Next, rinse the fixed cells three times with 2 ml/well of PBS 1×.

10.2.2.2 "Click chemistry": Probe labeling reactions

Prepare a moistened Whatman paper into a plastic box and place then a sheet of Parafilm on the top of the wet paper. With tweezers, transfer the coverslips from cell culture plates onto the Parafilm and wash the cells three times with ~200 μl/coverslip of PBS 1×.

To start the probe labeling reaction, incubate the fixed cells with 100 μl/coverslip of the permeabilizing solution (0.5% Triton X-100 in PBS 1×) for 10 min at room temperature (skip this step for methanol fixed cells). Then wash these cells three times with ~200 μl/coverslip of PBS 1×.

Label all cell lines with 100 μl/coverslip of click reaction solution (100 μM azido-545 fluorescent probe/50 μM $CuSO_4 \cdot 5H_2O$/300 μM BTTAA/2.5 mM ascorbic acid/100 mM HK_2PO_4, in water) for 45 min at room temperature in the dark under gentle shaking.

10.2.2.3 Immunolabeling

To start the immunolabeling, first wash the fixed and labeled cells twice with 100 μl/coverslip of 1.2% gelatin in PBS 1× for 5 min at room temperature in the dark. Then, rinse the cells three times with ~200 μl/coverslip of PBS 1×. Next, incubate the cells with 100 μl/coverslip of the blocking buffer (2% FBS, 2% BSA, 0.2% gelatin in PBS 1×) for 30 min at room temperature in the dark.

Finally, wash the cells three times with ~200 μl/coverslip of PBS 1×. Add either the primary polyclonal antibody against TGN46 with the primary monoclonal antibody against EEA1 or the primary polyclonal antibody against TGN46 with the primary monoclonal antibody against Lamp 2. All antibodies are diluted 1:100 in blocking buffer. Usually, 50 μl of the diluted antibodies is enough to cover the coverslip surface. Incubate the cells with the primary antibodies for 1 h at room temperature in the dark. Then wash the cells three times with ~200 μl/coverslip of PBS 1×. Next, add the secondary antibodies anti-rabbit Alexa Fluor 700 and anti-mouse Alexa Fluor 488 (diluted 1:600 in blocking buffer), and incubate for 1 h at room temperature in the dark.

10.2.2.4 Microscopic analysis of fluorescent labeling in cells

Wash the immunolabeled cells three times with ~200 μl/coverslip of PBS 1×. Then, for each coverslip, add a drop of Mowiol mounting medium (~6 μl) on microscope slides. With tweezers, dip the coverslips into deionized water several times to rinse them and immediately place the coverslips, cells side facing mowiol drop, on the microscope slides. Let dry at room temperature in the dark.

We use the Leica SP5 laser scanning confocal microscope to image the cells and to quantify the associated fluorescence to alkyne-tagged sialic acid incorporated into newly synthesized Golgi glycoproteins.

10.3 DISCUSSION

Over the past 30 years and thanks to improvements of chemical, physical, and enzymatic approaches, the chemistry of glycoconjugates has rapidly progressed, leading to the determination of the primary structure of several hundreds of glycans. Such analyses have provided the basis for a detailed knowledge of the mechanisms of glycan biosynthesis and catabolism and to the approach of the cellular function of these glycoconjugates. Distributed within cellular membranes, extracellular matrix, and biological fluids, protein *N*-glycosylation is certainly one of the most important post-translational modifications. In between the ER and Golgi compartments, this biosynthetic process not only requires an amazing number of enzymes, chaperones, and lectins whose functions delicately secure the fidelity of glycan structures but also needs to be tightly regulated. The crucial requirement for glycosylation regulation in the intricate design of life can dramatically be illustrated by a group of inherited human disorders named congenital disorders of glycosylation (CDG). To date, many different types of CDG have been defined genetically with defects impairing either the biosynthesis (CDG-I) or the addition and remodeling of *N*-glycans (CDG-II). These glycan structure changes are not only encountered in CDG but are found in many other different diseases, such as metabolic diseases, cancers, and autoimmune diseases. The identification of these glycan changes is difficult and relies either on the use of reporter glycoproteins such as transferrin or the use of HPLC and mass spectrometry (MS). *N*-Glycome fingerprinting by MS has indeed been very successful with CDG, ovarian and colon cancer, and liver fibrosis (Amano & Nishimura, 2010; Klein, 2008). Most of these analyses have been done on purified glycoproteins and whole-serum *N*-glycans after glycans' permethylation. While extremely powerful, the main limitation of such approach comes from the fact that these methods are almost exclusively performed on plasma samples and do not reflect the potential ER/Golgi glycosylation deficiency in cultured cells. The elucidation whether the cells are deficient or not in ER/Golgi glycosylation is, however, crucial to envisage and perform further biochemical and cell biological experiments. The two protocols outlined here, both based on metabolic labeling of cells, provide a detailed guide for analyzing both ER and Golgi glycosylation efficiency in a relatively short time

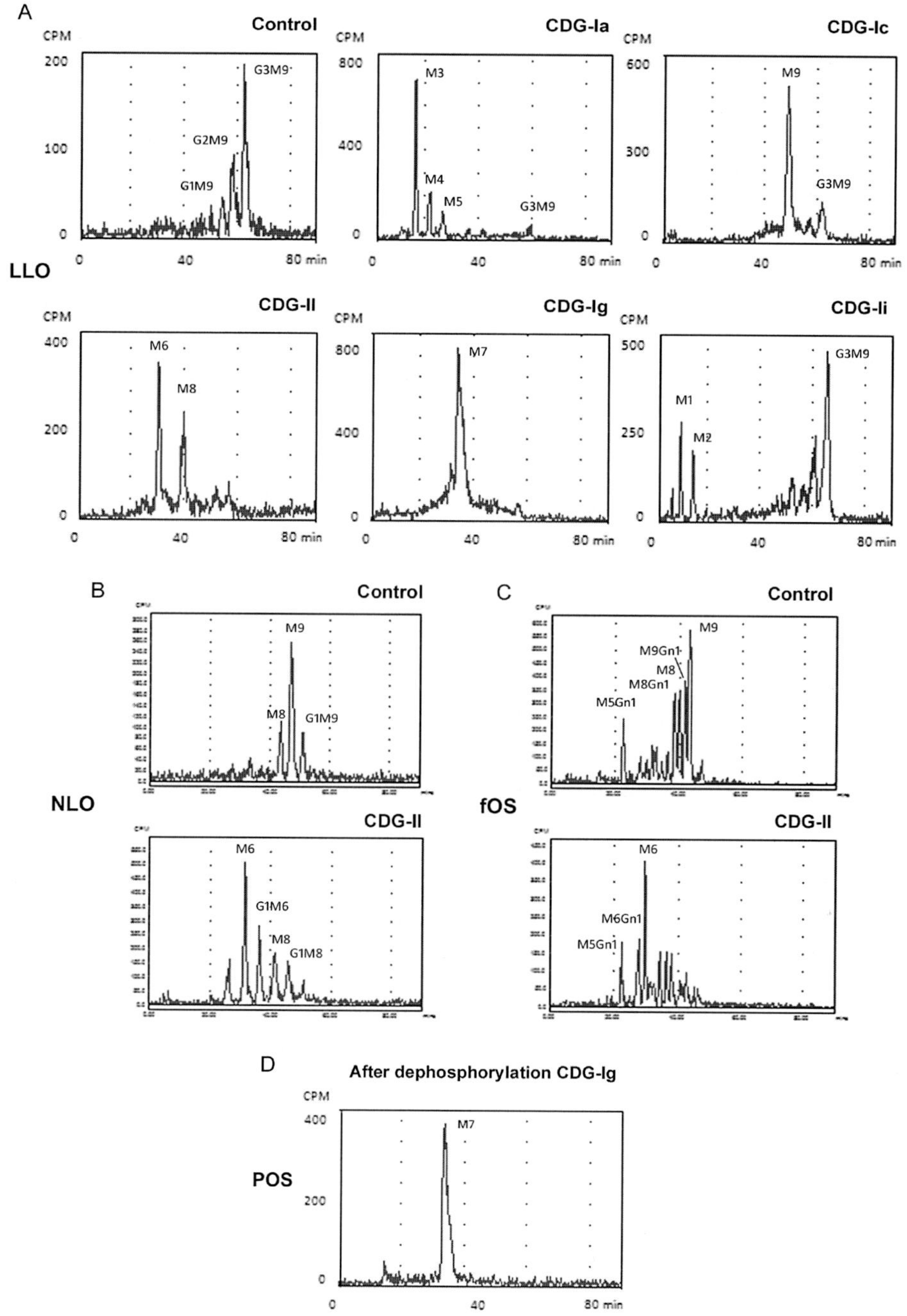

FIGURE 10.4

Example of HPLC separation and analysis of purified oligosaccharides. (A) For six different CDG-I patients' cells, LLO structures were analyzed after metabolic labeling and sequential extraction. HPLC analysis of LLO revealed mainly the fully assembled oligosaccharide precursor Glc3Man9GlcNAc2-PP-Dol in control cells. PMM2-CDG (CDG-Ia) cells accumulated Man3GlcNAc2-PP-Dol on LLO, while ALG6-CDG (CDG-Ic) cells accumulated the LLO intermediate Man9GlcNAc2-PP-Dol. ALG12-CDG (CDG-Ig) cells

Continued

and in cultured cells. To evaluate any ER glycosylation deficiency, the use of 2-[^{3}H]-mannose as sugar is extremely convenient since it is incorporated into lipid-linked oligosaccharides, *N*-linked glycoproteins, and soluble fOS. Considering the balance between their hydrophobic moiety (dolichol) and their hydrophilic part (glycan), the different compounds can be easily isolated using a sequential extraction procedure based on their solubility in organic solvent. The fractionation and analysis of free, protein- and/or dolichol-linked oligosaccharides on the basis of their retention time using HPLC can be very useful in the identification of glycosylation ER defects (Fig. 10.4). The use of amino bound HPLC columns to separate and differentiate oligosaccharides was introduced in the early 1980s (Mellis & Baenziger, 1981). To precisely determine the structures of the different oligosaccharide structures, sequential exoglycosidase treatment can be used. Analyses of the level and structure of the different oligosaccharide intermediates can identify the defects along the ER biosynthetic pathway of *N*-glycans. As fOSs are generated in the ER and cytosol during ER-associated glycoprotein degradation, their level and analysis may also reveal defects along the ER degradation pathway of *N*-glycans (Alonzi et al., 2013; Hirayama, Seino, Kitajima, Jigami, & Suzuki, 2010).

The second method offers a new and mostly unexplored way to better understand the molecular mechanisms of Golgi glycosylation and to quantitatively assess the Golgi glycosylation efficiency. The used SialNAl as an alkyne-labeled monosaccharide provided a highly specific approach for studying Golgi glycosylation efficiency, through *in vivo* metabolic labeling and chemoselective ligation. When supplied to cells, the SialNAl is incorporated into Golgi glycoconjugates. Thanks to the alkyne group, the sialic acid incorporation can be visualized and quantified using azido-activated reagents (Huisgen reaction) such as fluorescent probes. There are no doubts that the application of click chemistry to carbohydrates has tremendous implications for cell biological approaches. The protocol that we decided to share allows the identification of Golgi glycosylation deficiencies in any cultured mammalian cells. The use of the COG7-deficient CDG patient known to present strong sialylation deficiencies illustrates perfectly the power of the assay to assess and quantify the sialylation defect in cultured mammalian cells (Fig. 10.3). Altogether, this method is fast, reliable, easy to handle, quantifiable, and does not need large amount of cells.

FIGURE 10.4—Cont'd accumulated Man7GlcNAc2-PP-Dol on LLO. Finally, ALG2-CDG (CDG-Ii) cells accumulated the small LLO intermediates Man1GlcNAc2-PP-Dol and Man2GlcNAc2-PP-Dol, while ALG9-CDG (CDG-IL) cells accumulated Man6GlcNAc2-PP-Dol and Man8GlcNAc2-PP-Dol. (B) Example of HPLC analysis of the oligosaccharides linked to proteins (NLO) in control fibroblasts and in ALG9-CDG (CDG-IL) cells accumulating Man6GlcNAc2-PP-Dol and Man8GlcNAc2-PP-Dol. (C) Example of HPLC analysis of the free oligosaccharides in control fibroblasts and in ALG9-CDG (CDG-IL) cells. (D) Example of HPLC analysis of phosphorylated oligosaccharides (POS) structure formed in ALG12-CDG (CDG-Ig) cells after dephosphorylation. Symbols: M5–9, Man5–9GlcNAc2; G1–3M9, Glc1–3Man9GlcNAc2; M5–9Gn1, Man5–9GlcNAc1. (See color plate.)

SUMMARY

For the past decades, our knowledge on glycosylation has been progressing from the classic description of glycan structures to the understanding of not only their biological role but also of the intimate mechanisms regulating both their biosynthesis and regulation. *N*-Glycosylation occurs at the interface of the ER and the Golgi apparatus, and is the most widespread form of glycosylation responsible for the modification of secreted and transmembrane proteins. Highly diverse structures, *N*-glycans are indeed ideal and versatile molecules to ensure specific cellular and extracellular functions. Consequently, abnormal glycan structures are not only encountered in specific glycosylation disorders (CDG) but are also found in many other different diseases, such as metabolic diseases, cancers, and autoimmune diseases. Though, the identification of these glycan changes is difficult and relies either on the use of reporter glycoproteins such as transferrin or the use of HPLC and mass spectrometry. Nevertheless, such approaches do not reflect the potential ER/Golgi glycosylation deficiency in cultured cells, such as primary skin biopsies. Therefore, our two protocols outlined here both rely on the use of metabolic labeling of cultured cells, for a relatively fast and accurate way to quantitatively analyze ER and, mostly unexplored so far, Golgi glycosylation processes in cultured cells. We believe that these methods should be expanded to any platform and community willing to easily and efficiently enhance glycosylation deficiencies in mammalian cultured cells, and as well could serve for CDG diagnosis.

Acknowledgments

The authors wish to thank the Bioimaging Core Facility of Lille (Institut de Recherche Interdisciplinaire, CNRS). This work was supported by a grant from the Agence Nationale de la Recherche (ANR-JC-SWEET-CDG) to F. F., by grants from the Research Foundation (FWO) Flanders (G.0553.08 and G.0505.12) to G. M., and by Grant ERARE 11-135 (EURO-CDG) of the ERA-Net for Research Programmes on Rare Diseases Joint Transnational Call 2011 to F. F. and G. M. In addition, R. P. is a postdoctoral researcher (FWO Pegasus Marie Curie Fellow) of the FWO and J. V. got a grant from the Ministère de l'Enseignement Supérieur.

References

Alonzi, D. S., Kukushkin, N. V., Allman, S. A., Hakki, Z., Williams, S. J., Pierce, L., et al. (2013). Glycoprotein misfolding in the endoplasmic reticulum: Identification of released oligosaccharides reveals a second ER-associated degradation pathway for Golgi-retrieved proteins. *Cellular and Molecular Life Sciences*, *70*, 2799–2814.

Amano, M., & Nishimura, S. (2010). Large-scale glycomics for discovering cancer-associated N-glycans by integrating glycoblotting and mass spectrometry. *Methods in Enzymology*, *478*, 109–125.

Bardor, M., Nguyen, D. H., Diaz, S., & Varki, A. (2005). Mechanism of uptake and incorporation of the non-human sialic acid N-glycolylneuraminic acid into human cells. *The Journal of Biological Chemistry*, *280*, 4228–4237.

Hart, G. W., Holt, G. D., & Haltiwanger, R. S. (1988). Nuclear and cytoplasmic glycosylation: Novel saccharide linkages in unexpected places. *Trends in Biochemical Sciences*, *13*(10), 380–384, Review.

Hirayama, H., Seino, J., Kitajima, T., Jigami, Y., & Suzuki, T. (2010). Free oligosaccharides to monitor glycoprotein endoplasmic reticulum-associated degradation in Saccharomyces cerevisiae. *The Journal of Biological Chemistry*, *285*(16), 12390–12404.

Jaeken, J., & Matthijs, G. (2007). Congenital disorders of glycosylation: A rapidly expanding disease family. *Annual Review of Genomics and Human Genetics*, *8*, 261–278, Review.

Klein, A. (2008). Human total serum N-glycome. *Advances in Clinical Chemistry*, *46*, 51–85, Review.

Mellis, S. J., & Baenziger, J. U. (1981). Separation of neutral oligosaccharides by high-performance liquid chromatography. *Analytical Biochemistry*, *114*(2), 276–280.

Panneerselvam, K., Etchison, J. R., & Freeze, H. H. (1997). Human fibroblasts prefer mannose over glucose as a source of mannose for N-glycosylation. *The Journal of Biological Chemistry*, *272*, 23123–23129.

Prescher, J. A., & Bertozzi, C. R. (2005). Chemistry in living systems. *Nature Chemical Biology*, *1*, 13–21.

Rosnoblet, C., Peanne, R., Legrand, D., & Foulquier, F. (2013). Glycosylation disorders of membrane trafficking. *Glycoconjugate Journal*, *30*(1), 23–31.

Schachter, H. (2000). The joys of HexNAc. The synthesis and function of N- and O-glycan branches. *Glycoconjugate Journal*, *17*(7–9), 465–483, Review.

Theodore, M., & Morava, E. (2011). Congenital disorders of glycosylation: Sweet news. *Current Opinion in Pediatrics*, *23*(6), 581–587.

Vanbeselaere, J., Vicogne, D., Matthijs, G., Biot, C., Foulquier, F., & Guerardel, Y. (2013). Alkynyl monosaccharide analogues as a tool for evaluating Golgi glycosylation efficiency: Application to Congenital Disorders of Glycosylation (CDG). *Chemical Communications (Cambridge, England)*. [Epub ahead of print] http://dx.doi.org/10.1039/c3cc45914d.

PART

Imaging-Based Approaches

2

CHAPTER

Fluorescence-Based Analysis of Trafficking in Mammalian Cells

11

Gaelle Boncompain*,† and Franck Perez*,†

**Institut Curie, Paris, France*

†CNRS UMR144, Paris, France

CHAPTER OUTLINE

Introduction **180**

The Secretory Pathway and the Diversity of Golgi-dependent Trafficking Routes 180

Existing Tools to Monitor the Trafficking of Secretory Cargos 180

VSVGtsO45 181

Temperature Blocks and Reversible Inhibitory Molecules 181

Aggregation Systems (iDimerize, UVR8) 181

The RUSH System to Follow the Trafficking of Any Cargo in Physiological Conditions 182

Key Considerations for the Use of the RUSH System 182

11.1 Methods **184**

11.1.1 Generation of the RUSH Reporter of Interest 184

11.1.2 Generation of RUSH Stable Cell Lines 185

11.1.3 Analysis of the Trafficking of the Protein of Interest in End-point Experiments 187

11.1.4 Analysis of the Trafficking of One or Multiple Cargos in Real Time in Living Cells 188

11.2 Discussion **190**

Summary **192**

Acknowledgments **193**

References **193**

Abstract

Proteins destined for the secretory pathway start their journey in the endoplasmic reticulum and transit through the Golgi complex to be delivered to their destination compartment. Over the last decades, several fluorescence-based assays were developed to analyze the transport of proteins along the secretory pathway. In this review, we briefly introduce the existing tools. We provide detailed protocols to allow the

Methods in Cell Biology, Volume 118

ISSN 0091-679X
http://dx.doi.org/10.1016/B978-0-12-417164-0.00011-2

reader to use the newly developed secretory assay termed the RUSH system (Retention Using Selective Hooks). This assay enables the synchronous release of one to three cargos of interest from a donor compartment (the endoplasmic reticulum). Analysis of the transport steps of the cargos from the donor compartment to the acceptor compartment is accomplished by fluorescence-based methods.

INTRODUCTION

The secretory pathway and the diversity of Golgi-dependent trafficking routes

In mammalian cells, about one-third of the newly synthesized proteins are destined to be secreted. Conventional secretory proteins enter the secretory pathway at the level of the endoplasmic reticulum (ER), and they are then transported to the Golgi apparatus. From there, the secretory proteins are delivered to their final destination compartments (e.g., plasma membrane, extracellular medium, lysosomes) using post-Golgi transport carriers. During their transit through the Golgi complex, the proteins encounter posttranslational modifications such as glycosylation or proteolytic cleavage. The Golgi apparatus also ensures the sorting of proteins in the secretory pathway. An increasing number of regulators have been identified, such as Rab GTPases, SNAREs, or protein coats. The family of proteins named Golgins constitutes one of the key players at the level of the Golgi (Munro, 2011). These proteins ensure the sorting of secretory cargos and are specifically required for the trafficking of certain cargos. For instance, transport of E-cadherin out of the Golgi was shown to be dependent on Golgin-97 (Lock, Hammond, Houghton, Gleeson, & Stow, 2005). The trafficking of TNF is dependent on Golgin-245 but independent of Golgin-97 (Lieu et al., 2008). Other examples of the selective involvement of Golgins exist in the literature (Hicks, Horn, McCaffery, Zuckerman, & Machamer, 2006).

These data mean that it is crucial to follow the transport of the appropriate cargo. In the following sections, this review briefly describes the tools available for monitoring the trafficking of secretory cargos and explains in detail how to use the secretory assay developed by us, named the Retention Using Selective Hooks (RUSH) system. This assay enables monitoring of the trafficking of any protein of interest in physiological conditions and in a population of cells (Boncompain et al., 2012).

Existing tools to monitor the trafficking of secretory cargos

During the last decades, several powerful assays were developed to study the secretory pathway in mammalian cells.

VSVGts045

The thermosensitive mutant of VSVG (VSVGtsO45; Lafay, 1974) is one of the most used tools and has made an undeniable contribution to the field of membrane trafficking. VSVG is a viral glycoprotein that needs to trimerize to be secreted. The VSVGtsO45 mutant cannot trimerize at high temperature and is retained in the ER at the restrictive temperature of 40 °C. It can traffic toward the plasma membrane through the Golgi complex at the permissive temperature of 32 °C (Kreis & Lodish, 1986; Lafay, 1974). Fusion of VSVGtsO45 with fluorescent proteins (e.g., GFP, mCherry) allows real-time observation of its transport. In addition to the nonphysiological temperatures required for the retention and release of VSVGtsO45, the major drawback of this method resides in its capacity to monitor the transport of the VSVG protein only and not of a cargo of interest. A certain level of flexibility is, however, possible with respect to extension of these limits, as the exoplasmic domain of VSVG can be fused to other proteins to synchronize their transport (see Cole, Ellenberg, Song, DiEuliis, & Lippincott-Schwartz, 1998).

Temperature blocks and reversible inhibitory molecules

Methods based on temperature blocks enable accumulation of secretory proteins in a particular intracellular compartment. Incubation of cells at 15 °C leads to accumulation of proteins in a post-ER but pre-Golgi compartment while incubation at 20 °C blocks cargos in the *trans*-Golgi network (Griffiths, Pfeiffer, Simons, & Matlin, 1985; Matlin & Simons, 1983; Saraste & Kuismanen, 1984).

Several chemical compounds can also be used to study some steps of the anterograde trafficking of secretory cargos. The fungal metabolite brefeldin A induces relocation of the Golgi proteins in the ER and thus leads to disassembly of the Golgi complex (Lippincott-Schwartz, Yuan, Bonifacino, & Klausner, 1989). Washout of brefeldin A restores trafficking and building of the Golgi apparatus, allowing for the study of the early steps of the anterograde transport. Another drug, monensin, is employed to stop cargos in the Golgi apparatus. However, the mechanisms of action of such molecules are complex and affect several targets, leading potentially to side effects. In addition, when traffic resumes during washout, it is unclear whether the mechanisms of transport at work are identical to those responsible for steady-state trafficking.

In general, although powerful, both temperature blocks and chemical compounds place the cells in nonphysiological conditions that may perturb cell homeostasis.

Aggregation systems (iDimerize, UVR8)

More recently, a method based on the aggregation of the protein of interest in the ER was developed (Rivera et al., 2000). The cargo is fused to conditional aggregation domains that induce oligomerization of the protein at steady state after its translation and translocation in the ER. The aggregation is abolished by the addition of a small ligand, enabling its trafficking through the secretory pathway. In the same line, conditional aggregation reversible by light (using UV-B) was also developed very recently (Chen, Gibson, & Kennedy, 2013). Its application has so far been confirmed only for VSVG and for single-cell analysis. In addition to the aggregation domains, the cargos are also fused to fluorescent proteins to facilitate their observation in real-time.

This aggregation system has been used successfully by many labs and allows tight synchronization. The main drawback of these assays resides in the method of aggregation itself. If disaggregation is not complete, trafficking of oligomers of the cargo can occur but using pathways different from those of the unaggregated counterpart. In addition, aggregation at the cytoplasmic side of membrane proteins may induce aggresome formation or autophagy. Last, the aggregation domains are long and need to be present in several copies, which may perturb the transport of the tagged cargo.

The RUSH system to follow the trafficking of any cargo in physiological conditions

In order to overcome the drawbacks of the aforementioned methods and to be able to follow the trafficking of any protein of interest in physiological conditions for mammalian cells, we developed a novel assay named the RUSH system (Boncompain et al., 2012). This chapter provides the key considerations and methods for using the RUSH system.

This secretory assay is a two-state method based on the retention of the cargo (the reporter) in a donor compartment (e.g., the ER) without aggregation and its synchronous release. A hook protein that stably resides in the donor compartment ensures the retention of the reporter. The interaction between the hook and the reporter is mediated by streptavidin fused to the hook. A small peptide, the streptavidin-binding peptide (SBP), is fused to the reporter. At steady state without treatment, the streptavidin binds to the SBP. The interaction is released by the addition of the nontoxic vitamin biotin to the medium. The biotin binds to the streptavidin with high affinity and competes out the SBP tag (Fig. 11.1). The reporter is thus free to traffic to its acceptor compartment (usually the Golgi complex and then to the plasma membrane, extracellular milieu, lysosome, etc.). To facilitate visualization, a fluorescent protein (e.g., EGFP, mCherry) or fluorogenic tags (e.g., SNAP-tag, Halo tag) are fused to the reporter.

Because of the high synchronization and the simple pipetting of biotin as an inducer of trafficking, the RUSH system is amenable to automated screening approaches (Boncompain et al., 2012).

The RUSH assay provides the major advantage of the potential to use physiological conditions, in particular, physiological temperatures, for mammalian cells.

Key considerations for the use of the RUSH system

One great benefit of the RUSH system compared to most of the methods described earlier is its versatility. The RUSH assay is usable for virtually any cargo of the secretory pathway. However, the topology of the protein of interest and prior knowledge concerning its successful tagging (if available) are essential. Since this method relies on tagging, at least with the SBP (only 38 amino acids) and, eventually, with

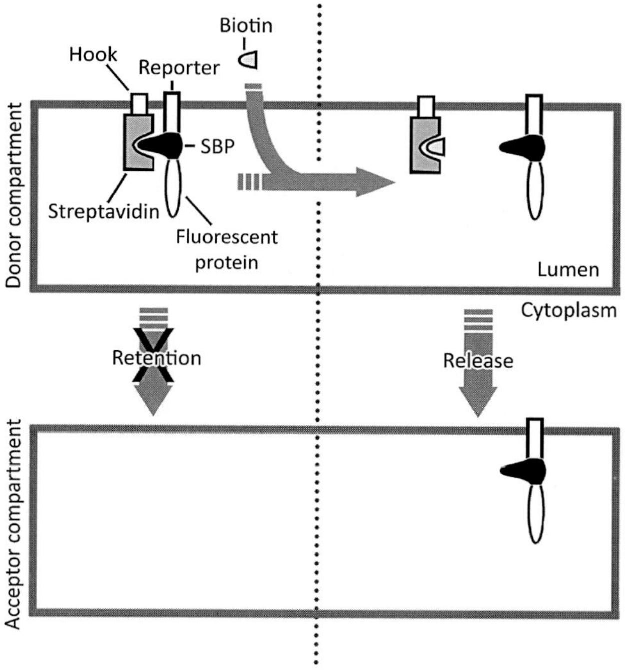

FIGURE 11.1

Principles of the RUSH (Retention Using Selective Hooks) system. At steady state, the reporter is retained in the donor compartment because of its interaction with the hook. Streptavidin fused to the hook and the SBP (streptavidin-binding peptide) fused to the reporter mediate the interaction between the two partners. This interaction is released upon addition of biotin allowing the reporter to traffic to its acceptor compartment.

fluorescent proteins for real-time observation in living cells, tagging the reporter in a way known to not affect the localization of the protein is crucial. Generally, we base our tagging strategies on previous studies using a GFP-tagged version of the cargo. If this has not been done before, and if an antibody is available, it may be worth checking the localization of the cargo at steady state.

The topology of the protein of interest is also a crucial point. Using the RUSH system, the reporter can be retained either at the luminal face of the donor compartment (e.g., the ER) or at its cytoplasmic face. Data on the topology of proteins can be found in the UniProt database (www.uniprot.org). It is absolutely necessary to take the topology into consideration before choosing the type of retention (luminal or cytosolic) and the right hook. If the SBP is placed in the luminal part of the protein, the chosen hook also has to have the streptavidin exposed to the luminal face of the ER to enable the interaction. In the same line, it is important to take into account the data available on the trafficking signals (e.g., signal peptide, motif in the cytoplasmic tail) borne by the cargo of interest in order to not hide them.

11.1 METHODS

11.1.1 Generation of the RUSH Reporter of interest

This method aims at obtaining a RUSH plasmid encoding for the reporter of interest that enables the synchronization of its trafficking using tagging with RUSH interaction domains and a fluorescent protein. The RUSH plasmid already bears the hook. The reporter is placed downstream of an IRES (Internal Ribosome Entry Site) to ensure co-expression of the two partners (Boncompain et al., 2012).

The RUSH reporter is constituted of four cloning cassettes. The two central cassettes (2 and 3) encode for the SBP and the fluorescent protein (see Fig. 11.2). Depending on its topology, the sequence coding for the reporter has to be inserted in cloning cassette 1 or cloning cassette 4.

MATERIALS

RUSH vectors (Boncompain et al., 2012)
cDNA encoding for the protein of interest flanked by convenient restriction sites (amplified by PCR or purchased using gene synthesis)
*Asc*I, *EcoR*I, *Fse*I, and *Pac*I restriction enzymes
Phosphatase and its buffer
T4 DNA ligase and its buffer
DNA electrophoresis equipment
E. coli competent cells
LB agar plates supplemented with ampicillin
Kits for extraction of DNA from agarose gel and from liquid cultures

A

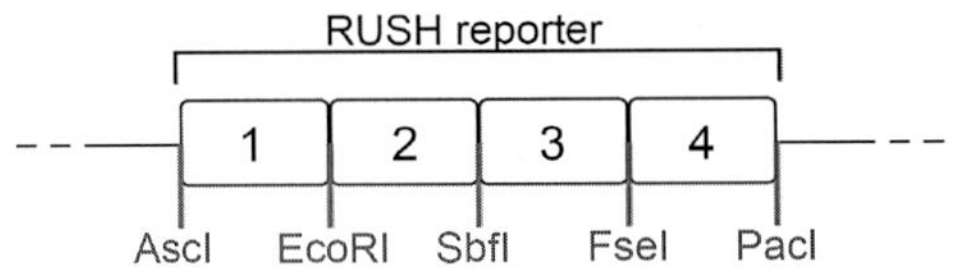

B

Type of retention	Topology of the protein X	Cloning cassettes 1	2	3	4
Luminal	Soluble secreted	Signal peptide	SBP	FP	X without signal peptide
		X with SP	SBP	FP	
	Type I	Signal peptide	SBP	FP	
	Type II	X	SBP	FP	
Cytosolic	Type I	X with SP	SBP	FP	
	Type II		SBP	FP	X

FIGURE 11.2

Cloning cassettes of the RUSH reporter. (A) The gene encoding for the reporter is divided in four cloning cassettes flanked by the indicated restriction sites. (B) The location of the different tags and cDNA is dependent on the topology of the protein of interest (X) and on the orientation of retention.

METHOD

Digest 5 μg of the RUSH vector adapted to the type of retention and to the topology of the protein of interest with the *adhoc* restriction enzymes (see earlier and Fig. 11.2) for 1 h at 37 °C.

Dephosphorylate the digested vector with alkaline phosphatase for 30 min at 37 °C. We use calf intestine phosphatase from Roche (#10 713 023 001).

Run the digested and dephosphorylated vector on an agarose gel and purify the right band from the gel. We generally use the NucleoSpin Gel and PCR Clean-up kit from Macherey–Nagel to perform this step.

Ligate the vector prepared as already described, with the insert coding for the protein of interest flanked by the restriction sites compatible with subcloning into the RUSH vector at the desired position. Use 50 ng of the vector and three times the molar amount of insert in diluted ligation buffer and T4 DNA ligase. Incubate either for 1 h at room temperature or overnight at 16 °C.

Transform *E. coli* competent cells and plate on LB agar plates supplemented with ampicillin.

Incubate the plates overnight at 37 °C to enable the growth of colonies.

Pick colonies and grow liquid cultures in LB containing ampicillin.

Extract the plasmid from these cultures using a kit, such as the NucleoSpin Quickpure from Macherey–Nagel, and according to the manufacturer's instructions.

Check the extracted plasmid by restriction with specific enzymes or by sequencing.

If the sequence of the plasmid corresponds to the expected sequence, a larger culture can be grown (150 mL) and DNA extracted using a kit (NucleoSpin Xtra Midi Plus from Macherey–Nagel).

At this step, the RUSH plasmid is ready to be tested in living cells (see below).

11.1.2 Generation of RUSH stable cell lines

The RUSH assay is usable in transiently transfected cells. However, for the quantitative analysis of the trafficking of a cargo over a population of cells (in particular, for cellular screens), it is advantageous to establish a stable cell line expressing the RUSH construct of interest. We found that lentiviral transduction systems are very efficient tools to construct stable cell lines. This method and the key considerations are described in the following paragraph.

One crucial parameter for the production of the viruses is the size of the DNA needed for encapsulation. It is necessary to keep the size of the insert below 4 kb. As described earlier, the RUSH plasmids for transient transfection encode for the hook and the reporter. The two cassettes are separated by an IRES to obtain co-expression of the two partners. Since it is difficult to get below the 4 kb limit using this bicistronic expression, we often establish stable cell lines in two steps. We first transduce the cells with lentiviruses encoding for the hook construct only. We then select a homogenous pool or a clone of cells stably expressing the RUSH hook.

These cells are then infected with lentiviruses encoding the RUSH reporter and pool, or clones with the desired expression levels are selected. The method that follows details the protocol to produce lentiviruses and infect model cells with these viruses. We describe the procedure only once. The same protocol is applied for generating cells stably expressing a hook (one step) or a hook and a reporter (two steps).

For lentiviral production, we use a pCDH vector from System Biosciences. The vector we chose bears a puromycin selection cassette. The genes encoding for the RUSH protein have to be subcloned into this vector.

To note: lentivirus production and infection of cells involve safety rules. We have first to know how to manipulate lentiviruses in a safe way and in the right environment, depending on the country and the institution.

MATERIALS

35-mm dish (or a well of a 6-well plate)
6-well plates
HEK-293T cells
Culture medium with serum
Packaging plasmids (pMD2.G and psPAX2)
Plasmid encoding the RUSH protein and suitable for lentivirus production
Transfection reagent (e.g., GeneJuice from Novagen)
Cells of interest
Puromycin
0.22 μm filter
Syringe without needle

METHOD

Seed the HEK-293T cells in a 35-mm dish or in one of the wells of a 6-well plate so as to be 50–80% confluent on the day of the transfection.
Transfect the HEK-293T cells with the three plasmids: plasmid encoding the RUSH protein, pMD2.G, and psPAX2. We use GeneJuice from Novagen to transfect HEK-293T cells, but other reagents are also suitable.
Incubate the dish or plate at 37 °C, 5% CO_2 for 72 h.
Harvest the culture supernatant and filtrate with a syringe and a 0.22 μm filter to remove cell debris. We did not find it necessary to concentrate the viruses.
Add 1 mL of this supernatant containing the viruses to the cells to infect those that were previously seeded. We successfully used a wide variety of cells (e.g., HeLa, RPE-1, MDA-MB231, MDCK, THP). We can prepare several wells to use different volumes of the virus solution to tune the expression levels.
Incubate for 24 h at 37 °C, 5% CO_2.
Remove the medium and replace it with medium containing puromycin. For HeLa cells, we classically use 4 μg/mL, but this has to be determined for every cell line. Incubate at 37 °C, 5% CO_2 for 48 h. The nontransduced cells should have died.

Split and expand the cells if necessary keeping puromycin in the culture medium. At this step, we can start to analyze the cells by checking whether they express the RUSH protein and whether the latter behaves as expected.
For the purpose of the experiment, we might need to work with a homogenous expression level over the population of cells. We can either sort cells with the desired expression level using a flow cytometer and/or isolate clones by limiting dilution.
As mentioned earlier, if we would like to express one or several RUSH proteins, we need to repeat the procedure, from the production of lentiviruses in HEK-293T cells to infection of the cells. In this case, the cells to be infected are the ones that already express one or more RUSH proteins.

11.1.3 Analysis of the trafficking of the protein of interest in end-point experiments

The method that follows is intended to evaluate the efficacy of synchronization of the RUSH reporter generated. As explained in the introduction, the release of the RUSH reporter from the hook is induced by the addition of biotin to the culture medium. Using increasing durations of incubation with biotin allows one to follow the transport steps of the cargos and their approximate kinetics.

MATERIALS

Adherent cells (we use HeLa or RPE-1 cells as model cell lines)
Glass coverslips
RUSH plasmid encoding the hook and the reporter of interest
Transfection reagent suitable for the cells (e.g., calcium phosphate, lipofectant)
Biotin (Sigma #B4501)
3% paraformaldehyde solution in PBS
Quenching solution (NH_4Cl 50 mM final in PBS)
Mounting medium (e.g., Mowiol)
Glass slides

METHOD

The day prior to transfection, the cells are seeded on glass coverslips at the adapted confluence to the transfection method to be used.
Transfect the cells with the RUSH plasmid encoding for the hook and cargo of interest. The method of transfection is not described here since it depends on the cells being used and on the material available in the lab.
For one coverslip, add biotin at 40 μM final diluted in culture medium. Prepare a stock solution of biotin at 4 mM (100 ×) diluted in the culture medium. This coverslip, maintained in the presence of biotin since the beginning of transfection, will allow analysis of the steady-state behavior of the cargo: since biotin is present, the hook and the reporter will not interact.

The day after transfection, keep one coverslip in medium without biotin to check the efficacy of the retention. In this condition, the reporter should be localized in the donor compartment (e.g., the ER).
Incubate one coverslip with biotin at 40 μM final for 15 min and another one for 60 min. Based on our experiments, these time points should be sufficient to visualize retention (no biotin), presence in the Golgi complex (15 min), complete or partial arrival at the destination compartment (60 min), and the steady-state localization (coverslip maintained with biotin) of the cargo. To define more precisely the different transport steps of the protein of interest, we can, of course, perform a time course with smaller time intervals. For cargos that traffic more slowly, a longer duration of incubation (2–3 h) with biotin will be required to recapitulate their complete trafficking pathway.
After the desired times of incubation of biotin, fix the cells with 3% paraformaldehyde for 15 min at room temperature. We routinely use paraformaldehyde to fix cells but other methods can be used (such as methanol) depending on one's habits and experiments.
Wash with PBS and neutralize the aldehyde solution using the quenching solution for 5 min at room temperature.
Wash with PBS.
Mount the coverslips on glass slides and let them dry before observation of the signal of the fluorescent protein fused to the reporter with a fluorescence microscope.
This is a basic protocol that aims at observing directly the localization of the reporter. A more complex analysis can be performed using immunofluorescence to label specific cell compartments. In particular, it is often interesting to stain the pool of proteins expressed at the cell surface. In this case, the immunostaining is done without permeabilizing the cells. When the reporter protein is tagged by GFP in its extracellular domain, anti-GFP antibodies can be used to stain the reporter present at the cell surface.

11.1.4 Analysis of the trafficking of one or multiple cargos in real time in living cells

The advantage of the use of a fluorescent protein in fusion with the RUSH reporter is the possibility to image its transport in living cells using time-lapse microscopy. The color palette of the fluorescent proteins available even offers the opportunity to perform multiple color analysis of the trafficking of different reporters at the same time in the same cell. To do so, we can fuse the diverse cargos with different fluorescent proteins following the method already described. In this case, we work in cloning cassette number 3 using *Sbf*I and *Fse*I restriction enzymes (see Fig. 11.2). Given the actual fluorescent proteins available, one can easily perform imaging up to three colors using a blue fluorescent protein (e.g. mTagBFP), a green fluorescent protein (e.g., EGFP, Venus), and a red fluorescent protein (e.g., mCherry, tagRFP, mKate2).

For time-lapse imaging of the transport and one or more cargos, the addition of biotin should be performed under the microscope. This will allow acquiring pictures of the retention state of the reporter prior to the addition of biotin and of the synchronous release of the cargo in the same cells upon addition of biotin. One of the most efficient ways to add biotin during the acquisition is to use a chamber connected with tubing. Using such a system, the culture medium is replaced by culture medium containing 40 μM of biotin ensuring good homogenization of biotin. We use a magnetic chamber with L-shaped tubing from Chamlide (see Fig. 11.3). In these chambers, the cells are seeded on glass coverslips. As an alternative, one can use disposable glass dishes for imaging (e.g., MatTek, Iwaki), but aspiration of the medium and addition of biotin can be more tricky. We now describe a method for real-time imaging using the Chamlide magnetic chamber connected with tubing. To prevent loss of focus during replacement of the medium or during long-term acquisition, we use an autofocus module (e.g., Perfect Focus System). The microscope has to be equipped with a thermostatic chamber or a heating stage to maintain the temperature at 37 °C. The temperature is a key parameter for the kinetics of transport in mammalian cells.

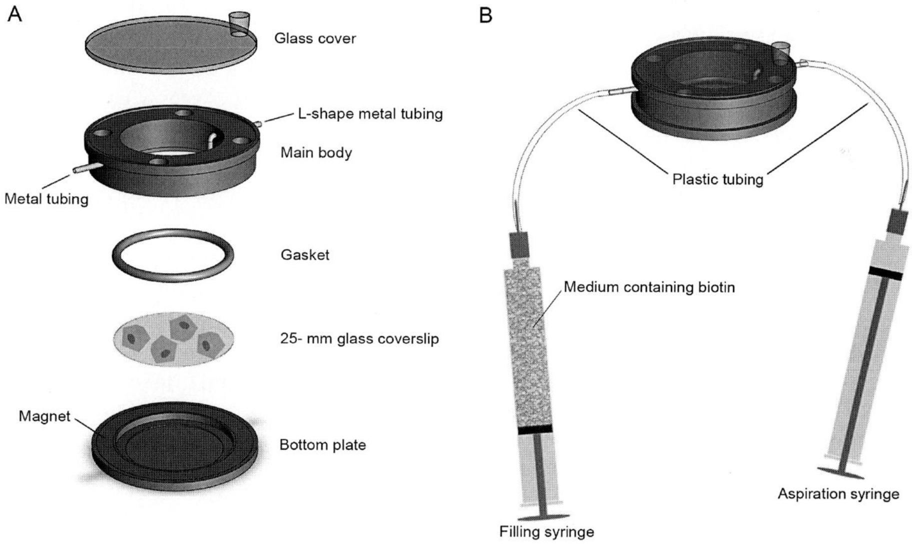

FIGURE 11.3

Schematic representation of the magnetic chamber (Chamlide) used for real-time imaging. (A) Components of the L-shaped Chamlide. (B) View of the assembled L-shaped Chamlide connected to tubing and syringes to perform the addition of biotin during the acquisition under the microscope.

Pictures adapted from www.chamlide.com (Live Cell Instruments).

MATERIALS

Adherent cells
25-mm glass coverslips
Plasmid encoding the RUSH reporter(s) fused to a fluorescent protein
Transfection reagent suitable for the cells
Chamlide with L-shaped tubing (see Fig. 11.3)
Imaging medium (e.g., Leibovitz's)
Imaging medium supplemented with 40 μM of biotin
Syringes and needles to connect plastic tubing
Immersion oil
Fluorescence microscope (we use a spinning disk microscope with inverted objective)

METHOD

Seed the cells on 25-mm glass coverslips the day prior to the transfection at the confluence required for the transfection method selected.
Transfect the cells with the method of your choice so as to introduce from one to three different RUSH plasmids (for one- to three-color imaging).
The day after, transfer the coverslips in the Chamlide chamber and connect the tubing, syringes, and needles (see Fig. 11.3). Add 1 mL of prewarmed imaging medium in the chamber. Fill a 1 mL syringe containing prewarmed imaging medium supplemented with 40 μM of biotin.
Add immersion oil to the inverted objective. Place the chamber on the stage of the microscope.
Set up the appropriate parameters for single or multiple wavelength acquisition with the fluorescent proteins chosen.
Adjust the focus and eventually save the positions if you are able to perform multistage positions.
Acquire pictures at appropriate time intervals and *z* steps.
After a few pictures and between two time points, carefully aspirate the medium in the chamber and replace it with medium containing biotin using the syringes connected to the plastic tubing. Be careful not to move the chamber or the focus.
Define the next picture as time zero. Since biotin has been added to the cells, release of the RUSH cargo(s) will start. Continue time-lapse imaging for the necessary time to allow transport of the cargo(s) to their destination compartment(s).
Reconstitute your movies using the single pictures. To do so, several software are available (e.g., ImageJ, MetaMorph).

11.2 DISCUSSION

- The RUSH system relies on the retention of a protein of interest by interaction with a hook stably located in a donor compartment. The interaction is released using biotin that is simply added to the culture medium. It is thus very important to check that the medium or the serum used does not contain too much biotin. If biotin is too concentrated in the medium, a fraction of the reporter will not bind to the hook. This will lead to leaking of the reporter of interest toward its destination compartment. We routinely use DMEM + 10% serum but have found that, for example, DMEM-F12, OptiMEM, and RPMI 1640 from Invitrogen contain too much additional biotin. To overcome this problem, we add streptavidin or the less costly avidin to the culture medium to trap external biotin. The quantity of avidin/streptavidin to be added has to be defined experimentally until the retention of the cargo is restored. For example, the typical concentrations that we use are 10^{-7} mol/L of avidin in OptiMEM and 0.4×10^{-8} mol/L in DMEM-F12 (both supplemented with 10% serum).

- Like any tagging system, one caveat of the RUSH system is that the reporters are tagged and exogenously expressed in cells. This caution is also applicable to other fluorescence-based secretory assays, such as VSVGtsO45-FP or aggregation methods. In some cases, tagging can affect the correct localization and/or the trafficking pathway of the RUSH cargo compared to the endogenous cargo. To lower the expression of the recombinant proteins in cells, we found that selection of a stable cell line enables working on a more homogenous and, in general, lower expression level. Of course, whenever possible, control experiments have to be performed to confirm that the steady-state localization of the tagged reporter is as expected.
- An important parameter when using two separate RUSH plasmids, in particular, for the generation of stable cell lines, is the relative ratio of the hook compared to the reporter. In the single plasmid used for transient expression, the hook is placed upstream of an IRES and the reporter downstream of it. The first expression cassette (the hook) is more expressed than the second one (the reporter) as a result of their positions relative to the IRES. This ensures that enough hooks are produced to retain all the reporter molecules translated. Thus when generating a stable cell line using a two-step procedure, the user has to select cells expressing the hook at a higher level than the reporter. If this is not the case, the non-retained reporter will be transported toward its destination compartment.
- The versatility of the RUSH system and multi-color imaging permit a comparative analysis of the trafficking of one up to three cargos in the same cell (Fig. 11.4). By using this approach, segregation in the ER, in the Golgi complex, or even in transport intermediates can be detected (Boncompain et al., 2012).

The RUSH system allows the fluorescence-based analysis of the trafficking of secretory cargos in mammalian cells in physiological conditions, that is, at physiological temperatures and using only the small nontoxic vitamin biotin as an inducer. This assay can be employed to describe the transport steps of the cargo of interest and to define its kinetics. It is also very useful for analyzing the trafficking of cargos under perturbation. For instance, at a small scale, specific pathway or regulators can be depleted using RNA interference (see Chapter 2). We also employed the RUSH system to evaluate the effect of infection with the pathogenic bacteria *Shigella* on the

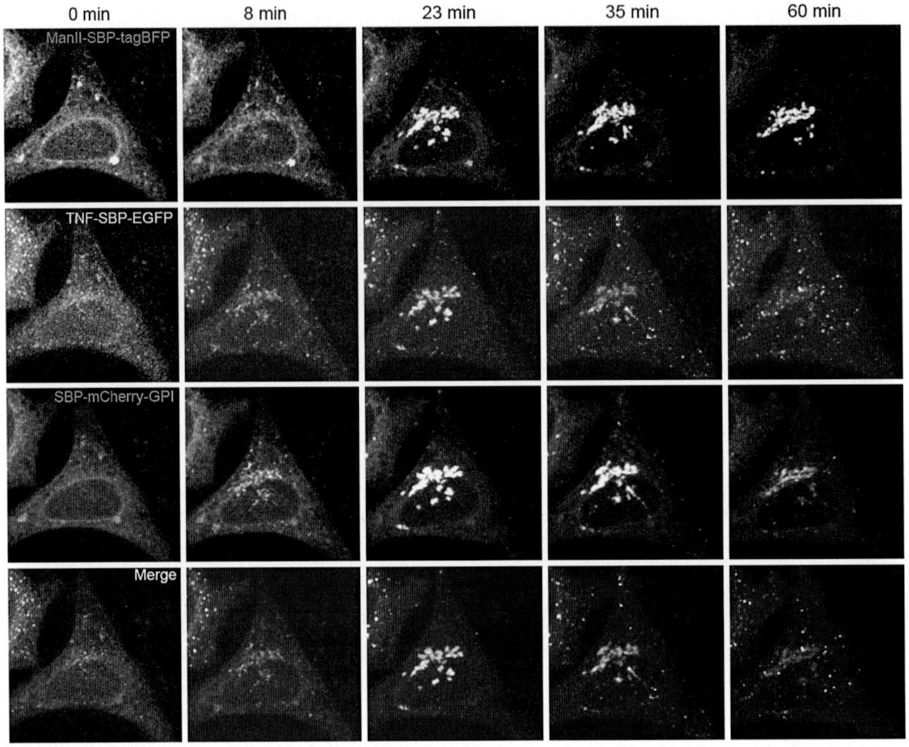

FIGURE 11.4

Three-color imaging of the simultaneous trafficking of three RUSH cargos. HeLa cells transiently expressing ManII-SBP-tagBFP (blue), TNF-SBP-EGFP (green) and SBP-mCherry-GPI (red) and streptavidin–KDEL as an ER hook. Release of the cargos was induced by addition of biotin at time 0 (0 min). (See color plate.)

membrane trafficking of the infected cell (Mounier et al., 2012). On a larger scale, regulators of the transport of a secretory protein can be identified by combining the RUSH system and high-content screening (see Chapter 17).

SUMMARY

To understand the dynamics and function of the secretory pathway and to start to draw comprehensive maps of associated regulatory networks, it is essential to study the diversity of Golgi-dependent trafficking routes. The RUSH assay, because it is quantitative and versatile, and usable in end-point and real-time experiments, represents a new tool that will help to advance faster in this understanding. Because it can be adapted to high-throughput screening, it will be instrumental in identifying novel regulatory proteins and inhibitory molecules that may be applied to therapeutic questions.

Acknowledgments

The work in the Perez Laboratory is supported by the Centre National pour la Recherche Scientifique (CNRS) and by the Institut Curie. Grants supporting our studies of intracellular trafficking have been obtained from the Agence Nationale de la Recherche (ANR), the Fondation pour la Recherche Medicale (FRM), and the Agence Nationale pour la Recherche contre le SIDA (ANRS).

References

Boncompain, G., Divoux, S., Gareil, N., de Forges, H., Lescure, A., Latreche, L., et al. (2012). Synchronization of secretory protein traffic in populations of cells. *Nature Methods*, *9*(5), 493–498.

Chen, D., Gibson, E. S., & Kennedy, M. J. (2013). A light-triggered protein secretion system. *Journal of Cell Biology*, *201*(4), 631–640.

Cole, N. B., Ellenberg, J., Song, J., DiEuliis, D., & Lippincott-Schwartz, J. (1998). Retrograde transport of Golgi-localized proteins to the ER. *Journal of Cell Biology*, *140*(1), 1–15.

Griffiths, G., Pfeiffer, S., Simons, K., & Matlin, K. (1985). Exit of newly synthesized membrane proteins from the trans cisterna of the Golgi complex to the plasma membrane. *Journal of Cell Biology*, *101*(3), 949–964.

Hicks, S. W., Horn, T. A., McCaffery, J. M., Zuckerman, D. M., & Machamer, C. E. (2006). Golgin-160 promotes cell surface expression of the beta-1 adrenergic receptor. *Traffic*, *7*(12), 1666–1677.

Kreis, T. E., & Lodish, H. F. (1986). Oligomerization is essential for transport of vesicular stomatitis viral glycoprotein to the cell surface. *Cell*, *46*(6), 929–937.

Lafay, F. (1974). Envelope proteins of vesicular stomatitis virus: Effect of temperature-sensitive mutations in complementation groups III and V. *Journal of Virology*, *14*(5), 1220–1228.

Lieu, Z. Z., Lock, J. G., Hammond, L. A., La Gruta, N. L., Stow, J. L., & Gleeson, P. A. (2008). A trans-Golgi network golgin is required for the regulated secretion of TNF in activated macrophages in vivo. *Proceedings of the National Academy of Sciences of the United States of America*, *105*(9), 3351–3356.

Lippincott-Schwartz, J., Yuan, L. C., Bonifacino, J. S., & Klausner, R. D. (1989). Rapid redistribution of Golgi proteins into the ER in cells treated with brefeldin A: Evidence for membrane cycling from Golgi to ER. *Cell*, *56*(5), 801–813.

Lock, J. G., Hammond, L. A., Houghton, F., Gleeson, P. A., & Stow, J. L. (2005). E-cadherin transport from the trans-Golgi network in tubulovesicular carriers is selectively regulated by golgin-97. *Traffic*, *6*(12), 1142–1156.

Matlin, K. S., & Simons, K. (1983). Reduced temperature prevents transfer of a membrane glycoprotein to the cell surface but does not prevent terminal glycosylation. *Cell*, *34*(1), 233–243.

Mounier, J., Boncompain, G., Senerovic, L., Lagache, T., Chretien, F., Perez, F., et al. (2012). Shigella effector IpaB-induced cholesterol relocation disrupts the golgi complex and recycling network to inhibit host cell secretion. *Cell Host and Microbe*, *12*(3), 381–389.

Munro, S. (2011). The golgin coiled-coil proteins of the Golgi apparatus. *Cold Spring Harbor Perspectives in Biology*, *3*.

Rivera, V. M., Wang, X., Wardwell, S., Courage, N. L., Volchuk, A., Keenan, T., et al. (2000). Regulation of protein secretion through controlled aggregation in the endoplasmic reticulum. *Science*, *287*(5454), 826–830.

Saraste, J., & Kuismanen, E. (1984). Pre- and post-Golgi vacuoles operate in the transport of Semliki Forest virus membrane glycoproteins to the cell surface. *Cell*, *38*(2), 535–549.

CHAPTER

Photobleaching Methods to Study Golgi Complex Dynamics in Living Cells

12

Erik Lee Snapp

Department of Anatomy and Structural Biology, Albert Einstein College of Medicine, New York, USA

CHAPTER OUTLINE

Introduction and Rationale **196**
12.1 Materials **197**
12.2 Methods **198**
12.2.1 Choice of FPs for Studies of the Secretory Pathway 198
12.2.2 Cell Culture and Transfection 200
12.2.2.1 Preparing the Imaging Chamber 201
12.2.3 Preparing the Imaging System 202
12.2.3.1 Defining the Capabilities of Your Confocal Microscope 202
12.2.4 FRAP Protocol 203
12.2.5 Inverse FRAP Protocol 206
12.2.6 Fluorescence Loss in Photobleaching Protocol 207
12.2.7 Data Analyses: Quantitation of Photobleaching Data 210
12.2.7.1 Mobile Fraction 210
12.2.7.2 Diffusion Analysis 210
12.2.7.3 $t_{1/2}$ Analysis 212
12.2.7.4 Calculation of Fractional Fluorescence 213
12.3 Discussion **213**
Summary **214**
Acknowledgments **214**
References **214**

Abstract

The Golgi complex (GC) is a highly dynamic organelle that constantly receives and exports proteins and lipids from both the endoplasmic reticulum and the plasma membrane. While protein trafficking can be monitored with traditional biochemical methods, these approaches average the behaviors of millions of cells, provide modest temporal

Methods in Cell Biology, Volume 118

ISSN 0091-679X
http://dx.doi.org/10.1016/B978-0-12-417164-0.00012-4

information and no spatial information. Photobleaching methods enable investigators to monitor protein trafficking in single cells or even single GC stacks with subsecond precision. Furthermore, photobleaching can be exploited to monitor the behaviors of resident GC proteins to provide insight into mechanisms of retention and trafficking. In this chapter, general photobleaching approaches with laser scanning confocal microscopes are described. Importantly, the problems associated with many fluorescent proteins (FPs) and their uses in the secretory pathway are discussed and appropriate choices are suggested. For example, Enhanced Green Fluorescent Protein (EGFP) and most red FPs are extremely problematic. Finally, options for data analyses are described.

INTRODUCTION AND RATIONALE

George Palade's Nobel prize-winning discovery that secretory proteins move between membranous cellular compartments raised several interesting questions. Do secretory proteins move from the endoplasmic reticulum (ER) to the Golgi complex (GC) to the plasma membrane or extracellular milieu in a unidirectional manner? Are the proteins compartmentalized into a single structure that matures into different types of compartments or do proteins move between distinct compartments? Within the compartments, are the proteins part of an immobilized matrix or do the proteins freely diffuse within a compartment. How and where are proteins sorted for different fates and destinations (such as the lysosome vs. the plasma membrane)? How fast do different proteins move through the secretory pathway? How are proteins retained or enriched in different compartments of the secretory pathway? All of these questions have been addressed to varying degrees with classical imaging techniques including immunogold transmission electron microscopy, immunofluorescence microscopy, biochemistry, and genetics.

However, the techniques are generally static (only capturing a single time point). Biochemistry experiments average the behaviors of millions of cells, which may be in different stages of the cell cycle, apoptotic, damaged, or differentiating. Fixed cell imaging experiments also do not illuminate how molecules behave within a cell, just that the molecules have a particular distribution.

Live cell imaging with dye-labeled proteins or organelle dyes provided new insights into the dynamics of cellular compartments. Yet, the cloning of green fluorescent protein (GFP) was the technological advance that now enabled investigation of the dynamics of proteins in cells, especially proteins in the early secretory pathway, as one cannot microinject dye-labeled proteins into the extremely small ER or GC. GFP-labeled proteins and the compartments they label can be visualized readily in live cell fluorescence microscope setups. Still, these proteins offer no insight into their organization or mobility within a compartment. The compartment is simply green.

Combining GFP-labeled proteins with laser scanning confocal microscopy opened up a new frontier and now with the application of photobleaching techniques, one could directly investigate the mobility and organization of proteins in cells

(Lippincott-Schwartz, Snapp, & Kenworthy, 2001). Better yet, the availability of other colors of fluorescent proteins (FPs) meant that multiple proteins can be monitored simultaneously, not just in solution, but in the actual native cellular environment. Photobleaching technology is possible on confocal microscopes available at most universities. In this chapter, a guide to successfully perform and interpret photobleaching experiments is described.

12.1 MATERIALS

1. Imaging medium: Phenol red free medium (such as RPMI or DMEM), 10% fetal bovine serum, 2 mM glutamine, 25 mM HEPES, pH 7.4. Other cell types may require different media. Note that phenol red is a chromophore that will absorb light and thus decrease the effective illumination of a sample. More importantly, phenol red can contribute to background and can even mark lysosomes in cells illuminated with near UV light. The important things are to keep cells supplied with energy (from media), growth factors (serum), and to avoid the negative effect of phenol red dye on fluorescence imaging. Media must not contain this dye.
2. Formaldehyde/PBS (phosphate-buffered saline) containing either 3.7% freshly diluted formaldehyde (from 37% stock) or 4% fresh paraformaldehyde.
3. Transfected eukaryotic cells expressing desired GFP chimeric protein or cells with plasmid plus transfection reagent or virus for expressing protein of interest.
4. Cell culture medium (i.e., DMEM with serum) for cell culturing and transfection.
5. Inverted Confocal Laser Scanning Microscope equipped with a filter set for GFP and a 488-nm excitation laser that is at least 10 mW and preferably 25 mW or more powerful laser and an acousto-optical transmission filter (AOTF). Suitable commonly available Zeiss LSM 510 or 710 with the physiology software package, Leica SP2, 5, and 7, Olympus FV1000 and FV1200 (Olympus Corporation of the Americas, Center Valley, PA), Nikon A1+ (Nikon Instruments, Inc., Melville, NY) or a microscope with a point scanning laser regulated by an AOTF. For example, some of the newer spinning disk confocal microscopes (UltraVIEW) have photobleaching laser modules (i.e., Perkin Elmer, Waltham, MA).
6. Imaging chamber (either a Nunc Lab-Tek chambered coverglass (Fig. 12.1) or MatTek dish (Ashland, MA)).
7. Immersion media (i.e., oil for oil microscope objectives).
8. Stage heater or environmental chamber (for cells that grow at temperatures above 28 °C).
9. A computer system capable of processing large image files (a 100+-GB hard drive, and at 2-GB RAM) equipped with image processing software (i.e., NIH Image, Matlab (TM), Volocity (TM), or Metamorph (TM)).

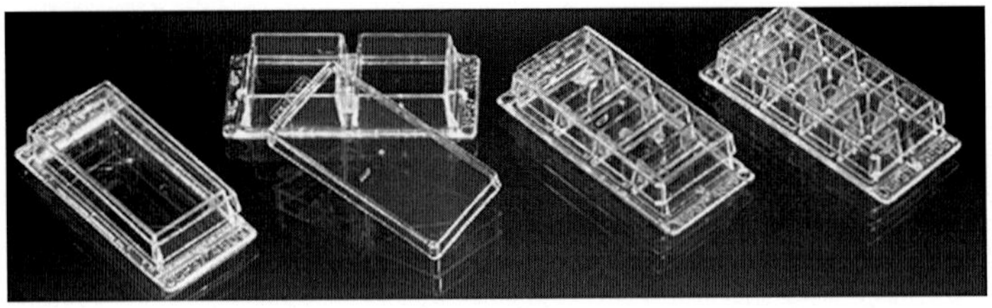

FIGURE 12.1

Lab-Tek chambered coverglasses. (See color plate.)

Image courtesy of Thermo Scientific.

12.2 METHODS

12.2.1 Choice of FPs for studies of the secretory pathway

Not all FPs are created equal. In fact, many FPs have very specific uses and are not interchangeable. While brightness, photostability, and folding time are all important properties for an FP, there are X key features that are often not considered. The FP must be suitable for the particular cellular environment. All natural FPs (such as GFP and DsRed) and most evolved FPs (i.e., mCherry) are from the cytoplasm of various eukaryotes and bacteria. This becomes a concern when putting an FP inside the secretory pathway. First, FPs often contain cysteines. In the oxidizing environment of the secretory pathway, the FPs can form interchain disulfide bonds that result in dark misfolded proteins (Aronson, Costantini, & Snapp, 2011; Jain, Joyce, Molinete, Halban, & Gorr, 2001). While a fluorescent signal may be present, 50% or more of the FP pool can be misfolded and dark. This dark pool will decrease overall fluorescent signal and will complicate the interpretation of biochemical assays using the secretory FP, as only part of the pool of fusion protein may actually be functional. Second, many FPs contain consensus sequences (N-X-S/T, where X is any amino acid except proline) for N-linked sugar additions. These sugars increase an FP's size and can influence its diffusive behavior (Costantini, Subach, Jaureguiberry-Bravo, Verkhusha, & Snapp, 2013) and cause a fusion protein to interact with the lectin chaperones calnexin and calreticulin. Third, FPs have a characteristic pK_a or pH at which the fluorescence intensity of the FP decreases by half and eventually becomes dark as the pH decreases further. In the neutral pH of the cytoplasm or the ER, this is not an issue. However, the successive compartments of the GC are increasingly more acidic, such that higher pK_a FPs (including EGFP 6.0, EYFP 6.9, Venus 6.0, and mKate 6.0; Shaner, Patterson, & Davidson, 2007) may be quite dim in the pH 6.0 trans GC (Paroutis, Touret, & Grinstein, 2004). Finally, regardless of whether the FP is in the lumen of the GC or associates with the GC surface in the cytoplasm, it is critical that the FP be truly monomeric. Many FPs are not and FPs that are labeled as monomeric may not be (i.e., TagRFP; Costantini, Fossati, Francolini, & Snapp, 2012). The ability of FPs to oligomerize can affect everything from causing false-positive signals in Förster Resonance Energy Transfer (FRET) experiments (Zacharias, Violin, Newton, & Tsien, 2002) to gross distortion of secretory organelles when FPs on apposing membranes interact (Snapp et al., 2003). In some cases, it is not even a matter of simple oligomerization. Some red FPs are relatively monomeric, but have a tendency to cluster or aggregate at high concentrations. This is particularly problematic when FPs are attached to membrane or membrane-associated proteins. By confining FPs to the two-dimensional surface of a membrane, the effective concentration substantially increases. Even weak dimerizing activity that is not problematic for soluble cytoplasmic proteins becomes a serious issue on a membrane surface.

Standard GC fluorescent reporters, such as the Galactosyl Transferase (GalT) FP construct, can adversely affect GC size and organization if a sticky or oligomerizing

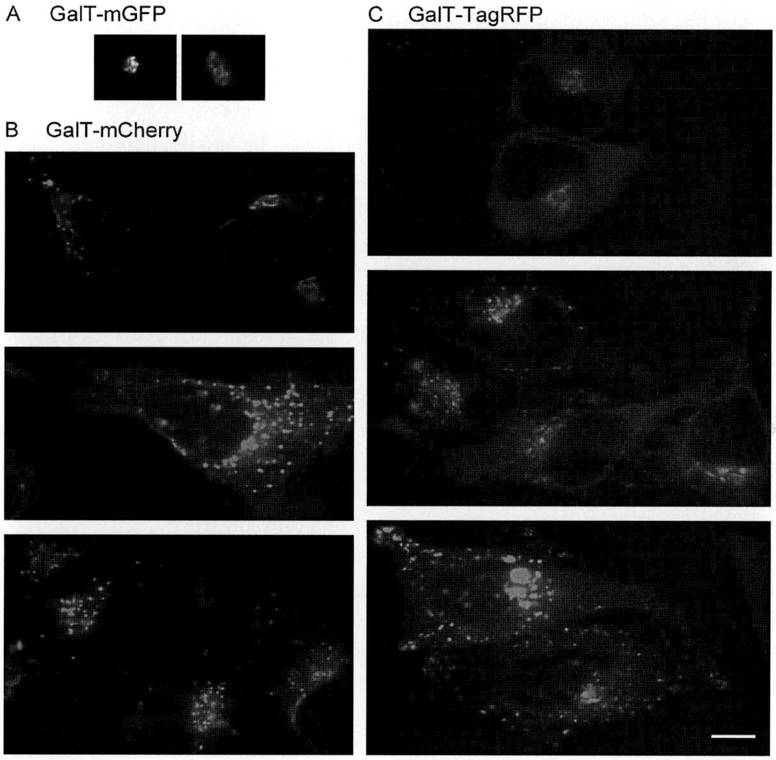

FIGURE 12.2

Effects of FP on GC structure. Fluorescence images of HeLa cells expressing GalT-mGFP (A), GalT-mCherry (B), and GalT (TagRFP) (C). In the lowest expressing cells in (B) and (C) (top panels), compact GC structures comparable to GalT-mGFP structures are observed. In modest to high expressing cells, the red constructs label dispersed punctate structures throughout the cell. For this reason, current red FPs are generally not suitable for secretory pathway membrane protein fusions. Scale bar is for all images and is equal to 10 μm. (See color plate.)

FP is incorporated (Fig. 12.2). Only an environmentally neutral FP should be used for a fusion protein, in general, and in particular for a GC membrane protein reporter. As with any FP reporter, it is important that the investigator establishes that the reporter does not alter the structure (unless is an experimental goal) or function of the GC.

To date, expression of red FPs typically distorts the GC at all but the lowest levels of expression. EGFP and related FPs do not distort the GC, but a large pool typically misfolds in the oxidizing lumen of the secretory pathway. The best behaved FPs we have identified are superfolder GFP (sfGFP) and its derivatives (Aronson et al., 2011; Pedelacq, Cabantous, Tran, Terwilliger, & Waldo, 2006). A blue FP suitable for the secretory pathway has been described (Costantini et al., 2013) and can be used in conjunction with sfGFP without significant crosstalk for most microscope filter sets. FPs with cysteines or N-linked glycosylation sites can be used on the cytoplasmic face of the GC, as long as the FPs are robustly monomeric.

Even with an optimal FP, the investigator is cautioned against selecting high expresser cells for imaging. Many cellular proteins are expressed at modest to very low (hundreds to thousands of molecules per cell). With a bright FP, some constructs, if expressed at near endogenous levels, may still not be easily detected. Within the tightly packed GC, this is often less of a concern. When a small number of molecules are concentrated in a tight space, the fluorescent signal can be excellent.

12.2.2 Cell culture and transfection

Before performing a photobleaching protocol, the investigator must ensure that there is sufficient fluorescent signal in the expressing cell relative to background noise after photobleaching. Thus, a sufficient number of cells expressing the fluorescent reporter must be ensured.

To image adherent cells, they are typically grown on coverglass discs or in multiwell chambers (Nunc Lab-Tek, #1 coverglass, especially for use with samples thicker than one layer of cells). We prefer the latter because cells retain their full three-dimensional shapes, while coverslips are often pressed against slides and the cells are squished. Plate the cells on the sterile coverglass in media of choice at least 16 h in advance. The next day, prewarm freshly prepared imaging media (to 37 °C for mammalian cells). Aspirate the growth media and replace with imaging media covering the cells generously, that is, 400 μl of imaging media per well in an eight-well Lab-Tek chamber.

To image suspension cells, the cells can be adhered to a coverglass that has been pretreated with concentrated poly-L-lysine. To successfully stick the cells to poly-L-lysine, the cells must be in a medium free of proteins, that is, 1 × PBS. Any negatively charged material can bind to the poly-L-lysine and compete with the cells for binding. The serum in cell media will completely prevent adherence of cells to poly-L-lysine.

First, add concentrated poly-L-lysine (P8920 from Sigma-Aldrich, St. Louis, MO) to coverglass and incubate for 5 min. Remove poly-L-lysine and save it, as it can be reused multiple times. Wash the coverslip 2× in dH_2O and allow to air dry. This can be done sterilely in a tissue culture hood if the cells are to be imaged for more than a few hours. Spin suspension cells at low centrifuge speed to gently pellet, resuspend in PBS, spin, and resuspend again in a low volume of PBS. Add to and cover the coverglass chamber, incubate 5 min at RT, wash 2 × gently with PBS, and then add imaging media.

If the cells are not stably expressing an FP construct, then the construct should be transfected at least 8–16 h in advance using cationic lipids (i.e., FuGENE6 (Roche, Indianapolis, IN) or LipofectAMINE 2000 (Invitrogen, Carlsbad, CA)) or a viral vector. Alternatively, electroporation can achieve high transfection efficiency. The optimal level of expression and timing of imaging should be determined empirically for each construct, cell type, and condition.

Transfection protocols will vary depending on the volume and cell type. Our lab typically uses Lipofectamine 2000 with good results for several tissue culture cell lines. Whichever transfection method is selected, the investigator needs several usable cells for each experiment. Note that for microscopy experiments, high transfection efficiency (70–100%) is not necessary and is often undesirable. Good images are often achieved at lower efficiencies so that individual cells can be readily distinguished. 10–40% efficiency is a good range for imaging single cells. Cells expressing low to modest levels of FPs are preferable for imaging. While high expressing cells are easier to image, the high expression levels may perturb GC structure and/or function. Note that stably expressing cells usually express a construct at comparably low levels of protein and this is also desirable. Higher expression levels can lead to GC structure artifacts, protein aggregation or titrate fusion protein cofactors.

12.2.2.1 Preparing the imaging chamber

1. When using an inverted microscope, adherent cells can be plated on Lab-Tek chambers (Nalgene, Naperville, IL). These chambers consist of wells with a cover glass bottom, which permits the use of high numerical aperture oil objectives for viewing. Prior to placing the cells on the microscope stage, the wells are filled to the rim with imaging media. This ensures that cells have sufficient nutrients during the imaging session. For experiments that will last longer than an hour, the top cover of the chamber should be sealed onto the chamber using petroleum jelly or silicon grease. These steps prevent both rapid evaporation of the media above the cells and decrease the alkalinization of the media by preventing room air from entering the chamber.

   Suspension cells can be grown in suspension and then adhered to coverslips or Lab-Tek chambers by precoating the chamber or clean coverslip with a concentrated (5–10 mg/ml in PBS) solution of poly-L-lysine (Sigma-Aldrich, St. Louis, MO). Incubate for 5 min. Wash twice with dH_2O. Suspension cells first must be washed three times in PBS, before adhering to the poly-L-lysine-coated surface. Incubate the cells on the surface for 2–5 min. Remove the nonadhering cells with PBS (wash twice). Then cover the cells in imaging medium.
2. Maintain cells at physiological temperature on the microscope stage using a stage heater, such as the Model ASI 400 Air Stream Stage Incubator (Nevtek,

Burnsville, VA), an environmental chamber or even a thermal collar objective heater (Bioptechs, Butler, PA). A temperature probe, such as a Thermolyne Pyrometer (Carl Parmer, Vernon Hills, IL), can be used to confirm maintenance of the temperature at the chamber.

12.2.3 Preparing the imaging system

Set up the confocal laser scanning microscope and its associated hardware. The investigator should be familiar with the basic operation of the confocal microscope. It is worth understanding both the concept and the operation of a pinhole, scan speed, zoom, detector gain, laser power, photobleach, and how the microscope collects a time series. The protocols in this chapter are guidelines and conditions will differ depending on laser power, scan speed, detector sensitivity, and sample properties. Understanding microscope principles will free the investigator to adapt protocols to his/her microscope.

12.2.3.1 Defining the capabilities of your confocal microscope

Before planning photobleaching experiments, it is vital to evaluate the status and abilities of the laser scanning confocal microscope. First, it is important to determine whether the microscope can deliver sufficient laser power for the FP of choice, that is, a 488-nm laser to photobleach sfGFP. This typically must be determined empirically. Even if a microscope has photobleaching capabilities, the laser may be too weak, may be out of alignment or may need to be replaced. Standard imaging experiments are often not a problem for a weak laser, as one can simply increase laser power to the sample. Photobleaching experiments demand substantial laser power that often starts with a laser of 10 mW power or higher. Lower powers can be used, but will be useful often only for slowly diffusing proteins or for FLIP experiments.

How can one assess the microscope's capabilities? An appropriate sample slide is needed with cells expressing the FP of choice. To avoid complexities of FP movements, the sample must be fixed. Wash the coverslip or Lab-Tek chamber once with 1 × PBS, aspirate, and add freshly prepared 3.7% formaldehyde in 1 × PBS and incubate in a dark drawer at RT for 15 min. Remove the formaldehyde and store in a chemical hazard container and add back 1 × PBS to the fixed cells.

Start up the confocal microscope according to protocol. Prewarm lasers for at least 5 min and then prepare to image. Select a high numerical aperture (NA) objective to maximize light collection. Excellent results can be obtained with a 60–63 × NA 1.4 objective. 40 × also works well and 100 ×, too. The 100 × objective typically causes samples to appear dimmer due to the relationship between magnification and brightness. For most imaging needs, the 100 × objective is rarely optimal.

Apply the appropriate medium to the objective, that is, oil, water, glycerol or other media and position the sample. Use the oculars to find and center a cell of interest. Switch to laser scanning mode and image the cell with low laser power, the highest speed that still maintains a sufficiently bright image (i.e., 15% or more of the intensity range of the image scale, 50 for an 8-bit image or 600 for a 12-bit image).

Avoid saturated pixels, as these are not suitable for quantitative imaging. Saturated pixels are literally intensity values higher than the detection scale and represent lost information. The pinhole should be open to maximize the sample signal. Note that the pinhole is positioned at the end of the light path, in front of the detector. Therefore, the pinhole has no effect on photobleaching efficiency. The key for most photobleaching experiments is quantitative signal, not necessarily attractive images.

Most photobleaching modules have a specific software program or macro. Open the program or macro and apply settings to define a photobleach region. Acquire a prebleach image for comparison. For maximum speed, use a horizontal box or region of interest (ROI). Set the photobleaching parameters for one iteration of photobleaching intensity laser power and attempt to photobleach the region of interest. If the region bleaches, then the conditions can be minimized until bleaching conditions achieve 70–90% bleach depth (completion).

If little or no bleaching is detected, modify parameters to increase bleaching. These conditions include the scan speed of the bleaching laser (slower bleaches more strongly), the zoom of the cell (higher zoom increases sample bleaching), number of scans, and activating other laser lines (i.e., sfGFP can be bleached by a 488-nm line, but also by the 514-, 456-, and 478-nm lines typically found on argon lasers. Similarly, mCherry photobleaches poorly with a 1-mW 543-nm laser, but has sufficient absorbance spectrum that even a 488-nm laser can help photobleach a sample.

The optimal bleaching conditions will depend on the type of microscope and its components. For a 63 × NA 1.4 objective on a Zeiss 510 with 40-mW 488-/514-nm Argon or 25-mW argon laser are 45–60% power with 100% transmission. For acquisition of recovery time points, use the same level of power with only 0.1–1.0% transmission (if possible), as regulated by the AOTF. For quantitative Fluorescence Recovery after Photobleaching (FRAP), the whole cell is usually scanned at scan speed 8–10 (0.798–3 s per 512 × 512 frame) with either two line averaging or no line averaging. For a Zeiss Duoscan with a 100-mW 489-nm laser, sfGFP is bleached using a 63 × NA 1.4 objective with one to three iterations at zoom 1, photobleaching laser scan speed of 6 with 100% transmission through an 80/20 beam splitter, with only 20% of laser power dedicated to the photobleaching scanner. Imaging is performed at 5 frames per second (fps) or faster with 0.5–1.5% transmission, and no frame averaging. For faster imaging, the cell of interest is often centered and a strip of only 100–200 pixels *Y*-axis of the normally 512 × 512 field is scanned.

If photobleaching cannot be performed rapidly or at all, it is often worth a visit from a service technician to align the laser (a common cause of low power) or the laser may need to be replaced. If everything is working to specifications, then the instrument may not be suitable for photobleaching experiments.

12.2.4 FRAP protocol

To perform data analysis of an FRAP experiment (Fig. 12.3), the user must collect at least one prebleach image of the cell, a series of postbleach images of the whole cell that extends from an immediate postbleach image to several images after the bleach

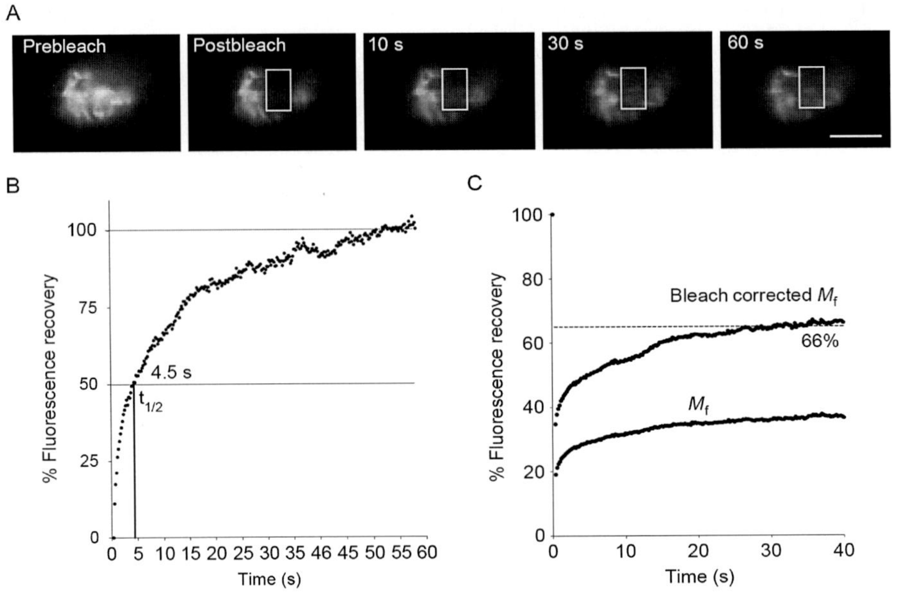

FIGURE 12.3

Golgi complex FRAP. (A) FRAP of a GC in HeLa cell expressing secretory sfGFP. (B) Plot of recovery data transformed for $t_{1/2}$ analysis. Note that prebleach intensity is not relevant to $t_{1/2}$ analysis. (C) Plots of M_f and bleach-corrected M_f data. The bleach-corrected M_f accounts for the substantial loss of total GC fluorescence and reveals a substantially higher value than the uncorrected M_f. (See color plate.)

ROI fluorescence intensity reaches a plateau, and a table of fluorescence intensity values including the photobleach ROI, the whole cell ROI, and a background ROI. The investigator also needs to know the corresponding time for each image and the image pixel size in microns. For the Zeiss 510 microscope, the physiology software package contains several windows that permit the user to determine the number of images to collect, laser intensities for bleaching and imaging, number of bleach iterations, how many images to collect before bleaching, and drawing of the photobleach ROI. Other microscopes may or may not contain their own software package or may require writing macros for the photobleach time series.

A quick note on software packages. It is important for the investigator to appreciate that not all software packages necessarily do the same thing. That is, some definitions of photobleaching experiments differ between companies. On the Leica SP2, for example, the FLIP wizard will only collect a prebleach image and an image at the end of the FLIP experiment. This protocol is not a true FLIP experiment and is of debatable utility. Some microscope company software packages also include analysis software. This author has tried three of them and, to date, has not found a diffusion analysis program that correctly calculates the diffusion coefficient (*D* value) for proteins of known mobilities. The take home message is not that it is impossible to calculate *D* values, but that investigators serious about quantitative microscopy should test any new system with standards with well-established properties. For example, cytoplasmic GFP diffuses at ~23 μm²/s (Lajoie & Snapp, 2010; Swaminathan, Hoang, & Verkman, 1997) and the membrane protein Vesicular Stomatitis Virus G protein fused to GFP (VSV G-GFP) diffuses at 0.45 μm²/s in the ER membrane (Nehls et al., 2000). GalT GFP, which uses a luminal GFP to replace the enzymatic domain of this GC integral membrane protein, diffuses at 0.3–0.5 μm²/s (Cole et al., 1996):

1. Collect multiple prebleach images to establish the prebleach fluorescence intensity and to confirm that the prebleach fluorescence intensities of the cell and the bleach ROI do not fluctuate significantly.
2. Photobleach the ROI with intense laser illumination. Continue to image the whole cell at low laser illumination (the same conditions as the prebleach images), until the recovery process has reached a steady state. These conditions must be determined quantitatively, as the human eye is incapable of distinguishing small differences in intensities. Typically, a 1- to 4-μm wide bleach ROI strip will recover within 30–120 s for most freely mobile soluble lumenal proteins and within 3–5 min for membrane proteins in the plasma membrane, ER, or GC. Cytoplasmic proteins may recover as rapidly as 1–20 s.

   In this way, the investigator can establish conditions to obtain data sufficient for estimation of $t_{1/2}$ and mobile fraction (M_f) (see below). The investigator should not attempt to refocus a cell during an FRAP experiment because the fluorescence intensities will shift and the recovery curves will no longer be smooth. Time series that contain a shift in focal plane should be discarded.

   If a protein has a high protein D value (i.e., ≥ 5 μm²/s), the investigator may not observe a defined bleach region following the photobleach, especially for slower confocal microscopes. Rapidly diffusing FPs may appear homogeneously distributed immediately following a photobleach. The investigator must either (1) increase the image collection rate, by changing the scan speed or reducing the size of the total frame to be collected, (2) increase the bleach ROI size, (3) increase the laser power or employ a combination of these suggestions.
3. Collect at least 10–20 data sets for each fluorescent reporter and treatment for robust statistical analyses. A fraction of the data sets are usually discarded because of problems that potentially bias imaging results (i.e., incomplete recovery, the focal plane shifted, cell movement, the recovery curve fitting method fails, etc.).

Note: For exceptionally rapidly moving molecules or structures (soluble proteins and vesicles), extremely rapid image acquisition speeds may be required. Newer spinning disk microscopes with sensitive electron multiplying Charge Coupled Device (emCCD) cameras or high speed acquisition microscopes, that is, the Duoscan (Carl Zeiss, Inc., Jena, Germany) or the Nikon LiveScan Swept Field Confocal microscope (Nikon Instruments, Inc., Melville, NY) can perform photobleaching operations and image at rates of tens of fps. For slower microscopes, an alternative approach is to image the photobleach ROI alone. The caveat of the method is that the cell may shift in focus or position during the course of imaging. Imaging the whole cell enables the advantage of visualizing any movement of the cell or focal plane, permitting the researcher to reject unusable data following visual inspection.

12.2.5 Inverse FRAP protocol

Inverse FRAP or iFRAP (Altan-Bonnet et al., 2006; Nichols et al., 2001) gets its name from the way in which everything *except* the region of interest is photobleached. Typically, this is accomplished by drawing an ROI over all parts of the cell excluding the region to be monitored, in this case, the GC. Because the ROI drawing programs only create continuous ROIs, the ROI will be drawn like a bagel with a hole in the center connected to the area outside of the cell by a narrow path (Fig. 12.4).

The iFRAP technique has three general uses. First, it can be used to perform a pulse-chase experiment. The ability of a population of proteins to exit the GC, the rate, and the types of structures that facilitate escape can be visualized (Fig. 12.4). Similar experiments can be performed with photoactivatable proteins (Lukyanov, Chudakov, Lukyanov, & Verkhusha, 2005; Shaner et al., 2007). The iFRAP technique permits the investigator to do all dynamics experiments with the same protein. Second, iFRAP helps reveal dim structures. In standard IF, all pools of a protein are labeled by an antibody. If a protein is mostly in the ER and a small

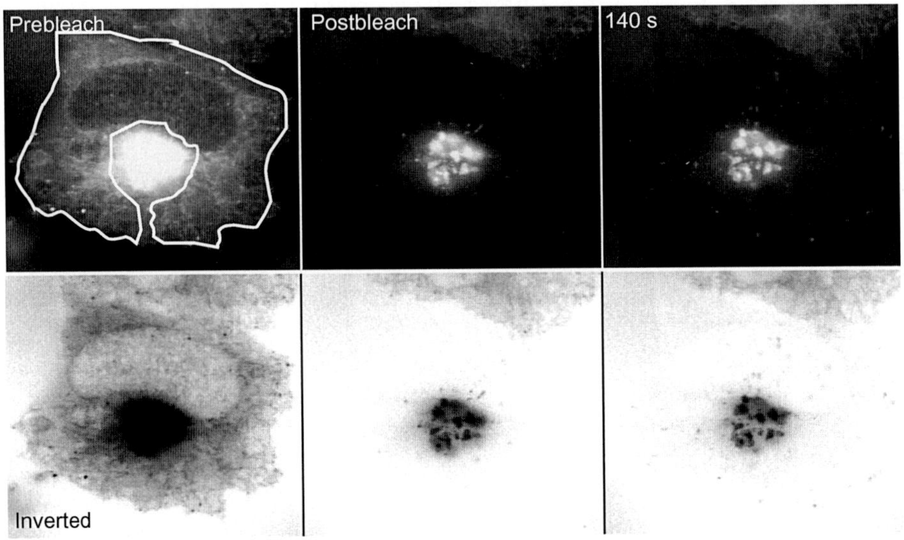

FIGURE 12.4

iFRAP of HeLa cell expressing VSV G-sfGFP. A "bagel bleach" ROI was drawn over the entire cell and excluded the GC. Immediately after the photobleach and especially after 140 s, dim vesicles and tubules can be visualized emerging from the GC. The image series also has been inverted to help visualize the number and variety of vesicles.

fraction is in vesicles, the bright signal can overwhelm the fluorescent signal of the vesicles. iFRAP enables the investigator to remove the ER signal and dramatically improve the signal to noise for very dim vesicles. Third, iFRAP compliments data from FRAP and FLIP experiments. While FRAP quantitates rate of recovery of fluorescence into an ROI, iFRAP can be used to measure rate of fluorescence loss from an ROI. FLIP helps visualize immobilized pools of FPs or discontinuous compartments.

Prepare sample and microscope as for FRAP. The key difference is that the photobleach ROI will be drawn to exclude the GC or other fluorescent structure of interest. Do not draw the ROI into the edge of the GC. The photobleaching laser is not always perfectly constrained to a single pixel. Include a small border around the GC (see Fig. 12.4). If visualizing dim structures is the goal of the experiment, it is often helpful to increase the image intensity (i.e., increased gain, more laser power, slower scan speed). After the photobleach, the dim structures will be often bright enough to detect against the otherwise dark background.

12.2.6 Fluorescence loss in photobleaching protocol

As in FRAP, fluorescence loss in photobleaching (FLIP) photobleaches an ROI with a high power laser. However, in FLIP, the ROI is repeatedly photobleached and the cell is alternatingly imaged to examine the behavior of the entire fluorescent pool throughout a structure or the cell. If the fluorescent molecules are completely mobile and have access to the photobleaching ROI, the entire fluorescent pool will be depleted (see Fig. 12.5). Thus, results from FLIP experiments can reveal the connectedness of structures containing FP reporters (Ellenberg et al., 1997; Nehls et al., 2000). FLIP can also be used to look at dynamics of proteins that bind and release from the GC surface, such as COP molecules (Presley et al., 2002).

In FLIP analysis, the user obtains a prebleach image, an immediate postbleach image following the first photobleach, images following successive photobleaches (often two to three images following each photobleach) until the entire cell or structure is depleted of fluorescence or several successive photobleaches do not further deplete cell fluorescence, the time for each image, pixel size, and fluorescence intensity values for the whole cell and the background intensity:

1. Prepare microscope and cells as in the FRAP protocol.
2. Identify the cell of interest on the confocal microscope. The optimal condition is to have two adjacent cells of similar fluorescence intensity in the imaging field. The unbleached cell serves as a control to insure that the imaging conditions do not cause nonspecific photobleaching of all fluorescence in the field of view. Imaging conditions must not significantly photobleach any adjacent cells during the experiment. The cell will often be imaged 100–1000 times during a FLIP experiment and that number of exposures will inherently cause photobleaching. Thus, it is important to image the cell with the fastest possible scan and lowest laser power to minimize photobleaching.

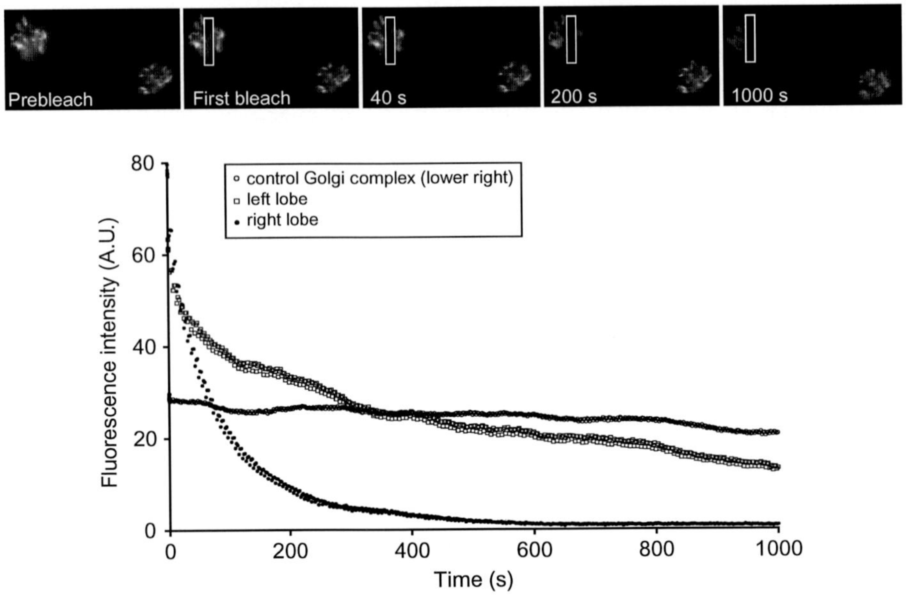

FIGURE 12.5

FLIP of GalT-GFP in HeLa cell. A bleach region of interest (ROI) was placed in the middle of the GC and bleached after every three images, collected with 2 s delay between images. The plot shows the different rates of loss of fluorescence for the two sides (lobes) of the GC, as well as the control GC in the lower right corner. Note that the control GC loses very little fluorescence, indicating that fluorescence loss in the GC of interest is specifically due to bleaching of the ROI. The two lobes lose fluorescence at different rates, suggesting that the reporter construct is either organized differently within domains of the GC and/or that there may be different rates of communication within this GC structure, that is, discontinuous cisternae.

Bring the desired cell into focus. Scan an image of the whole cell at the desired excitation light intensity, averaging, etc. Modify pinhole and detector gain for maximal fluorescence signal and minimal pixel saturation. Detector gain and offset will vary depending on the concentration of the fluorophore, the laser power, objective used, and the thickness of the fluorescently labeled organelle or region. The imaging parameters used for the prebleach image(s) should be used for the fluorescence recovery time series.

3. Define an ROI for the photobleach. The bleach ROI should be between 5% and 20% of the structure of interest. The photobleach ROI does not need to be a strip. A square or circle is equally appropriate.
4. Determine imaging and photobleaching conditions (i.e., scan speed, zoom, laser power, number of laser iterations required for photobleaching, microscope objective) that cause minimal photobleaching of the cell outside the photobleach

ROI. For example, a 40-mW 488-/514-nm argon laser can be used at 45–60% power with 10–20% transmission for photobleaching and 1% transmission for imaging on a Zeiss 510. For a Zeiss Duoscan, a 100-mW 489-nm laser is used at 100% transmission, scan speed 9, one bleach iteration. The photobleach does not need to be as strong as in an FRAP experiment. Molecules will continuously diffuse through the ROI and should eventually be photobleached even at the lower intensity. A benefit of this approach is that the cell should be less stressed by the laser over the course of the FLIP experiment.

Depending on the microscope, a field of cells (or a smaller imaging ROI if using a laser scanning confocal microscope) can be scanned at 0.1–2 s per image with two to four line or frame averaging (less averaging for series with increasing numbers of total exposures).

5. Collect 10 prebleach images to establish the prebleach fluorescence intensity and to confirm that the prebleach fluorescence intensity does not significantly fluctuate.
6. Photobleach the bleach ROI. Collect two images of the whole cell and photobleach the ROI again. Repeat the process until the fluorescent structure intensity is similar to background. Unlike FRAP, speed of imaging is not critical for most FLIP experiments. Bear in mind that the bleaches in the FLIP experiment are most useful when the bleach can further deplete fluorescence. If bleaches are performed too quickly in succession, there may be little recovery of fluorescence into the ROI and the next photobleach will have no productive effect. Typically, a delay of 1–5 s between images and the next photobleach works well. The delays allow unbleached molecules to diffuse into the photobleached ROI. Furthermore, collecting images between bleaches permits monitoring of FLIP progress and confirms that the cell remains in focus and has not moved. It is acceptable to refocus the cell during a FLIP experiment. Great care must be taken to return to the original focal plane. Shielding the stage from air currents in a room and minimizing temperature fluctuations on the stage can help maintain focus. Autofocus routines available for some microscopes can also help. Cell movement is more problematic. Extensive movement will mean the original ROI may be a significant distance away at the end of an experiment. Then the FLIP is not spatially restricted and other parts of the cell could be directly photobleached, which is counter to the goal of the FLIP experiment. If the cell moves significantly, the experimental results are unusable.

   For exceptionally slow proteins, the delay between images can be even longer. It is also acceptable to increase the size of the bleach ROI to decrease the time needed to deplete total structure fluorescence. It may require some experimentation to identify the best sized bleach area to maximize the number of data points for a well-defined fluorescence intensity data curve, while minimizing the time of imaging to prevent general photobleaching of the cell.
7. Collect at least 5–10 data sets for each FP reporter and treatment. Note that some data sets may be unusable due to focal plane shift, the cell movement, etc.

12.2.7 Data analyses: Quantitation of photobleaching data

After collecting photobleaching data, there are three general parameters that can be determined: the M_f of the FP reporter, its diffusion coefficient (D), and the $t_{1/2}$ of the fluorescence recovery or loss. Each parameter has potential utility. To make the most of the collected data, the investigator should perform calibration experiments with an FP reporter whose mobility properties have been previously characterized, especially when calculating D values.

12.2.7.1 Mobile fraction

M_f is the percentage of FPs capable of diffusing into a photobleached ROI during the time of the experiment. M_f, D, and $t_{1/2}$ are distinct parameters and must be understood as such. D is a characteristic of the *mobile* pool of FPs. The proteins in the *immobile* fraction do not diffuse or contribute to the D value. A low D value does not bear any relation to the fraction of immobile proteins.

An approximation of M_f can be calculated using the following equation, based on (Feder, Brust-Mascher, Slattery, Baird, & Webb, 1996) with the inclusion of a photobleaching correction (Lippincott-Schwartz, Snapp, & Kenworthy, 2001). Given the focus of this chapter, it is assumed that FRAP will be performed primarily on the GC. As diffusion will be relatively fast compared to rates of trafficking and protein synthesis, the fluorescence pool for recovery will be mostly confined to the GC:

$$M_f = 100 * \frac{\left(F_{\text{preGC}} - F_{\text{background}}\right)}{\left(F_{\infty\text{GC}} - F_{\text{background}}\right)} * \frac{\left(F_{\infty} - F_{\text{background}}\right)}{\left(F_{\text{pre}} - F_{\text{background}}\right)}$$

where F_{preGC} is the whole GC prebleach intensity, F_{pre} is the bleach ROI prebleach intensity, $F_{\infty\text{GC}}$ is the asymptote of fluorescence recovery of the whole GC, $F_{\text{background}}$ is the mean background intensity, and F_{∞} is the bleach ROI asymptote.

In the equation, the photobleach ROI and whole GC ROIs are background subtracted. Next, the photobleach ROI data are transformed such that the prebleach fluorescence intensity is defined as 100% fluorescence intensity. The equation includes a correction for the loss of total GC fluorescence (due to the photobleach of the ROI and bleaching of the GC during imaging). The correction is calculated by determining the prebleach fluorescence intensity of the whole GC ROI (F_{preGC}) and dividing it by the whole GC ROI intensity at time "t." Without the photobleach correction, the bleach ROI intensity can never recover to 100% of the prebleach fluorescence because the photobleach depletes 5–50% of the total GC fluorescence (see Fig. 12.3). To convert the fraction to a percentage, the data are multiplied by 100.

12.2.7.2 Diffusion analysis

D values reflect the mean squared displacement (usually $\mu m^2/s$ or cm^2/s, which can be multiplied by 10^8 to get $\mu m^2/s$) that a protein explores through a random walk (i.e., Brownian motion) over time. D values do NOT correspond to a linear process. The time required to cover increasing distances does not increase in a linear fashion. Time to sample increasing distances increases as the square of the distance covered divided by the D value. The primary constraints on protein diffusion are the viscosity of the environment (η), whether the protein is soluble or integrated into a membrane, and to a lesser extent, the hydrodynamic radius (R_h) of the protein. Protein–protein interactions and collisions with other molecules hinder free diffusion. By measuring D values, the investigator can obtain information related to a protein's environment and in some cases, whether the protein is interacting with large complexes.

Several equations and computer simulations have been developed to determine D values (Axelrod, Koppel, Schlessinger, Elson, & Webb, 1976; Gordon, Chazotte, Wang, & Herman, 1995; Saxton, 2001). Many available equations have been written for spot photobleaching or have been developed for highly specialized scenarios (i.e., the nuclear environment or the plasma membrane). Therefore, it is important to appreciate that many diffusion analytical methods are not necessarily appropriate for studies of the GC. The author has tried simulations developed for cells, in general, as well as simulations developed by microscope companies. Control protein D values were often incorrect by an order of magnitude or more. The problem is that most equations and simulations make important assumptions that may not hold true in the GC. In some cases, equations do not take into account the inhomogeneous diffusion that occurs in cellular compartments. The bottom line is one must validate a diffusion analysis method with robust control proteins. Do not simply accept values produced by an equation without establishing that the values are reasonable. For example, a membrane protein should not diffuse faster than 0.8–1.0 $\mu m^2/s$.

A robust method we have found for obtaining D value is through simulation of diffusive recovery into a bleached strip (Siggia, Lippincott-Schwartz, & Bekiranov, 2000). The Siggia simulation models inhomogeneous diffusion of unbleached proteins in the cell into the photobleach ROI. The simulation then compares the simulated recovery to the actual data to determine D (Siggia et al., 2000). The program has been used for ER membrane proteins (Ellenberg et al., 1997; Zaal et al., 1999) and has been used to calculate Ds comparable to values observed by other labs using different methods. Like other methods for determining D, simulation encounters difficulties when a significant fraction of the FP is immobilized (Siggia et al., 2000). In such cases, the simulation either fails to fit the simulation data to the experimental recovery data or the simulation cannot calculate D.

The Siggia simulation and a set of instructions can be obtained by contacting this author. The simulation has been written for UNIX systems and has been compiled for Apple computers pre-Intel chip (running OS $\times$ 10.4 and earlier). Recently, we were able to get the program successfully compiled for newer Apple computers.

After running any simulation or calculating a least squares fit to determine D, it is essential to compare the simulation or fit data to the experimental data by plotting the data sets in a spreadsheet program. Even if a program or equation produces D, a poor fit means that D is questionable at best. "Goodness of fit" describes how well the simulation or equation data overlaps the experimental data. The majority of the experimental data must overlap the simulation or equation plot. If the simulation misses the experimental data, the calculated D is questionable, at best. If there are any doubts about "goodness of fit," the user should consider another method of calculating the D.

Potential errors may occur in the course of data fitting. These errors are not unique to D calculations. They are part of the general problem of nonlinear regression analysis. Individuals interested in performing diffusion analysis can learn more about nonlinear regression analysis in an excellent (and free!) introduction to the theory and problems of nonlinear regression analysis at http://www.graphpad.com/guides/prism/6/curve-fitting/.

12.2.7.3 $t_{1/2}$ analysis

The mobility of several proteins depends on complex behaviors, such as binding and release or populations with multiple Ds. The architecture of a cellular structure can increase apparent D values by increasing the path length that a molecule must traverse (Sbalzarini, Mezzacasa, Helenius, & Koumoutsakos, 2005). Diffusion in the GC is further complicated by the extent of connectedness between cisternae (Patterson et al., 2008; Trucco et al., 2004). That is, when observing recovery, is it just from the remaining contents of a partially bleached cistern or also from a connected unbleached cistern. Such conditions complicate the analysis of D and may prevent fitting of data by traditional diffusion analysis.

When a diffusion equation or simulation is unable to fit fluorescence recoveries, the $t_{1/2}$ can be used to compare relative recovery rates between samples. The $t_{1/2}$ is the time in which the fluorescence intensity in the bleach ROI recovers to 50% of the asymptote or plateau fluorescence intensity. This value is *independent* of the prebleach ROI fluorescence intensity. While the $t_{1/2}$ can be a useful tool, it is only relevant for the user's system and conditions. The $t_{1/2}$ cannot be compared to $t_{1/2}$ values obtained by other investigators since the $t_{1/2}$ depends on the experimental setup.

To perform $t_{1/2}$ analysis, the data must have been collected using identical acquisition conditions—the same bleach ROI, number of bleach iterations, scan rate, frame rate, laser power, and even relative bleach ROI position (i.e., if the ROI is at the edge of the GC, all other ROIs must be similarly placed. If one ROI is on the edge and the next is in the middle, then one would receive fluorescence from one side and the other ROI would receive fluorescence from two sides, which should significantly impact the recovery rate).

The $t_{1/2}$ requires one data manipulation, conversion of the photobleach ROI recovery data into fractional fluorescence. The $t_{1/2}$ can be determined visually or by solving the following equation for each data set (Feder et al., 1996):

$$F(t) = \frac{\left[F_0 + F_\infty\left(\frac{t}{t_{1/2}}\right)\right]}{\left(1 + \left(\frac{t}{t_{1/2}}\right)\right)}$$

F_0 is the bleach ROI immediate postbleach intensity
F_∞ is the asymptote of the bleach ROI fluorescence recovery
t is the time for each ROI intensity value, usually in seconds or milliseconds
$t_{1/2}$ is the time required for the bleach ROI to recover to 50% of the asymptote.

12.2.7.4 Calculation of fractional fluorescence

To directly visualize and determine the $t_{1/2}$, transform the fluorescence intensity ($F(t)$) data to a 0–100% scale (see Fig. 12.4). The measurement is independent of the prebleach intensity and is not bleach corrected, as the relevant data occurs after the photobleach. The fluorescence recovery into the photobleach ROI must form a genuine plateau or asymptote or $t_{1/2}$ analysis cannot be performed. It is also important to have a large signal difference between the first postbleach intensity and the plateau. If the signal difference is too small, the plotted recovery will tend to be very broad and difficult to interpret. The following equation will convert ($F_2(t)$) the bleach ROI fluorescence recovery into fractional fluorescence data:

$$F_2(t) = 100 \times \frac{[F(t) - F_0]}{[F_\infty - F_0]}$$

$F(t)$ is the bleach ROI fluorescence intensity at time t
F_0 is the immediate postbleach bleach ROI fluorescence intensity
F_∞ is the asymptote of the bleach ROI fluorescence recovery
The $F_2(t)$ data are plotted versus time (in seconds) to determine $t_{1/2}$.

12.3 DISCUSSION

After acquiring and analyzing photobleaching data, one must still consider the problem of how to interpret the results. Few cellular proteins diffuse in a true random walk. Instead, diffusion barriers (i.e., the cytoskeleton, the nuclear pore, barriers within membranes, etc.), bind and release of proteins from membranes or relatively immobile proteins, and nondiffusive movements (i.e., motor proteins, vesicular traffic) all impact molecular movement relevant to the GC. Thus, standard diffusion analysis may be inappropriate for many investigations into GC function. While often noisy, $t_{1/2}$ measurements make no assumptions concerning the nature of molecular movement and can be used to make comparative measurements. Ultimately, most interesting photobleaching experiments concern comparative rather than absolute measurements. The effects of a treatment or mutant on the mobility of a protein primarily concern whether there is a change and how much change there is. Determining a D value certainly can be useful as it can be compared across a wide range of reported results, while $t_{1/2}$ values can only be directly compared for experiments with the identical microscope setup and imaging parameters.

With the availability of multiple spectrally distinct FPs, the investigator is encouraged to perform experiments with a control protein and the protein of interest in the same cell. Most confocal microscopes are equipped with multiple lasers and it should be possible to simultaneously photobleach two (or even three) different FP reporters within the same GC.

Today, the investigator is limited primarily by imagination in the study of GC proteins and membranes. Modern laser scanning confocal microscopes are sensitive enough and fast enough to image most GC proteins. The newer FPs ensure that reporters can behave more physiologically. The greatest limitation remains resolution of molecular movements at high speeds in individual cisternae. Superresolution microscopy techniques (Hell, 2007) suggest even this may be possible.

SUMMARY

Photobleaching of FPs in living cells is a powerful technology. As described here, the choice of FPs can adversely affect experimental results. Therefore, choosing sfGFP variants, truly monomeric proteins, and maintaining a healthy skepticism of untested FPs is critical for physiologic studies of the secretory pathway. Defining the capabilities of one's microscope is also critical for successful experiments. Insufficient laser power can mean the difference between an intentional FRAP experiment and an unintentional FLIP experiment. Finally, it is important to carefully consider the choice of quantitative analyses for different types of proteins and organelle structures. With the methods and considerations described in this chapter, investigators should be well positioned to exploit photobleaching microscopy techniques.

Acknowledgments

The author is supported by a grant to the Marion Bessin Liver Center (NIH/NIDDK 5 P30 DK041296) and is grateful for the use of microscopes in the Albert Einstein College of Medicine Analytical Imaging Facility and the resources of the Cell Structure and Imaging Core of the Marion Bessin Liver Center.

References

Altan-Bonnet, N., Sougrat, R., Liu, W., Snapp, E. L., Ward, T., & Lippincott-Schwartz, J. (2006). Golgi inheritance in mammalian cells is mediated through endoplasmic reticulum export activities. *Molecular Biology of the Cell*, *17*, 990–1005.

Aronson, D. E., Costantini, L. M., & Snapp, E. L. (2011). Superfolder GFP is fluorescent in oxidizing environments when targeted via the Sec translocon. *Traffic*, *12*, 543–548.

Axelrod, D., Koppel, D. E., Schlessinger, J., Elson, E., & Webb, W. W. (1976). Mobility measurement by analysis of fluorescence photobleaching recovery kinetics. *Biophysical Journal*, *16*, 1055–1069.

Cole, N. B., Smith, C. L., Sicaky, N., Terasaki, M., Edidin, M., & Lippincott-Schwartz, J. (1996). Diffusional mobility of Golgi proteins in membranes of living cells. *Science*, *273*, 797–801.

Costantini, L. M., Fossati, M., Francolini, M., & Snapp, E. L. (2012). Assessing the tendency of fluorescent proteins to oligomerize under physiologic conditions. *Traffic*, *13*, 643–649.

Costantini, L. M., Subach, O. M., Jaureguiberry-Bravo, M., Verkhusha, V. V., & Snapp, E. L. (2013). Cysteineless non-glycosylated monomeric blue fluorescent protein, secBFP2, for studies in the eukaryotic secretory pathway. *Biochemical and Biophysical Research Communications*, *430*, 1114–1119.

Ellenberg, J., Siggia, E. D., Moreira, J. E., Smith, C. L., Presley, J. F., Worman, H. J., et al. (1997). Nuclear membrane dynamics and reassembly in living cells: Targeting of an inner nuclear membrane protein in interphase and mitosis. *Journal of Cell Biology*, *138*, 1193–1206.

Feder, T. J., Brust-Mascher, I., Slattery, J. P., Baird, B., & Webb, W. W. (1996). Constrained diffusion or immobile fraction on cell surfaces: A new interpretation. *Biophysical Journal*, *70*, 2767–2773.

Gordon, G. W., Chazotte, B., Wang, X. F., & Herman, B. (1995). Analysis of simulated and experimental fluorescence recovery after photobleaching. Data for two diffusing components. *Biophysical Journal*, *68*, 766–778.

Hell, S. W. (2007). Far-field optical nanoscopy. *Science*, *316*, 1153–1158.

Jain, R. K., Joyce, P. B., Molinete, M., Halban, P. A., & Gorr, S. U. (2001). Oligomerization of green fluorescent protein in the secretory pathway of endocrine cells. *Biochemical Journal*, *360*, 645–649.

Lajoie, P., & Snapp, E. L. (2010). Formation and toxicity of soluble polyglutamine oligomers in living cells. *PLoS One*, *5*, e15245.

Lippincott-Schwartz, J., Snapp, E., & Kenworthy, A. (2001). Studying protein dynamics in living cells. *Nature Reviews. Molecular Cell Biology*, *2*, 444–456.

Lukyanov, K. A., Chudakov, D. M., Lukyanov, S., & Verkhusha, V. V. (2005). Innovation: Photoactivatable fluorescent proteins. *Nature Reviews. Molecular Cell Biology*, *6*, 885–891.

Nehls, S., Snapp, E. L., Cole, N. B., Zaal, K. J., Kenworthy, A. K., Roberts, T. H., et al. (2000). Dynamics and retention of misfolded proteins in native ER membranes. *Nature Cell Biology*, *2*, 288–295.

Nichols, B. J., Kenworthy, A. K., Polishchuk, R. S., Lodge, R., Roberts, T. H., Hirschberg, K., et al. (2001). Rapid cycling of lipid raft markers between the cell surface and Golgi complex. *Journal of Cell Biology*, *153*, 529–541.

Paroutis, P., Touret, N., & Grinstein, S. (2004). The pH of the secretory pathway: Measurement, determinants, and regulation. *Physiology (Bethesda)*, *19*, 207–215.

Patterson, G. H., Hirschberg, K., Polishchuk, R. S., Gerlich, D., Phair, R. D., & Lippincott-Schwartz, J. (2008). Transport through the Golgi apparatus by rapid partitioning within a two-phase membrane system. *Cell*, *133*, 1055–1067.

Pedelacq, J. D., Cabantous, S., Tran, T., Terwilliger, T. C., & Waldo, G. S. (2006). Engineering and characterization of a superfolder green fluorescent protein. *Nature Biotechnology*, *24*, 79–88.

Presley, J. F., Ward, T. H., Pfeifer, A. C., Siggia, E. D., Phair, R. D., & Lippincott-Schwartz, J. (2002). Dissection of COPI and Arf1 dynamics in vivo and role in Golgi membrane transport. *Nature*, *417*, 187–193.

Saxton, M. J. (2001). Anomalous subdiffusion in fluorescence photobleaching recovery: A Monte Carlo study. *Biophysical Journal*, *81*, 2226–2240.

Sbalzarini, I. F., Mezzacasa, A., Helenius, A., & Koumoutsakos, P. (2005). Effects of organelle shape on fluorescence recovery after photobleaching. *Biophysical Journal*, *89*, 1482–1492.

Shaner, N. C., Patterson, G. H., & Davidson, M. W. (2007). Advances in fluorescent protein technology. *Journal of Cell Science*, *120*, 4247–4260.

Siggia, E. D., Lippincott-Schwartz, J., & Bekiranov, S. (2000). Diffusion in inhomogeneous media: Theory and simulations applied to whole cell photobleach recovery. *Biophysical Journal*, *79*, 1761–1770.

Snapp, E. L., Hegde, R. S., Francolini, M., Lombardo, F., Colombo, S., Pedrazzini, E., et al. (2003). Formation of stacked ER cisternae by low affinity protein interactions. *Journal of Cell Biology*, *163*, 257–269.

Swaminathan, R., Hoang, C. P., & Verkman, A. S. (1997). Photobleaching recovery and anisotropy decay of green fluorescent protein GFP-S65T in solution and cells: Cytoplasmic viscosity probed by green fluorescent protein translational and rotational diffusion. *Biophysical Journal*, *72*, 1900–1907.

Trucco, A., Polishchuk, R. S., Martella, O., Di Pentima, A., Fusella, A., Di Giandomenico, D., et al. (2004). Secretory traffic triggers the formation of tubular continuities across Golgi sub-compartments. *Nature Cell Biology*, *6*, 1071–1081.

Zaal, K. J., Smith, C. L., Polishchuk, R. S., Altan, N., Cole, N. B., Ellenberg, J., et al. (1999). Golgi membranes are absorbed into and reemerge from the ER during mitosis. *Cell*, *99*, 589–601.

Zacharias, D. A., Violin, J. D., Newton, A. C., & Tsien, R. Y. (2002). Partitioning of lipid-modified monomeric GFPs into membrane microdomains of live cells. *Science*, *296*, 913–916.

CHAPTER

Photohighlighting Approaches to Access Membrane Dynamics of the Golgi Apparatus

13

Prabuddha Sengupta and Jennifer Lippincott-Schwartz
The Eunice Kennedy Shriver National Institute of Child Health and Human Development, National Institutes of Health, Bethesda, Maryland, USA

CHAPTER OUTLINE

Introduction **218**
13.1 Objectives and Rationale **219**
13.2 Materials and Instruments **219**
13.2.1 Reagents 219
13.2.2 Instrument 220
13.3 Methods **220**
13.3.1 Cell Plating and Transfection 220
13.3.2 Cell Imaging 220
13.3.3 Image Protocols and Acquisition 221
13.3.4 Experimental Strategies 224
13.3.4.1 Strategy 1: Measurement of Cargo Export Out of the Golgi .. 224
13.3.4.2 Strategy 2. Exchange of Golgi Enzymes Between the Golgi and ER 226
13.3.4.3 Strategy 3: Use of Photoactivation to Evaluate Exchange of Golgi Enzymes Between the Golgi and ER 230
Summary **232**
References **233**

Abstract

By providing quantitative, visual data of live cells, fluorescent protein-based microscopy techniques are furnishing novel insights into the complexities of membrane trafficking pathways and organelle dynamics. In this chapter, we describe experimental protocols employing fluorescent protein-based photohighlighting techniques to quantify protein movement into and out of the Golgi apparatus, an organelle that

Methods in Cell Biology, Volume 118

ISSN 0091-679X
http://dx.doi.org/10.1016/B978-0-12-417164-0.00013-6

serves as the central sorting and processing station of the secretory pathway. The methods allow kinetic characteristics of Golgi-associated protein trafficking to be deciphered, which can help clarify how the Golgi maintains itself as a steady-state structure despite a continuous flux of secretory cargo passing into and out of this organelle. The guidelines presented in this chapter can also be applied to examine the dynamics of other intracellular organelle systems, elucidating mechanisms for how proteins are maintained in specific organelles and/or circulated to other destinations within the cell.

INTRODUCTION

The Golgi apparatus functions as a carbohydrate factory at the crossroads of the secretory pathway, receiving proteins and lipids derived from the endoplasmic reticulum (ER) and sorting them into different plasma membrane (PM) intersecting pathways. During this process, proteins destined for the PM, endosomes, and lysosomes are separated from those to be retained in the Golgi and/or recycled back to the ER. How the Golgi is able to perform this selective filtration, including retaining its own specific proteins and lipids in the face of continuous outward trafficking of molecules, has been the subject of intense study and speculation (Altan-Bonnet, Sougrat, & Lippincott-Schwartz, 2004; Emr et al., 2009; Jackson, 2009; Lippincott-Schwartz, 2011; Nakano & Luini, 2010; Pelham & Rothman, 2000).

The development of fluorescent protein (FP)-based highlighting methodologies to directly visualize movement of proteins and organelles in live cells have provided important tools for addressing the dynamics of the Golgi apparatus (De Matteis & Luini, 2008; Lippincott-Schwartz & Patterson, 2003; Lippincott-Schwartz, Roberts, & Hirschberg, 2000; Lippincott-Schwartz, Snapp, & Kenworthy, 2001; Presley, 2005; Storrie, Starr, & Forsten-Williams, 2008). In these approaches, protein populations tagged with FPs in the Golgi are made visible either by photobleaching of molecules outside the Golgi region-of-interest (ROI) or through photoactivation of molecules inside it. Alternatively, FP-tagged molecules outside the Golgi are highlighted following photobleaching of the Golgi pool of fluorescent molecules. Thereafter, the highlighted proteins are tracked as they move into and out of the ROI in real time (Presley et al., 2002; Ward, Polishchuk, Caplan, Hirschberg, & Lippincott-Schwartz, 2001; Zaal et al., 1999). The spatial and temporal resolution achieved therein has enabled quantification of protein export rates from and import rates into the Golgi (Hirschberg & Lippincott-Schwartz, 1999; Patterson et al., 2008), as well as protein residence times within the Golgi (Altan-Bonnet et al., 2004; Storrie et al., 2008; Ward et al., 2001; Zaal et al., 1999). What has emerged from such measurements is a clearer picture of how membrane trafficking and sorting occur within the Golgi.

In this chapter, we describe these fluorescent highlighting protocols for precisely measuring export rates of proteins into and out from the Golgi. The use of these protocols can provide quantitative insights into the kinetics of Golgi membrane trafficking, either retrograde back to the ER or anterograde toward the PM. The imaging schemes can, furthermore, be used to measure dynamic associations of protein and lipids in other intracellular contexts in live cells (e.g., transport in/out of the nucleus). As such, they provide a universal toolkit for probing intracellular trafficking of molecules.

13.1 OBJECTIVES AND RATIONALE

Proteins that traffic to the Golgi apparatus are primarily of two types: cargo molecules that transit through the secretory pathway *en route* to different final destinations, and resident molecules such as Golgi enzymes that are retained in the Golgi for extended periods. Optical highlighting techniques using live cell imaging together with fluorescence recovery after photobleaching (FRAP) approaches or photoactivation have been valuable in teasing out the Golgi trafficking dynamics of these two classes of proteins, including the time frame of their association with the Golgi apparatus and their import and export kinetics from this organelle. This, in turn, has shed light on the mechanisms of protein sorting in the Golgi and the governing principles of Golgi maintenance.

Here, we discuss the operational principles of these optical highlighting techniques in the context of understanding the dynamics and steady-state organization of the Golgi apparatus. Using FRAP, a subset of fluorescent molecules is photobleached in a defined region inside a cell, either to measure the rate of movement of fluorescent molecules into the bleached, dark region, or to highlight a fluorescent region of the cell against a dark background followed by monitoring the movement of fluorescent molecules out of the highlighted region. Using photoactivation, a defined region inside a cell is made bright by switching on the fluorescent molecules in this region followed by the tracking of the molecules out of the photoactivated region. These two approaches (FRAP and photoactivation) can yield complementary data quantifying the kinetic parameters controlling cargo efflux out from and Golgi enzyme retention within the Golgi apparatus. In so doing, the methods can provide important new insights into the dynamic organization of the Golgi apparatus.

13.2 MATERIALS AND INSTRUMENTS

13.2.1 Reagents

Chinese Hamster Ovary (CHO) cells
COS-7 cells
HeLa cells
Plasmid constructs encoding galactosyl transferase (GalT)-Enhanced Green Fluorescent Protein (EGFP), Mannosidase II (ManII) Photoactivable monomeric cherry 1 (PAmCh), sialyl transferase (SiT)-EGFP, vesicular stomatitis virus glycoprotein with temperature sensitive ts045 mutation (VSVG)-green fluorescent protein (GFP), GFP-procollagen, and cleavable signal sequence (ss)-Yellow Fluorescent Protein (YFP)
Dublecco minimum essential medium (DMEM)

Fetal bovine serum
Glutamine
Penicillin
Streptomycin
Fibronectin
Cycloheximide
HEPES buffer

13.2.2 Instrument

Zeiss LSM 510 confocal microscope (or equivalent microscope)

13.3 METHODS

13.3.1 Cell plating and transfection

CHO and COS-7 cells are maintained in DMEM supplemented with 10% (v/v) fetal bovine serum, 2 mM glutamine, penicillin (100 U/mL), and streptomycin (100 μg/mL) at 37 °C in a 5%-CO_2 incubator. Ideally, the cells should be grown in phenol red free DMEM to minimize the background fluorescence during quantitative fluorescence imaging experiments. Cells with a flat morphology should be used for quantitative fluorescence microscopy experiments, so that the total fluorescence signal from the entire depth of the cell can be collected during image acquisition. The cells are plated on fibronectin-coated Labtek or Mattek chambers at a density such that they are 30–40% confluent after 24 h. They are transfected with plasmid constructs tagged with EGFP 24 h after plating.

13.3.2 Cell imaging

Prior to imaging, cells are incubated with fresh media containing cycloheximide (100–150 μg/mL) for 2–3 h at 37 °C for experiments requiring blocking of protein synthesis. Cycloheximide is an inhibitor of mammalian protein synthesis and prevents the synthesis of new copies of the FP-tagged protein of interest. This ensures that quantification of the time-lapse images is not complicated by the presence of newly synthesized proteins. The cells are then transferred to imaging medium, which is regular growth medium (with no phenol red) containing HEPES buffer (20 mM, pH7.4), with or without cycloheximide. Following photobleaching, slowly folding FP, reversible bleaching, or incomplete inhibition of protein synthesis can contribute to the recovery of fluorescence and lead to inaccurate quantification. In order to avoid such contributions, control experiments should be performed where the entire cell is photobleached and monitored for fluorescence recovery over an equivalent period of time. The imaging chamber should be kept humidified by the passage of humid air during image acquisition.

13.3.3 Image protocols and acquisition

The imaging protocols for the experimental strategies discussed in this chapter involve two distinct steps: (1) An initial photobleaching scheme when a user-defined ROI of a cell is selectively photobleached (Strategies 1, 2; Figs. 13.1 and 13.2) or

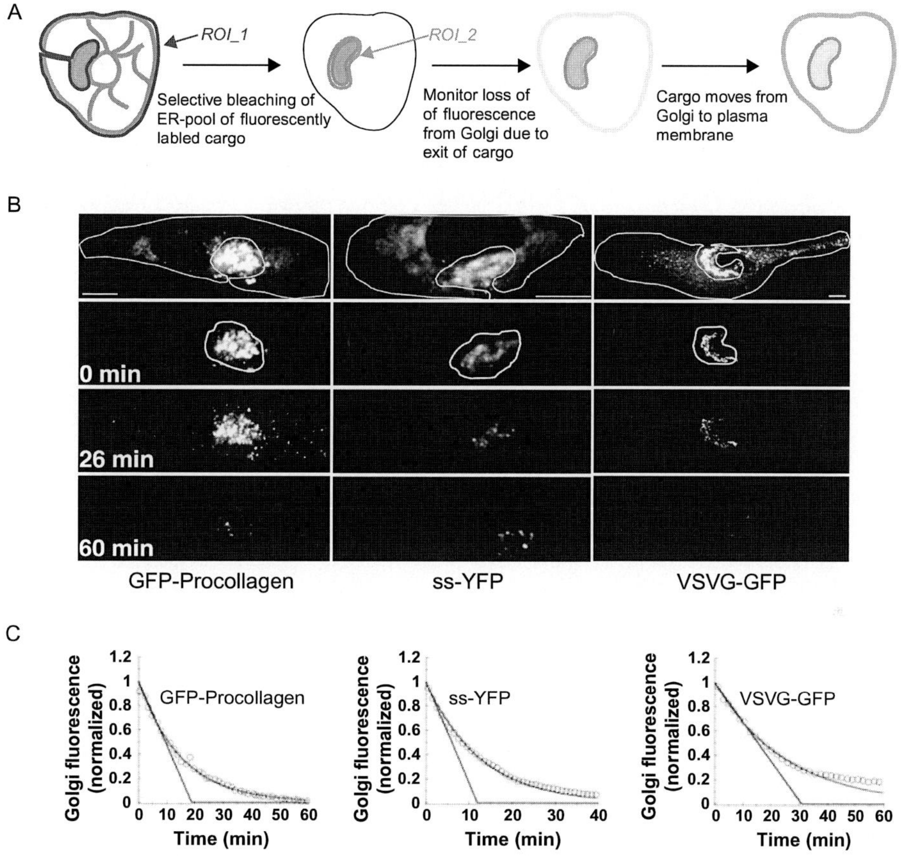

FIGURE 13.1

Workflow and results for Strategy 1. (A) Workflow for Strategy 1. (B) The fluorescent pool of three different cargo (left: GFP-procollagen, center: ss-YFP, right: VSVG-GFP) outside the Golgi apparatus is selectively photobleached, and the movement of cargo out of the Golgi is monitored by time-lapse imaging. (C) The plot of loss of fluorescence of cargo from the Golgi (black circles) shows an exponential decay for all the three cargo molecules. The red line represents the fit of the decay curve to a model involving rapid partitioning of cargo molecules between two different lipid environments in the Golgi and exit of cargo from the export domain according to first-order kinetics. The blue line represents the expected decay for a cisternal maturation model of cargo trafficking through the Golgi. (See color plate.)

Image adapted from Patterson et al. (2008).

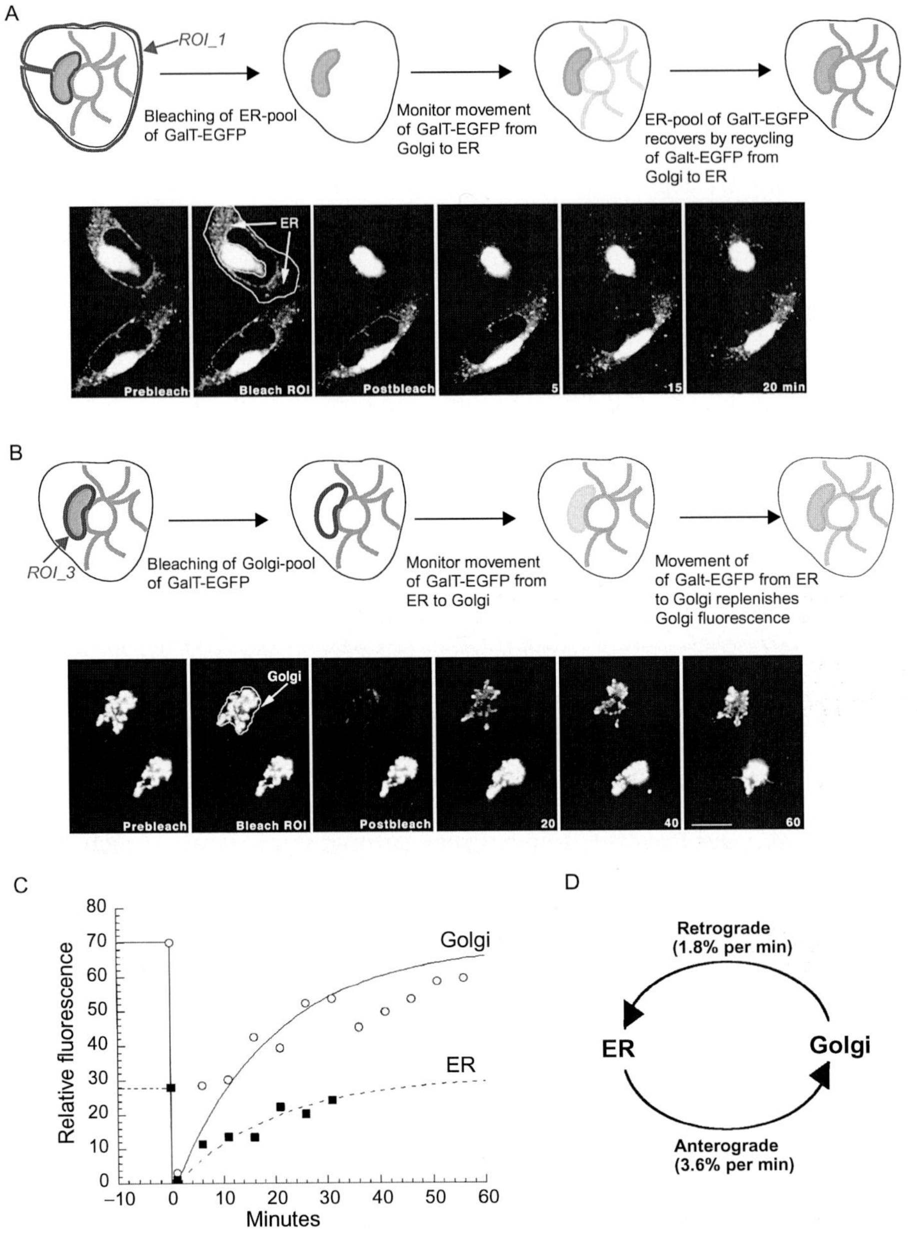

FIGURE 13.2

Workflow and results for Strategy 2. (A) The ER pool of GalT-EGFP is selectively photobleached by illuminating a defined ROI (*ROI_1*) comprising the entire cell except the perinuclear Golgi apparatus. Next, time-lapse images are collected to record the exchange of fluorescent GalT-EGFP between the Golgi and the dark ER network. (B) The GalT-EGFP present in the Golgi is photobleached using an ROI (*ROI_3*) that circumscribes the Golgi apparatus, leaving the ER pool of GalT-EGFP as the only source of fluorescent GalT-EGFP. Time-lapse images are acquired to evaluate the replenishment of fluorescent GalT-EGFP at the Golgi apparatus by movement of GalT-EGFP from the ER to the Golgi. (C) Plot of fluorescence intensity in the ER network (black squares) and the Golgi apparatus

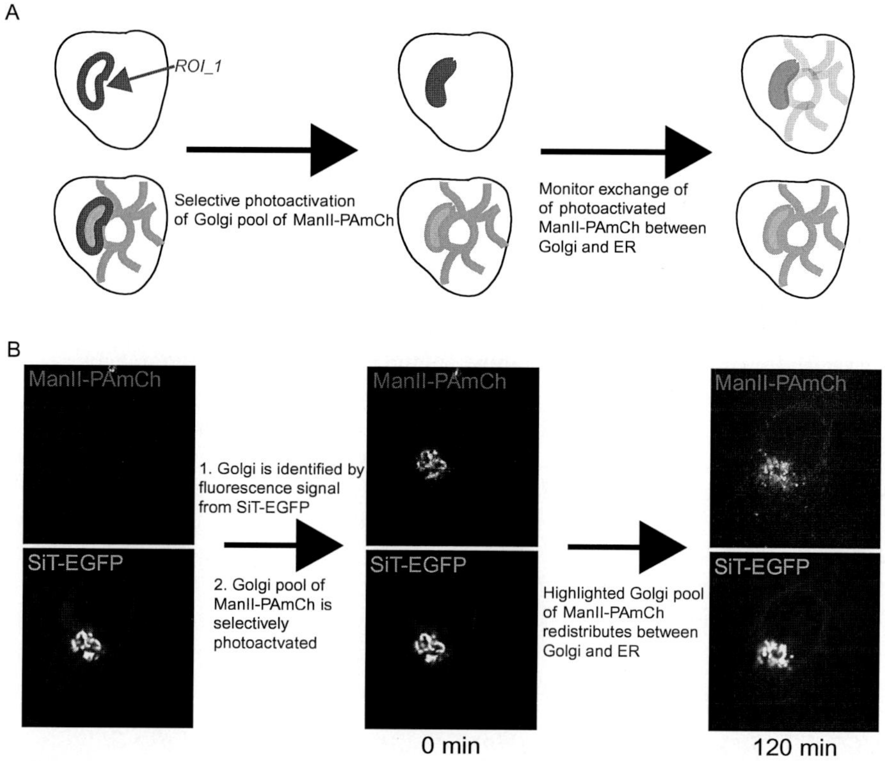

FIGURE 13.3

Workflow and results for Strategy 3. (A) Workflow for Strategy 3. (B) ManII-PAmCh is selectively photoactivated within the Golgi apparatus by illuminating an ROI encompassing the Golgi with a 405-nm laser. The fluorescence signal from SiT-EGFP highlights the Golgi in the green channel and is used to select the ROI. The exchange of photoactivated ManII-PAmCh between the Golgi and ER against a dark background is recorded by time-lapse imaging. (See color plate.)

photoactivated (Strategy 3; Fig. 13.3) and (2) collection of time-lapse images to monitor either the loss (Strategy 1, Fig. 13.1) or gain (Strategies 2 and 3, Figs. 13.2 and 13.3) of fluorescence within a specific area of the cell.

Imaging conditions are chosen upon careful calibration of laser power and exposure time such that there is minimal loss of fluorescence signal because of

(white circles) against time for time-lapse images following selective photobleaching of the ER pool and the Golgi pool of GalT-EGFP, respectively. (D) Rate constants for exchange of GalT-GFP between the ER and Golgi obtained by fitting the data from selective photobleaching experiments. (See color plate.)

Image adapted from Zaal et al. (1999).

photobleaching during the acquisition of the time series. For quantitative imaging, the total amount of fluorescence signal collected from the cell should stay constant, and should not vary as a result of redistribution of FPs between various organelles within the cell. In order to achieve this, the images acquired during the time-lapse should contain the fluorescence signal from the entire volume of the cell. In order to achieve this, images are acquired with a low magnification objective (20 × or 40 ×) and an open pinhole so that the collecting volume matches the *z*-dimensional thickness of the cell, and the fluorescence signal from the entire depth of the cell is collected.

In time-lapse imaging, the Definite Focus feature of the microscope can be used to compensate for any movement of the object being imaged in the plane-of-focus due to internal vibrations of the microscope. The Definite Focus operation uses the coverslip surface as a fixed reference plane and ensures that the images are captured at a user-defined distance from this reference plane.

Extra care should be taken to ensure that none of the pixels are saturated during the collection of images. The laser power, exposure, contrast, and black levels are adjusted so that the image pixel intensities stay within the measurable range during the course of the experiment (0–4095 gray levels for 12-bit image, 0–65,535 for 16-bit images), while sampling the maximum possible dynamic range for the camera.

13.3.4 Experimental strategies

13.3.4.1 Strategy 1: Measurement of cargo export out of the Golgi

13.3.4.1.1 Background and objective

Different models of intra-Golgi trafficking make different predictions regarding the export kinetics of cargo from the Golgi during secretory transport. For example, in the classic cisternal maturation model for Golgi transport, cargo is proposed to traverse the Golgi stack of cisternae while associated with a single cisterna, without leaving it until reaching the exit face of the cisternal stack (Bonfanti et al., 1998; Glick & Malhotra, 1998; Glick & Nakano, 2009). This predicts that cargo should leave the Golgi at a constant, linear rate, correlating with the pace at which an individual cisterna progresses through the Golgi cisternal stack to its exit face, where the cisterna disintegrates into transport vesicles. In models of Golgi transport invoking rapid vesicle or tubule mediated mixing of cargo prior to Golgi export (Lippincott-Schwartz & Phair, 2010; Patterson et al., 2008), by contrast, cargo export from the Golgi is predicted to be an exponential process (akin to radioactive decay), with the export kinetics dependent on cargo becoming concentrated in an export domain potentially present in every cisterna. Fluorescent highlighting approaches can provide valuable information for distinguishing between these two models of intra-Golgi transport owing to their ability to precisely quantify cargo export rates out of the Golgi during secretory transport.

One way to measure cargo export out of the Golgi using fluorescent highlighting is by photobleaching the entire pool of fluorescent cargo molecules outside a defined

Golgi ROI followed by recording of the loss of fluorescence from the ROI (Fig. 13.1A). The selective photobleaching of fluorescence from the entire cell outside the Golgi ROI highlights the Golgi pool of FP-tagged cargo molecules, which, because they have not been photobleached, are bright in comparison to the rest of the photobleached molecules in the cell. Subsequent time-lapse imaging permits visualization and quantification of the fluorescent cargo molecules as they move out from the Golgi apparatus. From the collected data, one can determine whether cargo proteins leave the Golgi at a constant rate (predicted by the classic cisternal progression model) or whether the rate is constantly changing over time as a function of the amount of cargo in the Golgi (predicted by the rapid mixing model).

Here, we describe Golgi export kinetics for three different classes of cargo that have been obtained using this experimental approach. The three classes of cargo include a transmembrane protein (temperature sensitive ts045 VSVG protein, VSVG-GFP); a large soluble cargo (GFP-procollagen); and a small soluble cargo (YFP with a cleavable signal sequence, ss-YFP). VSVG-GFP is retained as a misfolded protein in the ER at 40 °C before being released into the secretory pathway by the temperature shift to 32 °C, which induces the protein to properly fold and leave the ER (Hirschberg et al., 1998). Cells are transfected with one of each of the cargo molecules. Once expressed, the cargo protein is allowed to achieve a Golgi distribution. The Golgi pool is then pulse-labeled by selective photobleaching, and cargo export from the Golgi is measured over time (Fig. 13.1B).

13.3.4.1.2 Flow of experiment

1. Transfect COS-7 cells plated in three different Mattek dishes (or Labtek chambers) with VSVG-GFP, GFP-Procollagen, and ss-YFP, respectively. Maintain the cells at 40 °C for 20–24 hours. Shift cells to 32 °C for 30 min prior to the start of imaging. Perform the following steps with each of the dishes.
2. Mount the dish (or chamber) for imaging.
3. Choose a field of view with a spread out cell expressing the FP-labeled cargo.
4. Select an ROI (*ROI_1*) that includes the entire area of the cell, excluding the perinuclear Golgi network. The pool of the FP-tagged cargo molecule outside the Golgi is present within this ROI.
5. Select a complementary ROI (*ROI_2*) that encompasses the Golgi network.
6. Selectively bleach the FP-cargo fluorescence inside *ROI_1* using a few short pulses of high-intensity laser. Use a 488-nm laser for EGFP-tagged cargo molecules and a 514-nm laser for YFP-tagged cargo molecules.
7. Next, acquire time-lapse images with a low-intensity laser at 1-min intervals for 1 h to record the export of cargo molecules out from the Golgi.
8. Measure the background-subtracted fluorescence intensity of the *ROI_2* at each successive time point of the time series.
9. Plot the fluorescence intensities of *ROI_2* against time to obtain a graph of the loss of fluorescence from the Golgi over time caused by the exit of the FP-cargo molecules.

10. Examine whether the data best fit to a linear or exponential plot. The data will favor the classic cisternal progression model if the data fits a linear curve, and the rapid mixing model if it fits an exponential curve. For an exponential curve, the rate constant for cargo export, λ, can be calculated from $t_{1/2}$ ($\lambda = \ln(2)/t_{1/2}$) of the curve, whereas $1/\lambda$ gives an estimate of the residency time of the cargo in the Golgi.

13.3.4.1.3 Considerations

The fluorescence intensity in the Golgi region decreases during the time course of the experiment because of the exit of cargo from the Golgi. The imaging conditions at the start of the time-lapse imaging (immediately after the photobleaching step) should be set such that the intensity at the Golgi region is as close to the saturation value as possible (maximum recordable value by the camera), while ensuring minimal photobleaching. This enables utilization of a larger portion of the dynamic range of the camera or of the PMTs, enabling a higher detection sensitivity for changes in fluorescence intensity.

13.3.4.1.4 Results

The integration of background-subtracted fluorescence intensity in *ROI_2* measures the fluorescence of the FP-tagged cargo molecule within the Golgi and is directly proportional to the number of FP-cargo molecules present in this region. The fluorescence intensity of the *ROI_2* is measured at each successive time point and plotted against time to obtain a profile of the exit of the FP-cargo molecules from the Golgi.

In the experiment shown in Fig. 13.1, all the cargo varieties exhibited exponential exit kinetics, with each cargo type having a distinct Golgi export rate (i.e., VSVG-EGFP = 0.039 min^{-1}; GFP-procollagen = 0.065 min^{-1}; and ss-YFP = 0.07 min^{-1}; Fig. 13.1C). Therefore, the observed export kinetics supports a rapid mixing model for trafficking of these cargo proteins through the Golgi. For a more detailed discussion on how rapid mixing of cargo molecules within the Golgi can lead to selective export out of the Golgi, see Patterson et al. (2008), which posits an additional membrane partitioning step within the Golgi for export of transmembrane cargo proteins.

13.3.4.2 Strategy 2. Exchange of Golgi enzymes between the Golgi and ER

13.3.4.2.1 Background and objective

Even though the majority of Golgi enzymes reside within the Golgi at any particular moment within the cell, various lines of research have suggested that these enzymes undergo constitutive recycling back to the ER (Miles, McManus, Forsten, & Storrie, 2001; Storrie et al., 1998; Ward et al., 2001; Zaal et al., 1999). By photohighlighting fluorescently tagged Golgi enzymes in the Golgi and following their fate over time, it is possible to measure the length of time the enzymes remain within the Golgi prior to cycling back into the ER (Fig. 13.2A). Moreover, by photobleaching the entire pool of FP-tagged Golgi enzymes and watching the recovery of fluorescence into the Golgi from the nonbleached pool of molecules outside the Golgi in the presence of cycloheximide, one can estimate how quickly Golgi enzymes outside the Golgi (including those in the ER and transport intermediates) are retrieved back into the Golgi (Fig. 13.2B).

To perform these experiments, the entire population of fluorescently tagged Golgi proteins localized within (Fig. 13.2B) or outside the Golgi (Fig. 13.2A) is photobleached with a short, high-intensity laser pulse. Subsequently, fluorescence recovery (into or out of the Golgi) is measured by time-lapse imaging. This approach interrogates whether a Golgi protein cycles between the Golgi and ER during its normal trafficking itinerary, and can provide rates of exchange of the Golgi enzyme between the Golgi and the ER.

13.3.4.2.2 Flow of experiment

STAGE #1. DYNAMICS OF THE ER POOL OF GALT-EGFP

1. Transfect CHO cells with an FP-tagged Golgi enzyme, GalT-EGFP, and allow the protein to equilibrate throughout the secretory pathway by maintaining the transfected cells at 37 °C for 20–24 h. At steady state, GalT-EGFP is primarily localized in the Golgi, with a minor fraction in the ER.
2. Choose a field of view (Window 1) having two cells transfected with GalT-EGFP. One of the cells is subjected to the photobleaching protocol, while the second cell is used as a control to assess the amount of bleaching during the collection of time-lapse images.
3. Choose an ROI (*ROI_1*) in the first field of view (Window 1) that includes the entire cell, excluding the perinuclear Golgi network. The fluorescence in this ROI represents the GalT-EGFP present in the ER.
4. Record a few images (5–10) of the initial GalT-EGFP fluorescence distribution. Make sure that the pinhole is open.
5. Selectively bleach the entire ER pool of GalT-EGFP by illuminating ROI_1 with a few short pulses of a high-intensity laser at 488-nm.
6. Monitor the recovery of fluorescence in the ER by performing time-lapse images at 5-min intervals using a low laser power. Collect the time series for 1.5–2 h, or until you see recovery of fluorescence in the ER approach prebleach levels.
7. Choose a second field of view with one or two GalT-EGFP expressing cells (Window 2).
8. Choose an ROI that includes an entire cell excluding the Golgi apparatus (*ROI_2*).
9. Photobleach the complete pool of GalT-EGFP in the entire cell by illuminating the whole cell with short pulses of high-intensity laser at 488-nm.
10. Collect time-lapse images with low-intensity laser for 1 h to check if there is any recovery of fluorescence in *ROI_2*.

11. Calculate the total amount of fluorescence in the ER in each image of the time series by integrating the background-subtracted fluorescence over all the pixels inside *ROI_1* in the corresponding image.

$$I^i_{\text{ER}} = \sum ROI\_1^i$$

where I^i_{ER} represents ER fluorescence in image frame #i of the time series, and ROI_1^i is the *ROI_1* in image frame #i, and $\sum()$ represents the integration of fluorescence over all pixels in the defined region.

12. Calculate the fluorescence intensity of the Golgi pool of FP-labeled enzyme in each frame of the time-lapse images by subtracting the ER pool of fluorescence (I^i_{ER}) from the total cell fluorescence in the corresponding image.

$$I^i_{\text{Golgi}} = \sum \text{Cell}^i - I^i_{\text{ER}}$$

where I^i_{Golgi} and I^i_{ER} represent the Golgi pool and ER pool of fluorescence in image frame #i of the time series, respectively, and Cell^i is the total cell fluorescence in image frame #i.

13. Plot I^i_{ER} against time to obtain a graph of the recovery of fluorescence in ER with time.

STAGE #2. DYNAMICS OF THE GOLGI POOL OF GALT-EGFP

14. Next, choose a new region with two transfected cells in the field of view.
15. Choose an ROI (*ROI_3*) that encompasses the perinuclear Golgi network. The Golgi pool of GalT-EGFP resides inside this ROI.
16. Acquire a few images to record the initial distribution of the Golgi enzyme.
17. Bleach the fluorescence from the Golgi pool of GalT-EGFP by selectively illuminating *ROI_3* with a few pulses of high-intensity laser at 488 nm.
18. Record time-lapse images at 5-min intervals for 2 h to monitor the recovery of fluorescence in the Golgi network.
19. Measure the Golgi pool of fluorescence in each image by integrating the background-subtracted fluorescence intensity inside *ROI_3* in each image of the time series.

$$I^i_{\text{Golgi}} = \sum ROI\_3^i$$

where I^i_{Golgi} represents the ER fluorescence in image frame#i of the time series, and ROI_3^i is the *ROI_3* in image frame #i.

20. Calculate the fluorescence in the ER in each image frame by subtracting the Golgi pool of fluorescence from the total cell fluorescence in the corresponding image.

$$I^i_{\text{ER}} = \sum \text{Cell}^i - I^i_{\text{Golgi}}$$

where I^i_{Golgi} and I^i_{ER} represent the Golgi pool and ER pool of fluorescence in image frame# i of the time series, respectively, and Cell^i is the total cell fluorescence in image frame # i.

21. Plot I^i_{Golgi} against time to obtain a graph of the recovery of fluorescence in the Golgi with time.
22. Fit the measured values of I_{Golgi} and I_{ER} at each acquisition timepoint for the two stages of the experiments to a two-compartment model to obtain rate constants for the flow of GalT-EGFP from the Golgi to the ER (retrorograde) and from the ER to the Golgi (anterograde).

13.3.4.2.3 Considerations

Cells should be treated with a protein synthesis inhibitor such as cycloheximide (100–150 μg/mL) for 2 h before starting the image acquisition, and should also be maintained in media containing cycloheximide during the entire course of time-lapse imaging. This will ensure that the changes in fluorescence intensities do not arise from the synthesis of new FP-tagged proteins but reflect exchange of FP-tagged protein molecules between the Golgi and ER. In addition to newly synthesized proteins, reversible photobleaching or maturation of slowly folding FPs can contribute to the fluorescence recovery. The control experiment involving monitoring of fluorescence recovery following photobleaching of an entire cell (Steps #7–10) ensures that contributions from these processes are absent or negligible. For situations where protein synthesis inhibition can affect the results of the experiments (by affecting relevant physiological process), however, this strategy should not be used.

Golgi enzymes typically have a half-life of over 20 h or more; thus, the fluorescence intensities should be minimally affected by degradation of the FP-tagged Golgi enzymes. However, if the protein of interest has a shorter lifetime, then protein degradation can change the FP signal significantly. This has to be taken into account for quantitative interpretation of the time-lapse images.

For the second set of experiments (involving photobleaching the Golgi pool of FP-labeled protein), the movement of proteins from the Golgi to the ER will lead to an overall higher average fluorescence intensity in the Golgi region since the proteins are concentrated within a smaller volume at the Golgi than at the source organelle ER. Thus, the imaging conditions should be chosen such that the maximum fluorescence intensity at the ER immediately after photobleaching of the Golgi is significantly below the maximum pixel value that can be recorded by the camera. This will ensure that the increase in fluorescence intensity at the Golgi region stays within the linear range and does not saturate the pixels during time-lapse imaging.

13.3.4.2.4 Results

At steady state, the localization of GalT-EGFP is primarily restricted to two organelles, the Golgi and the ER. By measuring the recovery of GalT-EGFP fluorescence in the Golgi and ER independently following depletion of the fluorescent

GalT-EGFP in either of these compartments, this experimental protocol can provide information about the exchange of the enzyme between the two organelles.

Following selective bleaching of the GalT-EGFP present within *ROI_1*, most of the fluorescent signal in ER is recovered within 1 h (Fig. 13.2A). Following the photobleaching protocol, the only fluorescent pool of GalT-EGFP is present in the Golgi. In the absence of synthesis of new GalT-EGFP, recovery of fluorescence in the ER can occur only from recycling of GalT-EGFP from the Golgi to the ER. Furthermore, there is no fluorescence recovery in the ER when the entire cell is photobleached (*ROI_2* in Window 2), indicating that the fluorescence recovery in *ROI_1* is indeed caused by movement of GalT-EGFP back to the ER.

In the second stage of the experiment, the fluorescence in *ROI_3* (Golgi pool of GalT-EGFP) is recovered within 2 h and reaches the same percentage of total cellular fluorescence as recorded prior to the bleaching step (Fig. 13.2B). This recovery can happen only as a result of movement of GalT-EGFP from the only available source, the ER, indicating that the ER pool of GalT-EGFP is also in constant exchange with the Golgi apparatus.

The mean rate constant for retrograde and anterograde movement of GalT-EGFP obtained from this analysis is 1.8% per minute and 3.6% per minute, respectively (Fig. 13.2C). Taken together, these results indicate that the steady-state distribution of Golgi enzymes such as GalT arise by the continuous cycling of these proteins between these compartments, rather than by stable retention of the proteins in the lipid bilayer of the compartments.

13.3.4.3 Strategy 3: Use of photoactivation to evaluate exchange of Golgi enzymes between the Golgi and ER

13.3.4.3.1 Background and objective

A different strategy for measuring the flux of Golgi enzymes in and out of the Golgi apparatus is the use of photoactivable FPs (PA-FPs) to label a protein of interest and record its trafficking itinerary. Upon irradiation with light of a specific wavelength, photoactivable proteins exhibit dramatic changes in their spectral properties, manifested either by conversion from a dark state to a fluorescent state or emission in a different spectral window (Lippincott-Schwartz & Patterson, 2008). As a consequence, it is possible to selectively illuminate only a fraction of the total proteins within a defined region or an organelle inside the cell, and then track the spatial and temporal evolution of the fluorescent signal against a dark background (Fig. 13.3A). Since only the PA-FPs that are "switched on" during the initial photoactivation step are recorded during the imaging, experiments using PA-FPs are not affected by newly synthesized, dark PA-FP-tagged proteins. This permits quantitative imaging experiments without the need for protein synthesis inhibitor drugs. Consequently, experiments can be performed under more physiological conditions.

We utilized PA-FPs to evaluate whether Golgi enzymes concentrated in the perinuclear Golgi apparatus recycle back to the ER in HeLa cells. ManII, an enzyme involved in *N*-linked glycan processing within the Golgi, is labeled with photoactivable mCherry1 (ManII-PAmCh). Prior to photoactivation, ManII-PAmCh exists in a nonfluorescent dark state. However, upon irradiation with 405 nm light, ManII-PAmCh is converted to a fluorescent state, which absorbs maximally at 561 nm and emits between 590 and 650 nm. We selectively photoactivated ManII-PAmCh present in the Golgi, and then performed time-lapse imaging to find out whether ManII-PAmCh moves out of the Golgi into the dark ER network (Fig. 13.3B). Because it is challenging to identify the Golgi apparatus for selective photoactivation, we coexpressed a different Golgi enzyme, SiT-EGFP, to identify the Golgi region. It is this region that we then irradiate with 405-nm light to switch on the ManII-PAmCh molecules there.

13.3.4.3.2 Flow of experiment

1. Cotransfect cells with ManII-PAmCh and SiT-EGFP.
2. Use the excitation 488-nm laser and the green collection channel to find an expressing cell to image. This cell will show the SiT-EGFP signal only when imaged with 488-nm light. The image from the red channel (565-nm excitation) will have no fluorescence signal since ManII-PAmCh will be in the inactive, dark state.
3. Using the green channel image, select a region of interest (*ROI_1*) that circumscribes the entire Golgi network highlighted by the green fluorescence from SiT-EGFP.
4. Record a few images (~five) of the initial SiT-EGFP fluorescence distribution.
5. Selectively photoactivate the Golgi pool of ManII-PAmCh by illuminating *ROI_1* with a few short pulses of 405-nm laser.
6. Acquire a couple of images of the activated ManII-PAmCh in the red collection channel using a 561-nm laser line at low laser intensity.
7. Now collect time-lapse images at 10-min interval using the same laser power and imaging parameters as in Step5 to capture the loss of fluorescence in the Golgi and the concomitant increase in ER fluorescence because of exchange of fluorescent ManII-PAmCh molecules with dark ones. Continue collecting the time series for 1.5–2 h, or until the ManII-PAmCh fluorescence signal has equilibrated between the Golgi apparatus and the ER network.
8. Calculate the Golgi- and ER pool of ManII-PAmCh in each of the image frames of the time series as described earlier for **Strategy 2**, *Stage #1*.
9. Plot the measured Golgi- and ER pool of fluorescence in successive frames against time.

13.3.4.3.3 Considerations

There are many types of photoactivable proteins that can be used in these experiments. However, for optimal performance, a careful consideration should be made of their photophysical properties such as brightness, photostability, and contrast ratio (i.e., measure of the brightness of the activated or photoswitched form relative to the initial form) (Lippincott-Schwartz & Patterson, 2009; Sengupta & Lippincott-Schwartz, 2012). PAGFP, for example, has appreciable fluorescence in the inactive form, leading to a low contrast ratio. Thus, when possible, it is preferable to replace PAGFP with a different photoactivable probe. We recommend PAmCh or mEos3, each of which has a significantly higher contrast ratio than PAGFP.

We and other labs have noticed that mEos2, a widely used photoswitchable FP, often causes mislocalization of the protein of interest as a result of its tendency to form aggregates at physiological concentrations. Thus, PA-FP-labeled proteins should be tested for proper localization and function. The recently developed mEos3 (Zhang et al., 2012) is a brighter and more monomeric version of mEos2, and is recommended as a replacement for mEos2.

In addition to carefully choosing the photoactivable label, design the protein constructs so that the photoactivable protein is present in an environment where it can be efficiently activated and imaged. For example, certain photoactivable proteins such as PAGFP cannot be efficiently activated and imaged when they are present in the oxidizing environment of the ER lumen. Thus, for applications involving visualization of protein populations in the ER lumen, PAmCh or mEos2 should be used instead of PAGFP.

The laser power and gain should be carefully chosen for the photoactivation experiments. The photoactivation laser can also photobleach the photoactivated FP. Thus the illumination power and exposure time of the photoactivation laser power should be carefully calibrated to ensure optimal photoactivation with minimal photobleaching of the activated FP.

13.3.4.3.4 Results

Following photoactivation of ManII-PAmCh at the Golgi, fluorescence signal in the ER is detected, which progressively increases over the course of 2 h (Fig. 13.3B). Concomitantly, the ManII-PAmCherry fluorescence signal at the Golgi decreases by an equivalent amount. The photoactivation strategy ensures that the Golgi is the only source of fluorescent ManII-PAmCh, since the newly synthesized ManII-PAmCh molecules, which have not been photoactivated, are in a dark, nonfluorescent state. The buildup of fluorescent ManII-PAmCh in the ER is due to the retrograde movement of photoactivated ManII-PAmCh from the Golgi to the ER. Thus, these Golgi enzymes constitutively recycle, from the Golgi back to the ER in interphase cells.

SUMMARY

In this chapter, we have provided detailed experimental protocols for the detection and quantification of the dynamic association of fluorescently tagged proteins with intracellular organelles. Furthermore, these experimental strategies can be used to evaluate the exchange of proteins between two or more organelles within the cell. We have described how we could successfully use these photohighlighting strategies to characterize the temporal and spatial association of proteins with the Golgi apparatus and the ER, and obtain valuable insights into protein trafficking through the secretory pathway. Using photohighlighting, we show that different cargo molecules exit the Golgi with exponential kinetics, which is consistent with a rapid mixing model of cargo trafficking through the Golgi. We have also evaluated the exchange of Golgi enzymes between the Golgi and the ER, and found that Golgi enzymes undergo continuous recycling between these two organelles. Our results underscore the potential of these photohighlighting techniques for elucidation of the dynamic associations of proteins and organelles within the cell.

References

Altan-Bonnet, N., Sougrat, R., & Lippincott-Schwartz, J. (2004). Molecular basis for Golgi maintenance and biogenesis. *Current Opinion in Cell Biology*, *16*, 364–372.

Bonfanti, L., Mironov, A. A., Jr., Martinez-Menarguez, J. A., Martella, O., Fusella, A., Baldassarre, M., et al. (1998). Procollagen traverses the Golgi stack without leaving the lumen of cisternae: Evidence for cisternal maturation. *Cell*, *95*, 993–1003.

De Matteis, M. A., & Luini, A. (2008). Exiting the Golgi complex. *Nature Reviews Molecular Cell Biology*, *9*, 273–284.

Emr, S., Glick, B. S., Linstedt, A. D., Lippincott-Schwartz, J., Luini, A., Malhotra, V., et al. (2009). Journeys through the Golgi-taking stock in a new era. *The Journal of Cell Biology*, *187*, 449–453.

Glick, B. S., & Malhotra, V. (1998). The curious status of the Golgi apparatus. *Cell*, *95*, 883–889.

Glick, B. S., & Nakano, A. (2009). Membrane traffic within the Golgi apparatus. *Annual Review of Cell and Developmental Biology*, *25*, 113–132.

Hirschberg, K., & Lippincott-Schwartz, J. (1999). Secretory pathway kinetics and in vivo analysis of protein traffic from the Golgi complex to the cell surface. *FASEB Journal: Official Publication of the Federation of American Societies for Experimental Biology*, *13*(Suppl. 2), S251–S256.

Hirschberg, K., Miller, C. M., Ellenberg, J., Presley, J. F., Siggia, E. D., Phair, R. D., et al. (1998). Kinetic analysis of secretory protein traffic and characterization of golgi to plasma membrane transport intermediates in living cells. *Journal of Cell Biology*, *143*, 1485–1503.

Jackson, C. L. (2009). Mechanisms of transport through the Golgi complex. *Journal of Cell Science*, *122*, 443–452.

Lippincott-Schwartz, J. (2011). An evolving paradigm for the secretory pathway? *Molecular Biology of the Cell*, *22*, 3929–3932.

Lippincott-Schwartz, J., & Patterson, G. H. (2003). Development and use of fluorescent protein markers in living cells. *Science*, *300*, 87–91.

Lippincott-Schwartz, J., & Patterson, G. H. (2008). Fluorescent proteins for photoactivation experiments. *Methods in Cell Biology*, *85*, 45–61.

Lippincott-Schwartz, J., & Patterson, G. H. (2009). Photoactivatable fluorescent proteins for diffraction-limited and super-resolution imaging. *Trends in Cell Biology*, *19*, 555–565.

Lippincott-Schwartz, J., & Phair, R. D. (2010). Lipids and cholesterol as regulators of traffic in the endomembrane system. *Annual Review of Biophysics*, *39*, 559–578.

Lippincott-Schwartz, J., Roberts, T. H., & Hirschberg, K. (2000). Secretory protein trafficking and organelle dynamics in living cells. *Annual Review of Cell and Developmental Biology*, *16*, 557–589.

Lippincott-Schwartz, J., Snapp, E., & Kenworthy, A. (2001). Studying protein dynamics in living cells. *Nature Reviews Molecular Cell Biology*, *2*, 444–456.

Miles, S., McManus, H., Forsten, K. E., & Storrie, B. (2001). Evidence that the entire Golgi apparatus cycles in interphase HeLa cells: Sensitivity of Golgi matrix proteins to an ER exit block. *The Journal of Cell Biology*, *155*, 543–555.

Nakano, A., & Luini, A. (2010). Passage through the Golgi. *Current Opinion in Cell Biology*, *22*, 471–478.

Patterson, G. H., Hirschberg, K., Polishchuk, R. S., Gerlich, D., Phair, R. D., & Lippincott-Schwartz, J. (2008). Transport through the Golgi apparatus by rapid partitioning within a two-phase membrane system. *Cell*, *133*, 1055–1067.

Pelham, H. R., & Rothman, J. E. (2000). The debate about transport in the Golgi—two sides of the same coin? *Cell*, *102*, 713–719.

Presley, J. F. (2005). Imaging the secretory pathway: The past and future impact of live cell optical techniques. *Biochimica et Biophysica Acta*, *1744*, 259–272.

Presley, J. F., Ward, T. H., Pfeifer, A. C., Siggia, E. D., Phair, R. D., & Lippincott-Schwartz, J. (2002). Dissection of COPI and Arf1 dynamics in vivo and role in Golgi membrane transport. *Nature*, *417*, 187–193.

Sengupta, P., & Lippincott-Schwartz, J. (2012). Quantitative analysis of photoactivated localization microscopy (PALM) datasets using pair-correlation analysis. *BioEssays: News and Reviews in Molecular, Cellular and Developmental Biology*, *34*, 396–405.

Storrie, B., Starr, T., & Forsten-Williams, K. (2008). Using quantitative fluorescence microscopy to probe organelle assembly and membrane trafficking. *Methods in Molecular Biology*, *457*, 179–192.

Storrie, B., White, J., Rottger, S., Stelzer, E. H., Suganuma, T., & Nilsson, T. (1998). Recycling of golgi-resident glycosyltransferases through the ER reveals a novel pathway and provides an explanation for nocodazole-induced Golgi scattering. *The Journal of Cell Biology*, *143*, 1505–1521.

Ward, T. H., Polishchuk, R. S., Caplan, S., Hirschberg, K., & Lippincott-Schwartz, J. (2001). Maintenance of Golgi structure and function depends on the integrity of ER export. *The Journal of Cell Biology*, *155*, 557–570.

Zaal, K. J., Smith, C. L., Polishchuk, R. S., Altan, N., Cole, N. B., Ellenberg, J., Hirschberg, K., Presley, J. F., Roberts, T. H., Siggia, E., et al. (1999). Golgi membranes are absorbed into and reemerge from the ER during mitosis. *Cell*, *99*, 589–601.

Zhang, M., Chang, H., Zhang, Y., Yu, J., Wu, L., Ji, W., et al. (2012). Rational design of true monomeric and bright photoactivatable fluorescent proteins. *Nature Methods*, *9*, 727–729.

CHAPTER

Live Cell Visualization of Golgi Membrane Dynamics by Super-resolution Confocal Live Imaging Microscopy

14

Kazuo Kurokawa*, Midori Ishii*,†, Yasuyuki Suda*, Akira Ichihara* and Akihiko Nakano*,†

**Live Cell Molecular Imaging Research Team, Extreme Photonics Research Group, RIKEN Center for Advanced Photonics, Wako, Saitama, Japan*

†Department of Biological Sciences, Graduate School of Science, The University of Tokyo, Hongo, Bunkyo-ku, Tokyo, Japan

CHAPTER OUTLINE

Introduction 236
14.1 Instrumentation 237
14.2 Materials 237
14.3 Methods 238
14.4 Representative Results and Discussion 239
Acknowledgments 241
References 242

Abstract

Super-resolution confocal live imaging microscopy we developed provides cutting-edge high-speed live cell imaging at high space resolution. With this technology we are now able to observe details of membrane traffic events, including behaviors of small vesicles, cisternal maturation of the Golgi apparatus, and membrane segregation within a compartment.

Methods in Cell Biology, Volume 118

ISSN 0091-679X
http://dx.doi.org/10.1016/B978-0-12-417164-0.00014-8

INTRODUCTION

To solve many remaining problems of the Golgi apparatus, live cell imaging at super-high spatiotemporal resolution is a very powerful and promising approach (Emr et al., 2009; Glick & Nakano, 2009; Nakano & Luini, 2010; Matsuura-Tokita et al., 2006).

This approach has been successfully applied to observe the cisternal maturation process of the budding yeast *Saccharomyces cerevisiae* (Glick & Nakano, 2009; Losev et al., 2006; Nakano & Luini, 2010; Matsuura-Tokita et al., 2006; Papanikou & Glick, 2009; Suda & Nakano, 2012). An early cisterna labeled with a fluorescent protein changes its property to a later cisterna with another fluorescent protein over time. This change is *cis*-to-*trans* directional and involves dynamic mixing and resegregation of membrane components from both earlier and later cisternae (Matsuura-Tokita et al., 2006). Time and space resolutions are very important issues in this live imaging.

Demonstration of cisternal maturation in yeast (Losev et al., 2006; Matsuura-Tokita et al., 2006) gave a strong support for this mechanism in higher organisms that harbor more organized Golgi in the forms of stacks and ribbons (Nakano & Luini, 2010). Arguments continue regarding molecular basis of sorting of cargo and resident proteins (Patterson et al., 2008; Pfeffer, 2010). An intensive group discussion on this problem was held in 2009 at Barcelona, and one important message of this meeting was that advances of high-resolution microscopy would help to clarify disputed issues (Emr et al., 2009).

We had an opportunity to develop a high-speed and high-sensitivity confocal laser-scanning microscope by combining a spinning-disk confocal scanner and a high-sensitivity camera system (Matsuura-Tokita et al., 2006, Nakano, 2002, 2004, 2008; Nakano & Luini, 2010). With this system, collected images contain a massive amount of precise information so that mathematical processing (deconvolution) enables localization of fluorescence origins at very high accuracy. For example, acquisition of 50 optical slices over 5 μm gives 3D reconstruction of small vesicles and tubules of 50–60 nm in diameter after deconvolution (Nakano & Luini, 2010). Since the thickness of confocal slices is 1–2 μm, >10 times more data collection is regarded "oversampling" and thus contributes to the precision of deconvolution. This live imaging system has achieved space resolution beyond the diffraction limit of the classic optical theories, and now we have given it the name SCLIM, standing for super-resolution confocal live imaging microscopy (Ito et al., 2012; Okamoto et al., 2012).

Time and space resolutions are trading off each other because tiny vesicles moving around at high speed have a less probability of being captured during Z-stack image acquisition. So the desire for the high rate of image acquisition is limitless. To acquire image data at a higher rate, not only higher sensitivity but also higher signal-to-noise (S/N) ratio of cameras is critical for correct signal detection. Our current camera system utilizes a set of EM-CCD (electron multiplying charge coupled device) cameras and Image Intensifiers equipped with a custom-made cooling system, which give high-sensitivity (thousands-fold signal amplification) and high S/N ratio at video rate (30 frames/s). Higher acquisition rates are available with the latest EM-CCD cameras, and the new generation CMOS (complementary metal–oxide–semiconductor) devices will provide us even better performance in the near future.

14.1 INSTRUMENTATION

1. *Microscope*: Olympus model IX-71 inverted fluorescence microscope.
2. *Objective lens*: Olympus oil-immersion lens UPlanSApo (100 ×, NA 1.4).
3. *Piezo actuator*: Custom-made piezo actuator (Yokogawa Electric) equipped at the neck of the objective lens. It oscillates the *Z*-axis position of the lens top for up to 100 μm at high frequency (10–30 Hz).
4. *Confocal scanner*: Spinning-disk scanner with microlenses: Yokogawa Electric CSU-10, CSU-X1, or a custom-made model.
5. *Camera*: Hamamatsu Photonics EM-CCD camera C9100-13 (two cameras setup for green and red observation, more for multicolors).
6. *Image intensifier*: Hamamatsu Photonics Image Intensifier with a custom-made cooling system to achieve low S/N ratio amplification (one each for cameras).
7. *Spectroscopic unit*: A custom-made system utilizing dichroic mirrors, reflection mirrors, and band-pass filters (see below for details).
8. *Lasers*: Solid-state lasers with emission at 473 and 561 nm, Blues™ (CW 473 nm, DPSS, 50 mW, Cobolt) and Jive™ (CW 561 nm, DPSS, 50 mW, Cobolt).
9. *Other equipment*: A magnification lens (2.5–10 ×) is put in the light path between the confocal scanner and the spectroscopic unit. This lens is important for raising space resolution. Shutters are operated by the demand of computer. When necessary, the temperature of the sample during imaging is controlled by a thermocontrol stage (Tokai Hit).
10. *Software*: Collection of confocal fluorescence data is controlled by a custom-made system (Yokogawa Electric). Volocity (Perkin Elmer) is used to reconstruct and deconvolve 3D images. Fluorescence image data are quantified with Metamorph (Molecular Devices).

See Fig. 14.1 for the whole setup.

14.2 MATERIALS

1. Yeast cells are usually grown in MCD medium (0.67% yeast nitrogen base without amino acids, 0.5% casamino acids, and 2% glucose) with appropriate supplements at 23 or 30 °C. Cells are harvested at an early-to-mid logarithmic phase and resuspended in the MCD medium before microscopic observation.

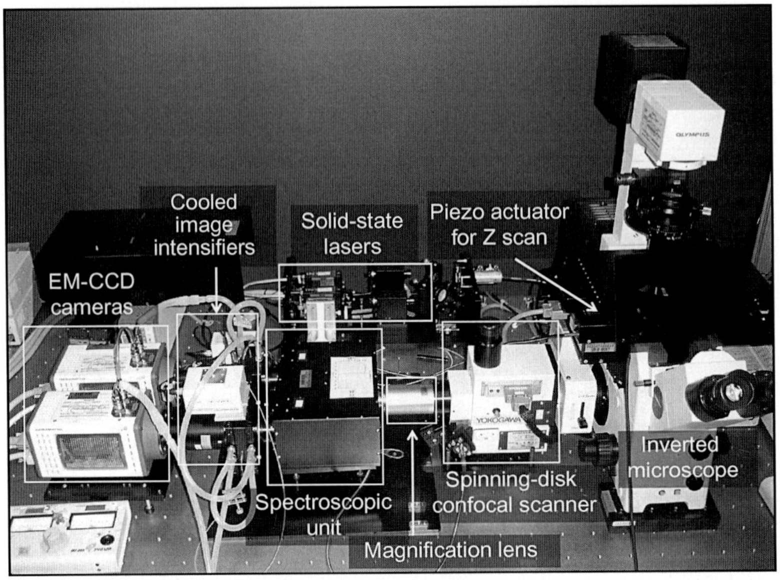

FIGURE 14.1

A setup of SCLIM (two-color system). A custom-made spinning disk confocal scanner is combined with a spectroscopic unit (containing a dichroic mirror and band-pass filters), cooled image intensifiers, and EM-CCD cameras. (See color plate.)

2. For expression of fluorescent proteins, fusion genes are constructed between an appropriate organelle marker or cargo and one of fluorescent protein genes (GFP, mRFP, etc.) usually under control of the constitutive *TDH3* or *ADH1* promoter on a low copy number plasmid pRS314-316 (Okamoto et al., 2012; Sato, Sato, & Nakano, 2001). Plasmids are introduced into cells by standard transformation methods (Burke, Dawson, & Stearns, 2000).
3. Coverslips (40 × 22 mm) are coated with 50 μl each of 0.1% Concanavalin A for 10 min, and then dried at room temperature. An aliquot of yeast liquid culture is sandwiched by a Concanavalin A-coated coverslip and another coverslip (18 × 18 mm), which are spaced with silicon grease. The yeast cells are then subjected to microscopic observation.

14.3 METHODS

1. Typically, yeast cells expressing two cisternal markers (e.g., GFP-Gos1 for medial and Sec7-mRFP for *trans*) are observed by SCLIM at room temperature.
2. 2D confocal fluorescence data (maximum size: 512 × 512 pixels) are simultaneously collected for the two spectral window regions defined by band-pass filters (green, 490–545 nm; red, 580–660 nm). An appropriate range of

Z-axis (e.g., 3 μm cover about 80% of yeast cell volume) is scanned into optical slices 0.1–0.5 μm apart by the piezo actuator equipped to the objective lens.

3. Raw data are once stored in the computer memory (in TIFF format) and read out later to reconstruct 3D images by Volocity. The 3D data are further subjected to deconvolution using point-spread function parameters optimized for the spinning disk confocal scanner by the maximum entropy method (Volocity).
4. Deconvolved 3D images are organized into 3D movies, which can be presented as views at any desired angles.
5. When quantitative analysis is necessary, regions of interest are manually selected and total fluorescent signal intensities in green and red channels are collected to calculate relative fluorescent intensity values using Metamorph.

14.4 REPRESENTATIVE RESULTS AND DISCUSSION

Figure 14.2 (Supplementary Video 1, http://dx.doi.org/10.1016/B978-0-12-417164-0.00014-8) is a recent version of a 3D movie showing cisternal maturation of yeast Golgi from medial (GFP-Gos1) to *trans* (Sec7-mRFP). Into a large structure of green (Gos1 cisterna), red small structures come in, which probably represent vesicles containing the *trans* component (Sec7). Upon increasing arrival of red component, the green structure begins to disassemble and finally dissipate into green small structures, which are again most probably vesicles. Comparison of this movie with our previous ones published in 2006 (Matsuura-Tokita et al., 2006) would let you recognize how much more details of the process we could learn from state-of-the-art live imaging.

The most important advantage of SCLIM is its high speed based on the high-sensitivity detection. Different fluorescent markers have to be imaged in an exactly simultaneous manner. As green and red fluorescent proteins have emission spectra well separated from each other, their simultaneous detection is rather easy. A combination of a conventional dichroic mirror and band-pass filters is good enough to separate green and red. However, limitation of the number of fluorescent markers has been a bottleneck of the experimental designs. For example, directionality of cargo tracking between two compartments can be unambiguously defined only when three markers are available: cargo, donor, and acceptor compartments. To separate multiple colors of fluorescence, several methods have been adapted. The filter wheel method rotating a set of band-pass filters is easy to set up, but mechanical limitation hinders complete simultaneousness. A diffraction grating is a very powerful way to provide spectroscopic information and has been utilized to simultaneously separate several colors of fluorescence, but its weakness is the time for data acquisition.

We have been struggling on this problem and longing for a simple and powerful method for simultaneous color separation at high speed and high efficiency. A great innovation is in progress in designing spectroscopic properties of filters, and we have finally succeeded in producing dichroic mirrors and band-pass

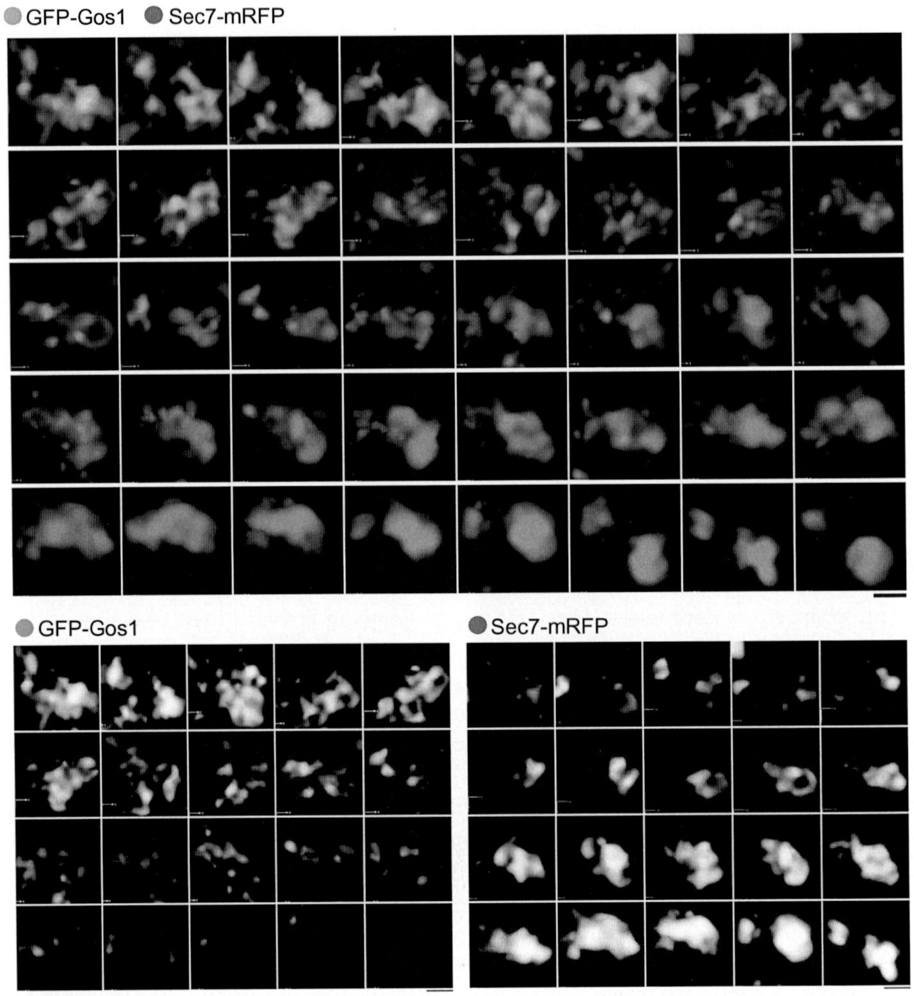

FIGURE 14.2

Cisternal maturation of yeast Golgi observed by the latest system of SCLIM. A yeast cell expressing GFP-Gos1 (medial-Golgi marker) and Sec7-mRFP (*trans*-Golgi marker) is observed by SCLIM. Time-lapse images of Supplementary Video 1 (http://dx.doi.org/10.1016/B978-0-12-417164-0.00014-8) taken at every 6.54 s are shown. Scale bar, 1 μm. (See color plate.)

filters of amazing performance. Figure 14.3 shows the specification of one of such test products, which transmits five narrow ranges of wavelengths in the visible light spectrum (yellow peaks) but efficiently cuts the other regions, and thus enables excitation of samples by five different lasers (e.g., 405, 491, 542, 594, and 671 nm) and collection of fluorescence in five windows (blue, green, orange,

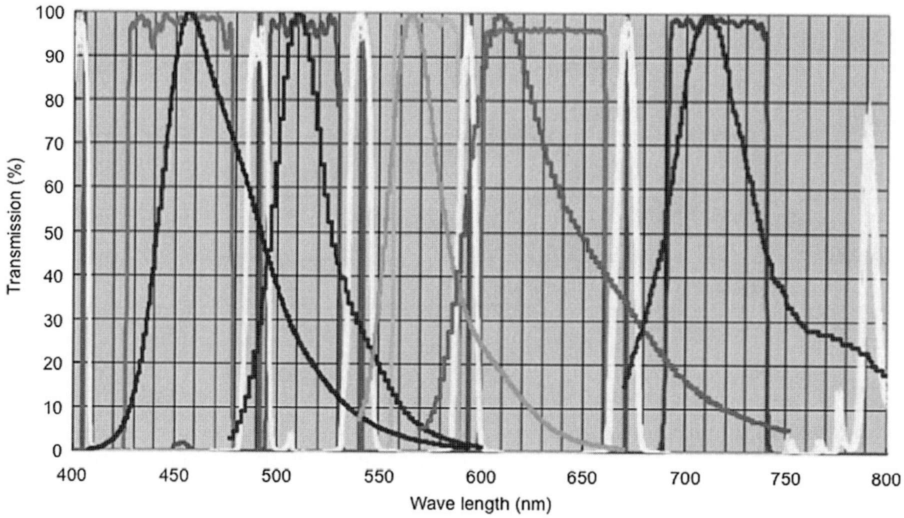

FIGURE 14.3

The spectral properties of a dichroic mirror/band-pass filter set designed for multicolor SCLIM test. Yellow (DM) line indicates transmission spectrum of a custom-made dichroic mirror and blue (BP1), green (BP2), orange (BP3), pink (BP4), and purple (BP5) lines indicate transmission spectra of five band-pass filters. Fluorescence emission spectra of representative fluorescent proteins (em1–em5) are also overlaid. (See color plate.)

pink, and purple spectra) in the combination with appropriate band-pass filters and mirrors. Because this color separation can be done in a completely simultaneous fashion, it provides us a wonderful opportunity of simultaneous multicolor confocal live imaging (multicolor SCLIM). High resolution in space can be achieved by oversampling and deconvolution equally well. Choice of fluorescent proteins with appropriate spectra that fit these windows is another problem to address, because fluorescence properties and behaviors in fusion proteins vary and depend on constructs. We will be able to find conditions to label cargo and several different compartments with different colors in the near future, which will greatly benefit understanding mechanisms of protein sorting within the Golgi apparatus.

Acknowledgments

We are grateful to all the members of the Nakano laboratory for development and evaluation tests of SCLIM. Studies mentioned in this chapter were supported by a Grant-in-Aid for Specially Promoted Research from the Ministry of Education, Culture, Sports, Science and Technology of Japan and by the Extreme Photonics and the Cellular Systems Biology Projects of RIKEN.

References

Burke, D., Dawson, D., & Stearns, T. (2000). *Methods in yeast genetics* (2000 ed.). New York: Cold Spring Harbor Laboratory Press.

Emr, S., Glick, B. S., Linstedt, A. D., Lippincott-Schwartz, J., Luini, A., Malhotra, V., et al. (2009). Journeys through the Golgi—Taking stock in a new era. *Journal of Cell Biology*, *187*, 449–453.

Glick, B. S., & Nakano, A. (2009). Membrane traffic within the Golgi apparatus. *Annual Review of Cell & Developmental Biology*, *25*, 113–132.

Ito, Y., Uemura, T., Shoda, K., Fujimoto, M., Ueda, T., & Nakano, A. (2012). *cis*-Golgi proteins accumulate near the ER exit sites and act as the scaffold for Golgi regeneration after brefeldin A treatment in tobacco BY-2 cells. *Molecular Biology of the Cell*, *23*, 3203–3214.

Losev, E., Reinke, C. A., Jellen, J., Strongin, D. E., Bevis, B. J., & Glick, B. S. (2006). Golgi maturation visualized in living yeast. *Nature*, *441*, 1002–1006.

Matsuura-Tokita, K., Takeuchi, M., Ichihara, A., Mikuriya, K., & Nakano, A. (2006). Live imaging of yeast Golgi cisternal maturation. *Nature*, *441*, 1007–1010.

Nakano, A. (2002). Spinning-disk confocal microscopy—A cutting-edge tool for imaging of membrane traffic. *Cell Structure and Function*, *27*, 349–355.

Nakano, A. (2004). Yeast Golgi apparatus—Dynamics and sorting. *Cellular and Molecular Life Sciences*, *61*, 186–191.

Nakano, A. (2008). Yeast Golgi apparatus. In A. Mironov & M. Pavelka (Eds.), *The Golgi Apparatus, State of the art 110 years after Camillo Golgi's discovery* (pp. 623–629). Wien: Springer-Verlag.

Nakano, A., & Luini, A. (2010). Passage through the Golgi. *Current Opinion in Cell Biology*, *22*, 471–478.

Okamoto, M., Kurokawa, K., Matsuura-Tokita, K., Saito, C., Hirata, R., & Nakano, A. (2012). High-curvature domains of the endoplasmic reticulum (ER) are important for the organization of ER exit sites in *Saccharomyces cerevisiae*. *Journal of Cell Science*, *125*, 3412–3420.

Papanikou, E., & Glick, B. S. (2009). The yeast Golgi apparatus: Insights and mysteries. *Federation of European Biochemical Societies Letters*, *583*, 3746–3751.

Patterson, G. H., Hirschberg, K., Polishchuk, R. S., Gerlich, D., Phair, R. D., & Lippincott-Schwartz, J. (2008). Transport through the Golgi apparatus by rapid partitioning within a two-phase membrane system. *Cell*, *133*, 1055–1067.

Pfeffer, S. R. (2010). How the Golgi works: A cisternal progenitor model. *Proceeding of the National Academy of Sciences of the United States of America*, *107*, 19614–19618.

Sato, K., Sato, M., & Nakano, A. (2001). Rer1p, a retrieval receptor for endoplasmic reticulum membrane proteins, is dynamically localized to the Golgi apparatus by coatomer. *Journal of Cell Biology*, *152*, 935–944.

Suda, Y., & Nakano, A. (2012). The yeast Golgi apparatus. *Traffic*, *13*, 505–510.

CHAPTER

Analysis of Golgi Complex Function Using Correlative Light-Electron Microscopy 15

Elena V. Polishchuk and Roman S. Polishchuk
Telethon Institute of Genetics and Medicine, Naples, Italy

CHAPTER OUTLINE

Introduction **244**
15.1 Materials **247**
15.2 Methods **249**
15.2.1 Cell Transfection, Observation, and Fixation 249
15.2.2 Immunolabeling 250
15.2.3 Embedding 251
15.2.4 Sectioning 252
15.2.5 Serial-section Analysis and 3D Reconstruction 253
15.3 Discussion **255**
Acknowledgments **256**
References **256**

Abstract

Though it has been studied for more than 100 years, the Golgi complex remains a fascinating organelle for cell biologists. This is because it is a unique intracellular structure with a list of well-known functions that continues to grow with the discovery of new genes and the development of novel advanced techniques. The Golgi apparatus has been carefully characterized in different organisms, tissues, and cell types both by light and by electron microscopy. However, the new quality step in Golgi research began with the development of correlative light-electron microscopy (CLEM). This innovative method allows visualizing the fate of the same organelle of interest at any moment of its life span first by video imaging in live cells and then by immunoelectron microscopy and 3D reconstruction. Here, we will present the historical overview of different Golgi-associated processes characterized by CLEM and describe the detailed protocol of this complex yet valuable technique.

Methods in Cell Biology, Volume 118

ISSN 0091-679X
http://dx.doi.org/10.1016/B978-0-12-417164-0.00015-X

INTRODUCTION

The Golgi complex, with its central position in the biosynthetic pathway, is an important organelle for the modification of newly synthesized proteins and lipids during their passage through the intracellular compartments. Moreover, the specific organization of the trans-Golgi network (TGN) at the exit pole of the Golgi stack allows Golgi to operate as a main sorting station that directs newly synthesized proteins towards their final post-Golgi destinations. As is often the case in biology, each decade is characterized by new revolutionary technologies. The mid-1990s was enlightened by the dawn of the use of green fluorescent protein (GFP) in intracellular trafficking. This new era started when GFP isolated from the jellyfish *Aequorea victoria* was tagged to cargo proteins with the goal of following the behavior of the molecule of interest through every compartment of the secretory pathway. Thanks to GFP, the Golgi became not only functional but also visible in living cells, allowing scientists to monitor endoplasmic reticulum (ER)-to-Golgi and Golgi-to-plasma membrane (PM) transport of the specific cargo molecule G protein of the vesicular stomatitis virus (VSVG) (Hirschberg et al., 1998; Presley et al., 1997). When the effectiveness of GFP-tagged properties in membrane trafficking was proven also by studying the endocytosis of functional epidermal growth factor receptor (Carter & Sorkin, 1998), GFP-based live-cell imaging became one of the most widely used and reliable tools for cell biologists in the membrane trafficking field.

However, the beauty of colored Golgi did not allow researchers to discriminate among its different subcompartments, thus presenting the question of how to understand the mechanisms of intra-Golgi trafficking. Moreover, due to close vicinity of fluorescence Golgi with neighboring organelles, the interaction between them was difficult to follow. Something new was needed to appreciate such dynamic changes in the tightly packed Golgi area for achieving a sufficiently high resolution that could be combined with advanced GFP technology. Fortunately, in 2000 a new method was developed that offered the advantage of analyzing the same biological structures using two complementary methods: dynamic live-cell imaging and high-resolution electron microscopy with subsequent 3D reconstruction. This innovative technique was called correlative light-electron microscopy (CLEM). It was set up during intense investigation of VSVG transport carriers operating between the Golgi and the PM (Polishchuk et al., 2000). The novelty of this study included a presentation of the technique as well as answers to an important and specific biological question. CLEM helped demonstrate that “classical vesicles,” which for many years have been considered to carry cargo molecules from the TGN to the PM, appeared as complex tubular–saccular structures. These findings confirmed the effectiveness and sensitivity of the technique and led to the discovery of a new mechanism of protein exit from the TGN.

Further, the high fidelity of CLEM helped characterize the mechanism of intra-Golgi trafficking of small (VSVG) and large (Procollagen-I (PC-I)) cargo molecules (Mironov et al., 2001). The authors demonstrated synchronous movement of both molecules through the entire Golgi stack, concluding that PC-I aggregates as well as VSVG could traverse the Golgi stack without leaving the cisternal lumen and without entering the small Golgi vesicles. To better understand how a cargo protein behaves in the Golgi stack upon arrival from the ER, a small pulse of VSVG–GFP was allowed to reach the Golgi stack. Then, using the accuracy of CLEM, it was possible to detect just-arrived VSVG exclusively in 1–2 Golgi cisternae across the stack, indicating that VSVG could move through the Golgi without diffusing through all Golgi cisternae of the stack under these conditions (Mironov et al., 2001). These important results helped outline the common mechanism by which both small and large cargo proteins traffic through intra-Golgi compartments. Numerous CLEM-based publications have now uncovered new findings in the field of intracellular trafficking. The mechanisms of formation of constitutive post-Golgi carriers from the TGN have been characterized in detail by both the immunoperoxidase and immunogold CLEM methods (both described later in the practical part of the chapter) (Polishchuk, Di Pentima, Luini, & Polishchuk, 2003). The trick of these experiments was to catch the very fast event of carrier formation from the Golgi. Tubular TGN precursors carrying VSVG were fixed before detachment of free post-Golgi transport carriers. Live-cell imaging allowed to follow the generation of such tubular precursors during their protrusion from the Golgi, while electron microscopy and 3D reconstruction confirmed the connectivity of these tubular structures with the Golgi and showed their complex tubular–reticular morphology (Polishchuk et al., 2003). Importantly, based on similarity with TGN membranes, the authors defined these tubular precursors as large TGN domains specifically implicated in the TGN exit of molecules destined for the constitutive post-Golgi trafficking route (Polishchuk et al., 2003).

A similar analysis was done to simultaneously characterize the fate of two cargo proteins: basolateral VSVG–CFP and apical GPI–YFP-myc. First, it has been shown that after leaving the Golgi, VSVG–CFP and GPI–YFP-myc are located inside the same carrier, which at the EM level appeared as a unique tubular structure with both proteins rather than a chain of vesicles carrying either VSVG–CFP or GPI–YFP-myc. CLEM revealed that the distribution of the aforementioned proteins was strikingly matched in both IF and EM images. Apparently, two proteins were segregated into distinct domains of the same structure when observed *in vivo*. In fact, this was confirmed when cells were fixed and labeled using an antibody to the myc epitope in GPI–YFP. An abundance of gold particles signified domains where GPI was concentrated along a single tubular carrier, while a few gold particles indicated GPI-scarce and VSVG-enriched domains. Taken together, these data indicated that in nonpolarized cells, VSVG- and GPI-anchored proteins move inside the same carriers from the Golgi to the PM (Polishchuk, Di Pentima, & Lippincott-Schwartz, 2004).

As previously mentioned, the Golgi serves as the main source for the generation of transport carriers operating in different post-Golgi transport pathways. One of these routes is utilized for delivery of mannose-6-phosphate receptors carrying lysosomal hydrolases from the TGN to the endosomal system. With the help of CLEM, the shape and behavior of TGN-derived carriers destined to endosomes was analyzed at three different stages: at the moment of their formation, during their travel through the cytosol, and finally at moment of their fusion with endosomes. Apart from typical clathrin-coated vesicles, three new classes of transport carriers of different morphology were shown to be operating in the Golgi-to-endosome transport route (Polishchuk, San Pietro, Di Pentima, Tete, & Bonifacino, 2006). Furthermore, an interesting finding of Golgi organization has been made in uroepithelial cells. Based on CLEM, it has been shown that in nondifferentiated cells of urinary bladder, the Golgi is located near the nucleus and represents a common ribbonlike morphology. Instead, in differentiated cells that have to assemble and deliver the most abundant bladder proteins, uroplakins, to the plasma membrane, the Golgi is organized as fragmented single stacks located near the apical plasma membrane. Such organization of the Golgi in these cells is very important for uniform delivery of uroplakins over the entire apical surface to maintain the blood–urine barrier (Kreft et al., 2010).

Further development of CLEM technology allowed researchers to examine the kinetics of the lysosomal membrane protein LAMP1 at the Golgi/lysosome interface. For this purpose, CLEM based on the Tokuyasu cryosectioning technique was developed (Oorschot, de Wit, Annaert, & Klumperman, 2002). CLEM showed that TGN-derived LAMP1-positive carriers fused directly with the dextran Alexa 568-containing lysosome. This was an important experiment that together with others helped the authors conclude that (i) a direct TGN-to-lysosome transport route exists, (ii) noncoated 70–200 nm carriers deliver newly synthesized lysosomal proteins from the TGN to lysosomes, and (iii) hVps41 and VAMP7 regulate trafficking through this direct pathway (Pols et al., 2013).

Often, very well-known functions of the Golgi helped researchers test new developing techniques. For example, the distribution of *N*-acetylgalactosaminyltransferase-2-GFP within the Golgi stack was used to test the reliability of GFP recognition after photobleaching (GRAB) method. After GFP bleaching and photooxidation of 3,3′-diaminobenzidine (DAB) into an electron-dense precipitate, the product of the reaction was analyzed by both electron microscopy and electron tomography. Correct Golgi localization of *N*-acetylgalactosaminyltransferase-2-GFP made GRAB one of the possible candidates among new advanced CLEM techniques (Grabenbauer et al., 2005). In fact, by using this GRAB method, the laboratory of Pavelka was able to follow the GFP photoconversion of five different target proteins and Bodipy ceramide lipid. Interestingly, by comparing the specific distribution of Bodipy ceramide inside the TGN, the authors performed a precise tomographic analysis of TGN/ER contact size. The complex helical organization of TGN/ER junction was revealed by the detailed 3D reconstruction (Meiblitzer-Ruppitsch et al., 2008).

Many conflicting reports regarding the dynamic changes of the Golgi and its relationship with the ER during mitosis coexist in literature (Rossanese & Glick, 2001). New insight into understanding this biological problem came from the work of the Ellisman group. To better understand the precise identity of Golgi membranes during mitosis, researchers in this group generated a new sort of CLEM (Gaietta et al., 2006). They produced a construct of alpha-mannosidase II (Man-II) tagged with GFP for live-cell imaging and a tetracysteine motif to avoid antibody labeling. Instead of antibody staining, a biarsenical compound ReAsH bound to tetracysteine component allowed for the FRET-based photoconversion of DAB and the subsequent visualization of Man-II in electron microscopy. As expected, the electron-dense product of DAB was found in medial trans-Golgi cisternae of interphase cells. Interestingly, during metaphase, Man-II positive vesicles situated along the metaphase plate were clearly separated from ER membranes, while those at the cell poles were often close to the ER. These results clearly indicate a potential interaction between the Golgi and ER membrane systems during mitotic metaphase (Gaietta et al., 2006).

The other peculiarity of the Golgi is its capacity to receive and modify either small or big cargo proteins. Whether or not such distinct molecules traverse the Golgi using the same mechanism has been the main question posed by different laboratories for years. In 2009, another interesting publication in *Traffic* by Perinetti et al. revealed how individual Golgi stacks containing big procollagen III aggregates might be helpful for understanding membrane trafficking by correlating their structure by both superresolution 4Pi light microscopy and CLEM. A detailed comparative analysis showed high morphological correlation of individual Golgi stacks induced by nocodazole (NZ) treatment and allowed the use of 4Pi microscopy as an accurate tool to study the movement of small cargo molecules. Subsequent transport experiments demonstrated that VSVG–GFP rapidly traverses the Golgi stack (Perinetti et al., 2009). Importantly, within the Golgi cisternae, VSVG–GFP is partially segregated from the Golgi-resident enzyme galactosyltransferase (GalT-venusYFP), confirming previously reported data on the cargo/enzyme segregation during the course of intra-Golgi trafficking (Patterson et al., 2008).

Therefore, CLEM, despite its technically demanding nature, proved to be a very efficient and fascinating tool for studying Golgi structure and function. Moving to the practical part of the chapter, we would like to ask readers not to be intimidated by the length of the protocol. We preferred to make each step of CLEM as clear as possible. We believe that compared to different variations of this method, our preembedding CLEM remains the cheapest and most robust technique for understanding fast cellular events and Golgi functions. Our experience has shown that even though it may appear difficult, it is nevertheless achievable, especially when the researcher knows the precise biological question to be answered. The outcome of CLEM experiments could be spectacular and even exceed the most optimistic expectations. On the other hand, it should be noted that CLEM is recommended when no other method can be applied to address a specific issue during functional analysis of intracellular organelles and molecules.

15.1 MATERIALS

1. 35 mm, glass grid-bottomed dishes (MatTek Corporation, P35G-2-14-C-grid).
2. HEPES: 200 mM, pH 7.3.
3. Paraformaldehyde (PFA): 4% solution in 200 mM HEPES, pH 7.3. Put just under 100 ml of 200 mM HEPES into a glass beaker with a magnetic stirring bar

and warm to about 70 °C. Weigh out 4.0 g PFA and add this to the beaker. Mix for a maximum of 10 min to have a clear/transparent solution; adjust the final volume to 100 ml with 200 mM HEPES. Using a funnel with a piece of soft filter paper, filter the solution. Store at 4 °C.

4. Mixture of 4% PFA and 0.05% glutaraldehyde (GA): to 10 ml of 4% PFA solution in 200 mM HEPES (pH 7.3) in a plastic container, add 62.5 μl of 8% GA to obtain a mixture with the required final concentrations.
5. Phosphate-buffered saline (PBS, pH 7.4) containing NaCl, KCl, Na_2HPO_4, and KH_2PO_4.
6. Blocking/permeabilizing solution: 0.5% bovine serum albumin (BSA), 0.1% saponin, 50 mM NH_4Cl in PBS. Weigh out 0.5 g BSA, 0.1 g saponin, 0.27 g NH_4Cl. Dissolve these in PBS to a final volume of 100 ml.
7. Primary polyclonal antibody against GFP (Abcam, Cat No. AB 290-50).
8. Antirabbit Fab fragment coupled to 1.4 nm gold particles (Molecular Probes, Cat No. 2004, antirabbit nanogold).
9. Gold enhancement (Nanoprobes, Cat No. 2113, GoldEnhancement for EM).
10. Phosphate buffer: 100 mM, pH 6.8: prepare solution A as 200 mM dibasic sodium phosphate. Weigh out 35.61 g $Na_2HPO_4{\cdot}2H_2O$ and dissolve it in water to 1.0 l. Prepare solution B as 200 mM monobasic sodium phosphate. Weigh out 31.21 g $NaH_2PO_4{\cdot}2H_2O$ and dissolve it in water to 1.0 l. Mix 24.5 ml solution A with 25.5 ml solution B and make up to a total volume of 100 ml with water.
11. Osmium tetroxide (OsO_4): 2% solution in water.
12. Potassium ferrocyanide ($K_4[Fe(CN)_6]{\cdot}3H_2O$): 3% solution in water.
13. Thiocarbohydrazide (CH_6N_4S): 1% solution in water.
14. Uranyl acetate ($UO_2(CH_3COO)_2{\cdot}2H_2O$): 0.5% solution in water.
15. Ethanol (C_2H_5OH): 100%.
16. Epoxy resin (EPON): add 33.6 g EPON (Fluka, Cat No. 45345), 21.0 g dodecenyl succinic anhydride (Fluka, Cat No. 45346), and 16.8 g methyl nadic anhydride (Fluka, Cat No. 45347) together in a test tube. Heat the tube in an oven for 2–3 min at 60 °C and mix gently. Add 0.96 g 2,4,6-tris (dimethylaminomethyl)phenol (DMP-30; Sigma, Cat No. 45348) and mix gently again. It is possible to freeze the EPON in aliquots at −20 °C and to store it for an extended time before use.
17. Slot grids covered with carbon–formvar supporting film Cu slots (Electron Microscopy Science, FCF 2010 Cu slots).
18. Pickup loop (Agar, Cambridge, England).
19. Tweezers.
20. Diamond knife.
21. Hydrofluoric acid: 40%.
22. Special EPON-made resin blocks. Prepare EPON resin blocks before starting the CLEM procedure. Fill the embedding capsules (Polysciences, Inc., Cat No. 08408-50) with fresh EPON and polymerize for 24 h in an oven at 60 °C. After polymerization, remove the EPON blocks from the capsules.

15.2 METHODS

15.2.1 Cell transfection, observation, and fixation

CLEM starts with seeding the cells of interest into the 35 mm Petri dishes (MatTek) with attached glass coverslip. Coverslip should contain a photo-etched grid on its surface. The coordinated grid (see example in Fig. 15.1) allows the cell of interest to be easily identified during the different steps of the CLEM procedure. It is worthwhile to note that cells at high confluence tend to mask the grid on the coverslip, and this, therefore, will complicate the detection of the cell of interest, which was observed

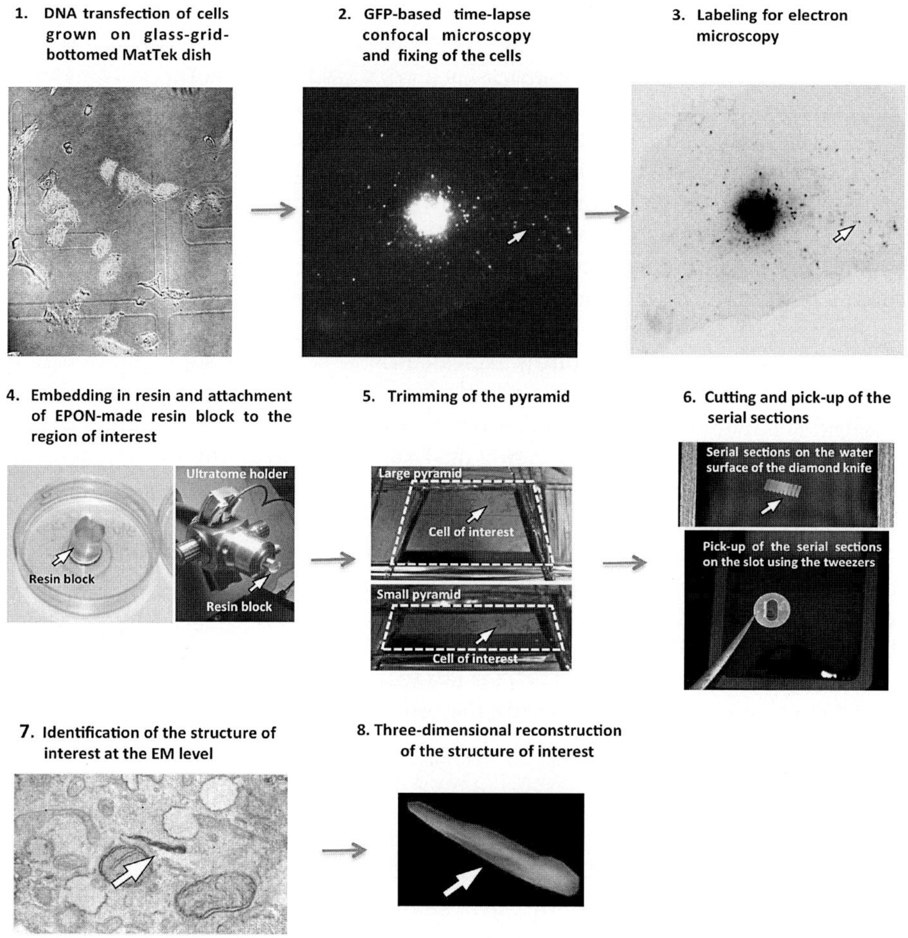

FIGURE 15.1

The stages of CLEM. The sequence of the main steps of the CLEM procedure is shown. Arrows between panels indicate the transition from one step of the technique to the other. For further details, see main text.

in vivo. Thus, quite a low cell density is recommended for the best observations (100,000 cells per 35 mm Petri dish are enough; e.g., for Cos7 or MDCK cells).

The following day, the cells have to be transfected with the GFP-tagged chimera of interest. CFP or YFP can be used instead of GFP for live-cell imaging. Red fluorescent proteins are not recommended due to a lack of the reliable antibody, which works for immuno-EM labeling. Instead of transfection, cells that stably express the GFP-fused protein of interest can be used for CLEM. GFP-expressing cells should then be observed under a fluorescent microscope equipped with a setup that can be used to take a time-lapse series of images. Before observation, it is very important to clean the bottom of the MatTek dish with 100% ethanol, as any dirt can affect the visualization of the coordinated grid. If the grid is barely visible under the 60 × oil-immersion objective, then switching to the 10 × or 20 × objectives can help determine exactly where the cell of interest is located. Draw the position of this cell on the grid map (available from MatTek) and acquire the image with the cell position on the MatTek grid in phase contrast (or DIC) and fluorescence channels.

Having found the structure of interest, video record it until the moment you want to capture for EM analysis. At this point, fix the cells by quickly adding 1 ml of the 4% PFA and 0.05% GA mixture prepared in 200 mM HEPES, pH 7.3. The fixative mixture can be added directly into the culture medium using a pipette. It is important to make sure that the cell of interest does not change its position in the Z-axis and that the structure of interest remains in focus during the fixation process. Focus drift during the fixation process usually happens and can even be caused by slight differences in the temperature between the cell culture medium and the fixative. In this case, the use of a hardware-automated focus device is recommended to keep the structure and the cell of interest in the focal plane during the fixation step. It is important to note that immediately after the addition of the fixative, the GFP fluorescence notably decays. Therefore, to keep the structure of interest visible, the CCD (or photomultiplier) gain has to be amplified at the moment the GFP signal starts to go down. As soon as the intracellular structures end any movement, stop grabbing time-lapse images and carefully change the solution, again adding 2 ml of the mixture of 4% PFA and 0.05% GA; then leave the cells for 10 min at room temperature. Next, "wash" the cells with 4% PFA once to remove the residual GA (as residual GA might affect the recognition of GFP by the antibody) and then postfix the cells with 4% PFA for 30 min.

15.2.2 Immunolabeling

In the past, we have used both immunogold and immunoperoxidase protocols to label structures of interest (Polishchuk et al., 2000, 2003) for its detection under the electron microscope. However, for the past few years, reliable secondary antibodies conjugated with horseradish peroxidase (HRP) have not been available on the market. Moreover, the immunoperoxidase method only allows the protein epitopes within the lumen of membrane-enclosed compartments to be labeled effectively, while the staining of the cytosolic protein domains frequently results in extensive diffusion of the electron-dense product from the place of the HRP-driven chemical reaction (Polishchuk et al., 2000). In contrast, the immunogold protocol with enhancement of ultrasmall particles provides reliable and reproducible labeling, regardless of the compartmentalization of antigen, and, in addition, it allows us to quantify the local density of proteins of interest. Therefore, we recommend the use of this immunogold EM protocol for the CLEM procedure.

To start the immunolabeling, wash the cells three times with PBS. Next incubate the cells with the blocking/permeabilizing mixture (0.5% BSA, 0.1% saponin, and 50 mM NH_4Cl) for 20–30 min and subsequently add the primary polyclonal antibody against GFP, diluted 1:250 in blocking/permeabilizing solution. It is worth noting that instead of using an anti-GFP IgG, an antibody against the protein of interest can be utilized to label the GFP-tagged chimera observed in living cells. However, such an antibody should be checked for its capacity to continue to recognize the antigen after fixation with the 4% PFA/0.05% GA mixture. Usually, 200 μl of the diluted antibody is enough to cover the center of the Petri dish where the gridded coverslip is attached. Incubate the cells with the primary antibody for 1 h at room temperature and then overnight at 4 °C. The following day, wash the cells six times with PBS and then add the secondary antibody, the antirabbit Fab fragment coupled to 1.4 nm gold particles (diluted 1:50 or 1:100 in blocking/permeabilizing solution), and leave for 2 h.

After this incubation with the antibodies, the gold-enhancement reaction has to be performed to increase the size of the 1.4 nm gold particles. The GoldEnhance mixture is prepared immediately before use, as outlined in the succeeding text.

GOLDENHANCEMENT REACTION

- mix 1 part (e.g., 3 drops) component A (enhancer, green cap) with 1 part (e.g., 3 drops) component B (activator, yellow cap) and wait for 10 min.
- Add 1 part (e.g., 3 drops) component C (initiator, magenta cap).
- Add 1 part (e.g., 3 drops) component D (buffer, white cap). Three drops of each solution are enough for one 35 mm Petri dish from MatTek.

Add the mixture to the cells and start observing them using a conventional bright-field light microscope. If you see the cells changing color after only a few minutes, immediately stop the reaction by washing the cells three times with PBS. The timely exposure of the cells in the enhance mixture is very important to achieve gold particles with a homogeneous size and to prevent them from aggregating into big clumps. The Gold-Enhancement procedure is successful when the cells became violet gray in color. At this point, the cells can be processed for embedding directly or washed three times in PBS and stored in 1% glutaraldehyde in HEPES for several days.

15.2.3 Embedding

After immunolabeling, the cells have to be embedded in resin to allow them to be further sectioned into serial slices for analysis under the electron microscope. Wash the immunolabeled cells three times with PBS and postfix the cells with a mixture of 2% OsO_4 and 100 mM phosphate buffer (pH 6.8) (1 part 2% OsO_4 plus 1 part 100 mM phosphate buffer) for 25–30 min on ice. Then wash the cells three times with water and incubate with 1% thiocarbohydrazide diluted in H_2O for 5 min. Wash the cells again, three times with water, and add a mixture of 2% OsO_4 and 3% potassium ferrocyanide (1 part 2% OsO_4 plus 1 part 3% potassium ferrocyanide) and leave for 25 min on ice. Following the OsO_4 postfixation, wash the cells three times with water and then incubate them overnight at 4 °C in 0.5% uranyl acetate diluted in H_2O. We prefer to stain the cells with 0.5% uranyl acetate before inclusion in the resin. Although uranyl acetate staining can also be done on sections, performing this procedure before the inclusion of cells in the resin allows reduced manipulation with slot-collected serial sections, which help to prevent their eventual damage. After the 0.5% uranyl acetate staining, wash the cells six times with water, and dehydrate the specimens by passing them through sequential solutions of 50%, 70%, and 90% ethanol, every 10 min, and then through 100% ethanol for 30 min, changing the 100% ethanol solution three times (every 10 min). Then add the mixture of EPON and 100% ethanol (1:1) to the cells and leave for 2–4 h at room temperature. Change the mixture for pure EPON and leave for 2–4 h at room temperature. We strongly recommend using EPON (and not other resins) for the embedding. Both Spurr and Araldite resins react with the plastic of the dish and are therefore unsuitable for the embedding procedure. Finally, incubate the specimens at 60 °C in an oven for 24 h to allow the EPON resin to polymerize. It is important to note that after the addition of EPON and the resin polymerization, the coordinated grid on the glass bottom of the MatTek dish is no longer visible. To observe your cells on the grid again, the glass needs to be removed with hydrofluoric acid. Take a plastic beaker with a few milliliters of 40% hydrofluoric acid in it and, using tweezers, place the MatTek dish inside the hydrofluoric acid for 30 min. Then take the MatTek dish out of the beaker (always using tweezers), wash it in 200 mM HEPES and then in water, clean it with a piece of soft paper, and dry it. A replica of the coordinated grid will be visible again on the surface of the resin at the bottom of the MatTek dish. In theory, liquid nitrogen can be utilized to detach the coverslip; however, we do not recommend using it because small remnants of the coverslip at the resin block might damage the knife during sectioning. There is also the risk that the thin layer of resin placed over the cells could crack in liquid nitrogen. After dissolving the glass, the special EPON-made resin block (see Section 15.1) has to be attached to the region where your cell of interest is located (see Fig. 15.1). First, use a conventional inverted microscope to locate the cell of interest on the grid. While observing the cell of interest under the microscope, put a mark near the cell on the surface of the resin in the MatTek dish using a STABILO OHPen universal marker. Then add the drop of resin on the top of the mark and stick the EPON block onto it, placing it at the top of the drop. Incubate this adjusted MatTek dish with the attached EPON block in an oven at 60 °C for an additional 24 h.

15.2.4 Sectioning

The sectioning is the trickiest part of the CLEM procedure and usually requires previous experience in the cutting of thin sections and in ultramicrotome use. To prepare the specimen for sectioning, check whether the EPON block and resin in the MatTek dish are fully attached. If so, break the wall of the MatTek dish with pliers and try to eliminate excess resin attached to the EPON block (always using the pliers). Be careful not to touch or damage the area of interest. Next, put the resin block into the holder of the ultramicrotome and examine it under a stereomicroscope. When the position of the cell is found, first, trim the block to a large pyramid (see Fig. 15.1) that has the cell of interest at its center. Then, by rotating the glass-knife stage, align the bottom edge of the pyramid parallel to the knife edge. The pyramid of optimal size should contain just a few cells with the cell of interest in the center (see Figs. 15.1 and 15.2). Bring the sample as close as possible towards the glass knife but do not touch the surface of the pyramid. Adjust the gap (which is visible as a bright band of shadow if all three lamps of the ultratome are switched on) between the knife edge and the surface of the sample. The gap must be the same width between the uppermost and lowermost edges of the sample during the up and down movement of the resin block.

When you are sure of the correct orientation of the pyramid, trim the excess resin to make the surface of the pyramid smaller (see Figs. 15.1 and 15.2). If the cell of interest is the only one present on the pyramid, this will help you to find it easily under the electron microscope. The length of the pyramid should be less than 0.5 mm and its height should be less than 0.1 mm. The small size of the pyramid will allow the collection of more serial sections on the slot. The top and bottom sides of the pyramid should be as parallel as possible to achieve a straight ribbon of slices during the sectioning procedure. Align the pyramid and the diamond knife again and begin cutting serial sections at 60–70 nm intervals, trying not to lose the first sections (see Fig. 15.1). Usually, these sections are very important for identifying of the structure of interest. Pick up the consecutive ribbons of the serial sections on the Cu slots (Electron Microscopy Science, FCF 2010, Cu slots), using the perfect loop or tweezers (see Fig. 15.1).

15.2.5 Serial-section analysis and 3D reconstruction

In our experience, identifying the structure of interest under the electron microscope is facilitated by its immunolabeling and by the presence of some fiduciary markers, namely,

- other GFP-labeled organelles that were observed *in vivo* together with the structure of interest,
- the cell nucleus,
- particular features of cell shape and/or cell protrusions.

Having the fluorescent and phase-contrast (or DIC) images of the cell also significantly helps to identify and track the structure of interest in the serial sections. To detect the structure of interest, place the slot grid with the first serial sections under the electron microscope and locate the cell of interest using the traces of the coordinates on the first few sections. After detecting the correct cell, try to find the

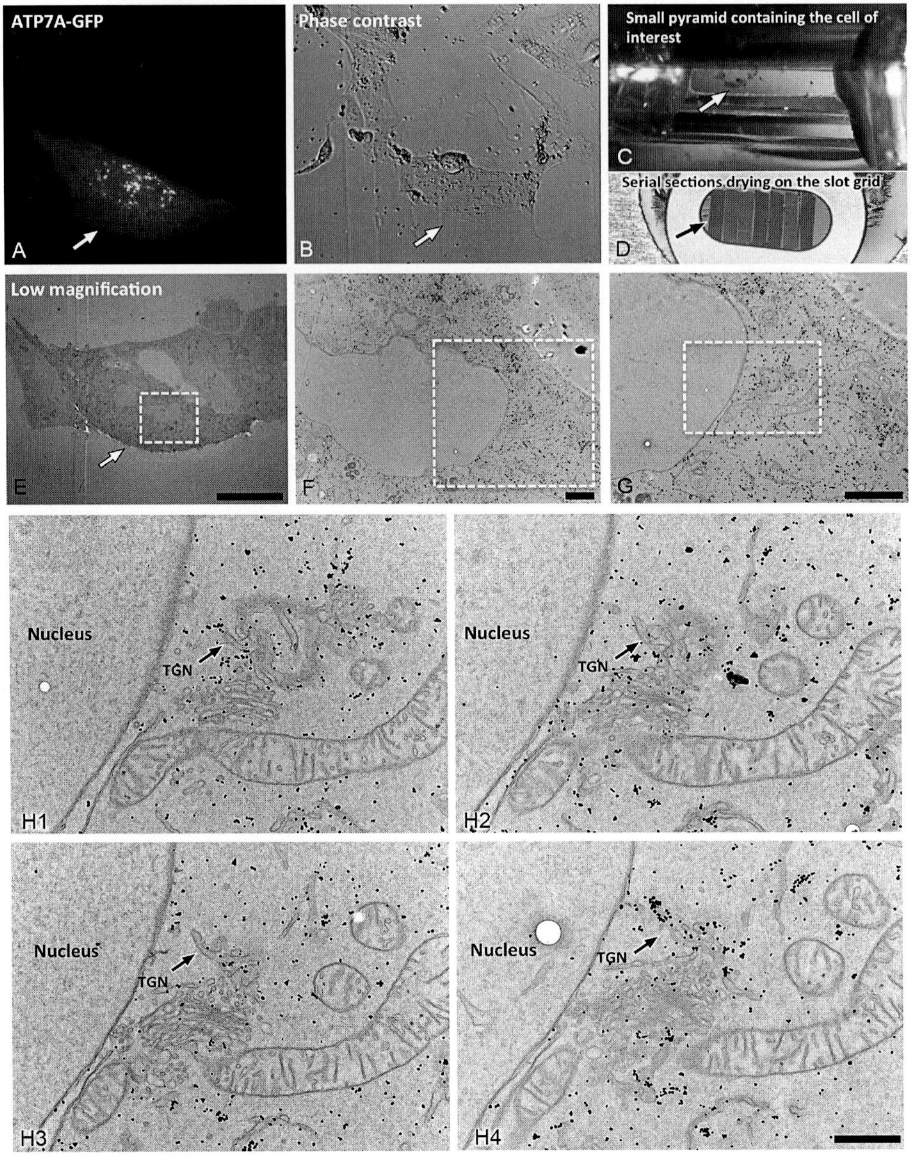

FIGURE 15.2

Example of CLEM analysis of the ATP7A-GFP distribution. (A) MDCK cells were transfected with a plasmid encoding ATP7A–GFP and imaged using confocal fluorescence microscopy. The arrow indicates the cell of interest. (B) The same cell of interest is shown under phase contrast (arrow). (C) The cells were then prepared for immunolabeling and cutting. The small pyramid exhibits the cell of interest on its surface (arrow). (D) Serial sections (indicated by arrow) are shown on the slot grid. (E) This low magnification EM image shows the same cell as in panels A, B, and C (arrow). (F) Higher magnification of the region indicated in panel E by dashed box. (G) The panel shows the area of the cell corresponding to the dashed box region in F. (H1–H4) Localization of ATP7A-GFP labeled with an anti-GFP antibody is shown in four serial thin sections from the region outlined by dashed box in G. The arrow indicates the tubular TGN subcompartment where ATP7A resides. Scale bar, 30 μm (E), 2 μm (F and G), and 500 nm (H1–H4).

structure of interest (that was observed under confocal microscope) and take images of this structure in the subsequent serial sections (see Fig. 15.2).

The image acquisition and overall analysis of the structure of interest can also be accelerated through the use of an automated XY stage in the electron microscope (which is usually available with modern electron microscopes). Such a stage allows to save the position of the structure of interest in each serial section during the EM observation and then to rapidly return to this position for image acquisition or evaluation of particular details of its morphology. Different designs of 3D reconstruction software (e.g., Amira, Volocity, and Imaris) can be used to further align the images of the organelle of interest and to build its 3D model.

If the electron microscope is equipped with a tomography stage, thicker (200–300 nm) serial sections can be used in combination with EM tomography to analyze the 3D architecture of the structure of interest. Usually such a tomography approach gives a more accurate reconstruction of the 3D organization of organelles (with 3–4 nm resolution) than the usual serial sections.

15.3 DISCUSSION

Despite its apparent complexity, we envision a bright future for CLEM in Golgi research. It is quite obvious that in a few years from now, new transport pathways to and from the Golgi will be discovered. Moreover, new emerging roles in basic metabolism, signaling and cell-fate decision were recently assigned to the Golgi (Wilson et al., 2011). Thus, a further understanding of how this complex organelle manages to coordinate such different functions in so compact space will require further CLEM development. In this context, the integration of CLEM with rapidly developing super-resolution light microscopy techniques will allow validation of their real resolution power and can produce new advantageous methods for looking at the Golgi at the molecular and ultrastructural level. Combining superior resolution of stimulated emission depletion microscopy or structured illumination microscopy (Toomre & Bewersdorf, 2010) in live-cell imaging with the power of EM should allow the development of new revolutionary CLEM approaches for the investigation of trafficking, sorting, and signaling events within the tightly packed space of the Golgi stack.

On the other hand, CLEM could also be very useful in Golgi-relevant translational research. In the past 20 years, many new diseases have emerged and new proteins playing key roles in well-known disorders (cancer, diabetes, etc.) have been identified. A variety of genetic, neurodegenerative, and metabolic disorders, as well as cancer, are associated with Golgi localized proteins, which therefore represent the primary focus of studies where CLEM could be effectively used. CLEM will allow scientists to analyze cells with different expression levels of such proteins and, therefore, to detect even tiny morphological variation induced by lower or higher amounts of the proteins present in the cell.

Sometimes, the simple analysis of GFP-tagged Golgi proteins involved in pathology becomes difficult when using constructs with low expression efficiency. Even though a few cells would be sufficient for *in vivo* observation, the electron microscopy analysis of a small population of cells definitely requires CLEM. An example of overcoming this problem with CLEM is shown in Fig. 15.2. ATP7A protein, which is affected in Menkes disease, plays a key role in Cu homeostasis (Mercer, 2001) and tumor resistance to platinum-based chemotherapy drugs (Safaei & Howell, 2005). Although ATP7A intracellular localization has an important role in the aforementioned processes, it remains poorly understood at the ultrastructural level due to the lack of an antibody that is suitable for immuno-EM. On the other hand, large size (1465 residues) and complex structure (8 transmembrane domains) allow the cloning of the ATP7A gene only in vectors with low expression efficiency (La Fontaine, Firth, Lockhart, Paynter, & Mercer, 1998). Therefore, the GFP-tagged ATP7A can only be seen in very few cells. These cells, therefore, can be captured for immuno-EM using CLEM procedure.

The CLEM approach (see detailed description in the text) revealed that ATP7A exhibits a particular distribution over the tubular subdomains of the TGN. Whether and how such a peculiar TGN subdomain is involved in Cu homeostasis and pathogenesis of Menkes disease remain to be elucidated.

In conclusion, we hope that future studies will continue to develop CLEM technology as well as create new opportunities for its use in improving our knowledge of Golgi structure and function in health and disease.

Acknowledgments

We would like to thank Ellen Abrams for critical reading of the chapter. This work was supported by Telethon Grants GTF08001 and TGM11CB4 and by AIRC Grant IG 10233.

References

Carter, R. E., & Sorkin, A. (1998). Endocytosis of functional epidermal growth factor receptor-green fluorescent protein chimera. *The Journal of Biological Chemistry*, *273*, 35000–35007.

Gaietta, G. M., Giepmans, B. N., Deerinck, T. J., Smith, W. B., Ngan, L., Llopis, J., et al. (2006). Golgi twins in late mitosis revealed by genetically encoded tags for live cell imaging and correlated electron microscopy. *Proceedings of the National Academy of Sciences of the United States of America*, *103*, 17777–17782.

Grabenbauer, M., Geerts, W. J., Fernadez-Rodriguez, J., Hoenger, A., Koster, A. J., & Nilsson, T. (2005). Correlative microscopy and electron tomography of GFP through photooxidation. *Nature Methods*, *2*, 857–862.

Hirschberg, K., Miller, C. M., Ellenberg, J., Presley, J. F., Siggia, E. D., Phair, R. D., et al. (1998). Kinetic analysis of secretory protein traffic and characterization of golgi to plasma membrane transport intermediates in living cells. *The Journal of Cell Biology*, *143*, 1485–1503.

Kreft, M. E., Di Giandomenico, D., Beznoussenko, G. V., Resnik, N., Mironov, A. A., & Jezernik, K. (2010). Golgi apparatus fragmentation as a mechanism responsible for uniform delivery of uroplakins to the apical plasma membrane of uroepithelial cells. *Biology of the Cell/Under the Auspices of the European Cell Biology Organization*, *102*, 593–607.

La Fontaine, S. L., Firth, S. D., Lockhart, P. J., Paynter, J. A., & Mercer, J. F. (1998). Eukaryotic expression vectors that replicate to low copy number in bacteria: Transient expression of the Menkes protein. *Plasmid*, *39*, 245–251.

Meiblitzer-Ruppitsch, C., Vetterlein, M., Stangl, H., Maier, S., Neumuller, J., Freissmuth, M., et al. (2008). Electron microscopic visualization of fluorescent signals in cellular compartments and organelles by means of DAB-photoconversion. *Histochemistry and Cell Biology*, *130*, 407–419.

Mercer, J. F. (2001). The molecular basis of copper-transport diseases. *Trends in Molecular Medicine*, *7*, 64–69.

Mironov, A. A., Beznoussenko, G. V., Nicoziani, P., Martella, O., Trucco, A., Kweon, H. S., et al. (2001). Small cargo proteins and large aggregates can traverse the Golgi by a common mechanism without leaving the lumen of cisternae. *The Journal of Cell Biology*, *155*, 1225–1238.

Oorschot, V., de Wit, H., Annaert, W. G., & Klumperman, J. (2002). A novel flat-embedding method to prepare ultrathin cryosections from cultured cells in their in situ orientation. *The Journal of Histochemistry and Cytochemistry: Official Journal of the Histochemistry Society*, *50*, 1067–1080.

Patterson, G. H., Hirschberg, K., Polishchuk, R. S., Gerlich, D., Phair, R. D., & Lippincott-Schwartz, J. (2008). Transport through the Golgi apparatus by rapid partitioning within a two-phase membrane system. *Cell*, *133*, 1055–1067.

Perinetti, G., Muller, T., Spaar, A., Polishchuk, R., Luini, A., & Egner, A. (2009). Correlation of 4Pi and electron microscopy to study transport through single Golgi stacks in living cells with super resolution. *Traffic*, *10*, 379–391.

Polishchuk, R., Di Pentima, A., & Lippincott-Schwartz, J. (2004). Delivery of raft-associated, GPI-anchored proteins to the apical surface of polarized MDCK cells by a transcytotic pathway. *Nature Cell Biology*, *6*, 297–307.

Polishchuk, E. V., Di Pentima, A., Luini, A., & Polishchuk, R. S. (2003). Mechanism of constitutive export from the golgi: Bulk flow via the formation, protrusion, and en bloc cleavage of large trans-golgi network tubular domains. *Molecular Biology of the Cell*, *14*, 4470–4485.

Polishchuk, R. S., Polishchuk, E. V., Marra, P., Alberti, S., Buccione, R., Luini, A., et al. (2000). Correlative light-electron microscopy reveals the tubular-saccular ultrastructure of carriers operating between Golgi apparatus and plasma membrane. *The Journal of Cell Biology*, *148*, 45–58.

Polishchuk, R. S., San Pietro, E., Di Pentima, A., Tete, S., & Bonifacino, J. S. (2006). Ultrastructure of long-range transport carriers moving from the trans Golgi network to peripheral endosomes. *Traffic*, *7*, 1092–1103.

Pols, M. S., van Meel, E., Oorschot, V., ten Brink, C., Fukuda, M., Swetha, M. G., et al. (2013). hVps41 and VAMP7 function in direct TGN to late endosome transport of lysosomal membrane proteins. *Nature Communications*, *4*, 1361.

Presley, J. F., Cole, N. B., Schroer, T. A., Hirschberg, K., Zaal, K. J., & Lippincott-Schwartz, J. (1997). ER-to-Golgi transport visualized in living cells. *Nature*, *389*, 81–85.

Rossanese, O. W., & Glick, B. S. (2001). Deconstructing Golgi inheritance. *Traffic*, *2*, 589–596.

Safaei, R., & Howell, S. B. (2005). Copper transporters regulate the cellular pharmacology and sensitivity to Pt drugs. *Critical Reviews in Oncology/Hematology*, *53*, 13–23.

Toomre, D., & Bewersdorf, J. (2010). A new wave of cellular imaging. *Annual Review of Cell and Developmental Biology*, *26*, 285–314.

Wilson, C., Venditti, R., Rega, L. R., Colanzi, A., D'Angelo, G., & De Matteis, M. A. (2011). The Golgi apparatus: An organelle with multiple complex functions. *The Biochemical Journal*, *433*, 1–9.

CHAPTER

Viewing Golgi Structure and Function from a Different Perspective—Insights from Electron Tomography

16

Brad J. Marsh* and Margit Pavelka†

*Institute for Molecular Bioscience, Centre for Microscopy and Microanalysis, The University of Queensland, Brisbane, Queensland, Australia

†Department of Cell Biology and Ultrastructure Research, Center for Anatomy and Cell Biology, Medical University of Vienna, Schwarzspanierstrasse 17, Vienna, Austria

CHAPTER OUTLINE

Introduction 260
The Golgi Apparatus: A Highly Complex and Dynamic Organelle 260
Electron Tomography: 3D Ultrastructural Imaging of the Golgi Apparatus 265
16.1 New Insights Into Golgi Architecture and Functional Organization Afforded by Electron Tomography 268
16.1.1 Stacks Versus Ribbons: Structure–function Variation Along the Golgi's Vertical Axis (*cis* to *trans*) Influences its Organization Along the Lateral Axis 268
16.1.2 The *cis*- and *trans*-Golgi: Zones of Dynamic Structural and Functional Transition 269
Summary 276
Acknowledgments 276
References 276

Abstract

Historically, ultrastructural investigations, which have focused on elucidating the biological idiosyncrasies of the Golgi apparatus, have tended towards oversimplified or fallacious hypotheses when postulating how the Golgi apparatus reorganizes itself both structurally and functionally to fulfill the plethora of cellular processes underpinned by this complex organelle. Key questions are still unanswered with regard to how changes in Golgi architecture correlate so reproducibly to changes in its

ISSN 0091-679X
http://dx.doi.org/10.1016/B978-0-12-417164-0.00016-1

functional priorities under different physiological conditions or experimental perturbations. This fact alone serves to highlight how the technical limitations associated with conventional two-dimensional imaging approaches employed in the past failed to adequately capture the extraordinary complexity of the Golgi's three-dimensional (3D) structure—now a hallmark of this challenging organelle. Consequently, this has hampered progress towards developing a clear understanding of how changes in its structure and function typically occur in parallel. In this chapter, we highlight but a few of the significant new insights regarding variations in the Golgi's structure–function relationships that have been afforded over recent years through advanced electron microscopic techniques for 3D image reconstruction, commonly referred to as electron tomography.

Abbreviations

2D	two dimensions/two-dimensional
3D	three dimensions/three-dimensional
CCD	charge-coupled device
CGN	*cis*-Golgi network
CLEM	correlative light–electron microscopy
EM	electron microscope/microscopic/microscopy
ER	endoplasmic reticulum
ERGIC	ER–Golgi intermediate compartment
ET	electron (microscope) tomography
GFP	green fluorescent protein
HPF/FS	high-pressure freezing/freeze substitution
keV	kilo electron volt(s)
LM	light microscope/microscopic/microscopy
nm	nanometer
SBFSEM	serial block-face scanning electron microscopy
SEM	scanning electron microscope/microscopy
TEM	transmission electron microscope/microscopy
TGN	*trans*-Golgi network

INTRODUCTION

The Golgi apparatus: A highly complex and dynamic organelle

Because of its extraordinary structural and functional complexity, the Golgi apparatus has presented an ongoing challenge to cell biologists. For over a century now, those who have embarked upon studies of the Golgi apparatus have had to continuously bear in mind the impressive scope of core cellular processes regulated by this enigmatic organelle including professional secretion of proteins and lipids, anterograde and retrograde membrane traffic, endocytosis, and signal transduction, among many others. Collectively, maintenance of these core cellular processes by the Golgi apparatus is what ultimately dictates our state of health at the level of an individual cell, of an organ, and/or of an entire organism. Moreover, it is now well understood that the complex internal mechanisms underpinning the Golgi's capacity to fulfill such a wide range of different functions materially influence its fine structure and gross architecture (for review, see Berger, 1997; Cancino & Luini, 2013; Chia et al., 2012; Donohoe et al., 2013; Farquhar & Hauri, 1997; Glick & Luini, 2011; Glick & Malhotra, 1998; Klumperman, 2011; Nakamura, Wei, & Seemann, 2012; Pfeffer, 2010; Wilson et al., 2011). A hallmark of this organelle is that its complex gross architecture encompasses different subcompartments, which undergo structural changes according to the functional state of a cell (e.g., stages of the cell cycle) and in response to variations in physiological demands upon the cell and/or upon exposure to pathological stimuli. Consequently, the main aim of Golgi researchers in the past and the present has centered on the continued development and application of improved methods/tools that help advance our current understanding about how key structure–function relationships change within this complex organelle under different biological conditions. Notably, working to better understand how gross changes in Golgi architecture may afford us reliable insights regarding its functional state for a given set of conditions remains an important objective in Golgi research. However, over recent decades, it has become increasingly important to refine our broader knowledge regarding how the nature of such gross changes in Golgi architecture that typically accompany changes in its functional priorities—at the organellar level—generally reflects the more specific alterations, which have taken place within each of the multiple subcompartments that it encompasses (each identified by a relatively unique set of structural and functional characteristics). Thus, more recent studies of the Golgi apparatus aim to elucidate its complex internal mechanisms with unprecedented precision in two, three, and even four dimensions through an array of powerful new techniques for mapping fundamental structure–function relationships within the Golgi at the molecular level. In particular, numerous advances in light microscopy (LM) and electron microscopy (EM)—including hybrid approaches for correlative light microscopy–electron microscopy (CLEM)—have yielded new insights regarding how the molecular residents within the Golgi's different subdomains act to cooperatively regulate the organelle's function under different conditions. Inevitably, as variations in Golgi structure are mapped at sufficiently high (i.e., macromolecular) resolution in three dimensions (3D) and four dimensions (4D), Golgi researchers will afford the powerful capacity to precisely identify the sites and molecules contributing to processing or trafficking defects within the Golgi that result in cellular dysfunction and the onset of pathological processes leading to chronic disease states such as diabetes and cancer.

As noted earlier, a variety of new imaging approaches had emerged within the field of Golgi research over the past several decades, all having the same basic objective in mind—that answering fundamental questions related to how the Golgi apparatus reorganizes itself according to different functional priorities requires mapping its structural complexity and architectural variations at high resolution.

However, over the years, a number of disparities have arisen between some key hypotheses formulated from the conclusions, which had been incorrectly drawn from image data that suffered from technical limitations associated with the imaging methods available at the time. The range of resolutions necessary for the most reliable image analysis of the Golgi's finer structural details (e.g., protein coats)—which are at the nanometer scale—has meant that from the very outset, EM has played a leading role (Porter, Claude, & Fullam, 1945). Until recently, however, most ultrastructural studies to examine structure–function variations of the Golgi apparatus have used traditional EM approaches that rely on interpreting the Golgi's 3D structure from conventional two-dimensional (2D) EM images (micrographs) taken from thin (50–80 nm) sections. As previously noted, any conclusions regarding the structural organization of the Golgi apparatus drawn by indirectly extrapolating its "predicted" 3D structure from 2D views must be treated with extreme caution, given that although membrane profiles are clearly visible in a conventional 2D image, the associated underlying structures and their extended architecture are not. Consequently, it became clear many years ago that gaining a more accurate understanding of how the Golgi's complex architectural changes in relation to different functional demands required the development and use of bona fide 3D imaging techniques, which would allow detailed ultrastructural investigation of much larger subcellular volumes than can be achieved by examining conventional thin sections. Originally, 3D studies were confined to stereoscopic investigations of the Golgi apparatus undertaken for a wide variety of different cell types (Rambourg & Clermont, 1997). However, despite substantially refining our understanding of the Golgi's 3D complexity, this method remained severely limited by the fact that 3D structural information within these images was superimposed relative to the *Z*-axis or thickness of the specimen. More recently, the technique now known as "electron tomography" (ET) has emerged as the "state-of-the-art" approach for 3D structure–function studies of the Golgi apparatus using the so-called "semithick" sections (i.e., 200–500 nm) and frequently encompassing multiple serial sections (Donohoe et al., 2013; Fridmann, Mader, Zwerger, Elia, & Medalia, 2012; Landinsky, Mastronarde, McIntosh, Howell, & Staehelin, 1999; Marsh, 2005; Marsh, Volkmann, McIntosh, & Howell, 2004; Martinez-Sanchez, Garcia, & Fernandez, 2011; Mogelsvang & Howell, 2006; Mogelsvang, Marsh, Ladinsky, & Howell, 2004; Storrie et al., 2012). Through their provision of important new insights into the Golgi's complex 3D structure at unprecedented clarity, these studies have demonstrated that many questions relating to the structural and functional organization of the Golgi apparatus will likely only be answered by exhaustive ultrastructural exploration of this organelle using ET.

The architecture of the Golgi apparatus has been described in numerous publications now spanning more than a century, since the basic principles of its organization were originally well outlined in the original drawings of Camillo Golgi (Golgi, 1898; for review, see Dröscher, 1998; Mazarello, 1999). The overall architecture of the Golgi apparatus comprises multiple subunits that are typically concentrated in the cell's center close to the nucleus but are sometimes redistributed to other areas of the cytoplasm; in either case, the subunits often seem to be connected. However, the precise structural details concerning the nature of such connections, for example, how and when they are formed and maintained, have only recently been resolved by 3D studies, which have employed high-resolution ET (Donohoe et al., 2013; Emr et al., 2009; Landinsky et al., 1999; Marsh et al., 2004; Storrie et al., 2012).

The Golgi apparatus' structure–function subunits (i.e., stacks) constitute a series of flat cisternae, which are stacked parallel to one another and generally exhibit unique characteristics allowing them to be allocated to either side (*cis* or *trans*) or the middle (*medial*) of the stack (Fig. 16.1). The *cis*-face of each Golgi stack tends to sit opposite the endoplasmic reticulum (ER) and serves as the main site of entry for newly synthesized molecules transported from transitional ER exit sites to the *cis*-most cisterna, presumably via an ER–Golgi-intermediate compartment (ERGIC) also referred to as vesicular–tubular clusters (VTC) (Appenzeller-Herzog & Hauri, 2006; Bannykh, Rowe, & Balch, 1996). After being taken up by the Golgi apparatus, these newly synthesized proteins and lipids are usually further modified as they progressively move forward through the stacked cisternae before being packaged at the *trans*-side of the Golgi stacks for delivery to their final destinations (for review, see Farquhar & Hauri, 1997). It is well known from conventional EM pictures (micrographs) of the Golgi apparatus that the profiles of flattened stacks of cisternae are typically accompanied by more irregularly shaped membrane profiles belonging to cisternae at either the *cis*- or *trans*-sides of the stack, tending to exhibit a convoluted or reticular appearance, leading to the terms *cis*-Golgi network and *trans*-Golgi network (CGN and TGN, respectively); these networks have already been shown in early 3D studies by serial sectioning and stereoscopic reconstitution (e.g., Mollenhauer & Morré, 1998; Rambourg & Clermont, 1997), whereas from conventional EM analysis, it is not possible to conclude that these are in fact "networks". Vesicle profiles with a "fuzzy" appearance that presumably correspond to COPI protein coats are routinely found adjacent to the membranes of *cis*-cisternae and at the dilated rims of either *cis*- or *medial*-cisternae; conversely, the characteristic "spiky" appearance of clathrin-coated budding profiles is dominant for *trans*-cisternae membranes, which are also frequently in close association with the membranes of specialized ER (*trans*-Golgi ER) (Fig. 16.1) (Marsh, 2005; for review, see Marsh & Howell, 2002). Reliably interpreting the 3D organization of Golgi membranes in these more complex regions of each stack especially is complicated further by the fact that this conventional arrangement of a "typical" Golgi stack is not always apparent in every cell type and may change extensively for different functional states of the cell, such as during increased protein secretion or endocytosis (e.g., Marsh et al., 2004; Pavelka, Neumüller, & Ellinger, 2008). Hence, among the burning questions that remain to be definitely answered by ET are those relating to the different structure–function relationships associated with the import and export of secretory/endocytic molecules; intra-Golgi trafficking; anterograde versus retrograde protein transport and membrane recycling (Cottam & Ungar, 2012;

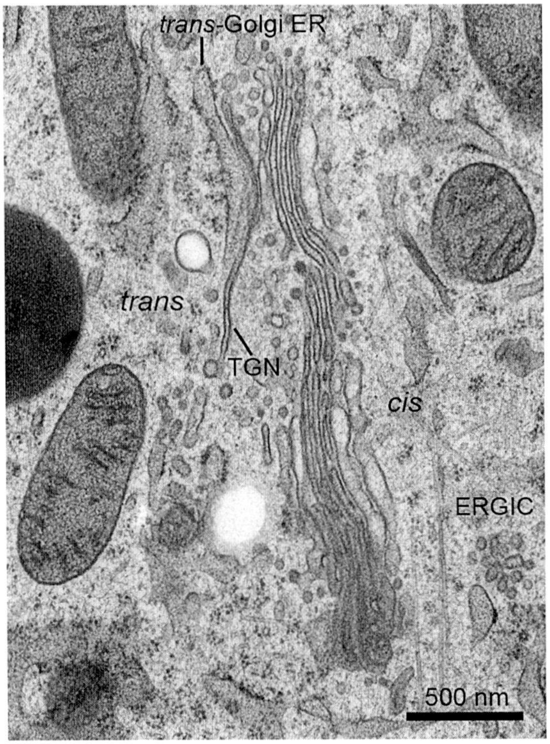

FIGURE 16.1

A Golgi "stack" as seen in a conventional (2D) micrograph of a thin (80 nm) section cut from a HepG2 hepatoma cell prepared by HPF/FS. The structural profiles of *medial*-cisternae are very similar when viewed in cross-section by EM, as shown here. Their limiting membranes are more closely oriented (in parallel) to one another and regularly spaced apart than for cisternae positioned elsewhere within the stack. In contrast to their counterparts at the *cis*- and *trans*-Golgi, the lumens of *medial*-Golgi cisternae tend to be relatively uniform with respect to each other and along their length. Moreover, they are often much narrower in width than either *cis*- or *trans*-cisternae, which are typically wider and exhibit a more irregular and convoluted architecture consistent with their roles in facilitating the entry and exit of membrane traffic to and from the Golgi. As shown in this example, vesicular–tubular clusters adjacent to the *cis*-Golgi region are visible, presumably belonging to the ER–Golgi intermediate compartment (ERGIC). At the *trans*-side of the stack, membrane profiles of the TGN partly attached to the stack are directly accompanied by *trans*-Golgi ER. Bar: 500 nm.

Glick & Luini, 2011; Nakano & Luini, 2010; Pfeffer, 2010; Rothman & Wieland, 1996); subcompartmentalization within the Golgi (Puthenveedu & Linstedt, 2005); packaging and formation of specialized transport carriers (De Matteis, Di Campli, & D´Angelo, 2007; De Matteis & Luini, 2008; Handa, Kumagain,

Tomishige, & Kawano, 2007; Valente et al., 2012); interactions between the Golgi apparatus and cytoskeletal elements of various kinds (Anitei & Hoflack, 2012); physiological and pathological reorganizations of the Golgi apparatus, including those during the cell cycle, in response to cellular stress, environmental changes, and drug treatments (for review, see Dinter & Berger, 1998; Meisslitzer-Ruppitsch et al., 2011; Tomas et al., 2012); and Golgi dissociation/reassembly versus *de novo* formation (Tang, Mar, Warren, & Wang, 2008). In this chapter, we will highlight by example why ET has provided such a powerful approach for shedding new light upon the 3D structural complexity of the Golgi apparatus.

Electron tomography: 3D ultrastructural imaging of the Golgi apparatus

Over the past two decades, in particular, the use of high-resolution ET for detailed structure–function investigations of the Golgi apparatus (i.e., how its membranes and associated tubules and vesicles are organized within mammalian cells under different physiological conditions and in different cell types) has convincingly demonstrated the power and accuracy of this approach of providing novel insights regarding membrane trafficking to, through, and from the Golgi apparatus and especially of providing a clearer understanding of the nature of structural events that occur at the site of cargo entry into the *cis*-Golgi and at its site of cargo exit at the *trans*-face of the Golgi (i.e., the *trans*-Golgi network (TGN)). High-resolution 3D reconstructions generated using ET, which have facilitated the precise ultrastructural dissection of the 3D image volumes encompassing the Golgi region through painstaking examination of the datasets (from one pixel-thick computed slice to the next and in any orientation in 3D), have helped to finally start providing answers to long-standing questions related to the Golgi's complex structure–function relationships.

Improved methods for more optimally preparing cells and tissue for EM (e.g., rapid cryopreservation) have been developed since the late 1970s leading to more reliable structural insights. Moreover, significant advances in computer hardware and software have dramatically improved the accuracy and efficiency with which large 3D image datasets are both acquired and analyzed. Electron microscopes specifically engineered for tomographic data acquisition have been commercially available for more than a decade. Typically, these more advanced microscopes fall into the category of either "intermediate voltage" or "high voltage" EMs since they routinely operate at higher accelerating voltages ($\geq$300 keV) than conventional EMs (i.e., 80–120 keV), enabling thicker than normal sections (e.g., 200–500 nm cf. 60–100 nm) cut from either plastic-embedded or frozen-hydrated cells and tissue to be imaged clearly. Moreover, these instruments offer precise (i.e., eucentric) tilting and repositioning of the sample (i.e., section) over a large angular range (e.g., $\pm$60–80°) relative to the incident electron beam through the use of specialized specimen stages not normally used in conventional EMs (see Fig. 16.2). Over recent years, improved electron optics and computerized control of microscope functions have led to the development and use of dedicated software packages for automatically acquiring the large image datasets (tilt series) needed for subsequent 3D (i.e., tomographic) reconstruction of the Golgi region at high (e.g., 4–8 nm) resolution. Semithick (400 nm) plastic sections cut from insulin-secreting pancreatic beta-cells prepared for ET by high-pressure freezing/freeze substitution (HPF/FS) and poststained with uranyl acetate and lead citrate are routinely imaged in an EM operated at $\geq$300 keV. Ideally, plastic sections are routinely coated with 5, 10, or 15 nm colloidal gold particles on both surfaces for subsequent image registration during tomogram computation, and they are also carbon-coated to minimize specimen instability during image acquisition, especially at high tilt angles (Marsh, 2005; Marsh et al., 2001, 2004). These days, tilt series datasets used for tomogram generation are typically collected using an intermediate voltage FEG–TEM operated at 300 keV. Generally, each section is "preirradiated" at low magnification in the EM for several minutes to promote more uniform physical changes within the thick plastic sections prior to subsequent collection of tilt series images (Marsh, 2005; Noske, Costin, Morgan, & Marsh, 2008). Precise alignment of the 2D image datasets (i.e., tilt series) collected as each section was serially tilted by fine (1–2°) angular increments over a range of ±65° is initially achieved by image cross correlation. For the so-called "dual-axis" tomography, the specimen is then rotated 90° before a second tilt series of exactly the same area is acquired around an orthogonal axis. Tilt series images first aligned by cross correlation are subsequently aligned much more precisely by tracking the positions of the gold fiducial markers across the entire set of images to bring them into register with one another as accurately as possible. Single-axis tomograms are then computed from each set of aligned tilts using the mathematical technique of "R-weighted back projection." The use of fiducial particles to generate a "fiduciary model" for more accurate alignment of individual tilt series images remains a very important step for optimal dual-axis tomographic reconstruction because the coordinates of the corresponding fiducial particles are used to compute the extent of 3D rotation, distortion, and shift between the two single-axis volumes. Tomograms calculated in this manner from each set of aligned tilts are then registered to each other in 3D space by nonlinear transformation using one tomogram as a reference volume and the two single-axis tomograms computationally combined to yield a single, high-resolution (dual-axis) 3D reconstruction (see Tomography Pipeline flowchart). Dual-axis reconstruction not only results in improved symmetry in all 3D but also benefits significantly from the fact that image "noise" artifacts resulting from technical limitations such as reduced tilt range and oversampling through lower tilt angles that are prominent when viewing each single-axis reconstruction are dramatically reduced after the two tomograms are combined in 3D space to produce a "new and improved" dual-axis tomogram.

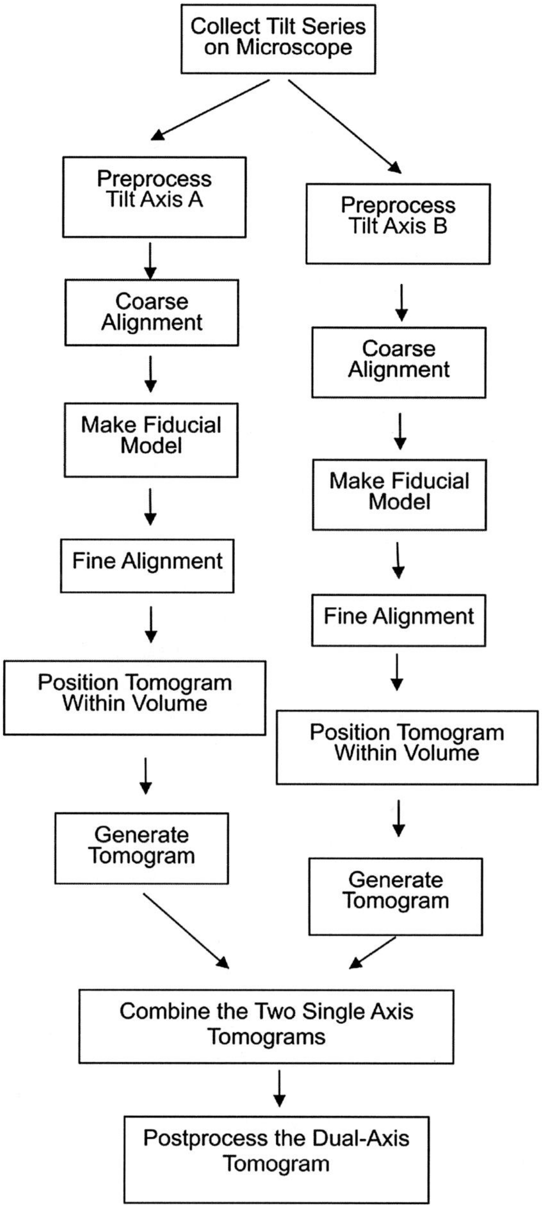

For "serial section tomography," dual-axis tomograms computed from serial semithick sections imaged and then reconstructed as outlined earlier are then brought into register with one another in 3D space to generate a single combined volume encompassing a very large cytoplasmic volume. Since the arrangement of Golgi membranes to form a "ribbon" typically extends over a considerable cellular area and in multiple orientations in 3D, limitations related to the field of view typically afforded by normal CCDs employed for digital imaging by EM can be overcome using a variety of methods including digital montaging of tilted image arrays and more advanced methods for ET reconstruction of entire mammalian cells albeit at slightly lower resolutions (Marsh, 2005; Marsh et al., 2004; Noske et al., 2008; Noske & Marsh, 2011).

16.1 NEW INSIGHTS INTO GOLGI ARCHITECTURE AND FUNCTIONAL ORGANIZATION AFFORDED BY ELECTRON TOMOGRAPHY

16.1.1 Stacks versus ribbons: Structure–function variation along the Golgi's vertical axis (*cis* to *trans*) influences its organization along the lateral axis

One of the most striking insights to come from large numbers of high-resolution ET studies of the Golgi region over recent years in our hands has been the range of structural and functional diversity in the Golgi region that is apparent among individual cells and even within any given cell being examined. In the past, due to the limited fields of view conventionally employed for ultrastructural investigations of the Golgi apparatus for both 2D and 3D studies, these differences have frequently been attributed to individual cells' differing responses to a particular physiological stimulus or experimental perturbation. However, as noted earlier, more recent advances in ET imaging, which allow for detailed 3D analysis of much more extensive cellular volumes and/or for relatively "high-throughput" analysis of high-resolution tomograms of the Golgi region collected from large numbers of different cells, have clearly demonstrated that the extent of structural variations observed can mostly be accounted for by structure–function variation along the length of the Golgi ribbon. That is, under different conditions leading to either stimulation or impairment of Golgi function, the ribbon gives rise to multiple and discrete functional domains asynchronously along its length, which indicate that activity along the Golgi ribbon occurs at different rates relative to individual stacks. Most recently, we have shown for molecular regulators of Golgi function such as the protein Rab6 that depletion of this molecule resulting in significant functional impairment of membrane traffic both to and from the Golgi has potent effects on structural organization of the Golgi apparatus both along its vertical (*cis* to *trans*) axis (i.e., inducing a significant increase in the number of cisternae within individual stacks) and along its lateral axis (i.e., inducing a significant increase in lateral continuity and cisternal length along the "backbone" of the Golgi ribbon per se) (Storrie et al., 2012).

This structure–function diversity from one stack to another is best demonstrated by viewing the 3D models for the Golgi apparatus produced following detailed image analysis to "segment" or delineate individual Golgi subcompartments (i.e., most notably, cisternae). Typically, different colors are used to define membrane contours that belong to a given either compartment or compartment type. In the examples provided here for Figs. 16.2–16.4, a particular color might be used to identify the membranes of a single continuous cisterna that occupies a certain position or hierarchical level in the stack (Fig. 16.3C and C′, D). The same color is also used to identify a cisterna determined to be either functionally equivalent or occupying an equivalent position in different stacks that are spatially distinct but belong to the same Golgi ribbon within a particular cell. Most notably, the relative differences as well as similarities for Golgi organization with respect to individual stacks versus the entire ribbon have only become possible to visualize ultrastructurally at the same scale, which is typically employed for LM investigations of the Golgi (see Fig. 16.3: compare panels A′ and B) through the development and implementation of cellular tomography approaches that finally allow 3D investigation of the Golgi ribbon *in toto* at the EM level (Noske et al., 2008; Noske & Marsh, 2011).

16.1.2 The *cis*- and *trans*-Golgi: Zones of dynamic structural and functional transition

One of the hallmarks of Golgi function has been the concept of different cargoes exiting the Golgi from different sites at its *trans*-face (including packaging different proteins for export from distinct *trans*-cisternae that collectively form the *trans*-Golgi network (TGN)) (Fig. 16.4). This notion has been based upon the idea that distinct *trans*-cisternae receive their respective cargoes labeled for redistribution via different cellular trafficking/export pathways (e.g., constitutive traffic to the apical vs. basolateral cell surface, the endosomal–lysosomal pathway or destined for the cell's "professional" regulated secretory pathway). Moreover, high-resolution ET data in particular have continued to provide strong evidence for a key role for specialized *trans*-Golgi-associated ER in regulating important events at the TGN (Marsh & Howell, 2002; Marsh et al., 2001, 2004). The close apposition of specialized ER membranes with *trans*-cisternae has been proposed as providing a mechanism for lipid exchange/modification at the ER/*trans*-Golgi face that would allow the direct transfer of lipids between these two organelles. Further supported by our own finding that specialized ER membranes are closely associated with those of *trans*-cisternae, these data lend weight to the idea that sphingolipid–cholesterol subdomains (i.e., lipid rafts) are directly involved in membrane modifications that determine cargo sorting at the TGN (Fig. 16.3).

In addition, high-resolution tomograms of the Golgi region in mammalian cells have revealed the extent of direct structural/spatial associations between the microtubule cytoskeleton and the Golgi cisternae at both *cis*- and *trans*-faces (Rios & Bornens, 2003). Previous high-resolution ET studies of the Golgi region have already provided novel insights regarding the distribution and nature of physical interactions between microtubules and the Golgi apparatus (Marsh et al., 2001; Rios & Bornens, 2003) (Fig. 16.4C′ and D). Among the most notable of these

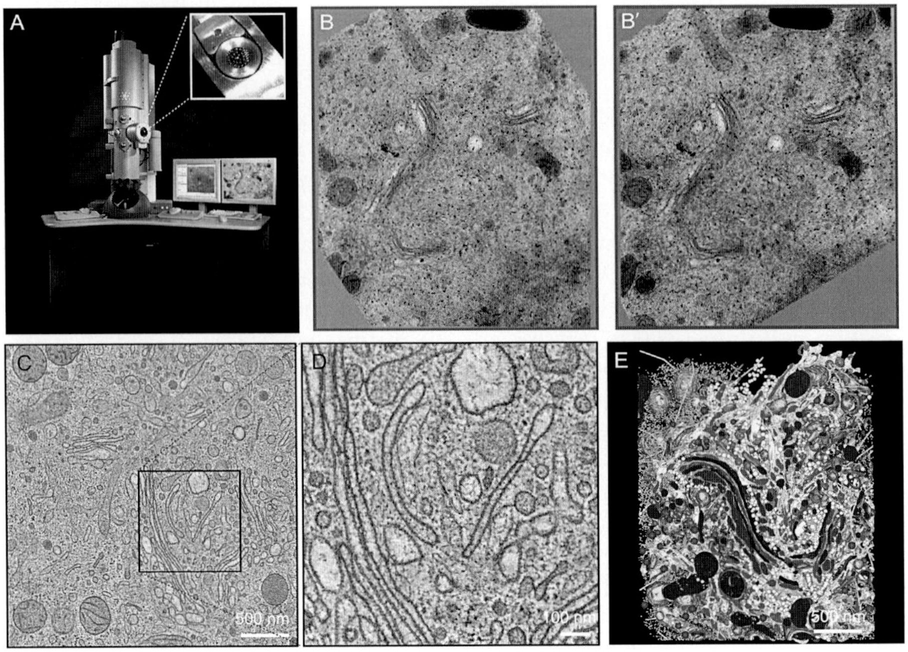

FIGURE 16.2

A figure panel illustrating some of the key steps required for high-resolution ET investigations of the Golgi apparatus. (Panel A) Image of a "Tecnai™" intermediate voltage EM optimized for automated tomographic data acquisition (reproduced courtesy of FEI Company). (Panels B, B′) Small (10–15 nm) gold fiducial particles adhered to both surfaces of each semithick (400 nm) section cut from a plastic-embedded cell prepared for ET prior to imaging afford improved accuracy during subsequent alignment of "tilt series" images. For the so-called "dual-axis" tomography, the specimen is rotated 90°, and a second tilt series of exactly the same area is acquired around an orthogonal axis (see also QuickTime movie Figure2ab.mov, http://dx.doi.org/10.1016/B978-0-12-417164-0.00016-1). (Panel C) Dual-axis tomograms computed from serial semithick sections are brought into register with one another in 3D space to produce a single 3D reconstruction encompassing a large cytoplasmic volume (see also QuickTime movie Figure2c.mov, http://dx.doi.org/10.1016/B978-0-12-417164-0.00016-1). (Panel D) This image reveals the subvolume highlighted in panel C (boxed area) at a typical working magnification (1 × zoom) as viewed during 3D ultrastructural analysis of the Golgi region (see also QuickTime movie Figure2d.mov, http://dx.doi.org/10.1016/B978-0-12-417164-0.00016-1). Each individual subcompartment and/or associated structure (e.g., membrane tubules and vesicles) within the tomogram is carefully delineated (i.e., segmented) from one pixel-thick slice to the next (along the *Z*-axis) so that precise morphometric information (e.g., membrane surface area, compartmental volume) and 3D spatial maps can be computed. (Panel E) Following detailed segmentation of the tomographic volume for the Golgi region, surface-rendered models were generated to allow 3D visualization and interrogation of the Golgi's complex architecture (Marsh, Mastronarde, Buttle, Howell, & McIntosh 2001) (see also QuickTime movie Figure2e.mov, http://dx.doi.org/10.1016/B978-0-12-417164-0.00016-1). Color coding: Golgi cisternae (*cis–trans*) (light blue, pink, cherry, green, dark blue, gold, and red); ER (yellow); membrane-bound ribosomes (blue); free ribosomes (orange); microtubules (bright green); insulin granules (blue); clathrin-negative vesicles (white); clathrin-positive compartments/vesicles (red); clathrin-negative compartments (purple); mitochondria (green). Bars: 500 nm (panels C, E); 100 nm (panel D). (See color plate.)

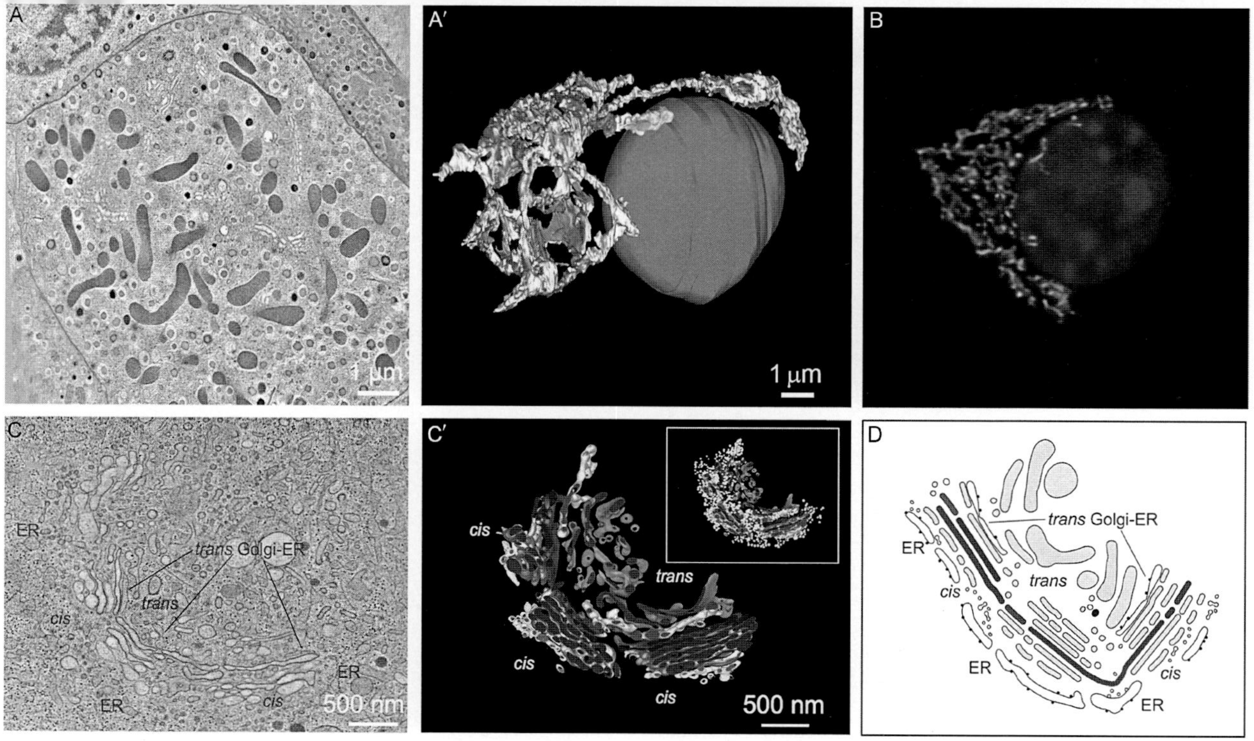

FIGURE 16.3

(Panel A) Shown is the *en face* (*X–Y*) view revealing a single image slice taken from a tomographic image volume reconstructed for an entire insulin-secreting cell prepared from an intact "islet of Langerhans" isolated from an adult mouse. 3D reconstruction of mammalian cells *in toto* (and *in situ*) at relatively high (~10 nm) resolution only recently became feasible through the development of more advanced ET approaches for cellular tomography (Noske et al., 2008; Noske & Marsh, 2011). (Panel A′) For the first time, the development and application of such methods has afforded 3D reconstruction and ultrastructural analysis of the entire Golgi ribbon at a similar scale to that commonly achieved through various LM techniques for 3D imaging of mammalian cells (see panel B for direct comparison; see also QuickTime movie Figure3a.mov, http://dx.doi.org/10.1016/B978-0-12-417164-0.00016-1). The main Golgi ribbon composed of *cis*- and *medial*-cisternae (gray) and the TGN (i.e., the *trans*-most Golgi cisternae (red) and penultimate *trans*-Golgi cisternae (gold)) is shown relative to the nucleus (yellow) in the cell's center. (Panel B) An equivalent view taken from a 3D deconvoluted LM image of a live insulin-secreting cell engineered to express green fluorescent protein (GFP)-tagged Golgi *N*-acetylgalactosamine:polypeptide *N*-acetylgalactosaminyltransferase 2 (GalNAcT2). Color coding: GFP-tagged GalNAcT2 (green), DAPI (blue), GM130 (red). (Panels C, C′) A dual-axis 3D tomogram of the Golgi region in a rodent HIT-T15 cell and its accompanying 3D model is shown (see also QuickTime movies Figure3c.mov and Figure3c′.mov, http://dx.doi.org/10.1016/B978-0-12-417164-0.00016-1, respectively). The accompanying QuickTime movie Figure3d.mov, http://dx.doi.org/10.1016/B978-0-12-417164-0.00016-1, provides further insight into the extent of structural complexity (relative to the *X–Y* axis for each pixel-thick tomographic slice) by revealing the color-coded contours, which delineate individual cisternae alone (as the movie progressively moves along the *Z*-axis through the depth of the image volume), followed by presentation of the resulting 3D surface-rendered model as it rotates around the vertical (*Y*) axis. (Panel D) Cartoon illustration highlighting the complexity of functional organization within the Golgi relative to its vertical (*cis–trans*) versus lateral aspects is shown. Functional continuity (based upon cisternal structure and function within each stack along their *cis–trans* axis) is shown in black (correlates to the cisternae color-coded in blue in panels C and C′). Color coding: *medial*-Golgi cisternae (gray); *trans*-most Golgi cisternae (red); penultimate *trans*-Golgi cisternae (gold); ERGIC (aquamarine); microtubules (bright green); Golgi-associated vesicles (white). Bars: 1 μm (panels A, A′); 500 nm (panels C, C′). (See color plate.)

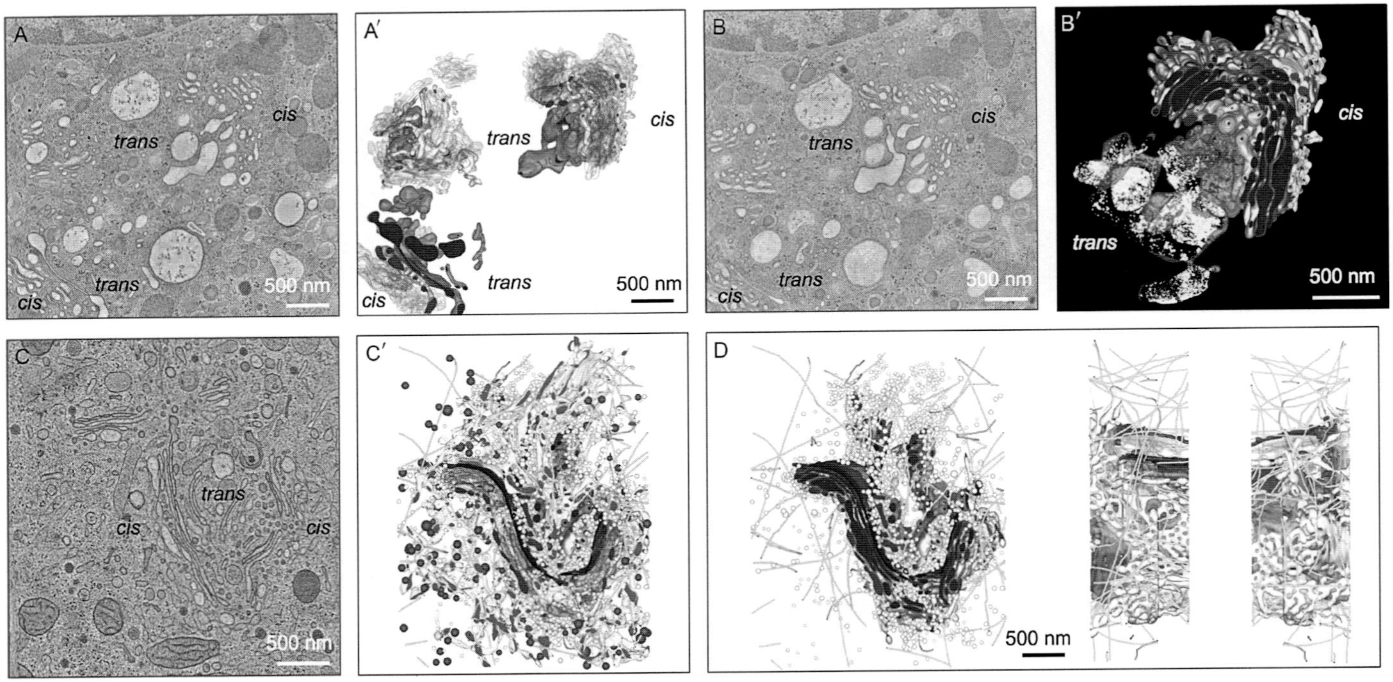

FIGURE 16.4

Illustrative tomographic examples highlighting the structural hallmarks of the specialized subcompartments at the *cis*- and *trans*-sides of the Golgi apparatus in insulin-secreting pancreatic beta cells. Panels A and A′ reveal the typical arrangement of ERGIC compartments (aquamarine) at the *cis*-face where new cargo enters the Golgi apparatus versus the typical organization of the TGN at the opposite *trans*-pole of each Golgi stack where the cargo exits (see also QuickTime movie Figure4a′.mov, http://dx.doi.org/10.1016/B978-0-12-417164-0.00016-1). (Panels B, B′) Newly synthesized proinsulin is packaged into nascent "immature" secretory granules (light blue) for transport to the cell surface for release into the bloodstream after progressing through the *trans*-most cisterna (red). In this example, the precise distribution and quantification of protein within each compartment has been determined by thresholding the particle density within their membrane boundaries (see also QuickTime movies Figure4b.mov and Figure4b′.mov, http://dx.doi.org/10.1016/B978-0-12-417164-0.00016-1). (Panels C, C′, D) Interactions of the microtubule cytoskeleton at the *cis*- and *trans*-sides of the Golgi apparatus play an important role in facilitating membrane traffic to and from the organelle (see also Rios & Bornens, 2003) (see also QuickTime movie Figure4c′.mov, http://dx.doi.org/10.1016/B978-0-12-417164-0.00016-1). Color coding: *medial*-Golgi cisternae (gray); *trans*-most Golgi cisternae (red); penultimate *trans*-Golgi cisternae (gold); ERGIC (aquamarine); microtubules (bright green); Golgi-associated vesicles (white); immature insulin granules (light blue); mature insulin granules (bright blue) (see also QuickTime movies Figure4e1.mov, Figure4e2.mov, and Figure3e3.mov, http://dx.doi.org/10.1016/B978-0-12-417164-0.00016-1, which highlight the 3D structural organization of the TGN at the level of an individual Golgi stack (Figure4e1.mov, http://dx.doi.org/10.1016/B978-0-12-417164-0.00016-1) vs. the entire Golgi ribbon (Figure4e2.mov and Figure3e3.mov, http://dx.doi.org/10.1016/B978-0-12-417164-0.00016-1)); these QuickTime movies correspond to the datasets presented in Figs. 16.2 and 16.3, respectively. Bars: 500 nm. (See color plate.)

findings was the extent of direct association between microtubules and membranes of the *cis*-most cisterna (Marsh et al., 2001). Since new cargo is delivered to the *cis*-face of the Golgi apparatus in mammalian cells via ER-derived membrane carriers referred to as either ERGIC (ER–Golgi intermediate compartment) elements or vesicular–tubular clusters these observations provided further direct support for the cisternal maturation/progression model, which proposes that new Golgi cisternae are formed following the docking and fusion of these carriers at the *cis*-face of the Golgi, using the preexisting *cis*-most cisterna as a template for the subsequent assembly of new cisternae (Emr et al., 2009; Marsh & Howell, 2002; Marsh et al., 2001). Interestingly, microtubules within the Golgi region did not display a typical radial distribution as if organized from the centrosome, but rather, they exhibited an arrangement similar to that observed in interphase epithelial cells (Marsh et al., 2001; Rios & Bornens, 2003).

The Golgi apparatus architecture is influenced not only by secretory transport but also by endocytic traffic as well (Gonatas, Kim, Stieber, & Avrameas, 1977; Pavelka et al., 1998, 2008; Sannerud, Saraste, & Goud, 2003; van Deurs, Tonnensen, Petersen, Sandvig, & Olsnes, 1986; Vetterlein, Ellinger, Neumüller, & Pavelka, 2002; Vetterlein, Niapir, Ellinger, Neumüller, & Pavelka, 2003); 3D studies by electron tomography considerably contributed to unravel the complex structures at the *trans*-Golgi side occurring in connection with retrograde transport. The detailed exploration of retrograde pathways towards, into, and across the Golgi apparatus and then further back to the ER and cytosol remains of outstanding interest with respect to the role of the Golgi apparatus in medicine, particularly for retrograde trafficking of toxins and drugs and, beyond this, for exploiting such knowledge as a basis towards developing new strategies for diagnosis, prevention, and therapies (Bachran & Fuchs, 2010; Engedal, Skotland, Torgersen, & Sandvig, 2011; Johannes & Popoff, 2008; Johannes & Römer, 2010; Sandvig, Garred, Prydz, Hansen, & van Deurs, 1992, Sandvig, Skotland, van Deurs, & Klokk, 2013; Tarrago-Trani & Storrie, 2007). Ultrastructural studies have shown that retrograde traffic from the cell surface to the Golgi apparatus is a complex multistep process that considerably influences the Golgi's architecture and leads to the new formation of extended endocytic *trans*-Golgi networks (endocytic TGN; Ellinger et al., 2010; Pavelka et al., 2008; Ranftler et al., 2013; Vetterlein et al., 2002, 2003). The 3D reconstructions by ET revealed complex contact structures and presented the endocytic TGN as being not a simple tubular or cisternal compartment but composed of globular and pot-like entities interconnected by bridges and crests. The 3D analysis of the Golgi apparatus by ET has revealed unexpected insights into the structures and dynamics of endocytic compartments and stimulated ongoing discussion of the novel mechanisms involved in reorganization of the Golgi apparatus during endocytosis. These questions, as well as a better understanding of Golgi connections to cytoskeleton and of general Golgi organization, will benefit from the development of more advanced techniques (Müller-Reichert et al., 2010; Figs. 16.5 and 16.6). Figure 16.6 highlights novel 3D imaging approaches complementing high-resolution ET investigations.

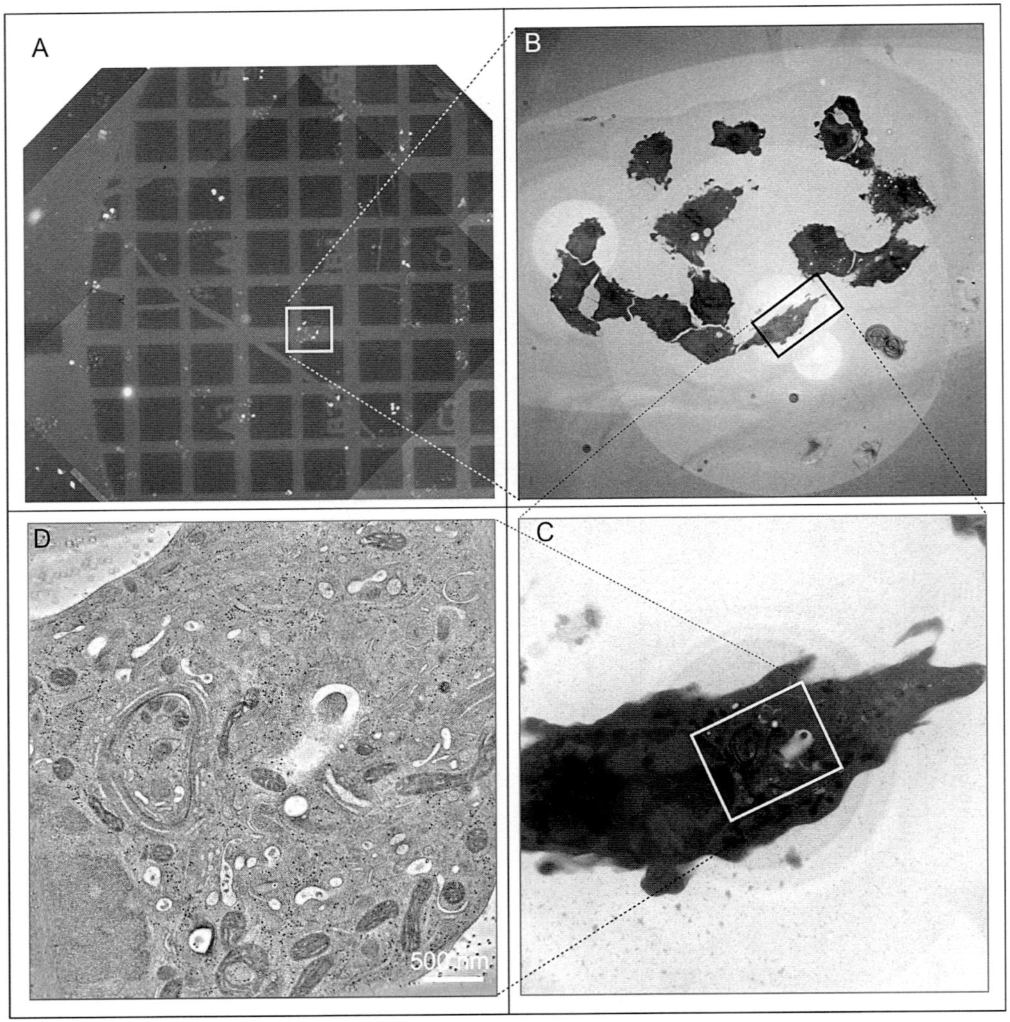

FIGURE 16.5

New approaches affording bona fide correlative imaging of live cells first by LM followed by EM analysis in 2D and 3D can now be routinely employed for more advanced structure–function investigations of the Golgi apparatus in mammalian cells (clockwise from top left). (Panel A) LM imaging of live insulin-secreting cells allows determination of which cells optimally express the marker being utilized (in this example, the GFP-tagged Golgi glycosyltransferase GalNAcT2, boxed area demarked in white). (Panel B) The exact area highlighted in panel A is next located at low magnification by conventional EM. (Panel C) Tilt series datasets are subsequently acquired by ET for the cell identified within the boxed region marked in panel B. Panel D shows the resulting dual-axis tomogram generated for the Golgi region highlighted within the white boxed area shown in panel C (see also QuickTime movie Figure5d.mov, http://dx.doi.org/10.1016/B978-0-12-417164-0.00016-1). Bar: 500 nm.

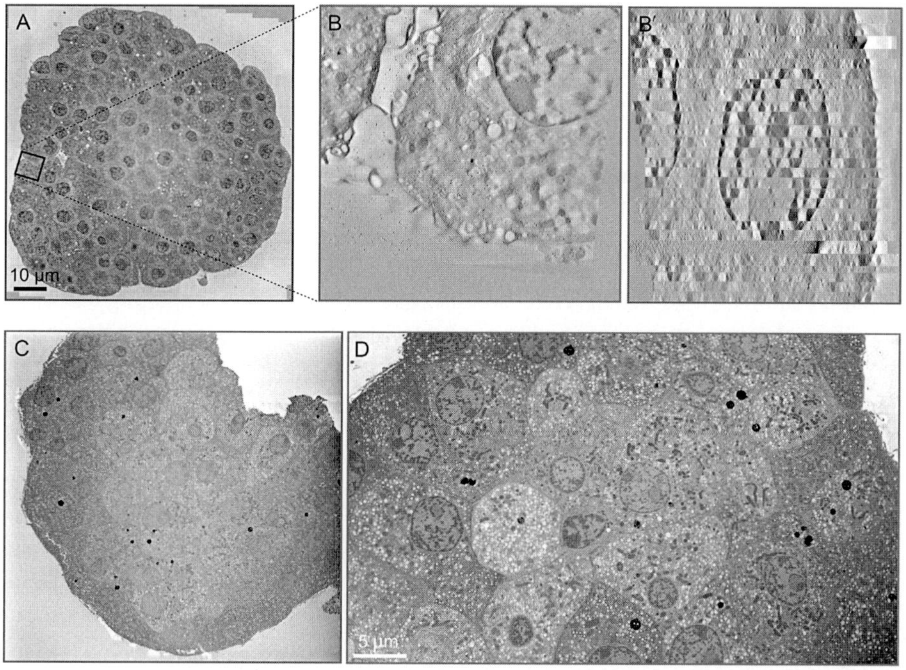

FIGURE 16.6

New 3D imaging approaches, which complement high-resolution ET investigations, now offer efficient and accurate solutions for 3D reconstruction of volumes on a much grander scale than has previously been possible (Müller-Reichert et al., 2010). Panel A shows the 2D overview of EM micrograph of a single, semithick section cut from a plastic-embedded pancreatic islet of Langerhans (see also Fig. 16.3 panels A and A′ for equivalent datasets). Although expedited approaches, which now allow 3D reconstruction of entire mammalian cells, have been developed and used routinely for "cellular tomography", the reconstruction and 3D analysis of such large tomograms still remain relatively slow and laborious compared to 3D imaging of an equivalent volume at the LM level (e.g., see Fig 16.3B). Panels B and B′ reveal the *X–Y* versus *Z*-axis views of a resulting cellular tomogram (see also QuickTime movies Figure6b.mov and Figure6b′.mov, http://dx.doi.org/10.1016/B978-0-12-417164-0.00016-1). That said, 3D imaging of the Golgi region by LM still typically suffers from both lower resolution and limited visualization of compartments of interest within the volume. New techniques such as that of serial block-face scanning electron microscopy (SBFSEM) now afford fully automated and efficient 3D reconstruction of entire tissues at relatively high resolutions (e.g., 5–50 nm) and can be employed as a hybrid approach that incorporates the same specimen for ultrastructural investigations using both methods (see also accompanying QuickTime movies Figure6c.mov and Figure6d.mov, http://dx.doi.org/10.1016/B978-0-12-417164-0.00016-1; movies reproduced courtesy of Gatan, Inc.).

SUMMARY

Over a century ago when Camillo Golgi provided the inaugural description of the enigmatic organelle that bears his name and that underpins a myriad of cellular processes essential to "life," little did he suspect that a complete understanding of its fundamental mechanisms would still constitute a significant challenge for cell biologists to this day. The extraordinary 3D complexity of the Golgi's architecture combined with its capacity to dynamically reorganize its structure in response to different intracellular and/or extracellular demands has hindered past ultrastructural studies constrained by conventional 2D imaging approaches. Here, we discuss how using the 3D imaging technique known as electron (microscope) tomography (ET) to precisely (i.e., at the nanoscale) map variations in Golgi architecture under different conditions has afforded new insights into how structure–function relationships change for this organelle.

Acknowledgments

This work was supported by grant funding (GNT0569535) to BJM from the National Health and Medical Research Council (NHMRC) of Australia and to MP from the Austrian Science Fund (FWF) project P20116. The authors gratefully acknowledge the generosity of FEI Company (Hillsboro, OR, USA) and Gatan, Inc. (Pleasanton, CA, USA) for permitting the use and reproduction of proprietary images and movies incorporated within this chapter and its accompanying online content. The authors cordially thank Mag. Beatrix Mallinger and Mrs. Regina Wegscheider for the excellent technical assistance and Mr. Thomas Nardelli and Mr. Ulrich Kaindl for their help with the artwork and the preparation of the 3D models.

References

Anitei, M., & Hoflack, B. (2012). Bridging membrane and cytoskeleton dynamics in the secretory and endocytic pathways. *Nature Cell Biology*, *14*, 11–19.

Appenzeller-Herzog, C., & Hauri, H. P. (2006). The ER-Golgi intermediate compartment (ERGIC): In search of its identity and function. *Journal of Cell Science*, *119*, 2173–2183.

Bachran, C., & Fuchs, H. (2010). Meeting report on the second targeted tumor therapies. *Molecular Cancer Therapy*, *9*, 17–23.

Bannykh, S. I., Rowe, T., & Balch, W. E. (1996). The organization of endoplasmic reticulum export complexes. *Journal of Cell Biology*, *135*, 19–35.

Berger, E. G. (1997). The Golgi apparatus: From discovery to contemporary studies. In E. G. Berger & J. Roth (Eds.), *The Golgi apparatus* (pp. 1–35). Basel, Boston, Berlin: Birkhäuser Verlag.

Cancino, J., & Luini, A. (2013). Signaling circuits on the Golgi complex. *Traffic*, *14*, 121–134.

Chia, J., Goh, G., Racine, V., Ng, S., Kumar, P., & Bard, F. (2012). RNAi screening reveals a large signaling network controlling the Golgi apparatus in human cells. *Molecular Systems Biology*, *8*, 629.

Cottam, N. P., & Ungar, D. (2012). Retrograde vesicle transport in the Golgi. *Protoplasma*, *249*, 943–955.

De Matteis, M. A., Di Campli, A., & D´Angelo, G. (2007). Lipid-transfer proteins in membrane trafficking at the Golgi complex. *Biochimica et Biophysica Acta*, *1771*, 761–768.

De Matteis, M. A., & Luini, A. (2008). Exiting the Golgi complex. *Nature Reviews Molecular Cell Biology*, *9*, 273–284.

Dinter, A., & Berger, E. G. (1998). Golgi-disturbing agents. *Histochemistry and Cell Biology*, *109*, 571–590.

Donohoe, B. S., Kang, B.-H., Gerl, M. J., Gergely, Z. R., McMichael, C. M., Bednarek, S. Y., et al. (2013). Cis-Golgi cisternal assembly and biosynthetic activation occur sequentially in plants and algae. *Traffic*, *14*, 551–567.

Dröscher, A. (1998). Camillo Golgi and the discovery of the Golgi apparatus. *Histochemistry and Cell Biology*, *109*, 425–430.

Ellinger, A., Vetterlein, M., Weiss, Ch., Meisslitzer-Ruppitsch, C., Neumüller, J., & Pavelka, M. (2010). High-pressure freezing combined with in vivo-DAB-cytochemistry. A novel approach for studies of endocytic compartments. *Journal of Structural Biology*, *169*, 286–293.

Emr, S., Glick, B. S., Linstedt, A. D., Lippincott-Schwartz, J., Luini, A., Malhotra, V., et al. (2009). Journeys through the Golgi—Taking stock in a new era. *Journal of Cell Biology*, *187*, 449–453.

Engedal, N., Skotland, T., Torgersen, M. L., & Sandvig, K. (2011). Shiga toxin and its use in targeted cancer therapy and imaging. *Microbiology and Biotechnology*, *4*, 32–46.

Farquhar, M. G., & Hauri, H. P. (1997). Protein sorting and vesicular traffic in the Golgi apparatus. In E. G. Berger & J. Roth (Eds.), *The Golgi apparatus* (pp. 1–35). Basel, Boston, Berlin: Birkhäuser Verlag.

Fridmann, K., Mader, A., Zwerger, M., Elia, N., & Medalia, O. (2012). Advances in tomography: Probing the molecular architecture of cells. *Nature Review Molecular Cell Biology*, *13*, 736–742.

Glick, B. S., & Luini, A. (2011). Models for Golgi traffic: A critical assessment. *Cold Spring Harbor Perspectives in Biology*, *a005215*, 1–15.

Glick, B. S., & Malhotra, V. (1998). The curious status of the Golgi apparatus. *Cell*, *95*, 883–889.

Golgi, C. (1898). Intorno alla struttura della cellula nervosa. *Archives Italiennes de Biologie*, *30*, 60–70.

Gonatas, N. K., Kim, S. U., Stieber, A., & Avrameas, S. (1977). Internalization of lectins in neuronal GERL. *Journal of Cell Biology*, *73*, 1–13.

Handa, K., Kumagain, K., Tomishige, N., & Kawano, M. (2007). CERT and intracellular trafficking of ceramide. *Biochimica et Biophysica Acta*, *1771*, 644–653.

Johannes, L., & Popoff, V. (2008). Tracing the retrograde route in protein trafficking. *Cell*, *135*, 1175–1187.

Johannes, L., & Römer, W. (2010). Shiga toxins—From cell biology to biomedical applications. *Nature Reviews. Microbiology*, *8*, 105–116.

Klumperman, J. (2011). Architecture of the mammalian Golgi. *Cold Spring Harbor Perspectives in Biology*, *a005181*, 1–19.

Landinsky, M. S., Mastronarde, D. N., McIntosh, J. R., Howell, K. E., & Staehelin, L. A. (1999). Golgi structure in three dimensions: Functional insights from the normal rat kidney cell. *Journal of Cell Biology*, *144*, 1135–1149.

Marsh, B. J. (2005). Lessons from tomographic studies of the mammalian Golgi. *Biochimica et Biophysica Acta, 1744*, 273–292.

Marsh, B. J., & Howell, K. E. (2002). The mammalian Golgi—Complex debates. *Nature Reviews. Molecular Cell Biology, 3*, 789–795.

Marsh, B. J., Mastronarde, D. N., Buttle, K. F., Howell, K. E., & McIntosh, J. R. (2001). Organellar relationships in the Golgi region of the pancreatic beta cell line, HIT-T15, visualized by high resolution electron tomography. *Proceedings of the National Academy of Sciences of the United States of America, 98*, 2399–2406.

Marsh, B. J., Volkmann, N., McIntosh, J. R., & Howell, K. E. (2004). Direct continuities between cisternae at different levels of the Golgi complex in glucose-stimulated mouse islet beta cells. *Proceedings of the National Academy of Sciences of the United States of America, 101*, 5565–5570.

Martinez-Sanchez, A., Garcia, I., & Fernandez, J. J. (2011). A differential structure approach to membrane segmentation in electron tomography. *Journal of Structural Biology, 175*, 372–383.

Mazarello, P. (1999). *The hidden structure. A scientific biography of Camillo Golgi.* New York: Oxford University Press, Inc.

Meisslitzer-Ruppitsch, C., Röhrl, C., Ranftler, C., Neumüller, J., Vetterlein, M., Ellinger, A., et al. (2011). The ceramide-enriched trans-Golgi compartments reorganize together with other parts of the Golgi apparatus in response to ATP-depletion. *Histochemistry and Cell Biology, 135*, 159–171.

Mogelsvang, S., & Howell, K. E. (2006). Global approaches to study Golgi function. *Current Opinion in Cell Biology, 18*, 438–443.

Mogelsvang, S., Marsh, B. J., Ladinsky, M. S., & Howell, K. E. (2004). Predicting function from structure: 3D structure studies of the mammalian Golgi complex. *Traffic, 5*, 338–345.

Mollenhauer, H. H., & Morré, D. J. (1998). The tubular network of the Golgi apparatus. *Histochemistry and Cell Biology, 109*, 533–543.

Müller-Reichert, T., Mancuso, J., Lich, B., & McDonald, K. (2010). Three-dimensional reconstruction methods for Caenorhabditis elegans ultrastructure. *Methods in Cell Biology, 96*, 331–361.

Nakamura, N., Wei, J. H., & Seemann, J. (2012). Modular organization of the mammalian Golgi apparatus. *Current Opinion in Cell Biology, 24*, 467–474.

Nakano, A., & Luini, A. (2010). Passage through the Golgi. *Current Opinion in Cell Biology, 22*, 471–478.

Noske, A. B., Costin, A. J., Morgan, G. P., & Marsh, B. J. (2008). Expedited approaches to whole cell electron tomography and organelle mark-up in situ in high-pressure frozen pancreatic islets. *Journal of Structural Biology, 161*, 298–313.

Noske, A. B., & Marsh, B. J. (2011). Mapping the β-Cell In 3D at the nanoscale using novel cellular electron tomography and computational approaches. In B. Booß-Bavnbek, B. Klösgen, J. Larsen, F. Pociot, & E. Renström (Eds.), *BetaSys—Systems biology of regulated exocytosis in pancreatic β-cells.* In S. Choi (Ed.), *Systems Biology*, Vol. 2, (pp. 147–183). New York: Springer Science+Business Media.

Pavelka, M., Ellinger, A., Debbage, P., Loewe, C., Vetterlein, M., & Roth, J. (1998). Endocytic routes to the Golgi apparatus. *Histochemistry and Cell Biology, 109*, 555–570.

Pavelka, M., Neumüller, J., & Ellinger, A. (2008). Retrograde traffic in the biosynthetic-secretory route. *Histochemistry and Cell Biology, 129*, 277–288.

Pfeffer, S. R. (2010). How the Golgi works: A cisternal progenitor model. *Proceedings of the National Academy of Sciences of the United States of America, 107*, 19614–19618.

Porter, K. R., Claude, A., & Fullam, E. F. (1945). A study of tissue culture cells by electron microscopy: Methods and preliminary observations. *Journal of Experimental Medicine, 81*, 233–246.

Puthenveedu, M. A., & Linstedt, A. D. (2005). Subcompartmentalizing the Golgi apparatus. *Current Opinion in Cell Biology*, *17*, 369–375.

Rambourg, A., & Clermont, Y. (1997). Three-dimensional structure of the Golgi apparatus in mammalian cells. In E. G. Berger & J. Roth (Eds.), *The Golgi apparatus* (pp. 37–61). Basel, Boston, Berlin: Birkhäuser Verlag.

Ranftler, C., Auinger, A., Meisslitzer-Ruppitsch, C., Neumüller, J., Ellinger, A., & Pavelka, M. (2013). Electron microscopy of endocytic pathways. In D. T. Taatjes & J. Roth (Eds.), *Cell imaging techniques: Methods and protocols* (pp. 437–447). New York/London: Humana Press/Springer.

Rios, R. M., & Bornens, M. (2003). The Golgi apparatus at the cell centre. *Current Opinion in Cell Biology*, *15*, 60–66.

Rothman, J. E., & Wieland, F. T. (1996). Protein sorting by transport vesicles. *Science*, *272*, 227–234.

Sandvig, K., Garred, O., Prydz, K., Hansen, S. H., & van Deurs, B. (1992). Retrograde transport of endocytosed Shiga toxin to the endoplasmic reticulum. *Nature*, *358*, 510–511.

Sandvig, K., Skotland, T., van Deurs, B., & Klokk, T. I. (2013). Retrograde transport of protein toxins through the Golgi apparatus. *Histochemistry and Cell Biology*, *140*, 317–326.

Sannerud, R., Saraste, J., & Goud, B. (2003). Retrograde traffic in the biosynthetic-secretory route: Pathways and machinery. *Current Opinion in Cell Biology*, *15*, 438–445.

Storrie, B., Micaroni, M., Morgan, G. P., Jones, N., Kamykowski, J. A., Wilkins, N., et al. (2012). Electron tomography reveals Rab6 is essential to the trafficking of trans-Golgi clathrin and COPI-coated vesicles and the maintenance of Golgi cisternal number. *Traffic*, *13*, 727–744.

Tang, D., Mar, K., Warren, G., & Wang, Y. (2008). Molecular mechanism of mitotic Golgi disassembly and reassembly revealed by a defined reconstitution assay. *Journal of Biological Chemistry*, *283*, 6085–6094.

Tarrago-Trani, M. T., & Storrie, B. (2007). Alternate routes for drug delivery to the cell interior: Pathways to the Golgi apparatus and endoplasmic reticulum. *Advanced Drug Delivery Reviews*, *29*, 782–797.

Tomas, M., Pilar-Marin, M., Martinez-Alonso, E., Esteban-Pretel, G., Diaz-Ruiz, A., Vazquez-Martinez, R., et al. (2012). Alcohol induces Golgi fragmentation in differentiated PC12 cells by deregulating Rab1-dependent ER-to-Golgi transport. *Histochemistry and Cell Biology*, *138*, 489–501.

Valente, C., Turacchio, G., Mariggiò, S., Pagliuso, A., Gaibisso, R., Di Tussio, G., et al. (2012). A 14-3-3 gamma dimer-based scaffold bridges CtBP1-S/BARS to PI(4)KIIIbeta to regulate post-Golgi carrier formation. *Nature Cell Biology*, *14*, 343–354.

van Deurs, B., Tonnensen, T. I., Petersen, O. W., Sandvig, K., & Olsnes, S. (1986). Routing of internalized ricin and ricin conjugates in the Golgi complex. *Journal of Cell Biology*, *102*, 37–47.

Vetterlein, M., Ellinger, A., Neumüller, J., & Pavelka, M. (2002). Golgi apparatus and TGN during endocytosis. *Histochemistry and Cell Biology*, *117*, 143–150.

Vetterlein, M., Niapir, M., Ellinger, A., Neumüller, J., & Pavelka, M. (2003). Brefeldin A-regulated retrograde transport into the endoplasmic reticulum of internalised wheat germ agglutinin. *Histochemistry and Cell Biology*, *120*, 121–128.

Wilson, C., Venditti, R., Rega, L. R., Colanzi, A., D'Angelo, G., & De Matteis, M. A. (2011). The Golgi apparatus: An organelle with multiple complex functions. *Biochemical Journal*, *433*, 1–9.

CHAPTER

High-Content Screening and Analysis of the Golgi Complex

17

George Galea and Jeremy C. Simpson

School of Biology and Environmental Science, Conway Institute of Biomolecular and Biomedical Research, University College Dublin (UCD), Dublin, Ireland

CHAPTER OUTLINE

Introduction **282**
17.1 Materials **284**
17.2 Methods **285**
17.2.1 Assay Concept and Design 285
17.2.2 Solid-phase Reverse Transfection 289
17.2.3 Cell Seeding, Fixing, and Staining 290
17.2.4 Image Acquisition 290
17.2.5 Image Analysis and Data Management 291
17.3 Discussion **292**
Summary **293**
Acknowledgments **293**
References **293**

Abstract

High-content screening (HCS) as a methodological tool has evolved relatively recently, largely driven by the demand for in depth spatial and temporal information from intact cells exposed to a range of chemical and/or genomic perturbations. The technology is based around automated fluorescence microscopy in combination with advanced imaging processing and analysis tools, which together can provide quantitative information as a first-level description of complex cellular events. HCS and high-content analysis are particularly powerful when combined with perturbation techniques such as RNA interference (RNAi), as this allows large families of genes to be interrogated with respect to a biological pathway or process of interest. In this methodology chapter, we describe an approach by which HCS can be applied to study the morphological state of the Golgi complex in cultured mammalian cells. We provide a detailed protocol for the highly parallel downregulation of gene

Methods in Cell Biology, Volume 118

ISSN 0091-679X
http://dx.doi.org/10.1016/B978-0-12-417164-0.00017-3

activity using RNAi in 384-well plates and describe an automated image analysis routine that could be used to quantify Golgi complex in a genome-wide RNAi context.

INTRODUCTION

Traditionally, biochemical assays and genetic methods have been the mainstay for determining the function and properties of proteins within cells. Although these techniques have been invaluable in terms of describing the intimate details about a particular protein under study, in many cases, they do not provide contextual "in cell" data. By contrast, fluorescence microscopy of intact cells is a key technique that can provide this level of information, although it potentially suffers from being low throughput and only qualitative. For these reasons, conventional fluorescence microscopy has not widely been used in large-scale systematic studies of cell function or in drug screens for particular phenotypes.

High-throughput screening (HTS), typically employing a simple colorimetric readout from an entire well, has however been used for several decades as a tool in industry for early stage drug discovery (Pereira & Williams, 2007). With increasing demand for more information, a progression in this technology was clearly needed. Through the 1990s, significant advances were being made in genome sequencing, chemical compound libraries were expanding at a dramatic rate, and there was a general move towards understanding cell function in a systematic way. High-content screening (HCS) was therefore a natural evolution from HTS, as it incorporated the automation, speed, and efficiency of HTS with the spatial and temporal subcellular information coming from wide-field fluorescence microscopy and latterly confocal microscopy (Taylor, 2010). HCS has thus been able to position itself as a technology to deliver high volumes of robust quantitative image data useful for both drug discovery and fundamental biomolecular research.

The term "HCS" was first brought to prominence by work of D. Lansing Taylor and colleagues in 1997. In this study, the group performed two automated screens to measure drug-induced transport of the human glucocorticoid receptor and also drug-induced apoptosis (Giuliano et al., 1997). These pioneering screens provided a variety of cellular measurements, including nuclear size and shape changes, nuclear DNA content, mitochondrial potential, and actin cytoskeletal rearrangement, all in a single series of experiments, thereby significantly extending the capability of HTS at that time. Since then, a large number of automated imaging platforms have been developed, making this technology and the ability to rapidly generate tens of thousands of images more widely accessible. Indeed, the large volumes of quantitative data that can potentially be extracted from fluorescence images of individual cells also present new challenges (Niederlein, Meyenhofer, White, & Bickle, 2009). HCS relies on the fact that all cells, in all wells across an entire screen, are imaged and analyzed in an identical manner. Of equal importance is that fact the software required to perform the analysis must be flexible, such that it can be easily tuned to various cell types and assays requiring a different readout. Open-source platforms, such as CellProfiler from the Broad Institute, have therefore been developed to provide tools by which a variety of biological questions can be quantitatively addressed (Carpenter et al., 2006; Eliceiri et al., 2012).

The Golgi complex is a unique and vital organelle, and due to its relatively consistent size and location in mammalian cells in particular, it could be seen as an ideal candidate for study using HCS. Being a central organelle in the secretory pathway, the physiological state of the Golgi complex is directly dependent on the bidirectional flow of membranes and proteins into and out of it, and as such any changes in flux through it are likely to cause a change in its appearance. Similarly, even at steady state, the shape of the Golgi complex can only be maintained through an extensive network of structural proteins (such as matrix and coat proteins) (Ramirez & Lowe, 2009) and interactions with the underlying cytoskeleton (de Forges, Bouissou, & Perez, 2012). One of the earliest attempts to use automated imaging in an HCS format to study Golgi complex function was reported back in 2003. The group of Rainer Pepperkok described the development of an automated microscope and associated assays, by which protein secretion and the integrity of the Golgi complex could be addressed (Liebel et al., 2003). This system was then applied to examine the effects of overexpression of 100 fluorescently tagged proteins in these two assays (Starkuviene et al., 2004). Specifically, a temperature-sensitive fluorescently tagged molecule (VSV-G tsO45), a variant of the vesicular stomatitis viral G-protein VSV-G, was used as a secretory cargo marker. This transmembrane protein has the feature of accumulating in the endoplasmic reticulum (ER) at 39.5 °C, but when induced to fold by reducing the temperature to 32 °C, it is rapidly transported through the secretory pathway, via the Golgi complex to the plasma membrane, where an antibody recognizing an external epitope can detect it. The premise of this assay, therefore, is that overexpression of any proteins associated with this pathway, including those physically located at the Golgi complex, should interfere with the delivery of this cargo to the cell surface. By incorporating an immunostaining for a Golgi marker (e.g., the *cis*-Golgi matrix protein GM130), the physical state of the Golgi complex itself was also assessed (Starkuviene et al., 2004).

More recently, RNA interference (RNAi) strategies, often at a genome-wide scale, have become the methodology of choice for probing protein function at the Golgi complex. The majority of these studies have so far concentrated on first identifying proteins associated with secretion and then analyzing Golgi complex morphology as a secondary screen. The first such study employed a chemiluminescence approach using horseradish peroxidase fused to a signal sequence as a measure of secretion (Bard et al., 2006). From this primary screen, 130 candidates were depleted by RNAi and the appearance of the Golgi complex (as judged by mannosidase II-GFP) was assessed. Although the effects on the Golgi complex were only manually annotated, the images that were presented would be highly amenable to automated HCS analysis. A subsequent secretion screen, also carried out in Drosophila S2 cells, followed a similar approach, but on this occasion, GM130 was used as the marker to assess Golgi complex physiology (Wendler et al., 2010). The only reported genome-wide screen for secretion in human cells was reported recently by our lab (Simpson et al., 2012). Although again, the focus of this work was secretion, in this case, the secondary analysis of the Golgi complex was semiautomated, with the effect of depletion of more than 550 genes on GM130 immunostaining being assessed.

All the studies described thus far have only assessed the Golgi complex in the context of its role in protein secretion and at best in a semiautomated manner. Very recently however, an RNAi screen of all kinases and phosphatases was carried out to specifically classify perturbations to Golgi complex morphology and thereby identify the key regulators of the maintenance of its architecture. The study used three different Golgi subcompartment markers (*Helix pomatia* lectin as a *cis*- marker, mannosidase II-GFP as a medial marker, and immunostained TGN46 as a *trans*- and *trans*-Golgi network marker) and applied a support vector machine learning approach to automatically identify and classify nine distinct Golgi complex phenotypes (Chia et al., 2012).

Although HCS is only now emerging as a mainstream cell biology tool in the academic environment, the initial signs are that it will become an increasingly important method for systematically dissecting a wide range of cellular processes and functions. In the protocol that we describe here, we present a robust method that can be used to systematically and quantitatively analyze the morphology of the Golgi complex from very large numbers of cells in a highly parallel manner. The protocol also describes how the cells for this assay can be prepared for RNAi experiments, as this is likely to be the perturbation most widely used in future studies aimed at understanding Golgi complex function.

17.1 MATERIALS

1. Lipofectamine 2000 transfection reagent (Life Technologies, Cat. No. 11668-019)
2. Opti-MEM reduced serum medium, no phenol red (Life Technologies, Cat. No. 11058-021)
3. Sucrose (Sigma-Aldrich, Cat. No. S0389)
4. 0.1% fibronectin solution from human plasma (Sigma-Aldrich, Cat. No. F0895)
5. Gelatin powder from bovine skin (Sigma-Aldrich, Cat. No. G9391)
6. RNase-free water (Fisher BioReagents, Cat. No. 7732-18-5)
7. Dulbecco's modified Eagle medium (DMEM), low glucose (Life Technologies, Cat. No. 31885-023), also containing additional supplements as required
8. Paraformaldehyde (PFA): 3% solution in 200 mM phosphate-buffered saline (PBS), pH 7.4. Prepared from fresh PFA and then stored at −20 °C in single-use aliquots
9. PBS (pH 7.4)
10. Hoechst 33342 nucleic acid stain (Life Technologies, Cat. No. H1399)
11. 96-well (v-shape) storage plate (Thermo Fisher Scientific, Cat. No. AB-1058); 384-well (shallow) plate (Thermo Fisher Scientific, Cat. No. 264704); 384-well

(deep storage) plate (Thermo Fisher Scientific, Cat. No. AB-1178); 384-well (optical-plastic bottom) Viewplate-384 FTC (PerkinElmer, Cat. No. 6007470)
12. Fully automated liquid handling system with individual pipetting channels and a 96- or 384-well head, a minimum of two cooled positions on the deck, and deck space for at least 18 multiwell plates, for example, Microlab STAR (Hamilton Robotics)
13. Centrifugal vacuum concentrator for drying down of small interfering RNA (siRNA) transfection mixtures into multiwell plates, for example, miVac Quattro (Genevac)
14. Handheld automated cell counter, for example, Scepter 2.0 (Merck Millipore)
15. Automated cell/liquid dispenser compatible with 96- and 384-well plates, for example, Multidrop 384 Reagent Dispenser (Thermo Fisher Scientific)
16. Semiautomated multichannel electronic pipette, ideally with a 96- or 384-well head, for example, VIAFLO (Integra Biosciences)
17. Fully automated HCS microscope with appropriate filter sets and objectives, for example, Scan^R (Olympus)
18. High-content analysis software, for example, CellProfiler (Broad Institute) or Columbus (PerkinElmer).

17.2 METHODS

17.2.1 Assay concept and design

The key to any HCS campaign lies in careful assay design and planning. Regardless of the complexity of the HCS assay in mind, it is essential to optimize each step of the protocol and typically carry out a series of pilot experiments to validate the complete pipeline. As most HCS screens involve processing very large sample numbers (cells, siRNA reagents, chemical compounds, antibodies, etc.), it is also important that particular batches of each are tested, and then sufficient quantities purchased or collected, such that an entire screen can be performed with a consistent set of materials all from an equivalent batch or lot number. This is vital to ensure reproducibility through a screen and to give confidence that cells analyzed from the first plate can be faithfully compared to cells analyzed from the final plate of the screen.

The protocol that we present here is based on the biological question of which genes play a role in maintenance of Golgi complex morphology in a human HeLa cell line. In order to do this, the genome is systematically probed by RNAi, specifically using siRNAs that target every gene in turn. The HCS pipeline is composed of a series of steps, as follows (Fig. 17.1). Multiwell optical quality imaging plates are first prepared by addition of a transfection mixture containing the siRNA of interest. These are dried down and can be stored in this highly stable format until needed (Fig. 17.1A). Cells stably expressing a GFP-tagged Golgi marker are then seeded into each well of the plate and incubated for the desired time required for depletion of the target protein. The cells are then fixed, and the nuclei counterstained in order to

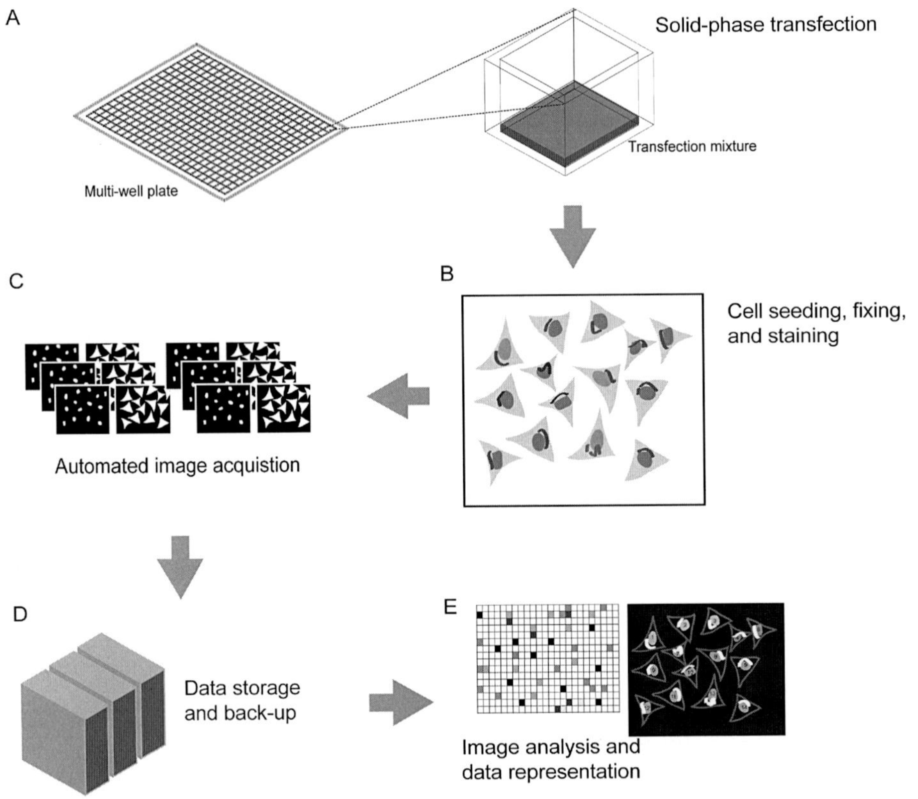

FIGURE 17.1

High-content screening pipeline. (A) Solid-phase reverse transfection plates are prepared in a 384-well plate format using a liquid handling robot. The plates are dried and stored until ready to use. (B) HeLa GalNAc-T2–GFP cells are plated at a density of 1000 cells per well in the 384-well plate and incubated for 48 h. The cells are fixed with 3% PFA, and the nuclei counterstained with Hoechst 33342. (C) Images from every well are acquired using an automated fluorescence microscope. (D) Image data are then duplicated and stored on a server, which is accessed by the (E) image analysis software, which segments the cells and extracts the relevant Golgi complex features. The image data and extracted features are then linked through a database, which also allows the data to be visualized.

provide a consistent marker for the autofocus in the imaging step (Fig. 17.1B). The cells are then imaged on an automated screening microscope, with every image across the entire screen being acquired in an identical manner (Fig. 17.1C). The image data are stored, duplicated, and transferred to the image analysis server (Fig. 17.1D). Finally, the images are analyzed in such a way that a quantitative measure of the Golgi complex morphology for every cell in every well across the screen can be made (Figs. 17.1E and 17.2).

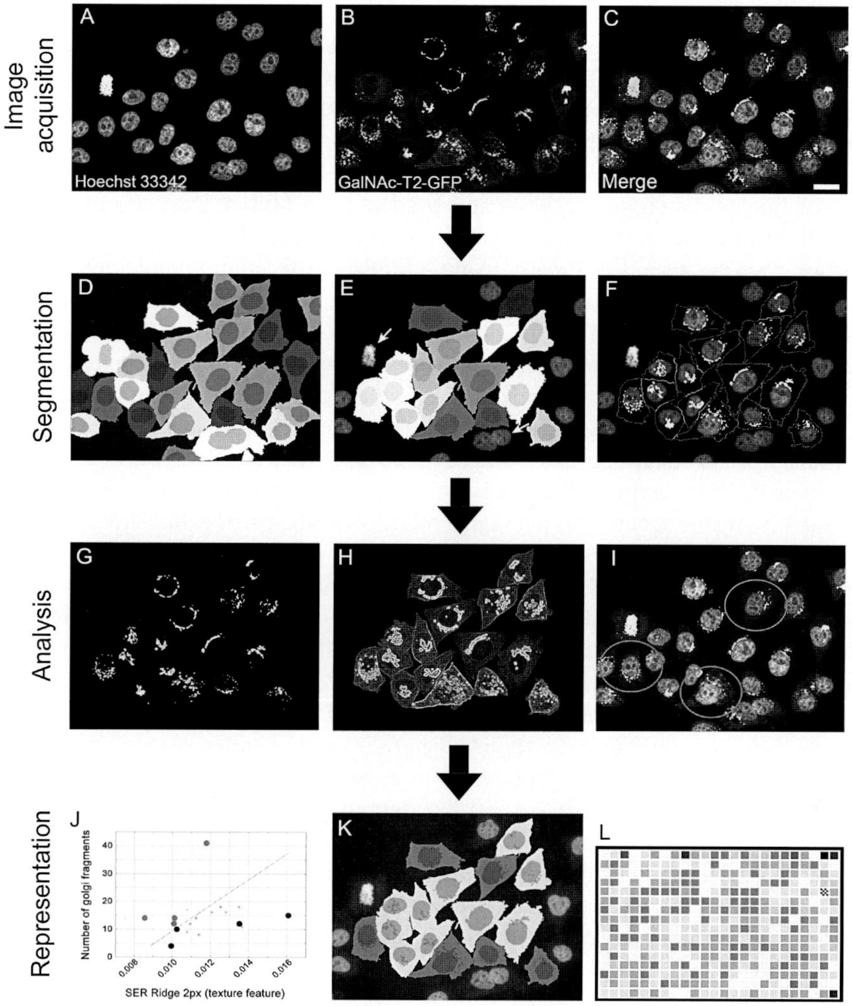

FIGURE 17.2

Image analysis strategy for quantifying Golgi morphology disruption. (A–C) Images of HeLa GalNAc-T2–GFP cells following 48 h transfection with siRNAs targeting Sec22B. Scale bar represents 20 μm. (D) The cytoplasm of each cell is segmented based on the fluorescence in the GFP channel. (E) Morphological and intensity measurements are then extracted from the segmented cells to distinguish between healthy and apoptotic or dividing cells that would skew the analysis. These cells are indicated with an arrow. Cells with a cytoplasm touching the border of the field of view are also removed at this point. (F) Golgi complex fragments present in each cell of the remaining population are next segmented. (G) Features such as Golgi complex total area, fragment area, and numbers of Golgi fragments per cell are extracted. (H) Texture feature analysis of the Golgi complex fragments identified in (F) and (G). (I) The cell population is manually inspected and cells showing either normal Golgi architecture or

During the pilot phase of the screen, it is essential to test and then select a series of appropriate negative and positive control siRNAs that will ultimately be added to every plate in the main screen. These controls will be used to assess the efficiency of transfection and the performance of assay (e.g., level of Golgi fragmentation) and also can be used for training the image analysis software machine learning algorithms. Although the controls will always to a certain extent be specific for the particular assay and cell line used, in our experience, it is prudent to first test at least 20 gene targets in order to identify the best performing 4–5 positive controls. In the case of the protocol that we describe here, specifically using a GFP-tagged Golgi complex enzyme as a readout for overall architecture of this organelle, we found that depletion of the ER–Golgi SNARE protein Sec22B, the Golgi SNARE proteins Syntaxin-16 and Ykt6, and the small GTPase Rab6A all gave robust but distinct phenotypic effects on the Golgi complex. A range of positive controls is also highly useful in the image analysis steps and for setting up linear classifiers. In addition to assay-specific positive controls, we also include several wells containing siRNAs targeting the INCENP gene. Depletion of this protein leads to the inability of cells to complete their cell cycle, with the result being the accumulation of large multilobed nucleated cells. This phenotype is straightforward to identify based on nuclear stain alone and provides an accurate measurement of siRNA transfection efficiency (Neumann et al., 2010; Simpson et al., 2012). Negative control siRNAs should give a similar Golgi complex phenotype to that seen in untreated cells. We would typically allocate a minimum of 20 wells to positive controls and 16 wells to negative controls across a 384-well plate.

In this protocol, we use a previously characterized HeLa cell line stably expressing GalNAc-T2–GFP in order to visualize the Golgi complex (Storrie et al., 1998). This was selected over other options such as antibody staining to reduce the costs across a large screen and also to minimize the postprocessing time of the plates prior to image acquisition. The choice of cell line and method of analysis are also relevant to the selection of plate type for the screen. Here, we use optical quality plastic 384-well plates. These are selected due to their relatively low cost compared to glass-bottomed imaging plates and their high robustness. Optical plastic has the added advantage that cells adhere better to it, thereby reducing the degree of cell loss during washing steps.

strongly fragmented phenotypes are selected and used as a training set for the linear classifier. Cells showing a fragmented Golgi complex and that are used for training the linear classifier are marked with a red circle. (J) Graphical output from the linear classifier. Morphological measurements and texture features are used to identify and classify the different phenotypes across the entire field of view. (K) Automated identification of two populations based on the linear classifier. Cells with nonfragmented Golgi complexes are shown in green, and cells with highly fragmented Golgi complexes in red. (L) Heat map indicating percentage of cells showing Golgi complex fragmentation in each well across the plate. Red indicates high levels of fragmentation, and green indicates low levels of fragmentation. (See color plate.)

17.2.2 Solid-phase reverse transfection

As discussed earlier, it is essential to invest time in optimizing all aspects of the screening pipeline. Similarly, accuracy and reproducibility in preparing the plates for solid-phase reverse transfection of siRNAs are also critical. The protocol given here is modified from a previously published protocol used for solid-phase transfection in a 96-well format (Erfle et al., 2008). Using 384-well plates has the advantage of significantly reducing the reagent costs and allows the siRNA library to be prepared on a smaller number of plates. A human genome-wide siRNA library containing ca. 22,000 targets can be prepared on between 60 and 70 384-well plates, depending on the number of control wells allocated. Libraries are available from a number of manufacturers, and siRNA duplexes can be either supplied individually or pooled for each target. In the protocol described here, we use a pool of three siRNA duplexes per target gene (per well), whereas, for example, using three duplexes per target but arraying them individually would result in the need for more than 180 384-well plates to be prepared. In general, we find that for large-scale HCS assays, solid-phase reverse transfection is superior to conventional liquid-based forward transfection for several reasons. Firstly, the plate preparation (even for an entire genome-wide library) can be done relatively quickly, meaning that the reagents are all treated and "aged" to a similar degree. Secondly, the plate preparation can easily programmed on a liquid handling robot, making all wells highly comparable. Thirdly, siRNA transfection efficiencies (typically >80%) tend to be higher than those obtained using liquid forward transfection. Finally, the plates can be stored in the dry condition for extensive periods of time following fabrication, meaning that experiments can be timed with the availability of the HCS imaging systems.

1. Add 0.205 g of sucrose to 1.5 mL Opti-MEM and heat at 37 °C until dissolved. Filter sterilize the solution through a 0.22 μm syringe filter.
2. Take 800 μL of the Opti-MEM/sucrose solution and add to it 900 μL of Lipofectamine 2000. Add 200 μL of this solution to each well in the first column only of a 96-well (v-shape) storage plate. Transfer this plate to an empty deck position on the liquid handling robot. Use the robot to distribute 3.8 μL of this solution to every well of a 384-well (shallow) plate.
3. Using the liquid handling robot, transfer 3 μL of each siRNA (6 μM) to each well of the transfection plate prepared in step 2, and incubate for 20 min at room temperature.
4. While incubating the transfection mixture, add 8.75 μL of fibronectin to 2.5 mL gelatin solution (0.032 g of gelatin in 40 mL of RNase-free water). Transfer 200 μL of this solution to each well in the second column only of the 96-well (v-shape) storage plate used in step 2. After the incubation period in step 3 is complete, use the liquid handling robot to transfer 4.16 μL of the gelatin/fibronectin solution to the transfection mixture.
5. Using the liquid handling robot, transfer 8.5 μL of the transfection mixture (containing the siRNA) into a 384-well (deep storage) plate previously filled with 166 μL of RNase-free water.

6. Use the liquid handling robot to aliquot 10.25 μL of the diluted transfection mixture to the corresponding well of the 384-well (optical) plates. Replace the lids on the 384-well plates. Up to 15 replicate plates can be prepared at this step.
7. Dry down the plates for 6 h at 37 °C using the centrifugal vacuum concentrator.
8. Store the plates in a sealed container containing a desiccating agent. The plates can be stored in this format for at least 1 year.

17.2.3 Cell seeding, fixing, and staining

In our protocol, cell seeding, fixing, and staining is performed in an automated or semiautomated manner. Cells are passaged routinely and used only up to passage number 10. Typically, 1000 cells are seeded per well in a total volume of 40 μL complete DMEM into the 384-well optical plates prepared as described earlier. Maximum reproducibility of cell number and distribution in the well is achieved when the cells are counted with an automated counting device (e.g., Scepter) and then seeded using a rapid dispensing system (e.g., MultiDrop 384 Reagent Dispenser). The cells are then incubated with the siRNAs for a total of 48 h prior to fixation. The cells are fixed using a semiautomated pipetting system (e.g., VIAFLO) with 3% PFA in PBS for 20 min, washed with PBS, and stained with 0.2 μg/mL Hoechst 33342. Although these pipetting steps can be carried out manually, instruments such as VIAFLO ensure that all the wells are treated at a similar time point, thereby increasing consistency. The cells are finally washed twice for 5 min each time with PBS and held at 4 °C overnight. This final step aids the uniform uptake of the nuclear stain prior to imaging.

17.2.4 Image acquisition

Achieving the highest possible image quality is an essential aspect of the screen, as ultimately not every image will be manually examined. HCS automated microscopes all tend to use a light source that is highly stabilized over time, meaning that the illumination given to wells across an entire screen should be consistent. The choice of objective is also critical and dependent on the subsequent analysis routine planned. For simple fluorescence intensity per cell measurements, a 10 ×/0.4 NA objective is likely to be sufficient, meaning that a field of view may contain in excess of 100 cells, and therefore perhaps no more than 2–4 fields of view per well need to be imaged. By contrast, when any advanced subcellular measurements are envisaged and resolution is important (as in this case), a 40 ×/0.95 NA objective should be used. In this scenario, no more than 25 cells may be present in a single field of view, and therefore upward of 20 fields of view may be required to obtain sufficient cells for the analysis. In the case of the analysis routine presented here, the image acquisition will use an Olympus Scan^R automated wide-field microscope, equipped with an Olympus 40 × Luc Plan FLN 0.6 NA objective. If screens are carried out on glass-bottomed plates, then super apochromatic objectives are recommended. In general, the vast majority of screens are carried out using air objectives. Although oil immersion objectives of the same magnification typically have superior optical performance, conventional mineral oil does not form droplets that are maintained across an entire plate during image acquisition. However, the latest silicone oil objectives show potential for use in HCS and clearly deliver superior quality images to air objectives. In the case of the assay for Golgi complex morphology described here, the air objective used appears to provide sufficient resolution to easily detect and quantify more than 50 individual Golgi fragments.

17.2.5 Image analysis and data management

Automated HCS imaging generates large volumes of data. In the assay described here, imaging a single 384-well plate, with 20 fields of view per well, and 2 color channels (Hoechst 33342 and GalNAc-T2-GFP) generates more than 15,000 image files of total size 40 GB. These data should be duplicated and archived on physically distinct storage devices.

The purpose of HCS is to generate quantitative single cell data in a completely unbiased manner. However, cells notoriously show a high degree of heterogeneity across a population, and this in itself can be problematic to automated analysis routines. It is therefore essential that the analysis procedure developed and applied can take account of possible artifacts wherever possible. This includes the removal of out-of-focus fields of view, cells on the border of fields of view, apoptotic cells, and cells undergoing mitosis. The choice of HCS image analysis software available is now very good, and we routinely use both open-source (e.g., CellProfiler, Broad Institute) and commercial (e.g., Columbus, PerkinElmer) solutions. Ideally, the software should be modular in nature, such that routines can be easily tailored to specific assays. It is also useful if they allow scripting within their environment.

The first task in the analysis routine is to segment the individual cells from the raw image data (Fig. 17.2A–C). In this assay, the cell line stably expresses GalNAc-T2–GFP, and although the majority of the fluorescence is concentrated at the Golgi complex, the small amount of fluorescence in the cytoplasm is sufficient to accurately identify the cytoplasm boundary for each cell (Fig. 17.2D). Morphological and intensity measurements are then extracted from the segmented cells to distinguish between healthy and apoptotic or dividing cells that would skew the analysis. At this point cells close to the border of the field of view are also removed from the analysis, allowing the Golgi complexes from only the remaining cells to be segmented (Fig. 17.2E and F). A variety of measurements can now be made from the segmented Golgi complexes of the remaining cells, including measurement of total area, fragment area, number of fragments per cell, and any texture features associated with the organelle (e.g., granularity) (Fig. 17.2G and H). The image analysis routine can be run across all wells and all plates in order to gather quantitative descriptors for every cell; however, this generates huge volumes of data that can be difficult to manage and interpret. At this point, therefore, an additional step can be introduced, specifically to use a linear classifier to select cells into populations based on particular features. In this case, cells showing either a wild-type or highly fragmented Golgi complex morphology are selected and used as a "training" set (Fig. 17.2I). The linear classifier then uses the features associated with this population of cells to classify the state of Golgi fragmentation across all cells in the well (Fig. 17.2J and K) and ultimately the entire plate (Fig. 17.2L). In this way, large volumes of highly complex image data become rapidly distilled down into a format that allows genes playing a role in Golgi fragmentation to be easily identified (Fig. 17.2L). The time required for such analyses is clearly dependent on the depth of information required and also on the computational power available. In the assay described here, a 2.4 GHz 16 core Xeon® processor takes between 5 and 6 h to analyze 15,000 images. If the protocol is based solely on morphological features (and not texture features), the analysis time can typically be reduced by half. The numerical data ultimately obtained can be normalized to both the negative and positive controls across the entire screen, and correlation studies can be carried out to test for reproducibility between replicate plates. Other statistical analyses may also be required, depending on the specific assay being carried out.

17.3 DISCUSSION

Traditionally, fluorescence microcopy has been long considered an appropriate method by which the general physiology of this organelle can be studied. Despite the advent of superresolution microscopy, light microscopy in general still suffers from fundamental limits in resolution. This is potentially problematic for the study of this organelle, as it is highly compact in nature, but yet proteomic techniques indicate that it is potentially "home" to several hundred proteins (Au et al., 2007). In recent years, there have been many elegant fluorescence microscopy techniques presented that can help define protein localization to this organelle (e.g., Dejgaard, Murshid, Dee, & Presley, 2007). However, the usual limitation of such approaches is that they are labor-intensive and not compatible for use with high numbers of samples or cells. Therefore, although automated fluorescence imaging and analysis of the Golgi will not solve the problems associated with resolution, as a technique it does provide accessibility in terms of systematically studying the relevance of all genes to the function of this organelle. To date, HCS on a genome-wide scale has already been applied to study many fundamental processes in the cell, for example, endocytosis (Collinet et al., 2010), secretion (Simpson et al., 2012), and cell division (Neumann et al., 2010), however not yet to the study of the physiology of a single organelle.

Although establishing a robust pipeline for high-content analysis of the Golgi complex is not a trivial task, we are now in a position that all the key tools to achieve this are available, and in this chapter, we present one potentially suitable strategy. In particular, it is worth commenting that advances in automated image analysis of fluorescence images have been significant in recent years. Automated classification of subcellular localization is now possible (Boland & Murphy, 2001; Chen & Murphy, 2006; Conrad et al., 2004; Hamilton, Pantelic, Hanson, & Teasdale, 2007), and indeed not only on a two-dimensional level but also on a three-dimensional one, whereby reconstruction of complex structures and automatic quantification of numerous three-dimensional descriptors can be made using intensity, morphological, and texture features from multiple stained structures (Dorval et al., 2010). Other examples of elaborate automated image analysis tools are too numerous to mention here but are comprehensively discussed in an article from the Murphy Lab (Shariff, Kangas, Coelho, Quinn, & Murphy, 2010). The Golgi complex is surely the first organelle that will be characterized in depth using a HCS approach; and these studies will undoubtedly reveal new and exciting features of how this central organelle is functionally integrated with all other cellular components.

SUMMARY

HCS and high-content analysis are an extremely powerful tool for the quantitative analysis of large volumes of image data. Although the infrastructure required for successful HCS experiments is both extensive and costly, it is a technique that can provide novel systems-level insight into complex biological questions. The protocols presented here provide a detailed guide for the systematic depletion of large numbers of target genes in a highly parallel manner, followed by quantitative analysis of the architecture of the Golgi complex in intact cells. We suggest that HCS will soon be applied in this way to gain a more thorough understanding of Golgi complex function.

Acknowledgments

We would like to thank all the members of the JCS Lab for excellent discussions and in particular Dr. Ciara Kelly the manager of the UCD Cell Screening Lab. The JCS Lab is supported by a Principal Investigator grant (09/IN.1/B2604) from Science Foundation Ireland (SFI) and by a Strategic Initiative award from the UCD College of Science.

References

Au, C. E., Bell, A. W., Gilchrist, A., Hiding, J., Nilsson, T., & Bergeron, J. J. (2007). Organellar proteomics to create the cell map. *Current Opinion in Cell Biology*, *19*, 376–385.

Bard, F., Casano, L., Mallabiabarrena, A., Wallace, E., Saito, K., Kitayama, H., et al. (2006). Functional genomics reveals genes involved in protein secretion and Golgi organization. *Nature*, *439*, 604–607.

Boland, M. V., & Murphy, R. F. (2001). A neural network classifier capable of recognizing the patterns of all major subcellular structures in fluorescence microscope images of HeLa cells. *Bioinformatics*, *17*, 1213–1223.

Carpenter, A. E., Jones, T. R., Lamprecht, M. R., Clarke, C., Kang, I. H., Friman, O., et al. (2006). Cell Profiler: Image analysis software for identifying and quantifying cell phenotypes. *Genome Biology*, *7*(10), R100.

Chen, S. C., & Murphy, R. F. (2006). A graphical model approach to automated classification of protein subcellular location patterns in multi-cell images. *BMC Bioinformatics*, *7*, 90.

Chia, J., Goh, G., Racine, V., Ng, S., Kumar, P., & Bard, F. (2012). RNAi screening reveals a large signaling network controlling the Golgi apparatus in human cells. *Molecular Systems Biology*, *8*, 629.

Collinet, C., Stöter, M., Bradshaw, C. R., Samusik, N., Rink, J. C., Kenski, D., et al. (2010). Systems survey of endocytosis by multiparametric image analysis. *Nature*, *464*, 243–249.

Conrad, C., Erfle, H., Warnat, P., Daigle, N., Lörch, T., Ellenberg, J., et al. (2004). Automatic identification of subcellular phenotypes on human cell arrays. *Genome Research*, *14*(6), 1130–1136.

de Forges, H., Bouissou, A., & Perez, F. (2012). Interplay between microtubule dynamics and intracellular organization. *The International Journal of Biochemistry & Cell Biology*, *44*, 266–274.

Dejgaard, S. Y., Murshid, A., Dee, K. M., & Presley, J. F. (2007). Confocal microscopy-based linescan methodologies for intra-Golgi localization of proteins. *Journal of Histochemistry and Cytochemistry*, *55*, 709–719.

Dorval, T., Ogier, A., Genovesio, A., Lim, H. K., Kwon do, Y., Lee, J. H., et al. (2010). Contextual automated 3D analysis of subcellular organelles adapted to high-content screening. *Journal of Biomolecular Screening*, *15*, 847–857.

Eliceiri, K. W., Berthold, M. R., Goldberg, I. G., Ibanez, L., Manjunath, B. S., Martone, M. E., et al. (2012). Biological imaging software tools. *Nature Methods*, *9*, 697–710.

Erfle, H., Neumann, B., Rogers, P., Bulkescher, J., Ellenberg, J., & Pepperkok, R. (2008). Work flow for multiplexing siRNA assays by solid-phase reverse transfection in multiwell plates. *Journal of Biomolecular Screening*, *13*, 575–580.

Giuliano, K. A., DeBiasio, R. L., Dunlay, R. T., Gough, A., Volosky, J. M., Zock, J., et al. (1997). High-content screening: A new approach to easing key bottlenecks in the drug discovery process. *Journal of Biomolecular Screening*, *2*, 249–259.

Hamilton, N. A., Pantelic, R. S., Hanson, K., & Teasdale, R. D. (2007). Fast automated cell phenotype image classification. *BMC Bioinformatics*, *8*, 110.

Liebel, U., Starkuviene, V., Erfle, H., Simpson, J. C., Poustka, A., Wiemann, S., et al. (2003). A microscope-based screening platform for large-scale functional protein analysis in intact cells. *FEBS Letters*, *554*, 394–398.

Neumann, B., Walter, T., Hériché, J. K., Bulkescher, J., Erfle, H., Conrad, C., et al. (2010). Phenotypic profiling of the human genome by time-lapse microscopy reveals cell division genes. *Nature*, *464*, 721–727.

Niederlein, A., Meyenhofer, F., White, D., & Bickle, M. (2009). Image analysis in high-content screening. *Combinatorial Chemistry & High Throughput Screening*, *12*, 899–907.

Pereira, D. A., & Williams, J. A. (2007). Origin and evolution of high throughput screening. *British Journal of Pharmacology*, *152*, 53–61.

Ramirez, I. B., & Lowe, M. (2009). Golgins and GRASPs: Holding the Golgi together. *Seminars in Cell & Developmental Biology*, *20*, 770–779.

Shariff, A., Kangas, J., Coelho, L. P., Quinn, S., & Murphy, R. F. (2010). Automated image analysis for high-content screening and analysis. *Journal of Biomolecular Screening*, *15*, 726–734.

Simpson, J. C., Joggerst, B., Laketa, V., Verissimo, F., Cetin, C., Erfle, H., et al. (2012). Genome-wide RNAi screening identifies human proteins with a regulatory function in the early secretory pathway. *Nature Cell Biology*, *14*, 764–774.

Starkuviene, V., Liebel, U., Simpson, J. C., Erfle, H., Poustka, A., Wiemann, S., et al. (2004). High-content screening microscopy identifies novel proteins with a putative role in secretory membrane traffic. *Genome Research*, *14*(10A), 1–11.

Storrie, B., White, J., Röttger, S., Stelzer, E. H., Suganuma, T., & Nilsson, T. (1998). Recycling of golgi-resident glycosyltransferases through the ER reveals a novel pathway and provides an explanation for nocodazole-induced Golgi scattering. *The Journal of Cell Biology*, *143*, 1505–1521.

Taylor, D. L. (2010). A personal perspective on high-content screening (HCS): From the beginning. *Journal of Biomolecular Screening*, *15*, 720–725.

Wendler, F., Gillingham, A. K., Sinka, R., Rosa-Ferreira, C., Gordon, D. E., Franch-Marro, X., et al. (2010). A genome-wide RNA interference screen identifies two novel components of the metazoan secretory pathway. *EMBO Journal*, *29*, 304–314.

PART 3

Emerging Studies [New Frontiers]

CHAPTER

(Re)Modeling the Golgi

18

Pierre Sens* and Madan Rao†

**Laboratoire Gulliver, CNRS-ESPCI, UMR 7083, 75231 Paris, France*

†Raman Research Institute and National Centre for Biological Sciences (Tata Institute of Fundamental Research), Bangalore, India

CHAPTER OUTLINE

Introduction: Coarse-Grained Modeling Based on Nonequilibrium Physics 300
18.1 Organelle Biogenesis and the Mechanistic Origins of Compartmentalization 300
18.1.1 Computational Models 301
18.1.2 Analytical Transport Models 302
18.2 Protein Transport of Golgi and Maintenance of Cisternal Identity 302
18.2.1 Quantification of Transport Through the Golgi Stack 303
18.2.2 Bulk Golgi Flow and Localization of Resident Enzymes 304
18.2.3 Regulation of Intercisternal Exchange 305
18.3 Kinetics of Budding and Cisternal Morphology 305
18.3.1 Kinetics of Coat-assisted Budding.. 306
18.3.2 Morphology Changes Induced by Active Fission–fusion 307
18.4 Future Outlook 307
Acknowledgments 309
References 309

Abstract

In this chapter, we summarize recent theoretical efforts to address a variety of issues in Golgi morphogenesis: *de novo* biogenesis of compartments with precise chemical identity, the transport of proteins through the Golgi, the maintenance of chemical identity, and the morphology of Golgi compartments, from the perspective of nonequilibrium physics.

Methods in Cell Biology, Volume 118

ISSN 0091-679X
http://dx.doi.org/10.1016/B978-0-12-417164-0.00018-5

INTRODUCTION: COARSE-GRAINED MODELING BASED ON NONEQUILIBRIUM PHYSICS

The need to segregate molecules for sequential and regulated chemical processing could have been the evolutionary basis for compartmentalization into pliable, membrane-bound organelles, of which a conspicuous example is the Golgi apparatus. Because of its identifiable structural and chemical signature, the Golgi of mammalian and plant cells offers the best arena in which to address fundamental questions in eukaryotic biology, namely, the robustness, precision, and biogenesis of organelle identity (both structural and chemical); the regulation of shape, size, and number of cisternae; and the connectivity between cisternae and other compartments.

While there is a rather detailed knowledge of the molecular processes involved in protein transport, membrane remodeling, and chemical processing, our understanding of the underlying physical principles at the core of these fundamental questions is still quite rudimentary. This, we argue, will require a confluence of quantitative high-resolution dynamic imaging of the Golgi and theoretical modeling, both computational and analytical. Theoretical models best suited for this are necessarily coarse-grained, described by a few generic rules and parameters and not overburdened by molecular detail. The evolution of specific protein machineries observed to be at work would then be viewed as embellishments on this organizing framework, leading to fine tuning, robustness, and efficiency.

The one incontrovertible feature of the Golgi organelle appears to be that they are dynamic membranous structures, subject to and driven by a continuous flux of membrane-bound material from the ER to the plasma membrane (PM). Thus, structure formation, chemical identity, and compartment morphology over large spatial and temporal scales emerge as steady states in the presence of this uninterrupted flux, suggesting the crucial role played by driven nonequilibrium physics. Here, we summarize recent theoretical studies that view the morphogenesis of the Golgi through the lens of nonequilibrium dynamics.

18.1 ORGANELLE BIOGENESIS AND THE MECHANISTIC ORIGINS OF COMPARTMENTALIZATION

There has been some debate about whether the Golgi organelle is constructed from a preexisting template or generated *de novo*, a result of self-organization, with recent experiments, reviewed in Glick and Nakano (2009), supporting the latter view. This opens the door to theoretical models, both computational and analytical, that take into account the essential ingredients of the dynamics at short scales to arrive at conditions under which the system can self-organize to form stable compartments with distinct chemical identity.

18.1.1 Computational models

Discrete-event Gillespie simulations (Gong, Guo, Linstedt, & Schwartz, 2010; Gong, Sengupta, Linstedt, & Schwartz, 2008) and continuous-time Brownian dynamics simulations (Kühnle et al., 2010) have been employed to study the biogenesis of compartmentalization. Discrete-event models (Gong et al., 2008, 2010) describe the stochastic time evolution of states defined by a list of membrane structures that come in three kinds: vesicles, tubular clusters formed by vesicle–vesicle fusion, and compartments. Each membrane structure is characterized by an instantaneous size and list of specific protein concentrations that come in three species: GEFs that initiate vesicle budding, vSNAREs that mark vesicles, and tSNAREs that mark compartments. Each kind of protein comes with a specific compartmental identity, which carries information about the structure that it resides in and the structure that it acts on. Thus, for instance, a GEF1 on a compartment will initiate the formation of a vSNARE1 marked vesicle, which in turn will target tSNARE1 marked compartment. Membrane and protein are exchanged among structures through budding + fusion of vesicles, where local budding rates depend on the concentration of GEFs in the compartment, and fusion probabilities depend cooperatively on vSNARE and tSNARE marker concentrations. These simulations can be used to determine parameters and initial conditions under which a desired number of compartments would emerge as stable steady states and the time evolution of sizes, number, and protein composition of the compartments following specific perturbations.

From a general physics viewpoint, the existence of a parameter/initial-condition regime for obtaining stable "compartments" within this model should not come as a surprise, since it contains a multitude of selective rules and provides sufficient constraints to ensure that there will be distinct compartments. These models are purely topological and contain no spatial information. There is no chemical maturation process. Most crucially, these are closed models, in which all material, proteins, and membranes are kept constant in time—there is no flux of material running through the system.

Brownian dynamics simulation models (Kühnle et al., 2010) restore spatial information, chemical maturation, and a steady flux of particles (vesicles) from discrete sources. Particles carrying two protein pools A and B are injected from the source (ER) with a fixed rate and allowed to diffuse within the simulation box and interact via short-range potentials. On encounter, particles can fuse with a probability that depends on the concentration of protein A (representing SNARES), which degrades with a rate r_A. Fusion of particles creates aggregates, whose shape, size, and fluidity are controlled so that the largest aggregates assume a flat, pancake-like structure, to resemble Golgi cisternae. Protein species B accumulates in each particle with a rate r_B and reflects the process of chemical maturation. Finally, above a threshold size n, the aggregate is allowed to eject particles into the PM. This model can be used to monitor the size, shape, and chemical composition of aggregates. Although this model exhibits the formation of a dynamically maintained stack structure, there is always only one such entity that is chemically partitioned (defined by a gradient in the concentration of protein B).

18.1.2 Analytical transport models

The above models only incorporate the elementary processes of fission, transport, and fusion. An alternate process, with some experimental support, is cisternal progression, where the entire compartment moves vectorially from the ER to the PM. This process can be easily incorporated and analyzed within a simplified one-dimensional transport model without sacrificing the essential principles, as done in Sachdeva, Barma, and Rao (2011, 2013, in preparation). Quite broadly, this transport model has the following processes: (i) localized injection of particles of A type, that is, injection of vesicles containing unprocessed biomolecules from the ER; (ii) transformation of particles from one species to the other, $A \rightarrow B \rightarrow C \rightarrow \cdots$, that is, chemical processing of the biomolecules by enzymes; (iii) transport of particles either by fission of a single particle (vesicle movement) or through movement of bigger aggregates (cisternal progression); and (iv) localized ejection of particles of any species, that is, removal of vesicles into the PM.

This model can be analytically solved in various limits, to determine the conditions under which one might obtain large, sharply peaked, well-separated, chemically distinct, and temporally stable compartments. The fact that the rules are kept fairly general, with no imposed constraints or selective rules, makes for analytical tractability. In the absence of cisternal progression, the most desirable compartmental features are obtained when the fission rate from an aggregate of size M is taken to scale as the aggregate area for small M with a crossover to a constant beyond a certain size, as is physically reasonable. Turning on a slight amount of cisternal progression typically maintains large finite structures but makes them less well separated. In the limit that cisternal progression dominates, transport proceeds through the movement of sporadic but large aggregates. If however, the fission rate is large, then transport occurs principally by a uniform flux of small aggregates and particles. There are many analytical predictions that emerge from these models and we encourage the reader to consult Sachdeva et al. (2011, 2013, in preparation).

18.2 PROTEIN TRANSPORT OF GOLGI AND MAINTENANCE OF CISTERNAL IDENTITY

A consensus on the mechanisms of protein transport through the Golgi stack has not emerged yet, but it is generally accepted that material must be exchanged between cisternae. In the vesicular exchange model, cisternae are stable "static" structures and transit proteins move along the stack by anterograde vesicular transport. In the cisternal progression model, the entire cisternae progress through the stack and Golgi resident enzymes undergo retrograde transport to remain in a fixed position. We review theoretical studies that provide (i) a general framework to quantify the transport of transit proteins across the Golgi stack and (ii) mechanisms for how compartments might maintain distinct biochemical identities in the face of constant exchange of membrane-bound material.

18.2.1 Quantification of transport through the Golgi stack

Although the principles governing the formation of the Golgi stack and the transport of proteins through it are undoubtedly intimately linked, precise quantification of transport requires a general framework independent of specific models of stack generation.

From a conceptual point of view, intra-Golgi transport can be seen as a one-dimensional movement from the *cis-* to the *trans*-Golgi face (Dmitrieff, Rao, & Sens, 2013). A minimal transport framework should include the two extreme mechanisms of Golgi transport, cisternal progression and intercisternal exchange. Figure 18.1 illustrates such a model that includes anterograde and retrograde transport of proteins (characterized by the rates k and k', respectively) and convection of entire cisterna with a velocity v_m (a rate corresponding to the inverse time needed for cisterna n to become cisterna $n+1$). In addition, proteins may exit any cisterna at a rate r. It was shown (Dmitrieff et al., 2013) that the concentration $C_n(t)$ of a given protein species in cisterna n at time t satisfies a diffusion–convection equation

$$\frac{\partial C_n}{\partial t} = \frac{\partial}{\partial n}\left(D_t \frac{\partial C_n}{\partial t} - (v_t + v_m)C_n\right) - rC_n \tag{18.1}$$

with

$$D_t = \frac{k+k'}{2}, \quad v_t = k - k'. \tag{18.2}$$

Here, the "maturation velocity" v_m is common to all species, while the velocity and diffusion coefficient v_t and D_t arising from intercisternal exchange may vary among species. An important implication of Eq. (18.1) is that while both cisternal

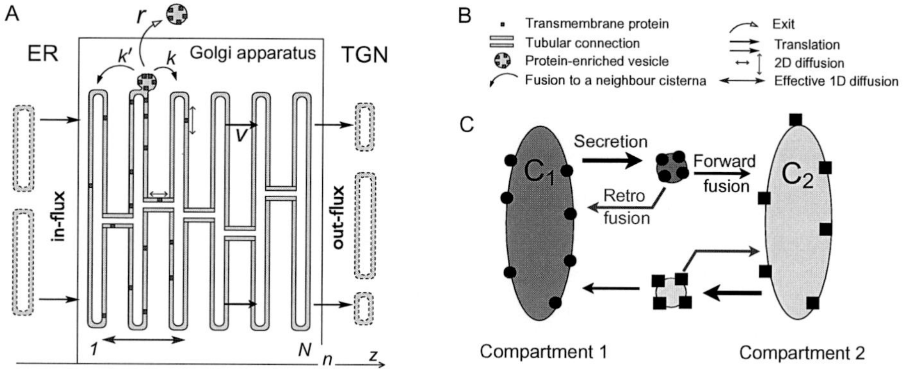

FIGURE 18.1

(A) Model for protein transport through the Golgi stack. (B) Sketch of the different processes contributing to intra-Golgi transport. (C) Sketch of the different steps permitting specificity of inter-compartment exchange. (See color plate.)

progression (v_m) and the difference between rates of anterograde and retrograde transport (v_t) result in the net progression of protein in the stack $v = v_m + v_t$, intercisternal exchange also gives rise to an effective diffusion of proteins along the *cis–trans* axis. The two transport coefficients v and D_t can *in principle* be characterized independently by analyzing the spatiotemporal evolution of the distribution of transit proteins in the Golgi. If it is found that $D_t \neq 0$ for a particular protein species, this implies that this species is exchanged between cisternae. In a stack containing N cisternae, for which the total transit time is $T = N/v$, the number of exchange steps experienced by a given protein is necessarily larger than $N \times 2D_t/v$, and the number of backward steps (toward the ER) is larger than $N \times (2D_t - v)/(2v)$. Quantifying cisternal progression is less straightforward, since the relative contribution of cisternal progression and biased vesicular transport to the net protein velocity v may not easily be distinguished. If a cargo is solely transported by anterograde vesicular transport ($k' = 0$ and $v_m = 0$), one expects $v = 2D_t = k$. Thus, measuring $v > 2D_t$ necessarily implies that cisternal progression contributes to the transport.

Quantitative kinetic data on Golgi protein transport are essentially of two kinds; fluorescence microscopy (FRAP) on entire Golgi gives access to the total quantity of protein in the stack with a good temporal resolution but no spatial information *within* the stack (Patterson et al., 2008), and electron microscopy data give access to the distribution of protein inside the stack but with poor temporal resolution (Bonfanti et al., 1998; Trucco et al., 2004). FRAP experiments mostly give information on the mechanism of protein exit from the Golgi. They have shown that the exit of the small membrane protein VSVG is not limited to the Golgi *trans* face and that the exit kinetics is inconsistent with a transport mechanism solely based on cisternal progression (Patterson et al., 2008). Analysis of EM data with Eqs. (18.1) and (18.2) leads to the conclusion that $D_t \simeq v = 0.2 - 0.3\ \text{min}^{-1}$ for both VSVG and the large protein aggregate procollagen. This shows that both cargoes are exchanged between cisternae and that they both undergo an average of $N/2 \simeq 3$ retrograde transport steps during their transit through the Golgi (Dmitrieff et al., 2013). This implies that transport may not be restricted to the conventional protein-coated vesicles of the COP type and could involve pleomorphic membrane carriers, including large cisterna fragments undergoing scission and fusion, as suggested by the recent cisternae progenitor model (Pfeffer, 2010).

18.2.2 Bulk Golgi flow and localization of resident enzymes

Resident proteins are often found to be localized in one or two Golgi cisternae (Donohoe et al., 2013); in the cisternal progression model, for instance, resident Golgi enzymes must remain localized against a bulk flow of material. It has been proposed (Glick, Elston, & Oster, 1997) that progression could be the consequence of retrograde transport, the *trans*-most cisterna disappearing when it is completely depleted of enzymes. Such a model is able to produce cisternae of different composition, if different enzymes compete for retrograde transport. Enzymes are then distributed sequentially according to their affinity to retrograde transport carriers.

Enzyme localization may also be achieved by recycling the protein back to their preferred location before they are convected by the bulk flow. Extensions of

Eq. (18.1) have shown that precise localization (within two cisternae) requires that proteins exit the Golgi stack at a rate faster than two to three times the net velocity or of order $r \simeq 1/2$ min (Dmitrieff et al., 2013).

18.2.3 Regulation of intercisternal exchange

Specificity during material exchange between organelles and between cisternae within the Golgi stack is commonly evoked as a crucial factor for compartmental identity. Theoretical modeling can inform us on the general principles permitting compartments constantly exchanging material to retain distinct biochemical identities. Heinrich and Rapoport (2005) showed that two nonidentical compartments can be generated in a model where vesicular exchange involves two kinds of coats and two pairs of SNARES. Vesicles formed by one type of coat preferentially package one type of SNARES pair, and vesicle fusion occurs preferentially with the compartment containing the highest concentration of matching SNARES. The authors showed that when the biases for preferential packaging or fusion reach threshold values, the dynamical system shows a bifurcation point and the symmetric state (equal concentration in both compartment) becomes unstable. The steady state of the system then becomes asymmetric, with the two different pairs of SNARES localized in different compartments. By making the budding and fusion rates dependent on the cisternae size, this mechanism of differentiation also allows for a regulation of the cisternae size. Finally, the distribution of particular cargo molecules can be biased toward one of the two compartments if it preferentially interacts with one of the two coat types.

This theoretical model illustrated the fundamental principles underlying the specificity of intracellular trafficking. In particular, it showed that a given species may be concentrated in a given compartment not by static retention, but through its affinity to molecules involved in transport. Indeed, molecules frequently escape their home compartment, but return to it rather than being exchanged. Generalization to a multicompartment system using the same general principles for establishing identity is straightforward and merely requires a multiplication of specific molecular players. An important improvement has been to consider an open system traversed by fluxes (Dmitrieff & Sens, 2011, 2012), as is the case *in vivo*. The specific recognition mechanisms discussed earlier allow for a concentration-dependent regulation of the rate of protein exchange between compartments. Under high concentration, the exchange is strongly reduced because most secreted vesicles undergo retro-fusion with the donor compartment. The residency time of a protein in the Golgi thus increases dramatically under conditions of high incoming flux, which might provide a functional advantage for efficient processing of all transit proteins by a limited amount of Golgi enzymes.

18.3 KINETICS OF BUDDING AND CISTERNAL MORPHOLOGY

Golgi compartments exhibit distinct morphologies at different scales—at the largest scale, they are flattened sacs or pancakes; at intermediate scales, they are ramified or folded with dynamic tubular extensions; and at smaller scales, the Golgi membranes can deform into buds of diameter 50–100 nm, often assisted by the action of coat proteins, COPI, II. In describing the morphology changes over a range of scales, it is necessary to appeal to different physical mechanisms; for instance, the dynamics at smaller scales is dictated by an interplay between coat polymerization kinetics and membrane forces and tension, while the dynamics at large scales is dominated by the nonequilibrium fluxing of vesicles that fission from and fuse onto the membrane compartments. On the other hand, membrane budding in the *trans*-Golgi is often not coat-assisted but involves enzymatic modification of lipids such as diacylglycerol and curvature inducing proteins that insert into the bilayer (Shemesh, Luini, Malhotra, Burger, & Kozlov, 2003); the sequential steps of membrane deformation leading to a large tubular carrier can be analyzed using an equilibrium elastic model of the membrane (Shemesh et al., 2003).

18.3.1 Kinetics of coat-assisted budding

The formation of a protein-coated transport vesicle of the COP or clathrin type involves several steps. The activation of a GTPase (Sar1 for COPII and Arf1 for COPI; McMahon & Mills, 2004) leads to its binding to the membrane and the recruitment of individual coatomer complexes (the monomers). The bound monomers diffuse in the membrane and aggregate into a growing coat of curved geometry that drives membrane deformation. Once completed, the mature coated vesicle eventually undergoes scission from the donor membrane and a transport vesicle is formed. Inactivation of the GTPase leads to the disassembly of the coat and the naked vesicle may then fuse with the receiving compartment.

This sequential and deterministic picture must be tempered by the fact that the exchange kinetics of coat components is typically much faster than the rate of vesicle secretion (Presley et al., 2002). This suggests that the process of coat formation must compete with the inactivation of coatomer complexes and their individual release from the membrane and that many futile activation/inactivation cycles accompany the formation of a coat. It has been shown theoretically that this apparent wasteful process allows for a precise regulation of vesicle secretion, coupled to the presence of cargo (Foret & Sens, 2008). Indeed, the competition between coat growth and the monomer recycling leads to the emergence of stationary coats of fixed area. The turnover of coat components allows for a highly sensitive switching mechanism between a quiescent membrane and a membrane producing vesicle, upon a slowing down of the exchange kinetics. Such slowing down may be due to the presence of secretory cargo, which have been shown to regulate the turnover of coat monomers (Forster et al., 2006). The competition between recycling and growth thus permits to avoid the formation of transport vesicles in the absence of cargo. Interestingly, the model (Foret & Sens, 2008) also highlights the importance of the mechanical properties of the membrane and in particular of membrane tension. Membrane deformation is penalized under conditions of high membrane tension, and the secretory switch just discussed may also be triggered by a decrease of the tension of the organelle's membrane.

18.3.2 Morphology changes induced by active fission–fusion

The typical rate at which material leaves the ER is estimated to be around 1000 transport vesicles per second (Wieland, Gleason, Serafini, & Rothman, 1987). Even if this overestimates the material flux through a single Golgi compartment, it suggests that the timescales of vesicle fission and fusion onto a compartment are at least comparable to membrane relaxation times, which for a micron-sized compartment is of the order of tens of seconds. Thus, the large-scale morphology of the Golgi compartments must be influenced by active out-of-equilibrium processes of fission and fusion of material.

The large-scale dynamics of a membrane subject to active fission and fusion can be theoretically described by a set of coarse-grained equations for the local membrane shape and lipid density (Girard, Julicher, & Prost, 2004; Rao & Sarasij, 2001). Because fission–fusion happens on the cytosolic side of the membrane, symmetry arguments imply that the local membrane velocity is proportional to curvature, a feature that can also be derived by analyzing the mechano-chemistry of the protein machinery at fission–fusion events. These studies predict novel shape fluctuations and instabilities; indeed, the prediction of tubular instabilities induced by negative tension has been verified in *in vitro* membrane systems (Solon et al., 2006; Staykova, Holmes, Read, & Stone, 2011).

These studies have been subsequently extended to closed vesicles (Gowrishankar, Ramakrishnan, Lakshmi, Sunil Kumar, & Rao, in preparation), where a perturbation analysis reveals three interesting features: (i) activity confined to a rim of the closed membrane, typically leads to a flattening of the structure; (ii) spatially asymmetric fission–fusion can provide the driving force for cisternal movement; and (iii) an imbalance between fission and fusion rates can lead to tubular instabilities and folded morphologies. These results have been confirmed by a dynamically triangulated Monte Carlo simulation of the closed membrane subject to curvature-dependent fission and fusion, which in addition shows a steady-state morphology strikingly reminiscent of Golgi compartments (Gowrishankar et al., in preparation).

18.4 FUTURE OUTLOOK

We have reviewed some recent modeling approaches to understand the underlying principles involved in Golgi morphogenesis that may be described as top-down, rather than the bottom-up approach of starting with a parts list of molecular players and their function. The aim of all these models is to be sufficiently minimalist without oversimplification. They achieve varying levels of coarse graining and varying successes. Computational models that incorporate many microscopic details and parameters are better suited for asking specifically molecular-based questions and predicting responses to specific molecular perturbations. However, since one cannot be certain about all the molecular processes in the elaborate network of interacting proteins and lipids, the response to specific perturbations may be more complex than envisaged by these models. Further, it is difficult to extract matters of principle from such detailed modeling. In contrast, coarse-grained models with a few effective parameters are more amenable to analytic and conceptual understanding; however, it is difficult to relate the effective parameters to specific molecular details.

We end with a list of questions that are amenable to the kind of coarse-grained modeling discussed here, to which only partial answers exist.

- Are the elementary nonequilibrium processes of fusion, scission, and selectivity-induced segregation sufficient to give rise to *de novo* self-assembly of the Golgi apparatus in all cases? What is the minimal selectivity that is required for robust, identifiable compartments?
- Since compartment biogenesis occurs by dynamic self-organizing principles, every component of the organelle must be subject to this dynamics. What is the distribution of lifetimes of the different components of a compartment?
- Is there chemical segregation within a compartment, both on the membrane and in the lumen? Are there active processes involved in this segregation or is it driven by equilibrium thermodynamic forces?
- Is the Golgi structure affected by what needs to be transported? Does the encapsulated cargo play a role in determining the size, shape, and chemical composition of budding vesicles or rates of fission? Do different cargoes take different routes in the secretory pathway?
- What fixes the timescale of protein transit through the Golgi: biochemical maturation, the making of transport intermediates, or physical transport?
- What is the spatial distribution of fission–fusion events at each cisterna?
- What is the physical driving force for cisternal progression?
- How does the size and number of cisternae depend on the flux of material through the Golgi?
- What are the factors that determine the spatial organization of the cisternae (e.g., distance from the ER and from each other)?
- What is the precise dynamics of reformation of the Golgi compartments from a fragmented state, both drug induced or after cell division? What is the dynamics of morphology changes upon blocking entry into the Golgi at different temperatures?
- What is the interplay between enzymatic kinetics responsible for chemical maturation in cisternae and the dynamics of compartmentalization?

A useful strategy to address these questions would be to construct abstract, predictive models and then see how the addition of selective microscopic features might lead to fine tuning, efficiency, and robustness. Such a strategy, combined with a quantitative analysis of the dynamics of Golgi morphogenesis using novel live-cell imaging techniques, would lead a much improved understanding.

Acknowledgments

We thank our colleagues M. Barma, S. Dmitrieff, L. Foret, V.S. Gayathri, N. Gov, K. Gowrishankar, V. Malhotra, S. Mayor, H. Sachdeva, Sarasij R.C., P.B. Sunil Kumar, and M.S. Turner for discussions and collaborations on some of the work reported here.

References

Bonfanti, L., Mironov, A., Jr., Martínez-Menárguez, J., Martella, O., Fusella, A., Baldassarre, M., et al. (1998). Procollagen traverses the golgi stack without leaving the lumen of cisternae: Evidence for cisternal maturation. *Cell*, *95*, 993–1003.

Dmitrieff, S., Rao, M., & Sens, P. (2013). Quantitative analysis of intra-golgi transport reveals inter-cisternal exchange for all cargo. *Proceedings of the National Academy of Sciences of the United States of America*, *110*, 15692–15697.

Dmitrieff, S., & Sens, P. (2011). Cooperative protein transport in cellular organelles. *Physical Review E*, *83*, 041923.

Dmitrieff, S., & Sens, P. (2012). Golgi apparatus: Homotypic fusion maintains biochemical gradients within the golgi and improves the accuracy of protein maturation. *International Journal of Biochemistry and Cell Biology*, *44*, 718–721.

Donohoe, B., Kang, B.-H., Gerl, M., Gergely, Z., McMichael, C., Bednarek, S., et al. (2013). Cis-golgi cisternal assembly and biosynthetic activation occur sequentially in plants and algae. *Traffic*, *14*, 551–567.

Foret, L., & Sens, P. (2008). Kinetic regulation of coated vesicle secretion. *Proceedings of the National Academy of Sciences of the United States of America*, *105*, 14763–14768.

Forster, R., Weiss, M., Zimmermann, T., Reynaud, E., Verissimo, F., Stephens, D., et al. (2006). Secretory cargo regulates the turnover of COPII subunits at single ER exit sites. *Current Biology*, *16*, 173–179.

Girard, P., Julicher, F., & Prost, J. (2004). Fluid membranes exchanging material with external reservoirs. *European Physical Journal E: Soft Matter and Biological Physics*, *14*, 387–394.

Glick, B., Elston, T., & Oster, G. (1997). A cisternal maturation mechanism can explain the asymmetry of the golgi stack. *FEBS Letters*, *414*, 177–181.

Glick, B. S., & Nakano, A. (2009). Membrane traffic within the golgi apparatus. *Annual Review Cell Development Biology*, *25*, 113–132.

Gong, H., Guo, Y., Linstedt, A., & Schwartz, R. (2010). Discrete, continuous, and stochastic models of protein sorting in the golgi apparatus. *Physical Review E*, *81*, 011914.

Gong, H., Sengupta, D., Linstedt, A., & Schwartz, R. (2008). Simulated de novo assembly of golgi compartments by selective cargo capture during vesicle budding and targeted vesicle fusion. *Biophysical Journal*, *95*, 1674–1688.

Gowrishankar, K., Ramakrishnan, N., Lakshmi, K. C., Sunil Kumar, P. B. & Rao, M., in preparation.

Heinrich, R., & Rapoport, T. A. (2005). Generation of nonidentical compartments in vesicular transport systems. *Journal of Cell Biology*, *168*, 271–280.

Kühnle, J., Shillcock, J., Mouritsen, O. G., & Weiss, M. (2010). A modeling approach to the self-assembly of the golgi apparatus. *Biophysical Journal*, *98*, 2839–2847.

McMahon, H. T., & Mills, I. G. (2004). Cop and clathrin-coated vesicle budding: Different pathways, common approaches. *Current Opinion in Cell Biology*, *16*, 379–391.

Patterson, G., Hirschberg, K., Polishchuk, R., Gerlich, D., Phair, R., & Lippincott-Schwartz, J. (2008). Transport through the golgi apparatus by rapid partitioning within a two-phase membrane system. *Cell*, *133*, 1055–1067.

Pfeffer, S. (2010). Recent advances in understanding golgi biogenesis. *F1000 Biology Reports*, *2*, 32.

Presley, J. F., Ward, T. H., Pfeifer, A. C., Siggia, E. D., Phair, R. D., & Lippincott-Schwartz, J. (2002). Dissection of copi and arf1 dynamics in vivo and role in golgi membrane transport. *Nature*, *417*, 187–193.

Rao, M., & Sarasij, R. C. (2001). Active fusion and fission processes on a fluid membrane. *Physics Review Letters*, *87*, 128101.

Sachdeva, H., Barma, M., & Rao, M. (2011). Multispecies model with interconversion, chipping and injection. *Physical Review E*, *84*, 031106.

Sachdeva, H., Barma, M., & Rao, M. (2013). Condensation and intermittency in an open-boundary aggregation-fragmentation model. *Physical Review Letters*, *110*, 150601.

Sachdeva, H., Barma, M., & Rao, M., in preparation.

Shemesh, T., Luini, A., Malhotra, V., Burger, K. N. J., & Kozlov, M. M. (2003). Prefission constriction of golgi tubular carriers driven by local lipid metabolism: A theoretical model. *Biophysical Journal*, *85*, 3813–3827.

Solon, J., Pecreaux, J., Girard, P., Faure, M., Prost, J., & Bassereau, P. (2006). Negative tension induced by lipid uptake. *Physical Review Letters*, *97*, 098103.

Staykova, M., Holmes, D. P., Read, C., & Stone, H. A. (2011). Mechanics of surface area regulation in cells examined with confined lipid membranes. *Proceedings of the National Academy of Sciences of the United States of America*, *108*, 9084–9088.

Trucco, A., Polishchuk, R., Martella, O., Di Pentima, A., Fusella, A., Di Giandomenico, D., et al. (2004). Secretory traffic triggers the formation of tubular continuities across golgi sub-compartments. *Nature Cell Biology*, *6*, 1071–1081.

Wieland, F., Gleason, M., Serafini, T., & Rothman, J. (1987). The rate of bulk flow from the endoplasmic reticulum to the cell surface. *Cell*, *50*, 289–300.

CHAPTER

Golgi Depletion from Living Cells with Laser Nanosurgery

19

Paolo Ronchi and Rainer Pepperkok

Cell Biology and Biophysics Unit, European Molecular Biology Laboratory, Heidelberg, Germany

CHAPTER OUTLINE

Introduction and Rationale **312**
19.1 Theory **312**
19.2 Equipment **313**
19.3 Materials **315**
19.4 Methods **315**
19.4.1 Microcontact Printing 315
19.4.1.1 Stamp Preparation 316
19.4.1.2 Printing on Glass Coverslips 316
19.4.1.3 Tips and Pitfalls 316
19.4.2 Laser Nanosurgery 319
19.4.2.1 Cell Preparation 319
19.4.2.2 Cell Dissection 319
19.4.2.3 Tips and Pitfalls 321
19.4.2.4 Preparation of Samples for Time-lapse Imaging 321
19.4.2.5 Tips and Pitfalls 322
19.5 Discussion **322**
Acknowledgment **323**
References **323**

Abstract

How Golgi biogenesis occurs in mammalian cells is a controversial problem. Can the Golgi complex (GC) form *de novo* from ER membranes or does it require a template? The method described in this chapter uses growth of cells on micropatterns to displace the GC from its juxtanuclear position and laser nanosurgery to subsequently deplete it from living cells. Golgi-depleted karyoplasts can be followed by time-lapse microscopy to address if and how the GC can be *de novo* synthesized from ER membranes.

Methods in Cell Biology, Volume 118 **ISSN 0091-679X**

http://dx.doi.org/10.1016/B978-0-12-417164-0.00019-7

Furthermore, the study of different processes in the absence of the GC can shed light on the role of this organelle in the intracellular signaling and homeostasis.

INTRODUCTION AND RATIONALE

Our understanding of the molecular mechanisms how the Golgi complex (GC) structure is achieved and maintained is still incomplete (Lowe & Barr, 2007). Numerous studies have investigated Golgi biogenesis in mammalian cells and, based on different experimental approaches, at least two alternative models for Golgi biogenesis have been suggested. One proposes that during mitosis, Golgi proteins are completely redistributed in the endoplasmic reticulum (ER), and upon exit from mitosis, cells resynthesize *de novo* a GC from ER membranes (Altan-Bonnet et al., 2006; Kasap et al., 2004; Lippincott-Schwartz, Yuan, Bonifacino, & Klausner, 1989; Puri & Linstedt, 2003; Zaal et al., 1999). The second model proposes that post-ER structures containing Golgi components are present throughout mitosis and equally partitioned into the daughter cells. These structures could act as templates and would be necessary for Golgi biogenesis (Axelsson & Warren, 2004; Jesch, Mehta, Velliste, Murphy, & Linstedt, 2001; Jollivet et al., 2007; Orci et al., 1993; Pelletier, Jokitalo, & Warren, 2000; Seemann, Jokitalo, Pypaert, & Warren, 2000; Shorter & Warren, 2002).

Despite the large amount of elegant studies, supporting either one of the two Golgi biogenesis models, none of the approaches used was able to deplete the GC or its remnants from intact cells, leaving the question of *de novo* Golgi biogenesis still open. The laser nanosurgery method described here allows physical removal of the GC and a substantial depletion of all its protein and lipid components from a living cell. The method was used to demonstrate that a structurally normal and functional GC reforms within 8–12 h after nanosurgery, providing evidence for a *de novo* Golgi biogenesis in mammalian cells (Tängemo et al., 2011). The need of resynthesizing Golgi components after their depletion by laser nanosurgery apparently slows down the Golgi biogenesis process and thus facilitates characterization of the process at the light and electron microscopy levels. In addition, laser nanosurgery allows the analysis of the behavior of living cells in the absence of a GC and offers the unprecedented possibility of studying the involvement of the GC in different cellular processes such as signal transduction, a question that is presently poorly understood.

In this chapter, we describe protocols for the generation of Golgi-depleted karyoplasts by laser nanosurgery and subsequent analyses of Golgi biogenesis by time-lapse microscopy.

19.1 THEORY

Laser nanosurgery has recently become a popular technique, because of the possibility to ablate biological structures inside living cells or organisms with submicrometer precision (Colombelli, Reynaud, & Stelzer, 2007). The technology has been applied to several fields of research in cell and developmental biology (see Colombelli et al., 2007; Kohli & Elezzabi, 2009; Ronchi, Terjung, & Pepperkok, 2012, for reviews).

We have recently introduced laser nanosurgery to deplete the GC from living cells, such that they still contain an intact nucleus and all the other cellular organelles, with the exception of the centrosome (Tängemo et al., 2011).

Depleting the GC from living cells by nanosurgery is challenging due to the fact that the GC is in close proximity to the cell nucleus in unperturbed cells, making any attempts to deplete the GC without severe cell damage difficult or even impossible. To overcome this limitation, we took advantage of recent observations by Pouthas et al. (2008) demonstrating that seeding BSC1 cells on micropatterned glass coverslips with 6 μm wide fibronectin lines surrounded by polyethylene glycol (PEG) results in the displacement of the GC from the nucleus in up to 10% of the cells (Fig. 19.1A; see also Pouthas et al., 2008; Tängemo et al., 2011). This displacement provides sufficient space between the nucleus and GC to perform the surgery without apparent damaging of the cells (Fig. 19.1A and C). Furthermore, cells move and stretch along the thin fibronectin lines such that microtubules and actin stress fibers redistribute parallel to the line direction (Pouthas et al., 2008; Fig. 19.1B). This considerably facilitates tuning the laser power to sever cytoskeleton fibers, but not membranes, which require higher energy due to their fluidity. The resulting release of the cytoskeleton elastic tension leads to the progressive separation of the cell in two parts, until the membranes are sealed generating a karyoplast and a Golgiplast (Fig. 19.1B and C; Tängemo et al., 2011).

19.2 EQUIPMENT

- Round coverslips (ø 15 mm) (Menzel Glaeser, Braunschweig, Germany)
- 35 mm culture dishes (Corning Inc., NY, United States) with a drilled hole (10 mm in diameter) in the center of the dish produced in-house
- Photolithography Si master for the production of poly(dimethyl siloxane) (PDMS) stamps with 6 μm wide lines spaced by at least 18 μm (GeSiM Gesellschaft fuer Silizium-Mikrosysteme mbH, Grosserkmannsdorf, Germany)
- Plasma generator (PlasmaPrep2, Gala Instrumente GmbH, Bad Schwalbach, Germany)
- Compressed nitrogen bottle
- Curved forceps (Fine Science Tools GmbH, Heidelberg, Germany)
- Filter forceps (Merck Millipore, Darmstadt, Germany)
- Nanosurgery device. We are currently using a wide-field Olympus Cell^R microscope (operated with the Excellence software) equipped with a pulsed 355 nm laser (DPSL-355/14, Rapp OptoElectronic, Hamburg, Germany). The pulse duration of this laser source is 1 ns with a repetition rate of 200 Hz. Cell dissection is typically achieved with power intensities in the range of 100–200 nJ/pulse. The laser beam is focused and moved on the sample with an integrated

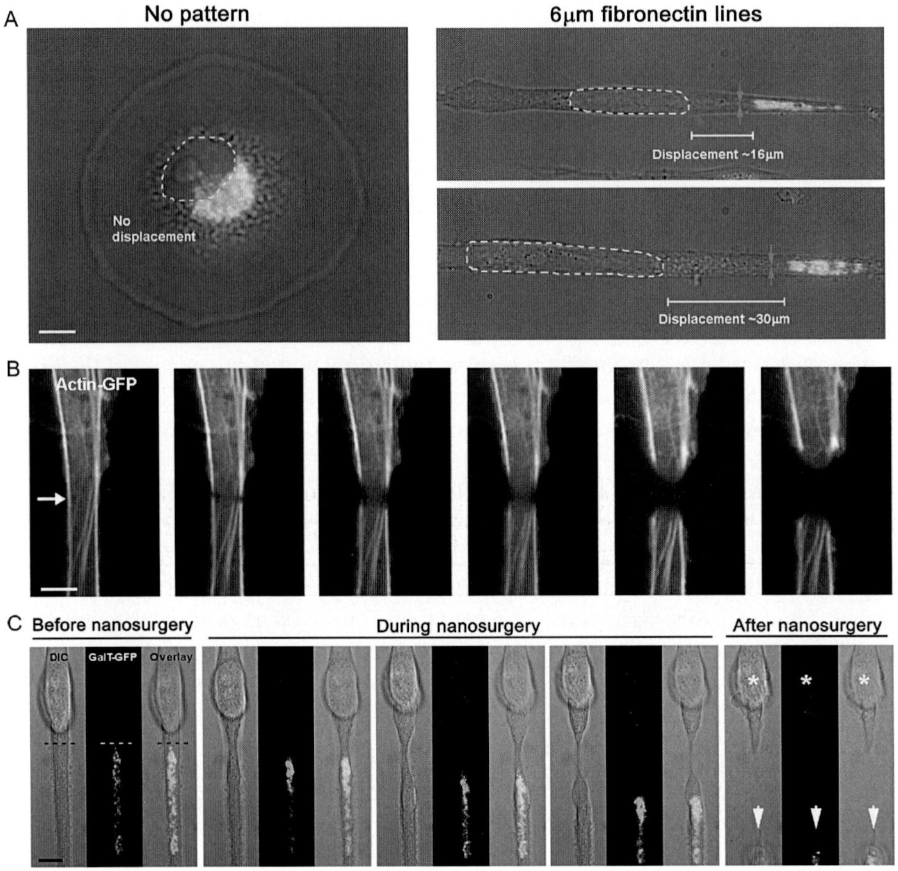

FIGURE 19.1

Laser nanosurgery of the Golgi complex. (A) BSC1 GalT-GFP cells were seeded on a nontreated glass surface (left) or on a micropatterned coverslip (6 μm wide fibronectin lines—right). Images were acquired in the GFP channel (showing the Golgi marker—green) and in the transmitted light channel using differential interference contrast (DIC). The two channels were then overlaid to observe the distance between the GC and the nucleus. The nucleus contour is highlighted by a dashed yellow line for better visualization. The optimal site of nanosurgery is indicated by the red arrows. Scale bar, 10 μm. (B) A BSC1 cell grown on 6 μm wide fibronectin lines and transfected with actin-GFP. Nanosurgery was performed along a linear pattern where indicated by the arrow. Subsequent time points show the severing of actin fibers and the consequent retraction of the two parts of the cell that leads to complete dissection. Scale bar, 10 μm. (C) A BSC1 GalT-GFP cell undergoes laser nanosurgery to deplete the "displaced" GC (visualized by fluorescence microscopy—in green) from the karyoplast (the nucleus localization is visualized in the transmission (DIC) channel). After nanosurgery, the cell is dissected into a nucleus containing a karyoplast (asterisk) and a Golgiplast (arrowhead). Scale bar, 10 μm. (See color plate.)

Micrographs in (B) and (C) have been reproduced and adapted from Tängemo et al. (2011).

UGA-40 scanner (Rapp OptoElectronic, Hamburg, Germany). Another system we have successfully used is an Olympus Fluoview 1000 Confocal Microscope (Olympus Deutschland, Hamburg, Germany) equipped with two scanning units, while one is continuously imaging, the second can operate the laser dissection (with a pulsed 375 nm laser—PDL 800D, PicoQuant, Berlin, Germany). In all cases, nanosurgery experiments have been carried out with a 60 ×/63 × water immersion objective. Other pulsed laser sources can be used and have been applied to different micro/nanosurgery experiments. For a comprehensive overview of the microscope setups and lasers that can be used for laser nanosurgery, the reader is referred to Ronchi et al. (2012). The temperature of the microscope stage needs to be stably controlled at 37 °C throughout the experiment. For this purpose, the microscope is equipped with an in-house built "environment box." Furthermore, an *XY* motorized stage is needed to store and find the positions of different cells in the same experiment.

19.3 MATERIALS

- Sylgard 184 Silicone Elastomer Kit (Dow Corning GmbH, Cat. no. 1673921)
- Fibronectin from bovine plasma (Sigma-Aldrich, Cat. no. F1141)
- Acetone
- Ethanol
- Deionized water
- 3-Aminopropyltriethoxysilane (APTS, Sigma-Aldrich, Cat. no. 440140)
- Phosphate-buffered saline (PBS, pH 7.4)
- Suberic acid bis-(3-sulfo-*N*-hydroxysuccinimide ester) sodium salt (BS3, Sigma-Aldrich Cat. no. S5799)
- Poly(L-lysine) hydrobromide (Sigma-Aldrich Cat. no. P8954)
- mPEG-SPA-2000 or mPEG-SPA-5000 (Shearwater)
- Silicone-free grease (Glisseal N, Borer Chemie, Zuchwil, Switzerland)
- Colorless nail polish (any manufacturer)
- BSC1 cells (ATCC CCL-26) stably transfected with a fluorescently tagged Golgi marker (e.g., GalT-GFP; Tängemo et al., 2011)
- Appropriate cell culture medium (Minimum Essential Medium—MEM, Gibco, Cat. no. 31095-029) supplemented with 1% fetal calf serum (FCS), L-glutamine, and penicillin + streptomycin
- CO_2-independent medium (Gibco, Cat. no. 18045-054) supplemented with 10% FCS, L-glutamine, and penicillin + streptomycin

19.4 METHODS

19.4.1 Microcontact printing

In order to give cells the desired shape that facilitates GC displacement from the nucleus, we use microcontact printing. This technique allows printing 6 μm wide fibronectin lines on a pretreated coverslip (Pouthas et al., 2008).

19.4.1.1 Stamp preparation

To achieve this pattern, we make use of an elastomeric stamp made of PDMS. To generate a stamp, the two components of the Sylgard 184 Silicone Elastomer Kit are mixed (follow the manufacturer's instructions) and poured on a custom-designed photolithography Si master (GeSim). Overnight incubation at 55 °C results in polymerization of a PDMS layer, which can be peeled off the master and cut in order to generate the individual stamps. The master reproduces the "negative" of the pattern of interest. The final stamp specifications are shown in Fig. 19.2A.

19.4.1.2 Printing on glass coverslips

The stamps prepared as described earlier are incubated with a 50 μg/ml fibronectin solution in PBS for 30–45 min (Fig. 19.2B). During the incubation time, round coverslips (ø 15 mm) are washed with acetone, ethanol, and deionized water and then dried with compressed nitrogen. Coverslips are then plasma treated for 30 s on each side under a plasma generator in order to clean and activate the surface. The glass surface is then silanized by incubating each coverslip for 30 s in a 5% APTS solution in acetone. Coverslips are then washed twice in PBS as shown in Fig. 19.3A before incubation with BS3 in order to later cross-link the silane groups to the fibronectin. For this step, each coverslip is laid on top of a 100 μl drop of 0.5 mg/ml BS3 solution in PBS for 10 min. Thereafter, the coverslips are washed twice in PBS as shown in Fig. 19.3A and then deposited on a parafilm with the BS3-treated side of the coverslip facing upward. The stamps are washed with water to remove excess fibronectin and dried with compressed nitrogen (Fig. 19.2C). One PDMS stamp is then laid on top of each coverslip and left in contact for 5 min to allow cross-linking of the fibronectin. With the use of filter forceps and curved forceps, the stamp is then carefully removed from the coverslip (Fig. 19.3B), leaving fibronectin lines cross-linked to the glass surface according to the pattern design (Fig. 19.2D). A drop of 1 mg/ml PLL–PEG solution in PBS is then deposited on top of the coverslip (Fig. 19.2E) and incubated for 30–45 min. PLL–PEG will be cross-linked to the glass in the areas left free from fibronectin, creating a repellent substrate for cells. Finally, the coverslips are washed again as shown in Fig. 19.3A to remove the noncross-linked PLL–PEG (Fig. 19.2F). The obtained micropatterned coverslips can be mounted on the bottom of a 35 mm cell culture dish with a drilled 10 mm ø hole using Glisseal N and nail polish (Fig. 19.3C).

19.4.1.3 Tips and pitfalls

- Design of the stamp can be critical: too deep grooves between the lines results in easier collapse of the lines upon use. If not deep enough, the printing results in a uniform fibronectin layer. The depth we use is 3 μm, which is a good compromise.
- Stamps can be used 5–10 times and have to be cleaned (ethanol and water, then dry. If available, wash using a sonicator) and stored in a dust-free environment.
- While washing the coverslips after each incubation, make sure that all the PBS is removed from the surface of the glass leaving only a small droplet at the rim

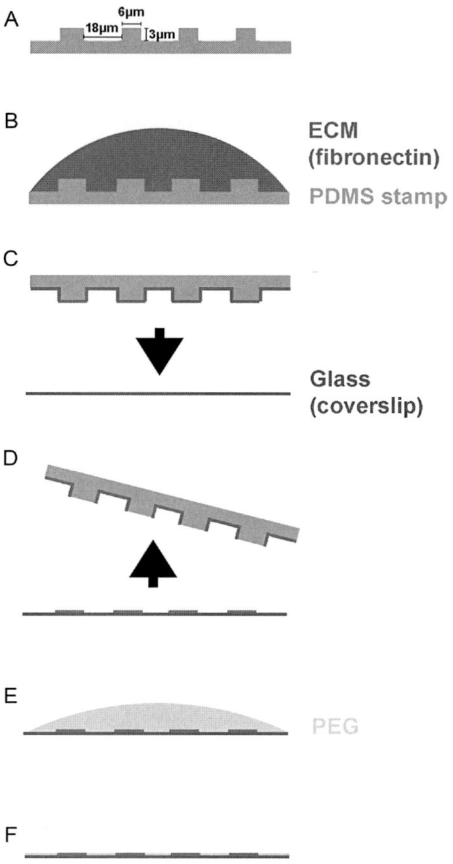

FIGURE 19.2

Microcontact printing procedure. (A) PDMS stamp design (lateral view). (B) PDMS stamps (in gray) are incubated with the fibronectin solution (red). ECM, extracellular matrix protein. (C) After removal of excess fibronectin with a nitrogen stream, a thin fibronectin layer on the surface of the stamp (red) is left. Stamps are then laid on clean coverslips (blue) previously treated as described in the main text to allow cross-linking of the fibronectin. (D) Removal of the stamps results in the deposition of a fibronectin pattern on the glass. (E) Coverslips are then incubated with a PLL–PEG solution (green). The PLL–PEG is cross-linked to the glass surface not covered by fibronectin, providing a hydrophobic surface, repellent for cells. (F) Excess PLL–PEG is removed and coverslips are washed, leaving the desired micropattern on the glass. (See color plate.)

(see Fig. 19.3A), otherwise salt precipitation in the center of the coverslip can affect the patterning. The APTS washing step is the most critical. Make sure that the coverslip is quickly transferred to PBS to avoid drying of the APTS–acetone solution. The dried chemical cannot be easily removed and the coverslip would have to be discarded.

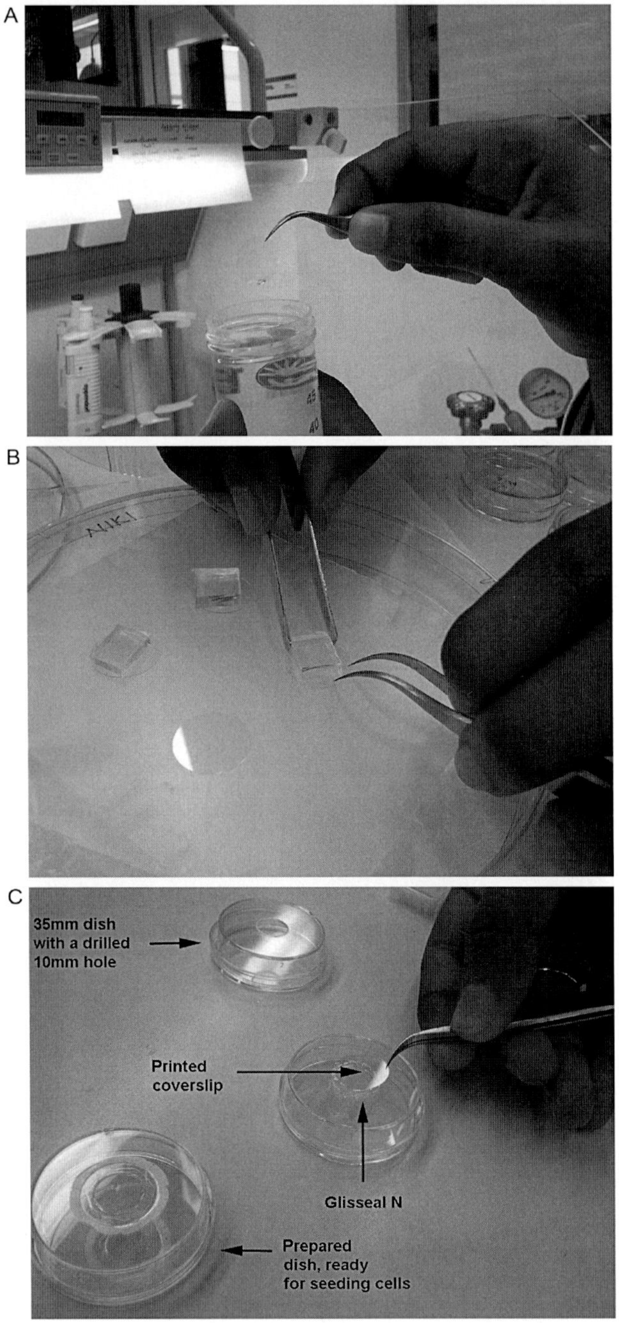

FIGURE 19.3

Critical steps during microcontact printing. (A) Washing coverslips after APTS, BS3, and PEG incubation. Dip the coverslip in PBS until the entire surface is washed, and the buffer flows over the surface only accumulating in a small droplet at the bottom of the coverslip. The PBS drop is cleaned with a tissue without touching the rest of the coverslip. (B) Removal of the PDMS stamp from the coverslip. (C) Mounting of the coverslips on 35 mm culture dish with a 10 mm hole in the center. (See color plate.)

- A simplified protocol can be used for the microcontact printing. Indeed, we noticed that cross-linking of the pattern to the glass is not strictly necessary. Non-cross-linked printed fibronectin and PEG are stable enough and do not dissolve in the medium while seeding cells. Nonetheless, the simplified procedure may result in a lower efficiency in the displacement of the Golgi from the nucleus of cultured cells.
- Printed coverslips can be stored at 4 °C for at least 2 weeks.

19.4.2 Laser nanosurgery

19.4.2.1 Cell preparation

Although there is no theoretical restriction to the use of a particular cell line for the described method (provided that cells are motile and express a fluorescently tagged Golgi marker), the micropattern described earlier has been optimized for the use of BSC1 cells. For the entire procedure, we will therefore use BSC1 cells expressing the Golgi marker GalT-GFP (Tängemo et al., 2011).

For an ideal concentration and occupancy of the micropattern, 70,000–80,000 cells/35 mm dish are seeded. Since they can attach only on the fibronectin lines and not on the PEG-coated surface, they will acquire a characteristic, elongated shape. Cells are kept in MEM supplemented with 1% FCS, L-glutamine, and antibiotics for 48–72 h. Low FCS concentration reduces cell division and avoids overcrowding of the cells on the fibronectin lines. 1 h before the nanosurgery experiment, the medium is replaced by CO_2-independent medium supplemented with 10% FCS, L-glutamine, and antibiotics. After the cells start to migrate along the lines, about 2–10% of them will show a displaced GC (Fig. 19.1A), thus being suitable for cutting. Cells are now ready to be moved to the prewarmed (37 °C) microscope stage.

19.4.2.2 Cell dissection

Laser nanosurgery to remove the GC has been successfully carried out in our laboratory with different devices (see Section 19.2). Nonetheless, we limit the description here to the use of a wide-field Olympus Cell^R microscope (operated with the Excellence software) with a pulsed 355 nm laser coupled in (DPSL-355/14, Rapp OptoElectronic, Hamburg, Germany). Cell dissections are performed with a high numerical aperture immersion objective. We prefer a 60×/1.2NA water immersion objective, which facilitates the subsequent switching to nonimmersion lenses for time-lapse imaging. An important prerequisite for a successful experiment is the precise spatial calibration of the cutting laser, which consists in the alignment of the scanning pattern defined by the software to the actual movements of the laser scanner. This procedure is performed in the "calibration" mode of the UGA control software. Thereafter, the first step is to look for cells suitable for nanosurgery (i.e., with sufficient nucleus–GC displacement). For this purpose, the Golgi position can be visualized by the GalT-GFP signal using the GFP illumination settings. The nuclear outline is observed using DIC illumination. Once the target cell is identified, an image in both channels (GFP and DIC) is captured using the Excellence software

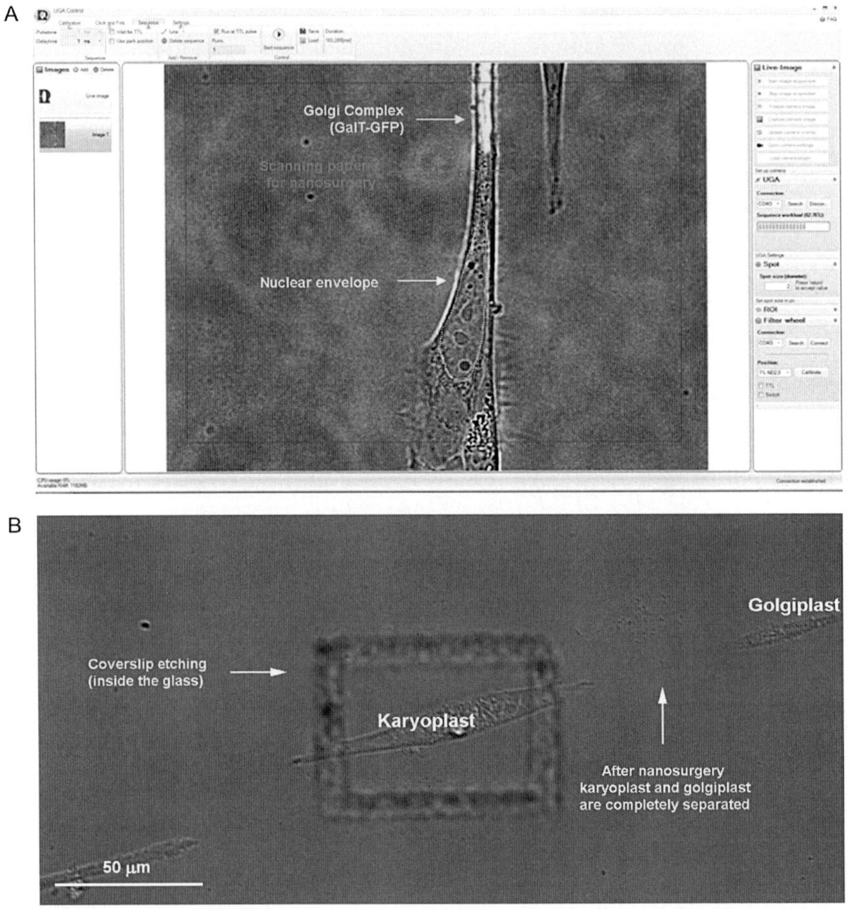

FIGURE 19.4

Cell dissection and glass etching. (A) Drawing of the pattern for nanosurgery (red line) on a cell showing GC displacement from the nucleus. This step is performed using the UGA control software on the DIC-GalT-GFP overlay image. The nuclear envelope is highlighted on the image with a yellow dashed line for better visualization. (B) Glass etching to mark the cell's position after Golgi removal. After nanosurgery, the cell is divided into karyoplast and Golgiplast. To retrieve the karyoplast position for subsequent studies, the same setup used for cell dissections can be used to generate a marking within the glass coverslip. The image shows a frame that is inscribed with the cutting laser into the glass around the karyoplast. (See color plate.)

and the overlay image is opened with the UGA control software. Making use of the interactive mouse interface, the scanning pattern for cutting (a line perpendicular to the cell axis spanning the entire cell between the GC and the nucleus) is manually drawn on the image (Fig. 19.4A). The laser output intensity in our case is controlled with a ND filter wheel and an additional diaphragm position along the laser path. The laser can be triggered with the UGA software, but to allow a better control over imaging conditions, we prefer to operate with the Excellence software. We therefore make use of an additional shutter on the laser path that is controlled by this software. While imaging the process live in the transmission channel, the shutter is opened and closed to obtain a cutting "cycle." We typically apply the laser for 2–4 s/cycle while in parallel manually moving the focus in the microscope from top to bottom (the laser will be continuously scanned along the designed line). This normally results in retraction of the two cell's parts. Typically 5–15 cycles are necessary to complete the dissection. The typical time between two cutting cycles is 1 min to allow the cell to recover between them. It is important to note that these guidelines will have to be adjusted by the operator taking into account the cell's reactions during nanosurgery. In our experience, due to cell variability, a successful removal of the GC can be obtained with different laser powers and number of laser "cycles" and the best conditions have to be experimentally determined each time.

19.4.2.3 Tips and pitfalls

- The position of the cutting line is crucial. Indeed, cutting as far as possible from the nucleus results in less cellular stress. On the other hand, the laser photobleaches the GFP signal. Small Golgi fragments and/or carriers can be bleached if the line is too close to the GC and the laser not properly calibrated. This may result in leaving part of the GC in the karyoplast, as the bleached material cannot be seen. The cutting pattern should therefore be designed at least 3–5 μm from the nearest Golgi object. Moreover, during the nanosurgery, the position of the GC should be occasionally monitored to ensure that fragments of it are not approaching the laser scanning path.
- Cutting perpendicular to the cell axis normally facilitates the procedure.
- If too high laser power is applied, cutting can result in a too fast separation of the two parts of the cell and eventually in cell death or in the formation of local blebbing. The two opposite behaviors can be likely explained by differences in the intrinsic tension of the cell that are more or less stretched on the linear pattern. In case of blebbing, longer time intervals between the cutting cycles could help the cell to recover.
- Even higher laser power can lead to glass surface etching that normally kills the cell.

19.4.2.4 Preparation of samples for time-lapse imaging

After Golgi removal, the karyoplasts can be followed with time-lapse microscopy, fixed at the stage of interest and immunolabeled for different cellular markers or processed for electron microscopy. For the identification of the cut cells after removing the dish from the microscope stage, a recognizable spatial reference is needed. In order to achieve this, it is possible to etch the glass inside the coverslip creating a pattern with the 355 nm "cutting laser" without damaging the living cells (Fig. 19.4B). For this purpose, a pattern of choice (e.g., a frame around the cell of interest) (Fig. 19.4B) can be designed with the UGA control software. The laser power needs to be increased to a value two to five times higher than the typical power needed to dissect cells. Due to the homogeneous nature of the glass coverslip, the laser settings necessary for this procedure are very reproducible and do not have to be determined for each experiment. For the etching, the system can be controlled with the Excellence software as for cell nanosurgery (see in the preceding text). Instead of focusing on the cell's plane, the focus is now moved several micrometers below the glass surface in order to etch inside the glass, without disturbing the surface.

Time-lapse imaging of the karyoplast is finally carried out using a lower magnification objective than used for nanosurgery (typically a long-distance objective with 20 × or 40 × magnification is used). To this end, the cell culture dish is first removed from the microscope stage to clean the coverslip from remaining water or immersion oil. The marking in the glass allows easy identification of the Golgi-depleted cells and repositioning of the culture dish for the following time-lapse imaging on any microscope system best suited for this purpose.

19.4.2.5 Tips and pitfalls

- Karyoplasts are more prone to cell death than neighboring cells, especially in the first hours after nanosurgery. Therefore, we advice starting time-lapse imaging not before 2 h after nanosurgery. Furthermore, the minimal possible excitation light intensity and exposure time should be used.
- Time intervals between subsequent image acquisitions depend on the aim of the experiments. We achieved time intervals down to 2 min without any apparent damage of cells and we could observe Golgi biogenesis, though with delayed kinetics.
- Frequently, karyoplasts move long distances along the linear pattern. Therefore, images of one or two frames in the neighborhood of the cell under view are acquired to be able to track the cell under observation when moving.

19.5 DISCUSSION

The nanosurgery technique presented earlier has allowed for the first time the depletion of the GC and all its molecular components from a living cell. It can be used to address important questions in the biology of this organelle, such as its biogenesis, the role of its proteins, and lipids in this process (e.g., when combined with siRNA-mediated knockdown or drug treatments) and the role of the GC in several cellular processes (e.g., ER export or endocytosis). The main advantage over other classical methods of investigation is the possibility to deplete the cell of the entire set of Golgi-localized molecules. As a consequence, the need to resynthesize its components makes *de novo* Golgi biogenesis as described here a long process. This facilitates characterization of the intermediates by light and electron microscopy. Moreover, the role of Golgi-localized molecules during Golgi biogenesis as well as in other cellular processes can be directly studied. For this purpose, the combination of laser nanosurgery with siRNA transfection has a great potential. Indeed, removing the Golgi-associated protein in cells where the siRNA blocks its resynthesis results in an acute depletion of the protein. This approach has been used to investigate the role the Golgi matrix protein GM130 in Golgi biogenesis (Tängemo et al. 2011) but could be potentially extended to the study of several other proteins.

Cellular dissection has been classically achieved with the use of micromanipulation needles (Poste, 1973). This method has been applied in the past to centrosome and Golgi biogenesis studies as well (Maniotis & Schliwa, 1991; Pelletier et al., 2000) and in combination with micropattern-mediated Golgi displacement from the nucleus could in principle lead to the same results that we have shown with laser nanosurgery. Nonetheless, the submicrometer precision of laser nanosurgery may allow in our view a more reproducible Golgi depletion with less cell damage compared to mechanical dissection methods. In addition, the possibility of glass etching and therefore the possibility of easily retrieving the cells of interest for subsequent studies are practical advantages.

The main limitations of the nanosurgery method are that it is technically demanding, time-consuming, and therefore low throughput. In a typical nanosurgery session (2–3 h), two to eight cells are successfully cut and approximately 50–80% survive and form a new GC. Nonetheless, for the unprecedented possibilities discussed earlier, we believe that nanosurgery-mediated Golgi depletion should be considered as an important tool to study the biology of the GC.

Acknowledgment

P. R. and R. P. are supported by an EU Systems Microscopy Network of Excellence grant: FP7/2007-2013-258068.

References

Altan-Bonnet, N., Sougrat, R., Liu, W., Snapp, E. L., Ward, T., & Lippincott-Schwartz, J. (2006). Golgi inheritance in mammalian cells is mediated through endoplasmic reticulum export activities. *Molecular Biology of the Cell*, *17*, 990–1005.

Axelsson, M. A., & Warren, G. (2004). Rapid, endoplasmic reticulum-independent diffusion of the mitotic Golgi haze. *Molecular Biology of the Cell*, *15*, 1843–1852.

Colombelli, J., Reynaud, E. G., & Stelzer, E. H. (2007). Investigating relaxation processes in cells and developing organisms: From cell ablation to cytoskeleton nanosurgery. *Methods in Cell Biology*, *82*, 267–291.

Jesch, S. A., Mehta, A. J., Velliste, M., Murphy, R. F., & Linstedt, A. D. (2001). Mitotic Golgi is in a dynamic equilibrium between clustered and free vesicles independent of the ER. *Traffic*, *2*, 873–884.

Jollivet, F., Raposo, G., Dimitrov, A., Sougrat, R., Goud, B., & Perez, F. (2007). Analysis of de novo Golgi complex formation after enzyme-based inactivation. *Molecular Biology of the Cell*, *18*, 4637–4647.

Kasap, M., Thomas, S., Danaher, E., Holton, V., Jiang, S., & Storrie, B. (2004). Dynamic nucleation of Golgi apparatus assembly from the endoplasmic reticulum in interphase hela cells. *Traffic*, *5*, 595–605.

Kohli, V., & Elezzabi, A. Y. (2009). Prospects and developments in cell and embryo laser nanosurgery. *Wiley Interdisciplinary Reviews Nanomedicine and Nanobiotechnology*, *1*, 11–25.

Lippincott-Schwartz, J., Yuan, L. C., Bonifacino, J. S., & Klausner, R. D. (1989). Rapid redistribution of Golgi proteins into the ER in cells treated with brefeldin A: Evidence for membrane cycling from Golgi to ER. *Cell*, *56*, 801–813.

Lowe, M., & Barr, F. A. (2007). Inheritance and biogenesis of organelles in the secretory pathway. *Nature Reviews Molecular Cell Biology*, *8*, 429–439.

Maniotis, A., & Schliwa, M. (1991). Microsurgical removal of centrosomes blocks cell reproduction and centriole generation in BSC-1 cells. *Cell*, *67*, 495–504.

Orci, L., Palmer, D. J., Ravazzola, M., Perrelet, A., Amherdt, M., & Rothman, J. E. (1993). Budding from Golgi membranes requires the coatomer complex of non-clathrin coat proteins. *Nature*, *362*, 648–652.

Pelletier, L., Jokitalo, E., & Warren, G. (2000). The effect of Golgi depletion on exocytic transport. *Nature Cell Biology*, *2*, 840–846.

Poste, G. (1973). Anucleate mammalian cells: Applications in cell biology and virology. *Methods in Cell Biology*, *7*, 211–249.

Pouthas, F., Girard, P., Lecaudey, V., Ly, T. B., Gilmour, D., Boulin, C., et al. (2008). In migrating cells, the Golgi complex and the position of the centrosome depend on geometrical constraints of the substratum. *Journal of Cell Science*, *121*, 2406–2414.

Puri, S., & Linstedt, A. D. (2003). Capacity of the golgi apparatus for biogenesis from the endoplasmic reticulum. *Molecular Biology of the Cell*, *14*, 5011–5018.

Ronchi, P., Terjung, S., & Pepperkok, R. (2012). At the cutting edge: Applications and perspectives of laser nanosurgery in cell biology. *Biological Chemistry*, *393*, 235–248.

Seemann, J., Jokitalo, E., Pypaert, M., & Warren, G. (2000). Matrix proteins can generate the higher order architecture of the Golgi apparatus. *Nature*, *407*, 1022–1026.

Shorter, J., & Warren, G. (2002). Golgi architecture and inheritance. *Annual Review of Cell and Developmental Biology*, *18*, 379–420.

Tängemo, C., Ronchi, P., Colombelli, J., Haselmann, U., Simpson, J. C., Antony, C., et al. (2011). A novel laser nanosurgery approach supports de novo Golgi biogenesis in mammalian cells. *Journal of Cell Science*, *124*, 978–987.

Zaal, K. J., Smith, C. L., Polishchuk, R. S., Altan, N., Cole, N. B., Ellenberg, J., et al. (1999). Golgi membranes are absorbed into and reemerge from the ER during mitosis. *Cell*, *99*, 589–601.

CHAPTER

Studying Intracellular Trafficking Pathways with Probabilistic Density Maps 20

Kristine Schauer*, Tarn Duong†,‡, Carina S. Gomes-Santos* and Bruno Goud*

**Molecular Mechanisms of Intracellular Transport, Unité Mixte de Recherche 144 Centre National de la Recherche Scientifique/Institut Curie, Paris cedex, France*

†Theoretical and Applied Statistics Laboratory (LSTA), University Pierre and Marie Curie—Paris, Paris, France

‡Institute of Translational Neurosciences (IHU-A-ICM), Pitié-Salpêtrière Hospital, Paris, France

CHAPTER OUTLINE

Introduction **326**
Intracellular Trafficking Pathways 326
Cell Confinement by Micropatterning Techniques 327
Probabilistic Density Mapping of Intracellular Compartments 329
Rationale 330
20.1 Material and Methods **330**
20.1.1 Specific Materials (Excluding Generic Products) 331
20.1.2 Methods 331
20.1.2.1 Preparation of Micropatterns with Photolithography 331
20.1.2.2 Cell Seeding, Treatment, and Fixation 332
20.1.2.3 Immunolabeling (Optional) 333
20.1.2.4 Image Acquisition of Many Single Cells and Preanalysis 333
20.1.2.5 Density Mapping 334
20.2 Results and Discussion **339**
Summary **341**
Acknowledgments **341**
References **341**

Abstract

The compartmentalization of cellular functions in complex membranous organelles is a key feature of eukaryotic cells. To cope with the enormous complexity of trafficking pathways that connect these compartments, new approaches need to be considered and introduced into the field of cell biology. We exploit the advantages of the "micropatterning technique," which is to bring cells to adopt a highly reproducible

Methods in Cell Biology, Volume 118

ISSN 0091-679X
http://dx.doi.org/10.1016/B978-0-12-417164-0.00020-3

shape, and probabilistic density mapping, which quantifies spatial organization of trafficking compartments, to study regulatory mechanisms of intracellular trafficking. Here, we provide a protocol to analyze and quantify alterations in trafficking compartments upon cellular manipulation. We demonstrate how this approach can be employed to study the regulation of Rab6-labeled transport carriers by the cytoskeleton.

INTRODUCTION

Intracellular trafficking pathways

The endomembranous system of eukaryotic cells comprises several diverse organelles and tubular–vesicular transport carriers that are dynamic and highly interconnected by multiple routes. Classically, intracellular trafficking routes are divided into the secretory (or anterograde) pathway and the endocytic pathway. The endoplasmic reticulum (ER), the intermediate compartment (IC), the Golgi apparatus, and the *trans*-Golgi network (TGN) are considered to belong to the secretory pathway, whereas early (sorting) endosomes, late endosomes, lysosomes, and the endocytic recycling compartment (ERC) are most often classified to the endocytic pathway. Although functionally distinct, the secretory and endocytic pathways have parallel functions (Saraste & Goud, 2007). These include membrane and receptor recycling to maintain the integrity of the ER and the plasma membrane, respectively (Griffiths, Back, & Marsh, 1989; Martinez-Menarguez, Geuze, Slot, & Klumperman, 1999; Sannerud, Saraste, & Goud, 2003). Furthermore, their respective molecular machineries resemble each other (Bonifacino & Glick, 2004; Chia & Gleeson, 2011). Importantly, both pathways seem to partly overlap and to be intertwined at various branch points (Ellis, Potter, Cresawn, & Weisz, 2006; Lipschutz, Lingappa, & Mostov, 2003; Sannerud et al., 2003; Toikkanen, Miller, Soderlund, Jantti, & Keranen, 2003). For example, the ERC can sort molecules to the TGN (Wilcke et al., 2000) and the IC can contribute to endocytosis (Becker, Volchuk, & Rothman, 2005).

Although allowing the segregation of cellular processes, cellular compartmentalization demands an efficient and regulated communication among those compartments. Small GTPases of the Rab family are key regulators of intracellular trafficking, present in both secretory and endocytic pathways. There are over 60 members in human and they are highly conserved (found in all eukaryotes) (Pereira-Leal & Seabra, 2001). In their active form (GTP-bound), Rabs recruit a diverse range of effector proteins, such as molecular motors, lipid kinases, tethering factors, and scaffolding proteins, controlling the formation of transport carriers from donor membranes, their movement along cytoskeletal tracks, and their tethering/fusion with target membranes (Stenmark, 2009). Because Rab proteins are restricted to specific organelles or membrane microdomains, their presence not only regulates the trafficking throughout compartments but also defines their identity (Hutagalung &

Novick, 2011). Rab proteins are themselves under a strict spatiotemporal control by a complex network of proteins. Guanine nucleotide exchange factors (GEFs) stimulate the exchange GDP for GTP and thus activate Rab proteins and GTPase-activating proteins (GAPs) facilitate GTP hydrolysis and thus deactivate Rabs. Recent findings indicate that Rab proteins not only regulate individual transport steps but also function in a coordinated fashion, orchestrating sequential steps along transport pathways. An upstream Rab can recruit a GEF for a downstream Rab. In addition, a downstream Rab can recruit a GAP that will inactivate the upstream Rab (Rab GTPase cascades). Such a mechanism appears to be crucial to ensure continuity and directionality in membrane traffic (Jean & Kiger, 2012; Markgraf, Peplowska, & Ungermann, 2007).

The establishment and dynamics of trafficking routes and their regulation by Rab proteins are still not fully understood. Major challenges are the dynamic nature of trafficking events. Additionally, many trafficking compartments are spatially dispersed. Unlike compact structures such as the Golgi apparatus in mammalian cells and the ERC, endosomes and various transport carriers are present in high numbers and are located throughout the cells. This dispersed localization makes it very challenging to detect alterations in the compartment due to experimental manipulation or in pathological situations such as cancer development. To tackle these challenges, it is crucial to develop new quantitative approaches. Here, we present a method that allies cell normalization by micropatterning with probabilistic density mapping to quantitatively study intracellular trafficking pathways. Figure 20.1 outlines the probabilistic density-mapping procedure for a single cell and a cell population.

Cell confinement by micropatterning techniques

The structured microenvironment of a tissue or organ imposes spatial restrictions on cells. These spatial cues force cells to take a specific, reproducible organization that is characteristic of tissues. Spatial cues have preponderant roles in a wide range of cellular functions, such as cell polarity, differentiation, division, and signaling (Gumbiner, 1996). Yet, classical *in vitro* cultured mammalian cells show strong cell-to-cell variations in morphology (Sigal et al., 2006). This heterogeneity has been attributed to unconstrained and dynamic cell shape, a consequence of the fact that cells move and constantly rearrange intracellular trafficking pathways. Furthermore, cells adapt to their local population context. Population size, local cell density, and position within a cellular islet have all been shown to determine cellular morphology and responsiveness (Snijder et al., 2009). The resulting heterogeneity hampers the understanding of the regulation of cellular trafficking pathways. In an attempt to transfer some of the *in situ* cues into classical cell culture conditions, various microfabricated substrates have been used in *in vitro* cell culture (Folch & Toner, 2000). Micropatterning technique comprises the fabrication of adhesive and nonadhesive regions that impose a defined geometry to cells upon adhesion. Throughout the years, the application of micropatterning technology has increased in cell biological studies, revealing important aspects of cell physiology. For instance, mesenchymal stem

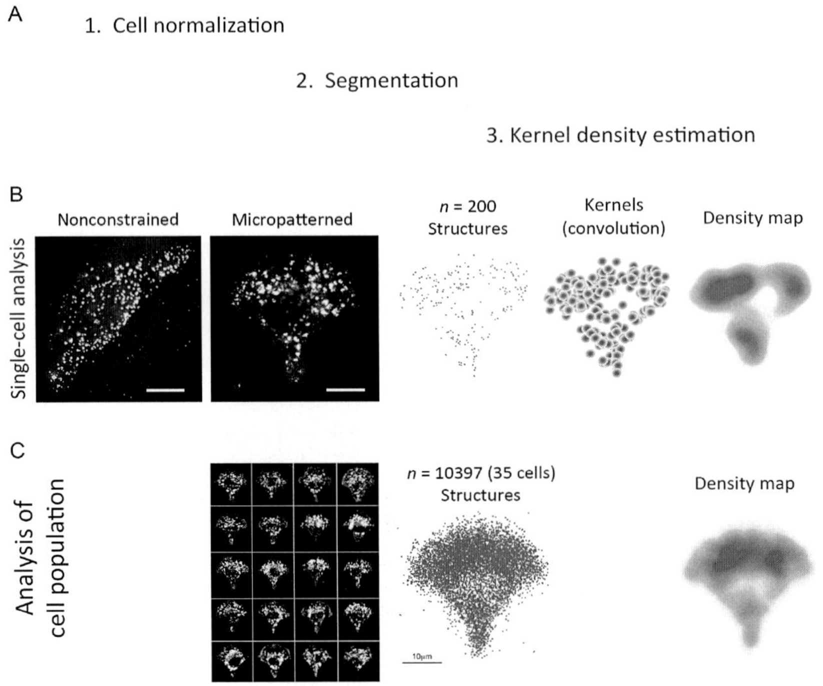

FIGURE 20.1

Outline of the probabilistic density-mapping procedure exemplified for one single cell and cell populations. (A) Probabilistic density mapping of intracellular compartments requires cell normalization by micropatterning, extraction of point coordinates of imaged intracellular structures by segmentation, and kernel density estimation. (B) Example of this analysis for one single cell: the shape of the cell is first normalized by adhesion on a micropatterned shape (such as a crossbow), fluorescently labeled intracellular structures are transformed into point coordinates, and each of them is replaced by a Gaussian function (kernels) with mean zero and an optimized variance. Kernels are summed revealing the underlying density throughout the cell, where "red" corresponds to the most dense cellular regions. (C) Typically, several thousands of coordinates from several tens of normalized cells are used to calculate the density map of a population of cells. (For interpretation of the references to color in this figure legend, the reader is referred to the online version of this chapter.)

cells plated on micropatterns differentiate either into adipocytes or osteoblasts, depending on the size and geometry of the pattern (McBeath, Pirone, Nelson, Bhadriraju, & Chen, 2004), revealing the importance of adhesive cues in cell fate. By applying controlled changes in micropattern geometry, it was also shown that cell adhesion determines the orientation of the mitotic spindle (Thery et al., 2005) and the orientation of the cell polarity axis (Thery et al., 2006).

These and other studies show that *in vitro* confinement of cell shape mimics the restriction of space that cells experience in body tissues (Gumbiner, 1996). Additionally, cell confinement on adhesive micropatterns of defined shapes normalizes cellular shape, inner organization (Fig. 20.1B and C), and responsiveness to a given manipulation. Consequently, cell normalization allows the reduction of the number of analyzed cells required for statistically significant results. We showed that subtle drug effects on intracellular compartment were assessable using less than 20 micropatterned cells (Duong, Goud, & Schauer, 2012; Schauer et al., 2010). Importantly, these effects could not be detected in conventionally grown unrestrained cells. Normalization additionally allows simplification and automatization of image analysis that is required for unbiased studies or high-throughput approaches.

Probabilistic density mapping of intracellular compartments

A primary method for studying cellular function is to examine fluorescently labeled proteins after a given manipulation. Despite of the central role of morphology comparisons in cell biological approaches, few statistical tools are available that allow to quantify the similarity or difference of fluorescent images containing multivariate data. Because direct visualization of a large number of structures is not able to reveal details of the underlying spatial organization, statistical methods such as density mapping are required to visualize this organization (Fig. 20.1B and C). Density mapping is based on kernel density estimation, a set of advanced statistical techniques that are well suited to analyze multivariate data. Kernel estimators are convolutions, as each data point is blurred, in an inverse operation of the more widely known deconvolution, which takes a blurry image and sharpens it. This blurring or smoothing is a mathematical representation that an observed structure should represent not only itself but also other nearby structures that were not observed. The level of smoothing is the most important parameter to estimate optimally and has been the subject of much research, as it is also the most mathematically difficult to resolve (Duong & Hazelton, 2003, 2005). An oversmoothing leads to estimates that obscure the detailed structure of the data, whereas an undersmoothing leads to estimates that are too noisy with spurious data artifacts (Fig. 20.2). The mathematical complexity of kernel density estimators has thus historically limited their use beyond

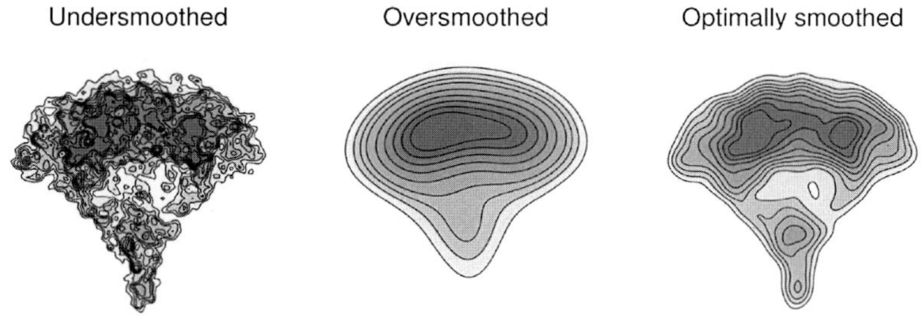

FIGURE 20.2

The smoothing parameter. The level of smoothing is the most important parameter to estimate optimally in a density map. An example is given of an oversmoothing that obscures the detailed structure of the data and an undersmoothing that leads to estimates that are too noisy with spurious data artifacts and optimal smoothing. (See color plate.)

the statistical computing community. As a result, kernel density maps have been underused in the quantitative analysis of biological data in comparison to histogram maps, despite the well-documented evidence that kernel maps possess many favorable statistical properties over histograms (Simonoff, 1996; Wand & Jones, 1995). Fortunately, the growing availability of the user-friendly, open-source statistical computing software R, and especially the user-contributed add-on packages, has brought advanced statistical analysis to a larger pool of users.

In order to create spatial density maps of intracellular compartments, fluorescent microscopy images of endomembranes are segmented and the 3D spatial coordinates of the structures are extracted. Typically, several tens to several hundred structures of a specific compartment are extracted from each cell. Then, coordinates are replaced by kernels and summed, revealing the underlying density of a protein throughout the cell. The reproducible cellular orientation induced by the micropattern allows the collation of several tens (or more) of cells, thus resulting in an average density map for a population of analyzed cells. In addition to a visual summary, density maps are used to derive statistics for quantitative analysis of cell morphology. Probability contours indicate the regions of the highest local density/concentration of endomembranes (minimal regions for a given fixed proportion). They are useful because they admit an intuitive probabilistic interpretation, for example, the 50% probability contour defines the smallest region in which 50% of the endosomes are located. Applying density mapping to Rab6, we have found that secretory vesicles, labeled by this GTPase, preferentially localize at the adhesive edge of the cellular periphery and are trapped there by the actin cytoskeleton.

Rationale

We will present how we exploit the advantages of the "micropatterning technique," which is to bring cells to adopt a highly reproducible shape, and probabilistic density mapping that measures spatial organization of intracellular structures to quantitatively study the regulation of Rab6-labeled compartment by the cytoskeleton. Rab6 is the most abundant Rab in the Golgi apparatus and it is also present on vesicular post-Golgi transport carriers. It was found to be involved in several trafficking steps, both in the secretory and retrograde pathways (Goud & Akhmanova, 2012). Rab6-labeled transport carriers move along microtubules from the Golgi to the cell periphery, where some of them interact with cortical proteins and subsequently fuse with the plasma membrane. We assess how Rab6-labeled transport carriers are affected upon depolymerization of the cytoskeleton using nocodazole or cytochalasin D, which, respectively, disrupt microtubules or actin filaments.

20.1 MATERIAL AND METHODS

The experimental procedure is outlined in Fig. 20.3. It includes the preparation of micropatterned coverslips, cell plating, visualization of Rab6-labeled compartment, imaging of several tens of cells, image segmentation, alignment, and density mapping.

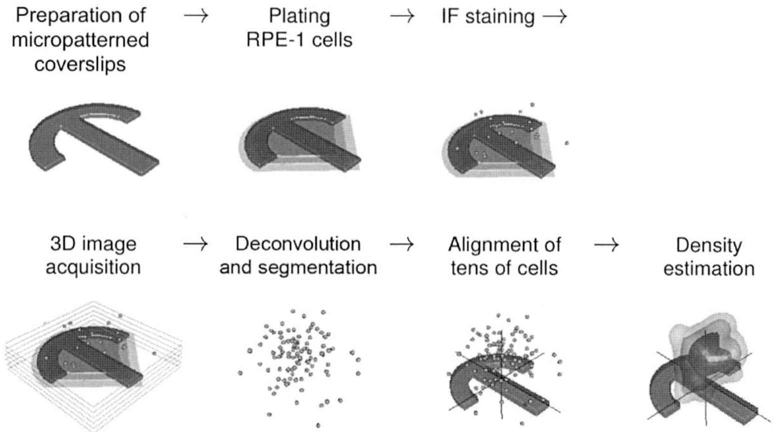

FIGURE 20.3

Schema of the experimental procedure. Micropatterned coverslips are prepared using photolithography. Cells are plated on micropatterns, structures of interest are visualized either by expressing fluorescently tagged proteins or indirect immunofluorescence, and 3D images are acquired. Images are deconvolved (if epifluorescence microscopy is used for 3D imaging) and positional information of fluorescently labeled structures is extracted from each single cell by segmentation analysis. Coordinates are aligned using the micropattern and the probability density map is calculated throughout the cell. To visualize the result, equally spaced probability contours are calculated that represent smallest areas in which a given percentage of structures are found. (See color plate.)

20.1.1 Specific materials (excluding generic products)

- hTERT-RPE1 cells stably expressing EGFP-Rab6
- PLL(20)-*g*[3.5]-PEG(2) (PLL-*g*-PEG, poly-L-lysine-*g*-poly(ethyleneglycol), Surface Solutions), keep a stock solution at 1 mg/ml in 10 mM HEPES (pH 7.3) at 4 °C
- Photomask (Delta Mask)
- UVO Cleaner (Model 342-220, Jelight Company)
- Fibronectin
- Alexa-labeled fibrinogen (Amersham or Molecular Probes)
- 16% formaldehyde
- PBS/saponin/BSA (10 × stock solution: 20 g BSA, 5 g saponin, and 1 l PBS 10 ×).

20.1.2 Methods

20.1.2.1 Preparation of micropatterns with photolithography

All steps have to be performed with the minimization of dust that will interfere with pattern quality. First, coverslips are cleaned by irradiation with UV light in the UV cleaner for 5 min. Coverslips are coated with PLL-g-PEG for 1 h, upside down in a wet chamber on parafilm, using 15 μl (12 mm ∅) or 30 μl (18 mm ∅) of 0.1 mg/ml PLL-g-PEG in 10 mM HEPES, pH 7.3. Coverslips are washed 2 × with PBS and

1 × with ddH_2O using a multiwell plate and dried with the PEG-treated (hydrophobic) surface up. Before printing the micropatterns, the photomask is first rinsed with ddH_2O and then with ethanol, dried with airflow, and eradiated with deep-UV light for 5 min. Then, 4 μl (12 mm ∅) or 6 μl (18 mm ∅) of ddH_2O is pipetted on the photomask and coverslips are attached upside down on the mask using the capillary pressure. Extra water is removed with a tissue, until coverslips do not move anymore, and coverslips are covered with a big, thin glass plate or plastic film (e.g., old Western blot film). The photomask with attached coverslips is carefully installed upside down under the UV lamp of the UV cleaner and eradiated for 6 min. The UV light (30–33 mW/cm^2, measured at 254 nm) passes through the micropattern shapes and changes the chemistry of the PLL-g-PEG layer (Azioune, Storch, Bornens, Thery, & Piel, 2009). Then, the photomask is removed from the UV cleaner and covered with plenteous water to remove the coverslips (coverslips will float). Coverslips are dried upright under a cover and stored for up to 4 weeks in multiwell plates until use.

20.1.2.2 Cell seeding, treatment, and fixation

Before cell plating, fibronectin is diluted to a final concentration of 50 μg/ml in ddH_2O using 20 μl for a 12 mm ∅ or 30 μl for a 18 mm ∅ coverslip. To visualize micropatterns, Alexa-labeled fibrinogen is added at a final concentration of 5 μg/ml to the fibronectin solution. Coverslips are placed with the patterned surface on the mix and incubated for 1 h in the dark, in a wet chamber on parafilm. Afterward, they are incubated in plain DMEM/F12 medium/20 mM HEPES in a multiwell plate for 5 min, at 37 °C. Meanwhile, cells are trypsinized (60–80% confluence, grown for 3 days) and resuspended in full medium (DMEM/F12, 10% FBS, and 1% penicillin/streptomycin) in order to inactivate the trypsin. To obtain single cells, they are passed several times through the pipette and then added dropwise onto the coverslips (~300,000 cells). Cells are incubated for 5 min at room temperature (RT) (RPE-1 cells) or for 20 min in the incubator (HeLa cells). Because the attachment time varies between different cell lines, this parameter needs to be adjusted accordingly. Generally, we wait until the cells are attaching (1 cell per pattern) and the rest are still floating. Cells are then washed 3–10 × with plain medium/20 mM HEPES in order to remove all unbound cells. As the surface of coverslips is more hydrophobic than under classical cell culture conditions, some medium should always be left in the well to avoid drying out of the coverslips, which will cause cell death. The cells are incubated at 37 °C for 2–5 h for full spreading: RPE-1 cells require at least 2 h, and HeLa cells require at least 4 h. Typically, 25% of micropatterns are covered by a single cell, whereas the other 75% are either empty or covered by more than one cell. Once cells are spread, drugs are added directly into the medium. In the case of drugs that depolymerize actin filaments, only low concentrations are used to prevent cell rounding (e.g., 0.5 μM cytochalasin D). To depolymerize microtubules, cells are placed at 4 °C for 1 h prior to the addition of nocodazole (10 μM), a treatment that does not change cell shape. Cells are fixed with 16% formaldehyde that is added directly to the cells to obtain a final concentration of 4% and incubated for 15 min at room temperature.

20.1.2.3 Immunolabeling (optional)

After cell fixation, formaldehyde is quenched with 50 mM NH_4Cl for 5 min. Cells are washed 3 × with PBS and permeabilized with PBS/saponin/BSA for 20 min at RT. Coverslips are then incubated upside down in a drop (20 μl for 12 mm ∅ coverslips) containing the primary antibody in 2% BSA in PBS, in a wet chamber, in the dark for 45 min. Then, coverslips are washed 3 × with PBS and incubated in the same way with the secondary antibody followed by three washes with PBS and a short dip in ddH_2O (to remove PBS crystals). Coverslips are mounted with mowiol/DAPI mix (ca. 5 μl for a 12 mm ∅ coverslips) on an ethanol-cleaned slide and care is taken to avoid air bubbles or moving of the coverslip, which may deform micropatterned cells.

20.1.2.4 Image acquisition of many single cells and preanalysis

Images are acquired with any conventional microscope using a magnification that resolves the structures of interest. In the case of Rab6-labeled secretory vesicles, a 63 × or 100 × objective is required. For a typical analysis, several tens of cells (>20) are imaged; hence, microscopes with an automated stage are advantageous. To obtain 3D density maps, *z*-dimension series are acquired every 0.2–0.3 μm and out-of-focus signals should be avoided (either by using confocal microscopy or performing deconvolution of epifluorescent images). To be able to process images in a batch mode, the images need to be numbered logically. For each cell, images are acquired for up to four fluorescence channels, including fluorescence staining of the micropattern (e.g., Cy5), the nucleus (e.g., Dapi), and the intracellular structures of interest (e.g., GFP). The nucleus staining is observed to verify that a single cell is analyzed, as several cells are frequently attached onto a single micropattern. Images of the structures of interest are segmented to extract their 3D spatial coordinates. This step translates the gray values of a camera chip into numbers that define intracellular structures. Several segmentation programs can be employed (Eliceiri et al., 2012), such as the "multidimensional image analysis" interface in MetaMorph (Universal Imaging Corporation) or "ParticleTracker.jar" in ImageJ (MOSAIC group, ETH, http://www.mosaic.ethz.ch/Downloads/ParticleTracker; Sbalzarini & Koumoutsakos, 2005). After segmentation, the coordinates (centroids) of all structures of interest from all cells are aligned using the micropattern. For this, coordinates of the center and the angle of rotation of the micropattern are measured, either automatically or manually. For automation, a macro can be established in ImageJ that chooses the most in-focus image of the 3D stack of fluorescently labeled patterns and applies the ImageJ plug-in "Hough Circles" to detect the center of a circle around the micropattern. Use a micropattern-shaped mask to determine the angle of rotation to orient cells in one direction. This preanalysis will provide a table containing (aligned) coordinates for each detected structure from several tens of cells. In supplementary data, you can find text files (.txt) containing coordinates from the small GTPase Rab6, monitored in 82 non-treated cells (Rab6NT82cells.txt), 50 cytochalasin D-treated cells (Rab6CD50cells.txt), and 47 nocodazole-treated cells (Rab6NZ47cells.txt), all normalized by micropatterning. These coordinate tables are required to calculate corresponding density maps for each condition.

20.1.2.5 Density mapping

The density estimation is performed using the ks library in R (R_Development_Core_Team, 2013). R is a free program and can be downloaded and installed from the CRAN (Comprehensive R Archive Network) Website (http://cran.r-project.org). It is available on most platforms including Windows, Mac OS, and Unix/Linux. The procedure for installing R is already well documented on this site. To visualize density maps, the extension libraries mvtnorm, rgl, miscd, and ks are required that are available from the CRAN Website. Binary files are compiled versions of the source code and are the easier option. For those who have the ability and desire to compile the source code, this requires certain libraries to be installed, including a gcc/Fortran77 compiler, X11 Windows server, and PNG libraries.

20.1.2.5.1 Program installation

1. Install the R base distribution from the R project Website http://www.r-project.org under the CRAN rubric, for example, R-3.0.1.pkg (for Macintosh) or R-3.0.1-win32.exe (for Windows).
2. Install the required extension packages. There are two main ways to do this.
   a. Automatic: If the computer is connected to the Internet, packages can be installed directly using the Package Manager interface inside R. This is the preferred version as the most updated versions of the libraries will be automatically installed. Note that networks with a proxy cache or firewall may not allow for downloading in batch mode. Start R.
      i. For Macintosh, select Packages & Data -> Package Installer. Make sure that At User Level is selected for Install Location. Choose the CRAN (binaries) under the Packages repository label. Click Get List.
      ii. For Windows, select Packages -> Install Packages. Choose the closest CRAN mirror site. This will bring up a list of all extension packages.

Scroll down to select ks, and press Install. Repeat for the rgl and misc3d libraries.

   b. Manual: If the computer is not connected to the Internet, then the following packages must be downloaded using another computer from CRAN: mvtnorm, rgl, miscd, and ks. Go to the R project Website -> CRAN. Choose a CRAN mirror site. Click on the Contributed extension packages link. This will bring up the list of all contributed packages. Look for mvtnorm. Download the appropriate binary file (.tgz for Macintosh and .zip for Windows) to the desktop (or other desired location). Repeat for the ks, rgl, and misc3d libraries. Once all files have been downloaded, start R. To install these extension packages, perform the following:
      i. For Macintosh, select Packages & Data -> Package Installer. Then select Local Binary Package under Packages Repository label. Select mvtnorm and press Install.
      ii. For Windows, select Packages -> Install Packages from local zip files.

Install the extension packages in this order: mvtnorm, ks, rgl, and misc3d.

20.1.2.5.2 Density estimation

Three simple commands in R produce a kernel density map: (i) `Hpi()` calculates the optimal smoothing parameter for a density map, using a plug-in algorithm (Duong & Hazelton, 2003); (ii) `kde()` computes the density map; and (iii) `plot()` creates a graphical visualization of the density map.

```
smooth.param = Hpi(x=x)
density.map = kde(x=x, H=smooth.param)
plot(density.map)
```

The following procedure will calculate the density maps of the Rab6-labeled compartment using the table of (aligned) coordinates for each detected structure from several tens of cells obtained from the preanalysis. The text files, containing coordinates of Rab6-labeled structures under different experimental conditions and the R script required for the analysis, are provided in supplementary data. Save these data into a dedicated folder that will be the working directory. For other experimental conditions, coordinate tables of structures of interest need to be obtained and the R script needs to be adjusted (e.g., exchange `rab6.NT`) in order to calculate corresponding density maps.

1. Start R and load the ks library.

This can be done by selecting Packages & Data -> Package Manager (Macintosh) or Packages -> Load package (Windows), and then ks. Or it can be loaded by typing library(ks) into the command line window.

2. Choose the working directory in which txt files for analysis are located.
   (a) In Mac: ->Misc->Change Working Directory
   (b) In PC: ->File->Change dir . . .
3. Open R script:

File->Open Document . . .

4. Copy from R script the following script and paste in R Console:

```
## Load libraries
library(ks)
## Define calculation space

xlim2 <- c(-300,300)
ylim2 <- c(-300,300)
xlim <- c(-400,400)
ylim <- c(-400,400)
zlim <- c(-50, 100)
lwd <- 3
xmin <- -c(400,400,100)
xmax <- c(400,400,100)
```

```
## Read in data files

rab6.NT <- read.table(file="Rab6NT82cells.txt", header=TRUE)
rab6.NZ <- read.table(file="Rab6NZ47cells.txt", header=TRUE)
rab6.CD <- read.table(file="Rab6CD50cells.txt", header=TRUE)

## Calculate density in 2D

H <- Hpi(rab6.NT[,1:2], binned=TRUE)
fhat2d.rab6.NT <- kde(x=rab6.NT[,1:2], H=H, xmin=xmin, xmax=xmax)
H <- Hpi(rab6.NZ[,1:2], binned=TRUE)
fhat2d.rab6.NZ <- kde(x=rab6.NZ[,1:2], H=H, xmin=xmin, xmax=xmax)
H <- Hpi(rab6.CD[,1:2], binned=TRUE)
fhat2d.rab6.CD <- kde(x=rab6.CD[,1:2], H=H, xmin=xmin, xmax=xmax)

## Calculate density in 3D

H <- Hpi(rab6.NT, binned=TRUE, bgridsize=rep(31,3))
fhat3d.rab6.NT <- kde(x=rab6.NT, H=H, xmin=xmin, xmax=xmax)
H <- Hpi(rab6.NZ, binned=TRUE, bgridsize=rep(31,3))
fhat3d.rab6.NZ <- kde(x=rab6.NZ, H=H, xmin=xmin, xmax=xmax)
H <- Hpi(rab6.CD, binned=TRUE, bgridsize=rep(31,3))
fhat3d.rab6.CD <- kde(x=rab6.CD, H=H, xmin=xmin, xmax=xmax)

## Create 2D plots
## 2D plotting functions

crossbow.3d <- array(0, dim=c(750, 750, 32))

crossbowt.3d <- crossbow.3d

contour.crossbow.2d <- function()
{
     x.cross <- tail(seq(-350, 350, length=751), -1)
     y.cross <- tail(seq(-350, 350, length=751), -1)
     contour(x.cross, y.cross, crossbowt.3d[,,5], drawlabels=FALSE,
lwd=4, col="grey", cex.lab=1.4, cex.axis=1.4, asp=1, xlim=xlim2,
ylim=ylim2, xlab="x", ylab="y")
     lines(c(-100-10000/64.5, -100), c(-300,-300), lwd=lwd)
     text(-100-10000/64.5/2,-300, expression(10*" "*mu*m), pos=3,
     cex=1.4)
}

## 2D plots

pdf(file="Rab6_NT.pdf")
par(mar=c(4,4,1,1), cex.main=1.4, lwd=lwd)
contour.crossbow.2d()
plot(fhat2d.rab6.NT, cont=25, col="red", add=TRUE, drawlabel=FALSE)
plot(fhat2d.rab6.NT, cont=50, col="orange", add=TRUE, drawlabel=FALSE)
```

```
plot(fhat2d.rab6.NT, cont=75, col="yellow", add=TRUE, drawlabel=FALSE)
legend("bottomright", title="Rab6 NT", legend=c("25%", "50%", "75%"),
lwd=lwd,
col=c("red", "orange", "yellow"), bty="n", cex=1.2)
dev.off()

pdf(file="Rab6_NZ.pdf")
par(mar=c(4,4,1,1), cex.main=1.4, lwd=lwd)
contour.crossbow.2d()
plot(fhat2d.rab6.NZ, cont=25, col="red", add=TRUE, drawlabel=FALSE)
plot(fhat2d.rab6.NZ, cont=50, col="orange", add=TRUE, drawlabel=FALSE)
plot(fhat2d.rab6.NZ, cont=75, col="yellow", add=TRUE, drawlabel=FALSE)
legend("bottomright", title="Rab6 NZ", legend=c("25%", "50%", "75%"),
lwd=lwd,
col=c("red", "orange", "yellow"), bty="n", cex=1.2)
dev.off()

pdf(file="Rab6_CD.pdf")
par(mar=c(4,4,1,1), cex.main=1.4, lwd=lwd)
contour.crossbow.2d()
plot(fhat2d.rab6.CD, cont=25, col="red", add=TRUE, drawlabel=FALSE)
plot(fhat2d.rab6.CD, cont=50, col="orange", add=TRUE, drawlabel=FALSE)
plot(fhat2d.rab6.CD, cont=75, col="yellow", add=TRUE, drawlabel=FALSE)
legend("bottomright", title="Rab6 CD", legend=c("25%", "50%", "75%"),
lwd=lwd,
col=c("red", "orange", "yellow"), bty="n", cex=1.2)
dev.off()

## Create 3D plots
## 3D plotting functions

vv3d <- function()
{
     par3d(userMatrix=matrix(c(0.87, -0.48, 0.04, 0, 0.10, 0.26, 0.96,
0, -0.48, -0.83, 0.29, 0, 0.00, 0.00, 0.00, 1), byrow=TRUE, ncol=4))
}

contour.crossbow.movie.3d <- function(add=TRUE, alpha=0.2, zoom=0.5)
{
     par3d(windowRect=c(0,0,500*16/9,500), zoom=zoom)
     ind <- seq(10, 750, by=10)
     x.cross <- tail(seq(-350, 350, length=751), -1)[ind]
     y.cross <- tail(seq(-350, 350, length=751), -1)[ind]
     z.cross <- seq(-2000/645*4, 2000/645*27, length=32)
     if (add)
     axis3d('z+-', at=seq(0, 100, length=5), pos=c(-100, 100, NA))
     axis3d('x+', pos=c(NA, 100, 0))
```

```
      axis3d('y+', pos=c(-100, NA, 0))
      text3d(-120, 120, 80, "z")
      text3d(0, 150, 0, "x")
      text3d(-130, 0, 0, "y")
      vv3d()
}

rotatez.movie.3d <- function(file="movie.mpeg", start=0, end=2*pi,
len=100, n=1)
{
      angle.step <- diff(seq(start, end, length=(n*len)))[1]
      for (i in 1:(n*len))
      {
      par3d(userMatrix=rotate3d(par3d()$userMatrix,angle.step,0,0,1))
      snapshot3d(file=paste(file, sprintf("%03d", i), ".png", sep=""))
   }

   if (.Platform$OS.type=="unix")
   {
#system(paste("convert -delay 10 ", file, "*.png -loop 0 ", file, sep=""))
   }
}

## 3d plots

plot(fhat3d.rab6.NT, col=c("yellow", "orange", "red"), xlim=xlim,
ylim=ylim, zlim=ylim, axes=FALSE, box=FALSE, asp=c(1,1,1), xlab="",
ylab="", zlab="")
contour.crossbow.movie.3d()
snapshot3d(file="Rab6_NT.png")

plot(fhat3d.rab6.NZ, col=c("yellow", "orange", "red"), xlim=xlim,
ylim=ylim, zlim=ylim, axes=FALSE, box=FALSE, asp=c(1,1,1), xlab="",
ylab="", zlab="")
contour.crossbow.movie.3d()
snapshot3d(file="Rab6_NZ.png")

plot(fhat3d.rab6.CD, col=c("yellow", "orange", "red"), xlim=xlim,
ylim=ylim, zlim=ylim, axes=FALSE, box=FALSE, asp=c(1,1,1), xlab="",
ylab="", zlab="")
contour.crossbow.movie.3d()
snapshot3d(file="Rab6_CD.png")
```

5. Additional options to save a movie of the 3D density map and to save and load the calculated densities are provided by the following commands:

```
## 3d movies

rotatez.movie.3d()
```

```
## save densities
save(list=ls(pattern="fhat*"), file="Rab6.RData")
## reload calculated densities
load("Rab6.RData")
```

20.2 RESULTS AND DISCUSSION

To demonstrate the potential of our method to detect alterations in intracellular trafficking pathways, we analyze the Rab6-labeled compartment under different experimental conditions as an example. Whereas treatment with nocodazole gives rise to a strong phenotype that has been previously described (Grigoriev et al., 2007; Racine et al., 2007), treatment with cytochalasin D leads to subtle morphological changes on Rab6-labeled transport carriers. Nocodazole treatment fragments the Golgi and redistributes Rab6-labeled transport carriers from the cell periphery to the cell center, a phenotype easily assessable in nonrestricted and micropatterned cells (Fig. 20.4C). On the contrary, disruption of actin filaments does not show any strong effect in nonconstrained cells or micropatterned cells as judged by visual estimation (Fig. 20.4B). Density mapping of the Rab6-labeled compartment in many normalized cells with disrupted actin shows however retraction of transport carriers from peripheral adhesive areas (Fig. 20.4B, right panel). This indicates that Rab6-labeled transport carriers are entrapped at the edge of the cell by actin filaments. Actin filaments are known to be organized close to the cell surface and to spread out trafficking vesicles (Nascimento, Roland, & Gelfand, 2003; Snider et al., 2004). To test for the significance of the difference between control cells and cytochalasin D-treated cells, density-based testing can be employed. Full automated testing can be performed using our multivariate two-sample test that is asymptotically normal under the null hypothesis allowing "black box" comparisons of multivariate data (Duong et al., 2012). The completely automatic testing procedure (including the parameter estimation and the computation of the test statistic and its *P*-value) is programmed in the ks library in R. Alternatively, resampling methods that compute the critical quantiles of the null distribution can be used requiring calibration for each data analysis at hand. For example, a distance function between pairs of points from different experimental conditions can be compared with a corresponding distance function from the same experimental condition (Schauer et al., 2010).

Alterations in intracellular trafficking are often revealed indirectly, mainly by the missorting of marker molecules, such as cargo proteins that are exchanged among the different intracellular compartments. Moreover, existing methods to study intracellular trafficking alterations do not provide a quantitative way to characterize the spatial distribution of the compartments, especially regarding those that are spatially dispersed, like transport carriers. By allying the micropatterning technique with probabilistic density maps, our method provides a direct and intuitive visualization of subcellular alterations in trafficking compartments. Therefore, it is extremely useful in the assignment of functional roles to trafficking components. In the future, it will be interesting to use this method to investigate whether patterned cells could be

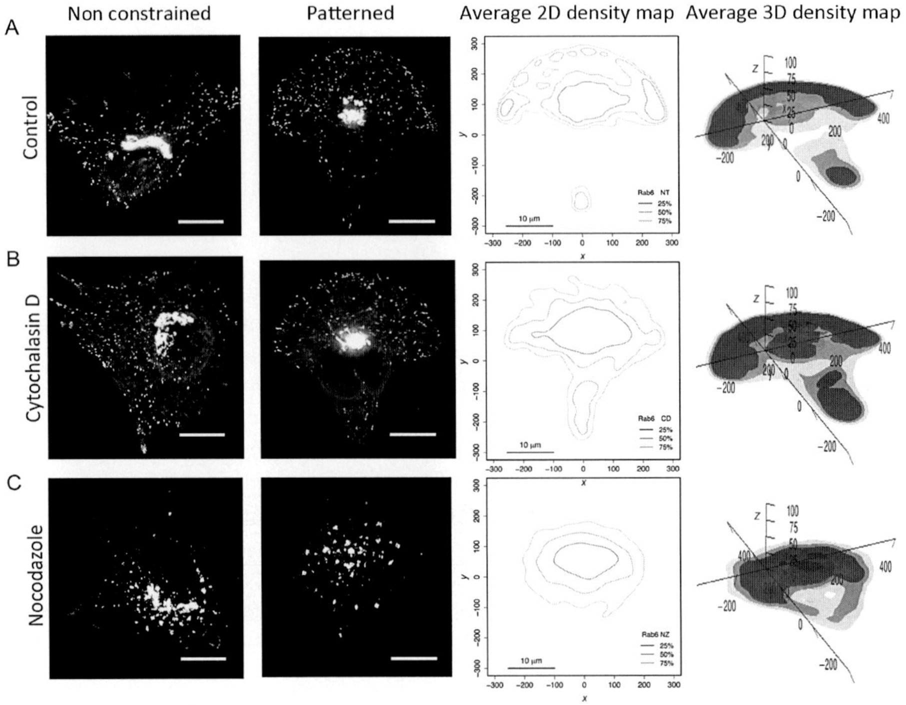

FIGURE 20.4

Changes of Rab6-labeled compartment upon cytoskeletal disruption. (A) Rab6-labeled compartment in nontreated cells. Nonconstrained and micropatterned cells and the corresponding 2D and 3D density maps of 82 normalized cells are shown. (B) Rab6-labeled compartment upon cytochalasin D treatment in nonconstrained and micropatterned cells and the corresponding 2D and 3D density maps of 50 normalized cells. (C) Rab6-labeled compartment upon nocodazole treatment in nonconstrained and micropatterned cells and the corresponding 2D and 3D density maps of 47 normalized cells. The 25% (red), 50% (orange), and 75% (yellow) probability contours represent the smallest area in which a given percentage of Rab6-labeled structures are found: for example, the 50% probability contour defines the smallest area in which 50% of structures are located. Scale bars, 10 mm. (See color plate.)

used as references to rapidly highlight pathological disorders, for example, during malignant development. Furthermore, this approach is a step towards the establishment of automated proceedings for high-content, high-throughput, phenotype-based screenings (Schauer et al., 2013). For instance, it could be used to identify scaffold, tether, or motor proteins that are regulated by a distinct Rab protein, for example, Rab6. As our method provides quantitative data from density-based analyses, we envision to use them for modeling approaches that could allow a better understanding of changes in intracellular organization of compartments in health and disease.

SUMMARY

Crossing the border between different scientific disciplines is required to quantify changes in cellular morphology in a robust, unbiased, and automated manner in high throughput. We present how an advanced mathematical method, density estimation, is employed to move to quantitative cell biology, allowing detailed analysis of complex cellular behavior.

Acknowledgments

We thank Laura Picas and Lena Oesterlin for critical reading of the chapter. K. S. received funding from the Fondation pour la Recherche Médicale en France and Association pour la Recherche sur le Cancer. T. D. received funding from the program "Investissements d'avenir" ANR-10-IAIHU-06. C. S. G. S. is supported by the Marie Curie Fellowship PIEF-GA-2011-299756. This project was further supported by grants from Agence Nationale de la Recherche (#2010 BLAN 122902), the Centre National de la Recherche Scientifique, and Institut Curie.

Conflict of interest statement: A patent (PCT n° EP2012/064239) has been filed on the density-based two-sample test.

References

Azioune, A., Storch, M., Bornens, M., Thery, M., & Piel, M. (2009). Simple and rapid process for single cell micro-patterning. *Lab on a Chip*, *9*, 1640–1642.

Becker, T., Volchuk, A., & Rothman, J. E. (2005). Differential use of endoplasmic reticulum membrane for phagocytosis in J774 macrophages. *Proceedings of the National Academy of Sciences of the United States of America*, *102*, 4022–4026.

Bonifacino, J. S., & Glick, B. S. (2004). The mechanisms of vesicle budding and fusion. *Cell*, *116*, 153–166.

Chia, P. Z., & Gleeson, P. A. (2011). The regulation of endosome-to-Golgi retrograde transport by tethers and scaffolds. *Traffic*, *12*, 939–947.

Duong, T., Goud, B., & Schauer, K. (2012). Closed-form density-based framework for automatic detection of cellular morphology changes. *Proceedings of the National Academy of Sciences of the United States of America*, *109*, 8382–8387.

Duong, T., & Hazelton, M. L. (2003). Plug-in bandwidth matrices for bivariate kernel density estimations. *Journal of Nonparametric Statistics*, *17*, 17–30.

Duong, T., & Hazelton, M. L. (2005). Cross-validation bandwidth matrices for multivariate kernel density estimation. *Scandinavian Journal of Statistics*, *32*, 485–506.

Eliceiri, K. W., Berthold, M. R., Goldberg, I. G., Ibanez, L., Manjunath, B. S., Martone, M. E., et al. (2012). Biological imaging software tools. *Nature Methods*, *9*, 697–710.

Ellis, M. A., Potter, B. A., Cresawn, K. O., & Weisz, O. A. (2006). Polarized biosynthetic traffic in renal epithelial cells: Sorting, sorting, everywhere. *American Journal of Physiology Renal Physiology*, *291*, F707–F713.

Folch, A., & Toner, M. (2000). Microengineering of cellular interactions. *Annual Review of Biomedical Engineering*, *2*, 227–256.

Goud, B., & Akhmanova, A. (2012). *Rab6 GTPase*. Danvers: Bentham Science.

Griffiths, G., Back, R., & Marsh, M. (1989). A quantitative analysis of the endocytic pathway in baby hamster kidney cells. *The Journal of Cell Biology*, *109*, 2703–2720.

Grigoriev, I., Splinter, D., Keijzer, N., Wulf, P. S., Demmers, J., Ohtsuka, T., et al. (2007). Rab6 regulates transport and targeting of exocytotic carriers. *Developmental Cell*, *13*, 305–314.

Gumbiner, B. M. (1996). Cell adhesion: The molecular basis of tissue architecture and morphogenesis. *Cell*, *84*, 345–357.

Hutagalung, A. H., & Novick, P. J. (2011). Role of Rab GTPases in membrane traffic and cell physiology. *Physiological Reviews*, *91*, 119–149.

Jean, S., & Kiger, A. A. (2012). Coordination between RAB GTPase and phosphoinositide regulation and functions. *Nature Reviews Molecular Cell Biology*, *13*, 463–470.

Lipschutz, J. H., Lingappa, V. R., & Mostov, K. E. (2003). The exocyst affects protein synthesis by acting on the translocation machinery of the endoplasmic reticulum. *Journal of Biological Chemistry*, *278*, 20954–20960.

Markgraf, D. F., Peplowska, K., & Ungermann, C. (2007). Rab cascades and tethering factors in the endomembrane system. *FEBS Letters*, *581*, 2125–2130.

Martinez-Menarguez, J. A., Geuze, H. J., Slot, J. W., & Klumperman, J. (1999). Vesicular tubular clusters between the ER and Golgi mediate concentration of soluble secretory proteins by exclusion from COPI-coated vesicles. *Cell*, *98*, 81–90.

McBeath, R., Pirone, D. M., Nelson, C. M., Bhadriraju, K., & Chen, C. S. (2004). Cell shape, cytoskeletal tension, and RhoA regulate stem cell lineage commitment. *Developmental Cell*, *6*, 483–495.

Nascimento, A. A., Roland, J. T., & Gelfand, V. I. (2003). Pigment cells: A model for the study of organelle transport. *Annual Review of Cell and Developmental Biology*, *19*, 469–491.

Pereira-Leal, J. B., & Seabra, M. C. (2001). Evolution of the Rab family of small GTP-binding proteins. *Journal of Molecular Biology*, *313*, 889–901.

R_Development_Core_Team (2013). *R: A language and environment for statistical computing*. Vienna, Austria: R Foundation for Statistical Computing.

Racine, V., Sachse, M., Salamero, J., Fraisier, V., Trubuil, A., & Sibarita, J. B. (2007). Visualization and quantification of vesicle trafficking on a three-dimensional cytoskeleton network in living cells. *Journal of Microscopy*, *225*, 214–228.

Sannerud, R., Saraste, J., & Goud, B. (2003). Retrograde traffic in the biosynthetic-secretory route: Pathways and machinery. *Current Opinion in Cell Biology*, *15*, 438–445.

Saraste, J., & Goud, B. (2007). Functional symmetry of endomembranes. *Molecular Biology of the Cell*, *18*, 1430–1436.

Sbalzarini, I. F., & Koumoutsakos, P. (2005). Feature point tracking and trajectory analysis for video imaging in cell biology. *Journal of Structural Biology*, *151*, 182–195.

Schauer, K., Duong, T., Bleakley, K., Bardin, S., Bornens, M., & Goud, B. (2010). Probabilistic density maps to study global endomembrane organization. *Nature Methods*, *7*, 560–566.

Schauer, K., Grossier, J.-P., Duong, T., Chapuis, V., Degot, S., Lescure, A., et al. (2013). A novel "Organelle Map" framework for high-content cell morphology analysis in high-throughput. *Journal of Biomolecular Screening*, 2013, Aug 16. [Epub ahead of print].

Sigal, A., Milo, R., Cohen, A., Geva-Zatorsky, N., Klein, Y., Liron, Y., et al. (2006). Variability and memory of protein levels in human cells. *Nature*, *444*, 643–646.

Simonoff, J. S. (1996). *Smoothing methods in statistics*. New York: Springer-Verlag.

Snider, J., Lin, F., Zahedi, N., Rodionov, V., Yu, C. C., & Gross, S. P. (2004). Intracellular actin-based transport: How far you go depends on how often you switch. *Proceedings of the National Academy of Sciences of the United States of America*, *101*, 13204–13209.

Snijder, B., Sacher, R., Ramo, P., Damm, E. M., Liberali, P., & Pelkmans, L. (2009). Population context determines cell-to-cell variability in endocytosis and virus infection. *Nature, 461*, 520–523.

Stenmark, H. (2009). Rab GTPases as coordinators of vesicle traffic. *Nature Reviews Molecular Cell Biology, 10*, 513–525.

Thery, M., Racine, V., Pepin, A., Piel, M., Chen, Y., Sibarita, J. B., et al. (2005). The extracellular matrix guides the orientation of the cell division axis. *Nature Cell Biology, 7*, 947–953.

Thery, M., Racine, V., Piel, M., Pepin, A., Dimitrov, A., Chen, Y., et al. (2006). Anisotropy of cell adhesive microenvironment governs cell internal organization and orientation of polarity. *Proceedings of the National Academy of Sciences of the United States of America, 103*, 19771–19776.

Toikkanen, J. H., Miller, K. J., Soderlund, H., Jantti, J., & Keranen, S. (2003). The beta subunit of the Sec61p endoplasmic reticulum translocon interacts with the exocyst complex in Saccharomyces cerevisiae. *Journal of Biological Chemistry, 278*, 20946–20953.

Wand, M. P., & Jones, M. C. (1995). *Kernel smoothing*. London: Chapman & Hall/CRC.

Wilcke, M., Johannes, L., Galli, T., Mayau, V., Goud, B., & Salamero, J. (2000). Rab11 regulates the compartmentalization of early endosomes required for efficient transport from early endosomes to the trans-golgi network. *The Journal of Cell Biology, 151*, 1207–1220.

CHAPTER

Methods to Study Signaling at the Golgi Apparatus

21

Veronika Reitere*,†, Francesco Baschieri*,†, Valentina Millarte*,† and Hesso Farhan*,†

*Biotechnology Institute Thurgau, Kreuzlingen, Switzerland

†Department of Biology, University of Konstanz, Konstanz, Germany

CHAPTER OUTLINE

Introduction and Rationale **346**
21.1 Materials **346**
21.2 Methods **347**
21.2.1 Lipid Signaling at the Golgi 347
21.2.1.1 Immunofluorescence 347
21.2.1.2 Fluorescent Reporters 348
21.2.1.3 Acute Deletion Using Recruitable Enzymes 349
21.2.2 GTPase Signaling at the Golgi 350
21.2.2.1 Ras Activity on the Golgi 350
21.2.2.2 Rho Family GTPases at the Golgi 351
21.2.3 FRET Reporter to Monitor PKC Activity 354
21.2.4 FRET Reporter to Monitor DAG Activity 354
21.2.5 Protein Phosphorylation as a Readout for Signaling Events 354
Concluding Remarks **356**
Acknowledgments **356**
References **356**

Abstract

Research on the secretory pathway in the past three decades accounts for our known knowledge about the composition and architecture of organelles and about the machinery that regulates membrane transport. An emerging topic in the past few years was the discovery that the secretory pathway is regulated by signaling, and in this regard, the Golgi apparatus received major attention. In the current chapter, we will highlight various techniques that are used by us and others to study signaling at the Golgi. We describe methods to study lipid and protein phosphorylation at the Golgi

Methods in Cell Biology, Volume 118 **ISSN 0091-679X**

http://dx.doi.org/10.1016/B978-0-12-417164-0.00021-5

and various techniques for studying spatial activation of GTPases at this organelle. We also discuss how combining these techniques and improving their limitations is important for gaining a better understanding of how the Golgi intersects with various signal transduction pathways.

INTRODUCTION AND RATIONALE

Mounting evidence in the past years indicates that the secretory pathway is regulated by various signal transduction pathways. In addition to being a receiver of signaling, organelles of the secretory pathway also house signaling molecules and modify their action, thereby contributing to the biologic outcome of a signaling cascade (Farhan & Rabouille, 2011; Sallese, Giannotta, & Luini, 2009). The increasing relevance of signaling for research on the secretory pathway becomes very evident from the fact that 20 years ago (in 1992), roughly 6% of all publications containing the term "secretory pathway" also contained the term "signaling." In 2012, this number rose to about 18% (source, NCBI). Among the organelles of the secretory pathway, signaling at the Golgi apparatus received great attention. Therefore, we aim in this chapter to describe various methods that are used to study signaling at the Golgi. We stress that there is no defined single "method" to study signaling, simply because signal transduction is a process and not a technique. Therefore, a combination of various methods and approaches has to be applied to investigate signaling pathway to and from the Golgi. In this chapter, we will try to cover key techniques that we and others have used. We will start by describing methods to study phosphorylation of lipids at the Golgi, then describe methods to study GTPase signaling, and finally review methods to study protein phosphorylation at the Golgi.

21.1 MATERIALS

Cell culture dishes (Greiner).
Glass coverslips (diameter 18 mm, thickness #1 (0.17 mm thick), Thermo Scientific).
Bovine serum albumin (BSA) (Sigma-Aldrich).
Phosphate-buffered saline (PBS) (137 mM NaCl, 2.7 mM KCl, 8.1 mM Na_2HPO4, and 1.5 mM KH_2PO4, pH 7.4).
SLO-buffer (20 mM HEPES (pH 7.2), 110 mM $Mg(OAc)_2$, and 1 mM dithiothreitol).
Paraformaldehyde (Riedel-de Haen) (to prepare a 3% PFA solution, dissolve 3 g PFA in 100 ml PBS and heat 60 °C under a fume hood and stir. Add dropwise NaOH until PFA dissolves. The pH of the solution has to be between 7.4 and 8).
Formaldehyde 30% stock solution (Polysciences).
Methanol (Applichem).
Digitonin (Sigma-Aldrich).

Goat serum (Santa Cruz).
HiPerFect siRNA transfection reagent (Qiagen).
Fugene6 transfection reagent (Promega).
Ludin live imaging chamber.
Confocal microscope (Leica, SP5).
Polyvinyl alcohol mounting medium with DABCO® antifading (Fluka).

21.2 METHODS

21.2.1 Lipid signaling at the Golgi

Lipids play an important role in regulating the architecture of the Golgi and trafficking to and from this organelle. A major role is assigned to phosphoinositides and metabolites thereof. These are especially important in the context of signaling because phosphoinositides are subject to regulation by phosphorylation and the resulting metabolites (such as inositol trisphosphate or diacylglycerol (DAG)) activate downstream signaling molecules such as members of the protein kinase C family. The most abundant phosphoinositide at the Golgi is phosphatidylinositol-4-phosphate (PI4P), which is formed by the action of different isoforms of PI4 kinases, namely, PI4K-IIα and PI4K-IIIβ. Later, we will describe the two main methods that are used to study PI4P at the Golgi.

21.2.1.1 Immunofluorescence

As a lipid component of Golgi membranes, PI4P (as other phospholipids) is very sensitive to detergent treatment; therefore its one critical step is the choice of method to permeabilize cells. We describe in the succeeding text a protocol based on a protocol from the group of Irvine (Hammond et al., 2012; Hammond, Schiavo, & Irvine, 2009) that we successfully apply to monitor PI4P using immunofluorescence:

- Grow adherent cells (e.g., HeLa cells) on glass coverslips. In case cells are poorly adherent, it might be necessary to coat the glass coverslips with collagen (rat tail collagen type I or mouse collagen, type IV or type I, BD Biosciences) or polylysine (Sigma-Aldrich). In case of HeLa cells, we plate cells in high-glucose DMEM (LONZA) supplemented with 10% FCS (Linaris) and antibiotics (penicillin–streptomycin 100 U/ml).
- In case RNAi transfection is desired, we typically use HiPerFect (Qiagen) for most epithelial cell lines. We typically transfect directly after plating the cells. We noticed that coating of the glass coverslips with polylysine affects the efficiency and homogeneity of the RNAi transfection. In this case, it might be better to transfect the cells several hours after cells have attached. Also, it is recommended to use a higher amount of siRNA and of transfection reagent. Typically, cells are analyzed 48–72 h after siRNA transfection.
- Cells are fixed with 2% formaldehyde for 15 min at room temperature. Alternative fixation methods are 3% paraformaldehyde for 10–20 min at room temperature.

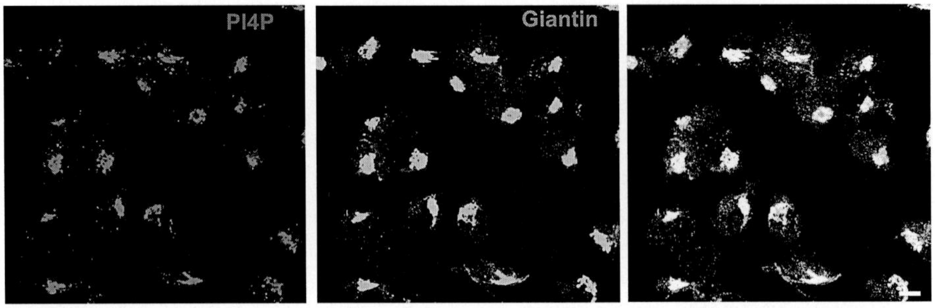

FIGURE 21.1

HeLa cells were plated on glass coverslips, fixed with PFA 2%, and immunostained for PI4P and Giantin as described in the protocol in the main text. Scale bar = 10 μm. (See color plate.)

Other reagents for fixation are ice-cold methanol at −20 °C for 5–10 min or acetone at −20 °C for 5–10 min.

- Cell permeabilization is achieved using incubation of fixed cells with low levels of digitonin (20 μM) for 5 min at room temperature. The low levels of digitonin were shown previously to permeabilize only the plasma membrane without affecting the integrity of endomembranes (Breuza et al., 2004). An alternative method for permeabilization is to subject fixed cells to a freeze–thaw cycle in the presence of 1 M sucrose (Wang et al., 2003). This freeze–thaw cycle only permeabilizes the plasma membrane (Tran et al., 1999), leaving the organelles intact.
- After washing the cells once with PBS, the anti-PI4P antibody is added (from Echelon), which is a mouse IgM antibody. The antibody is diluted 1:500 in PBS supplemented with 5% goat serum. Cells are incubated in the presence of primary antibody for 60 min at room temperature.
- To costain the Golgi, we use the polyclonal rabbit IgG anti-Giantin antibody (Covance) diluted at 1:1000 in the same solution with the anti-PI4P antibody.
- After washing three times with PBS, cells are incubated with the proper secondary antibody diluted in PBS with 5% goat serum.
- Cells are washed four times in PBS followed by mounting in polyvinyl alcohol with DABCO antifade.
- Colocalization of PI4P and the Golgi marker is performed using a LeicaSP5 confocal laser scanning microscope. An example for an immunofluorescence staining of PI4P and its colocalization with the Golgi is depicted in Fig. 21.1.

21.2.1.2 Fluorescent reporters

The PH (pleckstrin homology) domains are used by a variety of proteins to detect different phosphoinositides (Dowler et al., 2000) where different PH domains display high selectivity toward certain PI species. This specificity led to the use of isolated PH domains fused to a variant of the green fluorescence protein (GFP) to detect PI4P. One advantage of using reporters is that they permit visualization of PI4P in living cells, which further allows capturing dynamic spatiotemporal changes in the pool of this phosphoinositide. A major disadvantage is the exogenously expressed PH domain will be overexpressed at a level that exceeds the concentration of endogenous PH domain-containing proteins by severalfold. This will on one hand result in competition with endogenous proteins and alteration of their function, which is expected to alter cell homeostasis. On the other hand, the overexpressed PH domain will sequester PI4P, thereby also changing membrane composition and structure. Therefore, it is very critical to use moderately or low expressing cells and ignore the strong overexpressing cells as using these is likely to lead to erratic results. Especially when studying the Golgi, strong overexpression of PI4P reporters leads to structural distortion of the Golgi. One way to face this problem has been achieved through replacement of the CMV promoter of the reporter plasmids with a thymidine kinase promoter, which allowed fine-tuning of the expression levels (Szentpetery, Vàrnai, & Balla, 2010).

The most commonly used reporters for PI4P are the PH domain of FAPP1 (four-phosphate-adaptor protein) or of OSBP1 (oxysterol-binding protein) or of the yeast homologues of the latter OSH1.

The potential adverse effects of reporter overexpression, especially the effect on Golgi structure cell, can be overcome by applying these PH domains onto semi-intact cells. Cells are permeabilized using 20 μM digitonin for 5 min at room temperature. As mentioned earlier, this permeabilization method only permeabilizes the plasma membrane. Extensive washing (6–10 times) leads to a washout of the majority of the cellular cytosol. Alternatively, cells may be permeabilized using 1–2 U/ml of streptolysin O (SLO) diluted in SLO-buffer for 10 min at 4 °C. Afterward, fresh cytosol and an ATP-regeneration system together with the purified GST–FAPP–PH are added and cells are fixed after 10–30 min. The ATP-regeneration system consists of 1 mM ATP, 5 mM creatinine phosphate, and 7.3 U/ml creatine phosphokinase and is necessary because the binding of FAPP–PH domain to PI4P requires ATP, although it is not fully clear why this is the case (Godi et al., 2004).

Finally, the PI4P-sequestering effect of the FAPP–PH domain can be used to test the relevance of PI4P for certain biological processes. For instance, addition of a purified FAPP–PH domain to semi-intact cells was shown to inhibit export from the ER (Weixel, Blumental-Perry, Watkins, Aridor, & Weisz, 2005). A disadvantage of this method is that FAPP–PH is recruited by coincidence detection of PI4P and Arf1 (Godi et al., 2004) and thus is very unlikely to label PI4P on Arf1-free membranes. Another disadvantage is that this is only indirect evidence, and more recently, a better method was developed to locally deplete PI4P and therefore we will next discuss this method.

21.2.1.3 Acute deletion using recruitable enzymes

Acute depletion of PI4P from the Golgi might be achieved by using pharmacological inhibitors of PI4 kinases such phenyl-arsine oxide (Balla, Tuymetova, Barshishat, Geiszt, & Balla, 2002) or PIK93, which discriminates between PI4KIIIα and PI4KIIIβ (Knight et al., 2006). At low concentrations (250 nM), PI4KIIIβ is inhibited by 90%, whereas more than 10 μM are needed to inhibit PI4KIIIα by 50%. However, these inhibitors also inhibit PI4 kinases at other cellular locations, and as is the case with any chemicals, they suffer from nonspecific effects. In order to specifically determine the impact of selective and acute loss of PI4P from the Golgi, the group of Tamas Balla designed a version of Sac1, an enzyme that selectively hydrolyzes PI4P, which can be recruited to the Golgi (Szentpetery et al., 2010). The cytosolic portion of Sac1 was fused to mRFP (monomeric red fluorescent protein)–FKBP12 module. The mRFP part of the module is introduced to facilitate visualization of the construct. At the same time, cells were engineered to overexpress a CFP–TGN38–FRB construct. CFP-TGN38 is a cyan fluorescence protein tagged to TGN38, which localizes to the trans-Golgi network. FRB (FKBP12-binding protein) will bind to FKBP12 in the presence of rapamycin. Upon addition of rapamycin, the Sac1–mRFP–FKBP12 tandem is recruited to the Golgi, which leads to a rapid depletion of PI4P. In the succeeding text is a protocol for an experiment that we have performed:

- 2×10^5 HeLa cells were plated onto glass coverslips (diameter = 18 mm) in six-well cell culture plates and grown for 24 h.
- Cells were cotransfected with 0.5 μg of Sac1–mRFP–FKBP12 and 0.5 μg CFP–TGN38–FRB using Fugene6 (Promega).
- Cells are fixed and processed for PI4P staining as described earlier.
- Cells are treated with rapamycin followed by fixation and PI4P staining.

As shown in Fig. 21.2, addition of rapamycin for only 10 min leads to a massive depletion of PI4P from the Golgi.

21.2.2 GTPase signaling at the Golgi

21.2.2.1 Ras activity on the Golgi

The first sensor that was employed to study spatial aspects of Ras signaling was composed of a Ras-binding domain (RBD) of c-Raf N-terminally tagged with a variant of GFP, which was to detect activation of Ras at the Golgi upon EGF stimulation (Chiu et al., 2002). This is based on the fact that the RBD will only be recruited to locations where Ras is found in its GTP-bound state. Again, a major problem was that the levels of exogenously overexpressed RBD would exceed the levels of endogenous Ras by severalfold. Therefore, the RBD probe had to be coexpressed with Ras. The major drawback from this is that this reporter does not report activation of endogenous Ras but mainly of overexpressed Ras. This problem was overcome by using a method called bystander FRET (Quatela & Philips, 2006). This method requires overexpression of RBD–CFP and YFP-tagged CD8 protein. The latter construct illuminates the whole secretory pathway. Upon excitation at 458 nm (excitation wavelength of CFP), Chiu et al. (2002) detected a weak FRET signal in the Golgi region, which could be enhanced upon EGF stimulation. This signal resulted from the close proximity of CFP–RBD and CD8–YFP in the Golgi. This method compensates for the high background fluorescence of the CFP–RBD signal by focusing only on the FRET signal.

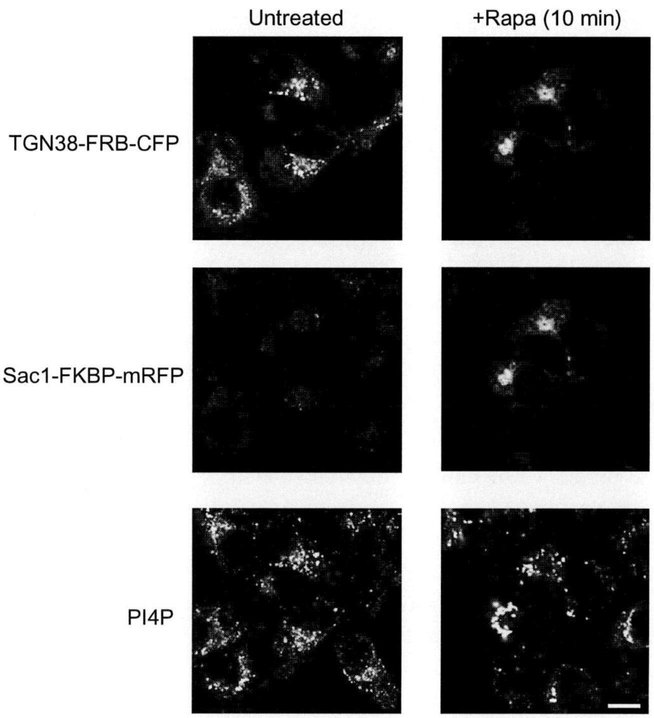

FIGURE 21.2

HeLa cells were plated on glass coverslips. After 24 h, cells were transfected with plasmids encoding the two parts of the recruitment system, TGN38–FRB–CFP and Sac1–FKBP–mRFP. After additional 24 h, one group of cells was fixed without further treatments (untreated), while another group was treated with 100 nM rapamycin for 10 min and then fixed. Cells were then immunostained for PI4P. Scale bar = 10 μm.

Another way to monitor endogenous Ras activity is the overexpression of a Ras–GEF (Caloca, Zugaza, & Bustelo, 2003); however, this biases Ras signaling toward the compartment where the GEF is mainly localized.

21.2.2.2 Rho family GTPases at the Golgi

The need to study the spatiotemporal aspects of Rho GTPases led to the construction of FRET-based sensors (Itoh et al., 2002; Kurokawa, Itoh, Yoshizaki, Nakamura, & Matsuda, 2004). Those sensors are composed by the small GTPase of interest (Rac, Cdc42, or Rho), the binding domain of an effector for the small GTPase (CRIB for Rac and Cdc42, the RBD of mDia, rhotekin, rhophilin, or PKN for Rho). These two modules are flanked by CFP and YFP. To target this probe to certain cellular location, different motifs are appended to the C-terminus of the probe. For instance, adding the lysine-rich motif of K-Ras targets the probe exclusively to the plasma membrane. On the contrary, adding the C-terminal–CAAx motif of the small GTPase of interest targets the probe to all locations where the endogenous GTPase is found. The probe containing Cdc42 and C-terminal motif of Cdc42 itself (Cdc42–Cdc42–CT) is normally enriched in a juxtanuclear region where it colocalizes with Golgi markers. When Cdc42 is stimulated, the GTPase binds to its effector, thereby bringing CFP and YFP into closer proximity. This change in confirmation can be measured by an increase in FRET efficiency.

Since the focus of this chapter is signaling at the Golgi, we will limit the method description to Cdc42 and its FRET Raichu probe as this probe most evidently localizes to the Golgi. The same techniques can however be applied to all the other Rho GTPases.

Here, we report a protocol that we are successfully using to measure Cdc42 activation at the Golgi:

- 2.5×10^5 HeLa cells are plated onto glass coverslips (18 mm diameter) in six-well cell culture plates. Cells can be transfected with a siRNA at this point.
- Cells are grown for 48 h.
- After 48 h, cells are transfected with 1 μg of Cdc42–Cdc42–CT using Fugene6 (Promega).
- Cells are grown for another 24 h, then growth medium is removed and exchanged with medium without FCS for at least 2 h.
- Fix cells to be able to measure the basal FRET signal in serum-starved cells.
- Cells are stimulated (for instance, with 10% FCS or any other desired stimulus). Note that the activation kinetics of Cdc42 are extremely fast and the peak of Cdc42 activation is found at 1 min or earlier. Therefore, it is critical to fix cells by directly adding the fixative into the medium. We add 200 μl of a 40% PFA solution into 2 ml of cell culture medium. Cells are incubated for 15 min.
- Cells are subsequently washed with PBS, dried, and mounted with polyvinyl alcohol with DABKO antifade.
- FRET is measured at the Golgi using a LeicaSP5 confocal laser scanning microscope. For users that have never performed FRET experiments before, we advise to use the FRET-AP Wizard that is provided together with the Leica software. FRET-AP uses the donor recovery after acceptor photobleaching method. Hereby, the acceptor (YFP) is bleached. In case the donor (CFP) molecule is in close proximity, bleaching of the acceptor will result in dequenching of the donor and an increase in its fluorescence.
- The FRET-AP wizard instructs first to acquire an image in the YFP channel, followed by an image using the CFP channel. Then, a region of interest is bleached using the laser setting of the acceptor channel at 100% laser intensity. Afterward, an image is acquired using the CFP channel.
- As shown in Fig. 21.3, an increase of CFP fluorescence is indicative of FRET.

A critical step is efficient bleaching of the acceptor. We typically exclude all measurements where the acceptor was bleached by less than 80%. To determine the

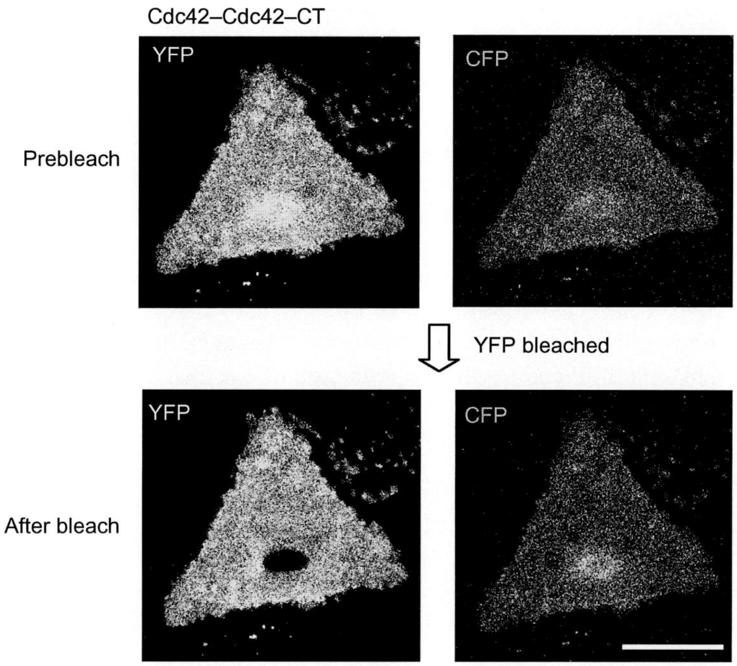

FIGURE 21.3

HeLa cells were plated on glass coverslips. After 24 h, cells were transfected with a plasmid encoding the FRET Raichu probe Cdc42–Cdc42–CT. After additional 24 h, cells are fixed as explained in the protocol and imaged before (upper two images) or after (lower two images) bleaching of YFP. Note the increase of CFP fluorescence in the region of bleached YFP. The increase of fluorescence is indicative of FRET. Scale bar = 20 μm. (See color plate.)

optimal laser parameters, cells can be transfected with a plasmid encoding CFP only or YFP only. First, the cells expressing YFP are bleached and the laser settings are regulated in order to get optimal bleaching. Then, the same bleaching parameters are applied to cells expressing only CFP and readjusted in order to get no alteration in the CFP fluorescence before and after bleaching.

Recently, the Raichu sensors were further improved with the addition of a longer linker sequence (Eevee, shortened EV) between the small GTPase and the effector. The EV linker reduces the basal activation level, thus significantly reducing the background (Komatsu et al., 2011). Raichu-EV probes have been developed to study also several other signaling cascades that can potentially take place on endomembranes, like ERK, Akt, and PKC signaling (Komatsu et al., 2011). However, it has to be kept in mind that neither the Raichu nor the Raichu-EV sensors are sensitive to GDIs.

21.2.3 FRET reporter to monitor PKC activity

Like the MAPK family, the protein kinase C family is a key player in diverse cellular functions (Parker & Murray-Rust, 2004; Toker, 1998). In order to investigate spatial regulation of PKC, Gallegos et al. (Gallegos, Kunkel, & Newton, 2006) constructed FRET-PKC reporters, which are targeted to distinct subcellular localization. The Golgi-CKAR (C kinase activity reporter) is targeted to the cytoplasmic face of the Golgi by fusing 33 amino acids of the N-terminal part of the endothelic nitric-oxid synthase. The reporter itself consists of CFP, YFP, a PKC substrate peptide, and an FHA2 phosphopeptide module. In an unphosphorylated state, the reporter shows FRET. Under conditions where PKC is activated, the substrate peptide becomes phosphorylated, leading to structural rearrangements in the reporter that displace CFP and YFP and result in a reduction of the FRET signal.

21.2.4 FRET reporter to monitor DAG activity

DAG is a signaling mediator that is produced downstream of several receptors including tyrosine kinase receptors and GPCRs. DAG is produced by members of the phospholipase C family that cleave phosphatidylinositol(4,5)bisphosphate to inositol trisphosphate and DAG. Importantly, DAG levels are high at the Golgi and are required for recruitment of protein kinase D to the TGN (Baron & Malhotra, 2002). Kunkel and Newton (2010) made use of a Golgi-targeted FRET reporter system to investigate DAG production at the Golgi. They targeted a CFP protein to the Golgi using a 35 aa targeting sequence from eNOS. As an FRET partner, they used the DAG-binding domain of PKCβII fused to YFP. To increase the affinity for DAG, they used not the wild-type of this domain, but a Y123W mutant, as this shows higher affinity to DAG. In case of DAG production at the Golgi, the YFP construct is recruited to the Golgi and therefore brought in close proximity to the Golgi resident CFP construct, thereby resulting in FRET. As DAG is produced downstream of PLC family members, this reporter can be used as an indirect measure to monitor the extent of PLC activation at the Golgi apparatus.

21.2.5 Protein phosphorylation as a readout for signaling events

Phosphorylation of proteins is the most common posttranslational modification and very often results in a change in protein conformation and/or function. Therefore, it is important to have means to demonstrate protein phosphorylation. In the best case, a specific antibody against the protein phosphorylated on a particular site exists. Unfortunately, the availability of such antibodies is rather the exception for a wide range of proteins studied. One approach is to generate such an antibody, which can be costly and labor-intensive. We will describe here a procedure that we recently used to study phosphorylation of the ER exit site protein Sec16 by ERK2, but we will try to keep the section as general as possible to be able to translate the protocol to other proteins.

Phosphorylation can be detected using antibodies recognizing phosphorylated tyrosine, serine, or threonine. However, these uses of these antibodies do not provide any insight into which kinase is responsible for the phosphorylation as these antibodies do not discriminate between different consensus motifs. The development of antibodies against different consensus motifs represents a major advantage (Phospho-(Ser/Thr) Kinase Substrate Antibody Sampler Kit #9920 Cell Signaling). In our particular case, we wanted to detect phosphorylation of a protein called Sec16, which is a component of ER exit sites. Therefore, Sec16 was immunoprecipitated followed by immunoblotting with an antibody that detects ERK1/2 consensus sites (MAPK substrate antibody) (Farhan et al., 2010).

- HeLa cells were seeded in 10 cm dishes (three dishes were used for each immunoprecipitation).
- The next day, cells were transfected with 3 μg GFP-Sec16 using Fugene6 as a transfection reagent.
- 24 h after transfection, cells were starved FCS to decrease ERK activation.
- 40 μl protein G-sepharose fast flow (50% slurry) was washed twice with IP buffer (50 mM Tris–HCl pH 7.4, 150 mM NaCl, 2 mM $CaCl_2$, and 1% Triton X-100) and incubated with 500 μl IP buffer and μl GFP antibody (Roche) for 2 h at room temperature.
- Cells were washed twice with 5 ml ice-cold PBS and collected in 800 μl of IP buffer supplemented with proteinase inhibitor and phosphatase inhibitor using a cell scraper and transferred into a 1.5-ml tube.
- Cell lysates were incubated for 10 min on ice.
- Insoluble material was pelleted by centrifugation at 4 °C, 16,000 × *g* for 10 min in a microcentrifuge.
- Supernatant was transferred into a fresh tube and 3% of the lysate as an input. This is used as a control whether the enrichment of the protein by immunoprecipitation worked. Afterward, protein G-sepharose with the adsorbed GFP antibody was added.
- The mixture was incubated at 4 °C in the cold room, end-over-end shaking overnight.
- The next day, samples were washed three times with IP buffer.
- Proteins were eluted by the addition of 30 μl 5 × reducing loading buffer and boiling at 95 °C for 3–5 min. Lysates were subjected to SDS-PAGE followed by blotting onto a nitrocellulose membrane using semidry transfer.
- Membranes were blocked in 5% milk in TBST (=TBS + 0.1% Tween 20) and incubated overnight with the MAPK substrate antibody (Phospho-(Thr) MAPK/CDK Substrate Mouse mAb #2321, cell signaling) diluted 1:1000 in TBST with 5% BSA.
- Blot was washed three times and incubated with HRP-conjugated anti-mouse secondary antibody diluted 1:5000 in TBST with 5% nonfat dry milk for 1 h at room temperature.
- Blot was washed three times and developed using ECL.
- Afterward, the blot was stripped and reprobed with Sec16 antibody to calculate the amount of phosphorylated protein compared to the total protein.

CONCLUDING REMARKS

Investigating the regulation of the secretory pathway in general and of the Golgi in particular is an emerging research area. Although some advance has been achieved in the past, we are still far from understanding how organelles of the secretory pathway are regulated by cellular signaling cascades. While many useful methods have been described to study this cross talk, there is certainly a large need to improve the methods to study the spatial details of the regulation of cellular signal transduction. In particular, work with the current fluorescent sensors needs careful experimentation in order to not pursue mere overexpression artifacts. Development of reporters that more closely resemble the physiological states will allow us to gain deeper insight into how signaling from and to endomembranes is regulated and how it regulates cell physiology. Such an integrative view of secretion and its role in cell homeostasis might be important to develop novel strategies that aim at targeting cellular processes that are deregulated in human pathologies.

Acknowledgments

This work was supported by the German Science Foundation, by the Canton of Thurgau, by the Swiss Science Foundation, and by a Young Scholar Fund from the University of Konstanz.

References

Balla, A., Tuymetova, G., Barshishat, M., Geiszt, M., & Balla, T. (2002). Characterization of type II phosphatidylinositol 4-kinase isoforms reveals association of the enzymes with endosomal vesicular compartments. *Journal of Biological Chemistry*, *277*(22), 20041–20050. http://dx.doi.org/10.1074/jbc.M111807200.

Baron, C. L., & Malhotra, V. (2002). Role of diacylglycerol in PKD recruitment to the TGN and protein transport to the plasma membrane. *Science*, *295*(5553), 325–328. http://dx.doi.org/10.1126/science.1066759.

Breuza, L., Halbeisen, R., Jenö, P., Otte, S., Barlowe, C., & Hong, W. (2004). Proteomics of endoplasmic reticulum-golgi intermediate compartment (ERGIC) membranes from Brefeldin A-treated HepG2 cells identifies ERGIC-32, a new cycling protein that interacts with human Erv46. *Journal of Biological Chemistry*, *279*(45), 47242–47253. http://dx.doi.org/10.1074/jbc.M406644200.

Caloca, M. J., Zugaza, J. L., & Bustelo, X. R. (2003). Exchange factors of the RasGRP family mediate Ras activation in the golgi. *Journal of Biological Chemistry*, *278*(35), 33465–33473. http://dx.doi.org/10.1074/jbc.M302807200.

Chiu, V. K., Bivona, T., Hach, A., Sajous, J. B., Silletti, J., & Wiener, H. (2002). Ras signalling on the endoplasmic reticulum and the Golgi. *Nature Cell Biology*, *4*(5), 343.

Dowler, S., Currie, R. A., Campbell, D. G., Deak, M., Kular, G., & Downes, C. P. (2000). Identification of pleckstrin-homology-domain-containing proteins with novel phosphoinositide-binding specificities. *Biochemical Journal*, *351*(1), 19.

Farhan, H., & Rabouille, C. (2011). Signalling to and from the secretory pathway. *Journal of Cell Science*, *124*(2), 171–180. http://dx.doi.org/10.1242/jcs.076455.

Farhan, H., Wendeler, M. W., Mitrovic, S., Fava, E., Silberberg, Y., & Sharan, R. (2010). MAPK signaling to the early secretory pathway revealed by kinase/phosphatase functional screening. *The Journal of Cell Biology*, *189*(6), 997–1011. http://dx.doi.org/10.1083/jcb.200912082.

Gallegos, L. L., Kunkel, M. T., & Newton, A. C. (2006). Targeting protein kinase C activity reporter to discrete intracellular regions reveals spatiotemporal differences in agonist-dependent signaling. *Journal of Biological Chemistry*, *281*(41), 30947–30956. http://dx.doi.org/10.1074/jbc.M603741200.

Godi, A., Di Campli, A., Konstantakopoulos, A., Di Tullio, G., Alessi, D. R., & Kular, G. S. (2004). FAPPs control Golgi-to-cell-surface membrane traffic by binding to ARF and PtdIns(4)P. *Nature Cell Biology*, *6*(5), 393.

Hammond, G. R. V., Fischer, M. J., Anderson, K. E., Holdich, J., Koteci, A., & Balla, T. (2012). PI4P and PI(4,5)P2 are essential but independent lipid determinants of membrane identity. *Science*, *337*(6095), 727–730. http://dx.doi.org/10.1126/science.1222483.

Hammond, G. R. V., Schiavo, G., & Irvine, R. F. (2009). Immunocytochemical techniques reveal multiple, distinct cellular pools of PtdIns4P and PtdIns(4,5)P2. *Biochemical Journal*, *422*(1), 23.

Itoh, R. E., Kurokawa, K., Ohba, Y., Yoshizaki, H., Mochizuki, N., & Matsuda, M. (2002). Activation of Rac and Cdc42 video imaged by fluorescent resonance energy transfer-based single-molecule probes in the membrane of living cells. *Molecular and Cellular Biology*, *22*(18), 6582–6591. http://dx.doi.org/10.1128/mcb.22.18.6582-6591.2002.

Knight, Z. A., Gonzalez, B., Feldman, M. E., Zunder, E. R., Goldenberg, D. D., & Williams, O. (2006). A pharmacological map of the PI3-K family defines a role for p110Î± in insulin signaling. *Cell*, *125*(4), 733.

Komatsu, N., Aoki, K., Yamada, M., Yukinaga, H., Fujita, Y., & Kamioka, Y. (2011). Development of an optimized backbone of FRET biosensors for kinases and GTPases. *Molecular Biology of the Cell*, *22*(23), 4647–4656. http://dx.doi.org/10.1091/mbc.E11-01-0072.

Kunkel, M. T., & Newton, A. C. (2010). Calcium transduces plasma membrane receptor signals to produce diacylglycerol at Golgi membranes. *Journal of Biological Chemistry*, *285*(30), 22748–22752. http://dx.doi.org/10.1074/jbc.C110.123133.

Kurokawa, K., Itoh, R. E., Yoshizaki, H., Nakamura, Y. O. T., & Matsuda, M. (2004). Coactivation of Rac1 and Cdc42 at Lamellipodia and membrane ruffles induced by epidermal growth factor. *Molecular Biology of the Cell*, *15*(3), 1003–1010. http://dx.doi.org/10.1091/mbc.E03-08-0609.

Parker, P. J., & Murray-Rust, J. (2004). PKC at a glance. *Journal of Cell Science*, *117*(2), 131–132. http://dx.doi.org/10.1242/jcs.00982.

Quatela, S. E., & Philips, M. R. (2006). Ras signaling on the Golgi. *Current Opinion in Cell Biology*, *18*(2), 162.

Sallese, M., Giannotta, M., & Luini, A. (2009). Coordination of the secretory compartments via inter-organelle signalling. *Seminars in Cell & Developmental Biology*, *20*(7), 801.

Szentpetery, Z., Vàrnai, P., & Balla, T. (2010). Acute manipulation of Golgi phosphoinositides to assess their importance in cellular trafficking and signaling. *Proceedings of the National Academy of Sciences*, *107*(18), 8225–8230. http://dx.doi.org/10.1073/pnas.1000157107.

Toker, A. (1998). Signaling through protein kinase C. *Frontiers in Bioscience*, *3*, D1134–D1147.

Tran, D., Stelly, N., Tordjmann, T., Durroux, T., Dufour, M. N., & Forchioni, A. (1999). Distribution of signaling molecules involved in vasopressin-induced Ca2+ mobilization in rat hepatocyte multiplets. *The Journal of Histochemistry and Cytochemistry*, *47*(5), 601–616. http://dx.doi.org/10.1177/002215549904700503.

Wang, Y. J., Wang, J., Sun, H. Q., Martinez, M., Sun, Y. X., & Macia, E. (2003). Phosphatidylinositol 4 phosphate regulates targeting of clathrin adaptor AP-1 complexes to the Golgi. *Cell*, *114*(3), 299.

Weixel, K. M., Blumental-Perry, A., Watkins, S. C., Aridor, M., & Weisz, O. A. (2005). Distinct Golgi populations of phosphatidylinositol 4-phosphate regulated by phosphatidylinositol 4-kinases. *Journal of Biological Chemistry*, *280*(11), 10501–10508. http://dx.doi.org/10.1074/jbc.M414304200.

CHAPTER

Golgi-Dependent Signaling: Self-Coordination of Membrane Trafficking

22

Jorge Cancino*,†, Anita Capalbo*,† and Alberto Luini*,†

*Telethon Institute of Genetics and Medicine, Napoli, Italy

†Istituto di Biochimica delle Proteine (IBP), Consiglio Nazionale delle Richerce (CNR), Napoli, Italy

CHAPTER OUTLINE

Introduction 361

22.1 Generation of Traffic-Dependent Golgi Signaling: ER-to-Golgi Traffic Synchronization Protocols 364

22.1.1 Synchronization of PC-I ER-to-Golgi Trafficking 365

22.1.2 VSVG Traffic Synchronization 366

22.1.3 ER-retention and Release of Cargo Proteins by Polymerization/Depolymerization 367

22.2 Morphological and Quantitative Analysis of Traffic-Generated Golgi-Based Signaling 368

22.2.1 Detection and Quantification of Traffic-induced Phosphorylation by IF Microscopy 368

22.2.2 Detection of Traffic-induced Phosphorylation by Western Blotting... 370

22.3 Analysis of KDELR-Dependent Signaling 371

22.3.1 Activation of the KDELR Signaling Cascade by Receptor Overexpression 372

22.3.2 Activation of the KDELR Signaling Cascade Using a Genetically Encoded Artificial Ligand 372

22.3.3 Activation of KDELR Signaling by Acute and Specific Ligand Incubation with Fluorescent Membrane-permeant Peptides 373

22.3.3.1 Labeling of Membrane-permeable Peptides 373

22.3.3.2 Activation of KDELR Signaling by KDEL-BODIPY Fluorescent Membrane-permeant Peptide 374

22.4 The KDELR Transduction Machinery 375

22.4.1 Detection and Quantification of Traffic-induced Phosphorylation Using Antibody-microarrays 375

22.4.2 Detection and Quantification of Traffic-induced Specific Phosphorylation Changes by Immunophosphoproteomic Mass Spectrometry 377

Methods in Cell Biology, Volume 118

ISSN 0091-679X
http://dx.doi.org/10.1016/B978-0-12-417164-0.00022-7

22.5 Discussion ... 378
Summary ... 379
Acknowledgment ... 379
References ... 379

Abstract

Multiple studies have shown that endomembranes can act as signaling platforms for plasma-membrane-originated signaling. In particular, the Golgi complex operates as a relay station for signaling, which is initiated by classical ligand–receptor systems at the plasma membrane, acting as a positive or negative regulator of these plasma-membrane signals. Thus, the Golgi complex has emerged as a hub for intracellular signaling. Furthermore, recent evidence has indicated that the Golgi complex can also trigger its own signaling cascades, which involve some of the molecular players that are classically engaged in signal transduction at the plasma membrane. This aspect of the Golgi complex, namely, the ability to generate autonomous signaling, has been experimentally addressed only in the last few years. These studies have revealed that the transport vesicles that leave the ER for the Golgi complex also carry signal molecules that can then be sensed by a receptor in the Golgi complex to coordinate secretory organelles. The receptor involved in the sensing of incoming traffic at the Golgi complex has been shown to be the KDEL receptor (KDELR), a proposed new G-protein-coupled receptor. Upon binding to a KDEL-containing ligand (a chaperone), the KDELR can activate a signaling cascade that regulates anterograde intra-Golgi trafficking. However, this Golgi-based signaling response is only partially understood to date. Here we report on several approaches that are suitable for the study of traffic-initiated and KDELR-dependent signaling.

Abbreviations

BODIPY	boron dipyrromethene
DMEM	Dulbecco's modified Eagle's medium
DPBS	Dulbecco's phosphate-buffered saline
DTT	dithiothreitol
ER	endoplasmic reticulum
FBS	fetal bovine serum
FRET	Forster resonance energy transfer
GPCR	G-protein-coupled receptor
HEPES	*N*-(2-hydroxyethyl)-piperazine-*N*′-2-ethanesulfonic acid
HRP	horse radish peroxidase
KDELR	KDEL receptor
MOPS	3-(N-morpholino)propanesulfonic acid
PBS	phosphate-buffered saline
PC-I	Pro-collagen I

PFA	paraformaldehyde
PKA	protein kinase A
SFK	Src family kinase
SILAC	stable isotope labeling by amino acids in cell culture
VSVG	G-protein of vesicular stomatitis virus

INTRODUCTION

Intracellular membrane transport involves a well-coordinated engagement of a series of organelles and machineries (Pulvirenti et al., 2008). This endomembrane system includes the endosomes, the endoplasmic reticulum (ER), and the Golgi complex, and in recent years, these have been emerging as novel signaling platforms (Sallese, Pulvirenti, & Luini, 2006). The Golgi complex in particular has received special interest due to its central position in the secretory pathway, as well as for its fascinating and complex structure (Lowe, 2011; Sengupta & Linstedt, 2011). Substantial advances have been made toward our understanding of Golgi function, but some aspects still remain beyond our comprehension (Wilson et al., 2010). Over the last few years, the Golgi complex has emerged as being even more complex than previously thought, with the finding that it is implicated in cellular processes other than traditional trafficking and biosynthetic pathways (Wilson et al., 2010). Thus, the Golgi complex might represent a hub for intracellular signaling (Cancino & Luini, 2013). One specific aspect of Golgi function is the response to the passage of variable amounts of membranes and cargo (Giannotta et al., 2012; Pulvirenti et al., 2008). This aspect of the Golgi, namely the ability to generate autonomous signaling in response to the arrival of cargo from the ER, has been addressed experimentally only in the last few years (Giannotta et al., 2012; Pulvirenti et al., 2008). These studies have revealed that transport vesicles that leave the ER for the Golgi complex also carry signal molecules that can then be sensed by a receptor in the Golgi. Activation of this receptor initiates signaling pathways that activate anterograde trafficking (Pulvirenti et al., 2008). Thus, the functional significance of this regulatory mechanism appears to be to maintain the homeostasis of the Golgi complex, and also to activate intra-Golgi trafficking upon demand; that is, when a traffic load reaches the Golgi. Some of the molecular components of this sensing/signaling system have been identified (Giannotta et al., 2012; Pulvirenti et al., 2008). The receptor involved in sensing the incoming traffic at the Golgi complex has been shown to be the KDEL receptor (KDELR), a seven-transmembrane-domain protein (Scheel & Pelham, 1998; Singh, Tang, Wong, & Hong, 1993) that cycles between the ER and the Golgi and mediates the return to the ER of chaperones that have "escaped" from the ER (Lewis & Pelham, 1990; Semenza, Hardwick, Dean, & Pelham, 1990). Recognition between chaperones and the KDELR is mediated by the KDEL peptide at the C-terminus of the ER chaperones; thus, when chaperones reach the Golgi complex via transport carriers, they can bind to the KDELR and return with it to the ER (Semenza et al., 1990).

Based on molecular modeling data, it has been proposed that the KDELR has a three-dimensional fold that is similar to that of the G-protein-coupled receptors (GPCRs) (Giannotta et al., 2012). The members of this large protein superfamily can couple to heterotrimeric G-proteins at the plasma membrane. Upon binding of their specific ligands, the GPCRs are activated, and can initiate a large variety of signaling pathways via the activation of the G-proteins. It has also been shown that the KDELR interacts (presumably directly) with two major G-proteins at the Golgi complex, G_q and G_s (Giannotta et al., 2012), and that upon binding to a KDEL-containing ligand (a chaperone), the KDELR can activate G_q (and presumably also G_s). G_q, in turn, initiates a tyrosine-phosphorylation cascade that leads to the activation of a Golgi pool of the Src family kinases (SFKs) (Giannotta et al., 2012). Through this SFK activation, there is acceleration of anterograde intra-Golgi trafficking.

It is unlikely that the roles of these signaling pathways triggered by the KDELR are restricted to the regulation of Golgi traffic. Instead, they are probably as complex as those of the GPCRs that operate at the plasma membrane, and they can probably control a similarly vast array of cellular functions (Cancino & Luini, 2013). However, despite their physiological importance, the organizational principles, the full molecular composition, and the flow of information through the endomembrane-based signaling pathways remain poorly understood.

To study these novel endomembrane regulatory circuits, transport synchronization protocols are needed. Transport synchronization assays are based on both endogenous and exogenous synchronizable secretory cargo; for example, pro-collagen I (PC-I) and the G-protein of the vesicular stomatitis virus (VSVG), respectively (Bonfanti et al., 1998; Giannotta et al., 2012; Mironov et al., 2001; Pulvirenti et al., 2008). When cells are exposed to a "restrictive" temperature (40 °C), these proteins cannot fold completely, which results in their ER accumulation and their clearance from the rest of the secretory pathway. This block can be reversed by incubation at the "permissive" temperature (32 °C), which results in correct protein folding, thus allowing these proteins to leave the ER. This "pulse" of secretory cargo then crosses the secretory pathway synchronously. Alternatively, membrane trafficking can be synchronized according to the polymerization/depolymerization of the FK506 binding protein (FKBP)-domain mutant (FM) (Rivera et al., 2000; Volchuk et al., 2000), which is modulated by rapamycin and/or rapamycin analogs, for example, see Bond, Peden, Kendrick-Jones, Sellers, and Buss (2011).

Here, we report several experimental approaches to establish posttranslational modifications, and in particular, phosphorylation (Fig. 22.1). These methods are suitable to resolve intracellular localization and semiquantitative phosphorylation changes of specific signaling molecules. Moreover, quantitative changes of hundreds to thousands of phosphorylated proteins can be achieved through phosphoproteomic methodologies. When combined, all of these methods are helpful to examine this novel Golgi-based endomembrane-initiated signaling, and in particular, KDELR-dependent signaling.

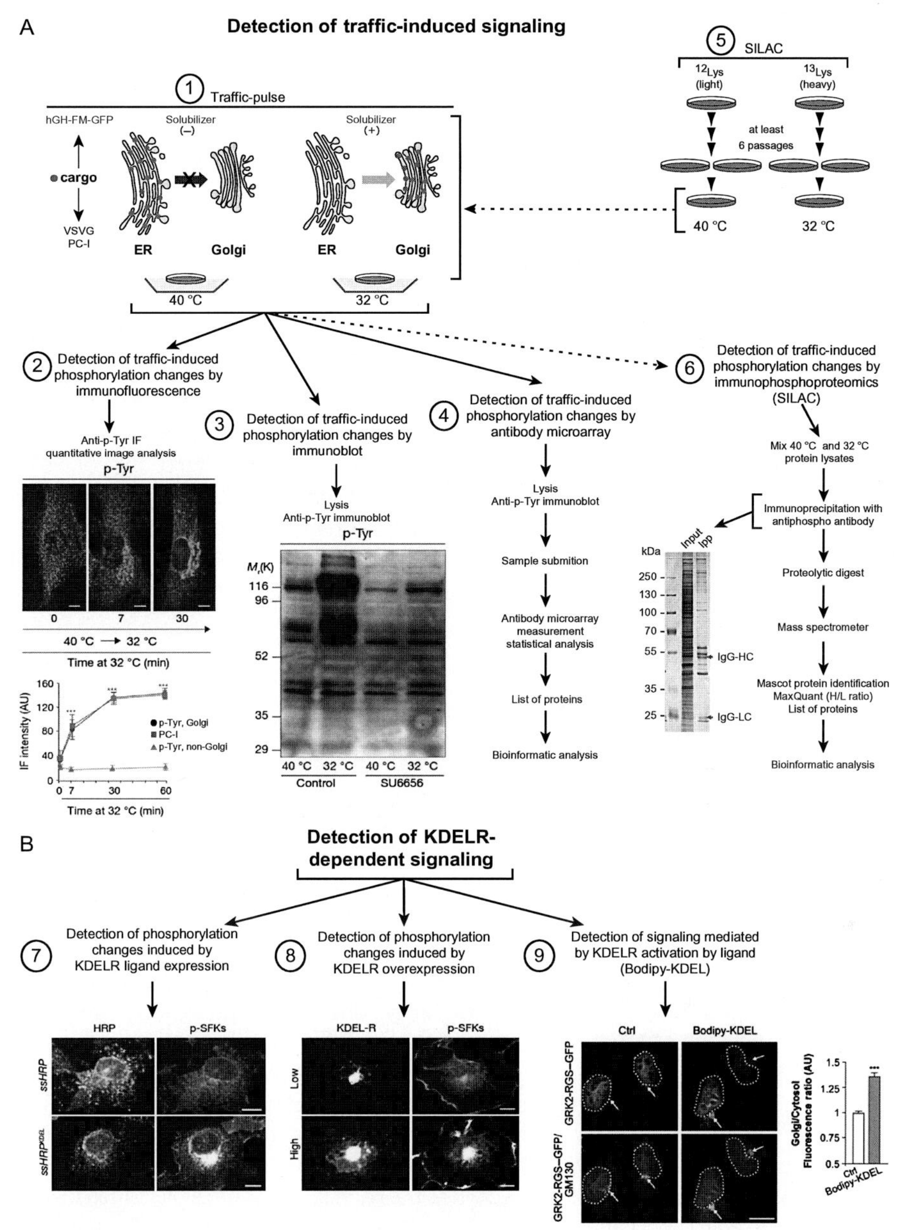

FIGURE 22.1

Workflow of traffic- and KDELR-dependent signaling detection. (A) ER–Golgi transport synchronization can be performed using either sensitive temperature cargoes (VSVG; PC-I) or polymerization/depolymerization conditions of hGH-FM-GFP construct. (1) At 40 °C or in polymerization condition (in the absence of solubilizer), these proteins are not able to fold properly and accumulate within the ER. ER release can be triggered with temperature shift to 32 °C, or by the addition of the D/D solubilizer, resulting in a traffic-pulse from the ER to the Golgi apparatus. Traffic-induced phosphorylation can be detected and quantified with different methods: (2) Immunofluorescence analysis is used to follow signaling activation with

Continued

22.1 GENERATION OF TRAFFIC-DEPENDENT GOLGI SIGNALING: ER-TO-GOLGI TRAFFIC SYNCHRONIZATION PROTOCOLS

The generation of traffic-dependent signaling (a traffic-pulse) at the Golgi is based on the synchronization of the trafficking of specific ER-to-Golgi cargo (Gallione & Rose, 1985; Giannotta et al., 2012; Hirschberg et al., 1998; Pulvirenti et al., 2008) (Fig. 22.1A1). Historically, two main cargo proteins have been used to study the different secretion steps, based on their temperature-dependent folding: VSVG and PC-I/II/IV (Bruckner & Eikenberry, 1984; Gallione & Rose, 1985; Mironov et al., 2001). Recently, it has been shown that traffic pulses can also be generated by controlled secretion of genetically modified cargo proteins through the control of their polymerization/depolymerization states, such as FM, the FKBP-domain mutant (Bond et al., 2011; Giannotta et al., 2012; Pulvirenti et al., 2008; Trucco et al., 2004) (Fig. 22.1A1). The methods mentioned above are detailed in this section.

FIGURE 22.1—Cont'd phospho-specific antibody (e.g., p-Tyr) during temperature shift (e.g., in green in the Golgi area). Quantitative image analysis can be performed to establish relative changes on Golgi phospho-level (Pulvirenti et al., 2008). (3) Immunoblot analysis is used to look for increased phosphorylation pattern after traffic-pulse 32 °C with the identification of proteins, which specifically responds to the cargo arrival to the Golgi (Pulvirenti et al., 2008). (4) Total cell lysate from a traffic-pulse can be analyzed by antibody microarray analysis. Phosphorylation levels of hundreds of important regulatory protein such as kinases, phosphatases, and transcription factors are measured in response to traffic. Specific phosphorylation changes can be assessed by affinity chromatography (Immunoprecipitation, IP) and phosphoproteomics approach. (5) Cells are grown for stable isotope labeling by amino acids in cell culture (SILAC). Two populations of cells are grown separately in either isotopically labeled "light" (^{12}C) or "heavy" (^{13}C) L-lysine till near 100% of isotope incorporation (at least 6 passages) is reached. (6) The cells are subjected to an ER-block and traffic-pulse as shown in (1). 40 °C (^{12}C) and 32 °C (^{13}C) cell lysates are produced and mixed to equal protein quantity. Immunoprecipitation is performed using a phospho-specific antibody in order to enrich samples in phosphoproteins. Immunoprecipitated phosphoproteins are eluted, then digested with trypsin and analyzed by mass spectrometry. Protein identification and H/L ratios are assessed by Mascot and MaxQuant software analysis, respectively. (B) KDELR-dependent signaling activation can be detected upon KDELR activation. (7) Expressing KDELR ligand such as KDEL-bearing soluble secretable HRP (ssHRP-KDEL) construct increases p-SFKs on Golgi (Pulvirenti et al., 2008). (8) Overexpressing the KDELR using KDELR-GFP construct results in p-SFKs-dependent phosphorylation on Golgi (Pulvirenti et al., 2008). (9) Direct KDELR activation by a cell membrane-permeable KDEL-BODIPY peptide resulting in a Gq activation on Golgi (Giannotta et al., 2012). Whatever is the way to activate KDELR, the KDELR-dependent signaling can be detected by quantitative IF or by any of the other methods described above in the text. (See color plate.)

GENERAL MATERIALS

1. DPBS, Phosphate-buffered saline (PBS, pH 7.4) containing NaCl, KCl, Na_2HPO_4, KH_2PO_4 (10 ×, Invitrogen, Cat. No. 14200)
2. PFA 4% (PFA 8%, Electron Microscopy Sciences, Cat. No. RT 157-8) diluted in PBS-CM
3. Secondary alexa-conjugated antibodies (Molecular Probes, Alexa Fluor® 488 Donkey Anti-Mouse IgG (H+L) Cat. No. A21202; Alexa Fluor® 488 Donkey Anti-Rabbit IgG (H+L) Cat. No. A21206; Alexa Fluor® 568 Donkey Anti-Mouse IgG (H+L) Cat. No. A10037; Alexa Fluor® 568 Donkey Anti-Rabbit IgG (H+L) (Cat. No. A10042)
4. DMEM (1 ×), liquid (4.5 g/l D-glucose) (Invitrogen, Cat. No. 11960)
5. Cycloheximide (Fluka, Cat. No. 01810)
6. fetal bovine serum (FBS) (Invitrogen, Cat. No. 160000)
7. Glutamine (Euroclone, Cat. No. ECB3000D)
8. Penicillin plus Streptomycin (Euroclone, Cat. No. ECB 3001D)
9. HEPES (Sigma-Aldrich, Cat. No. H3375)
10. Mowiol (Calbiochem, Cat. No. 475904) or Fluoromount-G (Electron Microscopy Sciences, Cat. No. 17984-25) No. 610823)
11. Saponin (Sigma, Cat. No. 407036)
12. Bovine Serum Albumine (Sigma, Cat. No. A2153)
13. Ammonium Cloride (Sigma, Cat. No. A9434)

22.1.1 Synchronization of PC-I ER-to-Golgi trafficking

Human skin fibroblasts produce and secrete large amounts of PC-I. As described above, PC-I is a temperature-sensitive protein that can be synchronized in a temperature-dependent manner to load up the different organelles through the secretory pathway (Bonfanti et al., 1998; Mironov et al., 2001) (Fig. 22.1A1). For this reason, human skin fibroblasts are a good cell model to study membrane trafficking because they can be easily synchronized and analyzed with different techniques (e.g., immunofluorescence (IF) (Fig. 22.1A2); Western blotting (Fig. 22.1A3); phosphoproteomics (Fig. 22.1A6)).

SPECIFIC MATERIALS

1. Human Skin Fibroblast, obtained from GDB biobank IRCCS Casa Sollievo della Sofferenza, San Giovanni Rotondo, FG 71013, Italy
2. Anti-Collagen I (Calbiochem, Cat. No. CP17)
3. Ascorbic Acid (Sigma-Aldrich, Cat. No. A4544)

To start with, plate human skin fibroblasts (if they are not immortalized cells, plate them between passages 3 and 8) on glass coverslips (10 mm^2; 50,000–70,000 cells) in 24-well plates in DMEM supplemented with 10% FBS, 4 mM L-glutamine, 100 U/ml penicillin, and 100 μg/ml streptomycin, at 37 °C in a 5% CO_2 atmosphere. When the cells are sufficiently confluent (90% confluence), wash them three times with

0.5 ml DMEM, prewarmed to 37 °C, containing 20 mM HEPES, 4 mM L-glutamine, 100 U/ml penicillin, and 100 μg/ml streptomycin, pH 7.4 (DMEM–HEPES) to remove the serum. Then incubate the cells with DMEM–HEPES at 40 °C in a water bath for 3 h, to accumulate newly synthesized PC-I in the ER (Gallione & Rose, 1985; Giannotta et al., 2012; Hirschberg et al., 1998; Pulvirenti et al., 2008). To release the PC-I from the ER, change the medium to 0.5 ml DMEM–HEPES (prewarmed to 32 °C) containing 10% FBS, 100 μg/ml ascorbic acid, and 100 μg/ml cycloheximide. Ascorbic acid acts as a cofactor of the enzyme 4-prolyl-4-hydroxylase by maintaining its catalytic iron atom in the active +2 redox state (Peterkofsky & Udenfriend, 1965); in this way, ascorbic acid supplementation promotes PC-I folding and exit from the ER (Bonfanti et al., 1998). Cycloheximide allows single and synchronized secretion to be followed by blocking new cargo synthesis. These cells can be incubated for the desired times, noting that PC-I usually reaches the Golgi complex in 10–30 min at 32 °C. At the desired times during the cargo release at 32 °C, place the cell plate on ice and wash the cells twice with DPBS (1 ×), pH 7.2, precooled to 4 °C. For IF, fix the cells by adding 0.5 ml 4% PFA in DPBS (1 ×) for 12 min at room temperature. Wash the cells three times with DPBS (1 ×) to completely remove the PFA, and incubate the cells in blocking/permeabilization solution (0.05% saponin, 0.5% BSA, 50 mM NH_4Cl, in DPBS (1 ×)) for 45 min at room temperature. Incubate the cells overnight at 4 °C with an anti-PC-I antibody (HC-I; Daichii Pure Chemicals) diluted 1:100 in blocking/permeabilization solution. To monitor PC-I arrival at the Golgi complex, co-stain the cells for a Golgi marker protein (e.g., an antibody against the *cis*-Golgi marker GM130). The day after, wash the cells with DPBS (1 ×) three times, and then incubate them for 1 h at room temperature with an appropriate fluorophore-conjugated secondary antibody (e.g., Alexa Fluor 488-conjugated or Alexa Fluor 568-conjugated secondary antibodies; diluted 1:400 in blocking/permeabilization solution; Molecular Probes). Wash the cells three times with DPBS (1 ×) and mount the coverslips on microscope slides in mowiol or fluoromount. We suggest that cell staining is to examine using an inverted confocal microscope. For biochemical studies, add cell lysis buffer after the DPBS (1 ×) washing step as described in Section 22.2.2.

22.1.2 VSVG traffic synchronization

The VSVG protein can be expressed in cells upon infection with the vesicular stomatitis virus (VSV). Biosafety Level 2 practices and facilities are recommended for activities utilizing laboratory-adapted strains of demonstrated low virulence. A temperature-sensitive mutant VSV strain (ts045) is widely used to study membrane trafficking in mammalian cells (Beckers et al., 1987; Bergmann, 1989; Doms, Keller, Helenius, & Balch, 1987; Hirschberg et al., 1998; Kreis and Lodish, 1986). The VSVG produced by the infected cells does not fold properly in the ER at the "restrictive" temperature (40 °C), which results in a block of ER-to-Golgi transport. At the "permissive" temperature (32 °C), the VSVG can fold and leave the ER, thus moving to the Golgi complex and from the Golgi complex to the plasma membrane (Doms et al., 1987; Hirschberg et al., 1998). The VSVG synchronization protocol is quite similar to the PC-I synchronization protocol.

SPECIFIC MATERIALS

1. Anti-VSVG P5D4 antibody (Sigma; Cat. No. V4888)
2. tsO45 (V) mutant of VSV

Plate the cells on glass coverslips (10 mm^2; 50,000–70,000 cells) in a 24-well multiwall plate. The cell confluence is crucial for efficiency of infection. If the cells are too confluent, they will be poorly infected, although with cells at low confluence, they can detach during the temperature shift from 40 to 32 °C. Remove the FBS completely from the cells by washing them 3–5 times with 0.5 ml DMEM–HEPES, prewarmed to 32 °C. Removal of the FBS is necessary because it can interfere with the viral infection, with 32 °C allowing good cell infection. Infect the cells by the addition of VSV ($>10^8$–10^9 PFU/ml) diluted 1:20 in DMEM–HEPES. This dilution is generally sufficient to ensure 90% infection efficiency, considering that the VSV has a high titer. Incubate the cells with 0.5 ml of this VSV diluted in DMEM–HEPES for 1 h at 32 °C. After infection, completely remove the VSV by washing the cells 3–5 times with 0.5 ml DMEM–HEPES with 10% FBS prewarmed to 40 °C, and incubate the cells for 3 h to accumulate VSVG in the ER. To allow the VSVG to fold, with its consequent ER export and Golgi arrival, incubate the cells in 0.5 ml DMEM–HEPES containing FBS prewarmed to 32 °C for 10–30 min. In this time range, the cargo reaches the Golgi complex. The VSVG transport can be followed by IF using the IF protocol described in the Section 22.1.1. To stain the VSVG, incubate the cells with the anti-VSVG antibody (P5D4; Sigma; catalog number, V4888) diluted 1:100 in blocking/permeabilization solution. Another way to express the VSVG is through VSVG cDNA transfection. A GFP tag is added to the cytoplasmic tail of VSVG, which makes it also suitable for transport studies in living cells (Presley et al., 1997).

22.1.3 ER-retention and release of cargo proteins by polymerization/depolymerization

ER-retention can be achieved by polymerization/ depolymerization of genetically modified cargo proteins. This method is based on a fusion of cargo proteins with FM, the FKBP-domain mutant (Volchuk et al., 2000) (Fig. 22.1A1). The transfected FM–cargo fusion protein polymerizes and cannot leave the ER until the cells are treated with the FM ligand, which is a synthetic derivative of rapamycin that acts as a solubilizer (D/D solubilizer). Addition of the solubilizer allows the synchronized release of the selected cargo protein from the ER.

SPECIFIC MATERIALS

1. D/D Solubilizer (Clontech, Cat. No. 635054)
2. TransIT®-LT1 Transfection Reagent (Myrus, Cat. No. MIR 2300)
3. Plasmid encoding the hGH-FM–GFP fusion protein

Plate HeLa cells at 70,000 cells/cm^2 in a 24-well multiwall dish. The day after, the cells are transfected using the TransIT®-LT1 Transfection Reagent (according to the manufacturer instructions) with 0.5 μg plasmid encoding the FM–cargo fusion protein. The transfected cells are incubated at 37 °C with 5% CO_2 for 24/48 h. Then wash the cells three times with DMEM–HEPES, replace this with DMEM–HEPES, and leave the cells to equilibrate in a water bath to 37 °C. To release the cargo protein, change the culture media for prewarmed (37 °C) DMEM–HEPES with 50 μg/ml cycloheximide and 50 μM (DD solubilizer). As an example, hGH-GFP-FM reaches the Golgi in 5–10 min. This arrival of hGH-GFP-FM at the Golgi complex is monitored by IF as detailed in the Section 22.1.1.

22.2 MORPHOLOGICAL AND QUANTITATIVE ANALYSIS OF TRAFFIC-GENERATED GOLGI-BASED SIGNALING

Signaling cascades that are generated by the arrival of a traffic-pulse at the Golgi complex need to be analyzed by morphological approaches to establish their intracellular response localization. Thus, the phosphorylation state and localization are usually detected using fluorescence microscopy (Fig. 22.1A2) and biochemical analysis (Fig. 22.1A3). However, large-scale identification of signaling cascades and their substrates requires identification and quantitative evaluation of their phosphorylation state. This phosphorylation state can now be addressed by quantitative post-translational protein modification using an antibody microarray (Fig. 22.1A4) (Ivanov et al., 2004) and/or by mass spectrometer analysis; for example, using the stable isotope labeling by amino acids in cell culture (SILAC) method (Fig. 22.1A5 and A6) (Ong et al., 2002) (see below).

GENERAL MATERIALS

1. DMEM (1 ×), liquid (4.5 g/l D-glucose) (Invitrogen, Cat. No. 11960)
2. Cycloheximide (Fluka, Cat. No. 01810)
3. FBS (Invitrogen, Cat. No. 160000)
4. Glutamine (Euroclone, Cat. No. ECB3000D)
5. Penicillin plus Streptomycin (Euroclone, Cat. No. ECB 3001D)
6. HEPES (Sigma-Aldrich, Cat. No. H3375)
7. DPBS, PBS (pH 7.4) containing NaCl, KCl, Na_2HPO_4, KH_2PO_4 (10 ×, Invitrogen, Cat. No. 14200)
8. Bovine Serum Albumin (BSA, Sigma, Cat. No. A9418)
9. Saponin (Sigma, Cat. No. 47036)

22.2.1 Detection and quantification of Traffic-induced phosphorylation by IF microscopy

The following procedures use a Leica SP5 inverted confocal microscope system (Leica, Germany), although they apply to any confocal microscope system. The quantitative fluorescence analysis is performed on a workstation running the MetaMorph or ImageJ software (Fig. 22.1A2).

SPECIFIC MATERIALS

1. PBS-CM, PBS (pH 7.4) containing NaCl, KCl, Na_2HPO_4, KH_2PO_4, 0.1 mM $CaCl_2$, 1 mM $MgCl_2$
2. PFA 4% (PFA 8%, Electron Microscopy Sciences, Cat. No. RT 157-8) diluted in PBS-CM
3. Anti-p-SFK (Cell Signaling, Cat. No. Y416)
4. Anti-p-Tyr-100 (Cell Signaling, Cat. No. 94115)
5. Anti-GM130 (BD Transduction Laboratories; Cat. No. 610823)
6. Secondary alexa-conjugated antibodies (Molecular Probes, Alexa Fluor® 488 Donkey Anti-Mouse IgG (H+L) Cat. No. A21202; Alexa Fluor® 488 Donkey Anti-Rabbit IgG (H+L) Cat. No. A21206; Alexa Fluor® 568 Donkey Anti-Mouse IgG (H+L) Cat. No. A10037; Alexa Fluor® 568 Donkey Anti-Rabbit IgG (H+L) Cat. No. A10042)
7. Mowiol (Calbiochem, Cat. No. 475904) or Fluoromount-G (Electron Microscopy Sciences, Cat. No. 17984-25)

At the desired times during the traffic-pulse (from 10 to 30 min), the cells can be processed for IF, by washing them two times with PBS-CM, and then fixing them by incubation with 4% PFA in PBS-CM for 25 min at room temperature. Then wash the cells with PBS-CM three times, and incubate them with the blocking buffer (5% BSA, 0.1% saponin) for 15 min at room temperature. Incubate the cells with the desired antibody (against p-Src, p-Tyr) diluted in blocking buffer for 1–2 h at room temperature; for example, using an anti-p-Tyr antibody (1:100, v/v). To use the Golgi area as a reference for quantification, co-stain the cells for the Golgi marker GM130. Then incubate the cells with the appropriate secondary Alexa-conjugated antibodies, and mount the coverslip using mowiol or fluoromount as the mounting media. After the mounting, examine the cells under 63 × oil immersion, with the inverted confocal microscope system. For image acquisition and to collect the maximum fluorescent signal set the pinhole completely open. This will reduce the variations in the total fluorescence intensity due to different focal planes between individual cells on the field. Acquire the unsaturated images (pixel fluorescence below 255 arbitrary units), using the maximum intensity obtained in a positive control experiment as the reference. Use 1024 × 1024 pixels image resolution, and 4 averages scanning. Use the same acquisition settings for all of the samples, to allow direct comparisons between them. Acquire images from 30 to 50 cells per time point per experiment. Use the acquired images to quantify the IF signal of interest in the Golgi area by selecting the region of interest according to the reference labeling (e.g., GM130, giantin). Use the threshold function to select the specific signal and to eliminate any background, and then apply the same threshold level to all of the images. To monitor the saturated pixels, use the upper threshold tool to highlight saturated areas, and avoid these in the quantification. Export the average and total intensity data to an Excel sheet using the export function. The total intensity (or integrated intensity) is the more accurate mode for determining increases/decreases in the phosphorylation state because it is independent of morphology changes. Average the total fluorescence intensity across all of the cells analyzed (Fig. 22.1A2).

22.2.2 Detection of traffic-induced phosphorylation by Western blotting

Protein phosphorylation is one of the most important posttranslational modifications in nature. The addition/detachment of a phosphate group on a protein is the regulatory signal for many important cellular processes, such as metabolism, homeostasis, and cell communication and proliferation. Unfortunately, the protein phosphorylation state in cells is very unstable, and for this reason it is very difficult to study. Inhibition of proteases and phosphatases together with the low temperature are considered to be good ways to preserve as many of the phosphate groups on a peptide as possible. In this context, if there is the need to detect phosphorylation biochemically, the preparation of total lysates becomes a crucial step (Fig. 22.1A3).

SPECIFIC MATERIALS

1. 3-(*N*-morpholino)propanesulfonic acid (MOPS Sigma, Cat. No. M9381) (1 M, pH 7.0, stock)
2. Ethylenediaminetetraacetic acid (EDTA Sigma, Cat. No. E4884) (500 mM stock)
3. Ethylene glycol tetraacetic acid (EGTA Sigma, Cat. No. E3889) (500 mM stock)
4. Triton X-100 (Sigma, Cat. No. X100)
5. Sodium Fluoride (NaF) (Sigma-Aldrich, Cat. No. 450022)
6. Sodium Ortovanadate (Na_3VO_4) (Sigma-Aldrich, Cat. No. 450243)
7. β-Glycerophosphate (Sigma-Aldrich, Cat. No. G9422)
8. Dithiothreitol (DTT) (Invitrogen, Cat. No. D1532)
9. Protease cocktail inhibitors (Roche, Cat. No. 05 056 489 001)
10. PhosSTOP (Roche, Cat. No. 04 906 837 001)
11. Luminata crescendo (Millipore, Cat. No. WBLUR0500)
12. Anti-rabbit HRP-conjugated (Calbiochem; Cat. No. 401215)
13. Anti-Phospho-(Ser/Thr) PKA Substrate (Cell Signaling, Cat. No. #9621)
14. Potran nitrocellulose, 0.45 μm pore size (Whatman, Cat. No. BA85)
15. Pierce-BCA Protein Assay Kit (Pierce, Cat. No. #23225)
16. Sodium dodecylsulfate (SDS Sigma, Cat. No. L3771)
17. Glycerol (Sigma, Cat. No. G5516)
18. Tris(hydroxymethyl)aminomethane (Tris, Sigma Cat. No. 252859)
19. Bromophenol blue (Sigma, Cat. No. B0126).

Protease and phosphatase inhibitors must be used, by adding them into the lysis buffer and into all of the procedures that follow, while all of the procedures must be carried out strictly at 4 °C. During cargo accumulation prepare cell lysis buffer, we are having very good results using MOPS 1 M at pH 7.0 at the final concentration of 20 mM; EGTA 0.5 M at the final concentration of 2 mM; EDTA 0.5 M at the final concentration of 5 mM; 1% Tryton, anyway whatever is the lysis buffer used, add phosphatase and protease inhibitors immediately before to use: 30 mM Sodium Fluoride (NaF); 5 mM, 60 mM β-glycerophosphate, 1 mM of DTT, phosphatases cocktail inhibitors, PhosSTOP.

At the end of the traffic-pulse, place the cell plate on ice and wash the cells three times with DPBS (1 ×), prechilled to 4 °C. Add the cold lysis buffer with all of the phosphatase and protease inhibitors, and detach the cells from the plate using a cell scraper. Collect the cell lysate into a low protein binding 1.5-ml microcentrifuge tube (Eppendorf, Cat. No. 022431081) and disrupt the cells with sonication for three times for 10 s each at 50% power (Misonix, Microson XL Cell Disruptor). Any delays during this cell harvesting and sample processing are deleterious to the accurate and consistent detection of phosphorylation. Centrifuge the cell lysate at 14,000 rpm for 5 min (microcentrifuge) at 4 °C, and carefully transfer the supernatant into a new 1.5-ml low protein-binding tube. Take an aliquot of the supernatant for the determination of protein concentration (e.g., using BCA protein assay kits (Pierce), according to the manufacturer instructions). Dilute the Laemmli buffer (5 ×: 10% SDS, 50% glycerol, 300 mM Tris–HCl, pH 6.8, 0.05%, w/v, bromophenol blue), into each sample, and incubate at 99 °C for 3 min. Run the required SDS-polyacrilamide gel electrophoresis (PAGE) gels, and transfer the protein to Protran nitrocellulose hybridization transfer membranes (0.2-μm pore size; Whatman). To saturate the nitrocellulose membranes, incubate them with 5% (w/v) bovine serum albumin in 150 mM NaCl, 10 mM Tris–HCl, pH 7.4, 0.05% Tween-20 (saturation solution) for 1 h at room temperature. The use of fat-free milk saturation needs to be avoided here as it contains phosphorylated casein, which interferes with phosphoprotein detection. Incubate the membranes with the anti-phospho-(Ser/Thr) PKA substrate antibody, or any phospho-antibody, at 1:1000 dilution in saturated solution, overnight at 4 °C. The day after, incubate with the secondary HRP-conjugated anti-rabbit antibody (1:10,000), for 1 h at room temperature. To detect the HRP signal use any ECL system; we obtain good results using Luminata crescendo (Millipore). For quantitative analysis, digital image acquisition (using ChemiDoc) is recommended. Use several acquisition times and select nonsaturated images for quantification. The gel analysis tools provided by ImageJ (free software from National Institutes of Health, USA) provide useful and easy methods of quantification.

22.3 ANALYSIS OF KDELR-DEPENDENT SIGNALING

The main difficulty for the study of KDELR-dependent signal-transduction mechanisms is that the KDELR ligand-binding domain is in the lumen of the secretory organelles. Overexpression of the KDELR leads to its autoactivation and redistribution to the ER, with the consequent SFK activation in the Golgi area (Giannotta et al., 2012; Pulvirenti et al., 2008) (Fig. 22.1B8). Alternatively, overexpression of a synthetic KDEL-containing polypeptide (e.g., ssHRP^{KDEL}) can stimulate the KDELR (Connolly et al., 1994). ssHRP lacking the KDEL motif provides the control (Fig. 22.1B7). Moreover, recent studies have demonstrated that KDEL peptides conjugated to boron dipyrromethene (BODIPY) can cross biological membranes and overcome this limitation (Giannotta et al., 2012; Pap, Dansen, & Wirtz, 2001) (Fig. 22.1B9). The selection of the activation protocols will depend on requirements of either chronic (overexpression, ssHRP^{KDEL}) or acute (KDEL-BODIPY) KDELR activation.

GENERAL MATERIALS

1. DMEM (1 ×), liquid (4.5 g/l D-glucose) (Invitrogen, Cat. No. 11960)
2. DPBS, PBS (pH 7.4) containing NaCl, KCl, Na_2HPO_4, KH_2PO_4 (10 ×, Invitrogen, Cat. No. 14200)
3. PFA 4% (PFA 8%, Electron Microscopy Sciences, Cat. No. RT 157-8) diluted in PBS-CM
4. FBS (Invitrogen, Cat. No. 160000)
5. Glutamine (Euroclone, Cat. No. ECB3000D)
6. Penicillin plus Streptomycin (Euroclone, Cat. No. ECB 3001D)
7. HEPES (Sigma-Aldrich, Cat. No. H3375).

22.3.1 Activation of the KDELR signaling cascade by receptor overexpression

SPECIFIC MATERIALS

1. Plasmid encoding the KDLER-GFP
2. TransIT®-LT1 Transfection Reagent (Myrus, Cat. No. MIR 2300).

Plate HeLa cells as described in Section 22.1.3. Transfect the cells with a GFP-tagged KDELR1 construct, and incubate for 16–24 h. Quantify the selected p-antibody levels by IF in cells with high levels of the KDELR in the ER. The KDELR in the ER is recognized by bright green fluorescence in a ring around the nucleus, as the nuclear envelope. This pattern of KDELR distribution is a *bona fide* indicator that a portion of the total KDELR has been activated (Fig. 22.1B8).

22.3.2 Activation of the KDELR signaling cascade using a genetically Encoded artificial ligand

SPECIFIC MATERIALS

1. Plasmid encoding the ssHRP control
2. Plasmid encoding the ssHRP-KDEL
3. HRP monoclonal antibody (Abcam; Cat. No. ab2110)
4. Anti-KDELR monoclonal antibody (clone KR-10; Stressgen Biotechnology; Cat. No. ADI-VAA-PT048-D)
5. TransIT®-LT1 Transfection Reagent (Myrus, Cat. No. MIR 2300)
6. Saponin (Sigma, Cat. No. 47036)
7. Bovine serum albumin (BSA, Sigma, Cat. No. A9418)
8. Ammonium chloride (Sigma, Cat. No. A9434).

Transfect HeLa or Cos7 cells with ssHRP (control) or $ssHRP^{KDEL}$, and incubate them at 37 °C for 16–24 h; then process the cells for IF as described in Section 22.1.1. Stain the HRP to identify transfected cells, using an anti-HRP antibody diluted 1:100 in blocking/permeabilization solution (0.05% saponin, 0.5%

BSA, 50 mM NH_4Cl, in DPBS (1 ×)), for 2 h at room temperature. Stain the cells using the selected primary antibody. Stain a parallel set of cells for the KDELR using an anti-KDELR monoclonal antibody (Stressgen) diluted 1:100 in blocking/permeabilization solution (0.05% saponin, 0.5% BSA, 50 mM NH_4Cl, in DPBS (1 ×)), for 2 h at room temperature. This procedure is necessary to monitor $ssHRP^{KDEL}$-dependent KDELR redistribution. Quantify the selected p-antibody level changes in the Golgi area by IF (Fig. 22.1B7).

22.3.3 Activation of KDELR signaling by acute and specific ligand Incubation with fluorescent Membrane-permeant peptides

Several fluorescent BODIPY derivatives are available, and the selection of the specific wavelength excitation/emission will depend on the experimental requirements. In our experience, $BODIPY^{630/650}$ (BODIPY® 630/650-X, SE, 6-(((4,4-difluoro-5-(2-thienyl)-4-bora-3a,4a-diaza-s-indacene-3-yl)styryloxy)acetyl)aminohexanoic acid, succinimidyl ester) is a good choice, due to the longer linker (additional seven-atom aminohexanoyl spacer, X) between BODIPY and the conjugated peptide, the high fluorescent quantum yield, and the long wavelength excitation/emission, which avoids fluorescent signal "bleeding" with the commonly used wavelength (488–568 nm). The $BODIPY^{630/650}$ derivative also allows Forster resonance energy transfer (FRET) measurements with the most commonly used FRET couple (CFP/YFP) and mcherry-VSVG recording.

SPECIFIC MATERIALS

1. BODIPY630/650,6-(((4,4-difluoro-5-(2-thienyl)-4-bora-3a,4a-diaza-s-indacene-3-yl)styryloxy)acetyl)aminohexanoic acid, succinimidyl ester (630/650 nm) (Invitrogen, Cat. No. D10000)

22.3.3.1 Labeling of Membrane-permeable peptides

The KDEL peptide conjugation with BODIPY derivatives is made as follows.

SPECIFIC MATERIAL

1. CFFKDEL (KDEL) and CFFKDEA (KDEA) peptides
2. *N,N*-Dimethylformamide (DMF; Sigma Cat. No. 227056)
3. Toluene (Sigma, Cat. No. 650579)
4. Acetonitrile (CH_3CN, Sigma, Cat. No. 34998)
5. Trifluoroacetic acid (Sigma, Cat. No. 91707)
6. Potassium carbonate (Sigma, Cat. No. 367877).

Two peptides are needed: the KDEL peptide (NH_2-CFFKDEL-COOH), and the signaling inactive KDEA peptide (NH_2-CFFKDEA-COOH). These KDEL and KDEA peptides are synthesized using standard automated solid-phase methods, then purified by reverse-phase HPLC, and analyzed by matrix-assisted laser-desorption/ionization–time-of-flight (MALDI-TOF) mass spectrometry. The labeling reaction is performed as follows: 1.0 mg of each lyophilized peptide (KDEL and KDEA) is weighted out and dissolved in 0.100 ml anhydrous DMF. Weight out 27.6 mg K_2CO_3 and dissolve it in 2.0 ml anhydrous DMF, for a 100 mM solution. Then add 20 μl (2 base equivalents) of this solution to freshly prepared peptide solutions that were made beforehand. A small amount of the reactive BODIPY ester derivative (dye) is picked from the vial with a pipette tip and dissolved in 20 μl anhydrous DMF. The concentration of this BODIPY dye stock solution in DMF is measured by standard UV–VIS measurements, diluting 1 μl of this DMF solution of fluorophore in 0.5 ml PBS. The absorbance is measured at 630–650 nm, and the concentration is calculated using the molar extinction coefficient provided by the manufacturer (Invitrogen). Add 4.5 μl of the reactive dye (33 μmol, stoichiometric ratio peptide:fluorophore 1:30) rapidly from the stock DMF solution to each reaction mix containing the base and peptide. The 30-fold high molar excess of fluorophore ensures the completeness of the labeling reaction. The mixture is allowed to react overnight in the dark at room temperature, with occasional vortexing. The solvent is dried by rotary evaporation and vacuum azeotropic codistillation with toluene. The separation of the unbound dye and the nonfluorescent peptide from the labeled peptide is performed by HPLC chromatography, using a reverse-phase C18 analytical column with a linear CH_3CN/H_2O/trifluoroacetic acid gradient, (flow rate, 1 ml/min; 50 min CH_3CN 0% → 100%). We use a Dionex Ultimate 3000 HPLC chromatographic system equipped with a multichannel UV reader. The elution is monitored by absorbance at 220 nm (the peptide bond) and 588 nm (the dye absorption maximum).

Note: The first peak that absorbs at 588 nm usually has the same retention time as the free dye, and thus this is discarded. The next peak at 588 nm (with a longer retention time) is collected and dried under vacuum centrifugation. The last peak at 588 nm is discarded, as this is attributed to the double-labeled peptide.

The conjugated peptide concentration is calculated on the assumption that the fluorescent conjugate keeps the same extinction coefficient as the free dye.

22.3.3.2 Activation of KDELR signaling by KDEL-BODIPY fluorescent membrane-permeant peptide

SPECIFIC MATERIAL

1. KDEL-BODIPY peptides
2. KDEA-BODIPY peptides

Plate the HeLa cells as described in Section 22.1.3. Prepare 3 μM solutions of KDEL-BODIPY or KDEA-BODIPY (inactive) (Giannotta et al., 2012) peptides in DMEM–HEPES, and warm the mixture to 37 °C. Wash the cells three times with prewarmed (37 °C) DMEM–HEPES, and replace this with KDEL-peptide- or KDEA-peptide-containing medium. Incubate the cells at 37 °C for the desired times. Generally, 15–30 min are sufficient for activation of the KDELR signaling cascade (Giannotta et al., 2012). Fix and process the cells for IF, and follow KDELR activation by monitoring its redistribution into the ER.

22.4 THE KDELR TRANSDUCTION MACHINERY

Over the last few decades, great efforts have been made to identify and characterize posttranslational modifications, among which phosphorylation is probably the most important. Because of its dynamics, phosphorylation events occur in a sub-stoichiometric manner, making its detection difficult to reveal and quantify. Phosphoproteomics is the branch of proteomics that uses the dynamics of phosphorylation/dephosphorylation to quantitatively unravel phosphorylation events triggered by kinases in a signaling network. As detailed above, KDELR-dependent signaling can be triggered by chronic and acute activation. Analysis of the downstream signaling pathways can be addressed through high-content analysis, by looking for phosphorylation changes of signaling molecules or substrates. Two methods are available that allow the analysis of important signaling pathways and/or specific kinase cascades. Antibody microarrays can be used to analyze well-known signaling molecules, and this provides the relative quantitative results with information about posttranslational modifications of well-known regulatory phosphorylation sites on these molecules (Ivanov et al., 2004) (Fig. 22.1A4). On the other hand, SILAC (Fig. 22.1A5) (Ong et al., 2002) can be used to determine quantitative changes in the posttranslational modification (phosphorylation) of substrates in specific pathways when it is combined with phospho-specific kinase motif antibody substrate immunoprecipitation (Fig. 22.1A5 and A6). Both of these methods need to be combined with bioinformatics tools for pathway reconstruction, such as the Keck pathway, DAVID ontology, and/or gene-set enrichment analysis. Different techniques are used to study the global phosphoproteome, and we will concentrate on those mentioned here.

GENERAL MATERIALS

1. DMEM (1 ×), liquid (4.5 g/l D-glucose) (Invitrogen,Cat. No. 11960)
2. DPBS, PBS (pH 7.4) containing NaCl, KCl, Na_2HPO_4, KH_2PO_4 (10 ×, Invitrogen, Cat. No. 14200)
3. FBS (Invitrogen, Cat. No. 160000)
4. Glutamine (Euroclone, Cat. No. ECB3000D)
5. Penicillin plus Streptomycin (Euroclone, Cat. No. ECB 3001D)
6. HEPES (Sigma-Aldrich, Cat. No. H3375).

22.4.1 Detection and quantification of Traffic-induced phosphorylation using antibody-microarrays

In recent years, Kinexus (Canada) has provided an interesting antibody microarray that can give information about the expression and phosphorylation levels of the main regulatory and signaling molecules. Proteins from cell lysates are labeled with different dyes, and the antibody chip is incubated with these dye-labeled proteins. Traffic-induced (Fig. 22.1A1, B7, B8, and B9) or KDELR-dependent phosphorylation changes can be determined by combining the experimental approaches mentioned above with quantitative phosphorylation analysis (Fig. 22.1A4–A6).

SPECIFIC MATERIALS

1. 3-(*N*-morpholino)propanesulfonic acid (MOPS Sigma, Cat. No. M9381) (1 M, pH 7.0, stock)
2. Ethylenediaminetetraacetic acid (EDTA Sigma, Cat. No. E4884) (500 mM stock)
3. Ethylene glycol tetraacetic acid (EGTA Sigma, Cat. No. E3889) (500 mM stock)
4. Triton X-100 (Sigma, Cat. No. X100)
5. Sodium Fluoride (NaF) (Sigma-Aldrich, Cat. No. 450022)
6. Sodium Ortovanadate (Na_3VO_4) (Sigma-Aldrich, Cat. No. 450243)
7. β-Glycerophosphate (Sigma-Aldrich, Cat. No. G9422)
8. DTT (Invitrogen, Cat. No. D1532)
9. Protease cocktail inhibitors (Roche, Cat. No. 05 056 489 001)
10. PhosSTOP (Roche, Cat. No. 04 906 837 001)
11. Luminata crescendo (Millipore, Cat. No. WBLUR0500)
12. Anti-rabbit HRP-conjugated (Calbiochem; Cat. No. 401215)
13. Anti-Phospho-(Ser/Thr) PKA substrate (Cell Signaling, Cat. No. #9621)
14. Potran nitrocellulose, 0.45 μm pore size (Whatman, Cat. No. BA85)
15. Pierce-BCA Protein Assay Kit (Pierce, Cat. No. #23225)
16. Sodium dodecylsulfate (SDS Sigma, Cat. No. L3771)
17. Glycerol (Sigma, Cat. No. G5516)
18. Tris(hydroxymethyl)aminomethane (Tris, Sigma Cat. No. 252859)
19. Bromophenol blue (Sigma, Cat. No. B0126).

We plate 2×10^6 HFs cells on 10 cm, this step differs along different cell lines so plate the cells in a way that the day after they reach 80%/90% of confluence. Traffic-pulse or KDELR activation is performed as described in Section 22.1.1 and in Section 22.3 (Fig. 22.1A1, B7, B8, and B9), then we prepared cell lysate according to the precise manufactures instruction:

- Remove medium from culture dishes containing;
- Rinse the cells twice with ice-cold PBS to remove medium residue (serum must be completely removed from cells);
- remove as much PBS as possible after the last rinse;
- Add 100 μl ice-cold lysis buffer as described in Section 22.2.2 to 10 cm culture dish per sample (more lysis buffer can be added if cells are concentrated);
- Scrape the cells in lysis buffer, collect the cell suspension from the dishes and transfer it into a 1.5-ml microcentrifuge tube;
- Sonicate four times for 10 s each time with 10–15 s intervals on ice to rupture the cells and to shear nuclear DNA;
- this step is crucial and cannot be omitted;
- Centrifuge the homogenate at $90{,}000 \times g$ or higher for 30 min at 4 °C in a Beckman Table Top TL-100 ultracentrifuge, Beckman Airfuge or equivalent;
- Transfer the resulting supernatant fraction to a 1.5-ml microcentrifuge tube;
- Adjust the protein lysate concentration, which should be concentrated 2 mg/ml.

Hundred microgram of the protein lysate concentrated 2 mg/ml is frozen at −80 °C and kept separate. Part of the cell lysates are monitored for increased phosphorylation in response to treatment, using anti-p-Tyr or anti-PKA-substrate antibodies, by Western blotting as described in Section 22.2.2. After sample submission, measurement, and statistical analysis, the proteins with increased percentages of changes and with Z-ratio values ≥ 1 are selected for further tests, according to normalized and statistically relevant changes. The pitfalls using this method include false positives and false negatives; chemical cleavage of the protein sample can reduce these. All the results obtained should also be confirmed by Western blotting.

22.4.2 Detection and quantification of traffic-induced specific phosphorylation changes by immunophosphoproteomic mass spectrometry

The SILAC approach is based on the metabolic incorporation of "light" or "heavy" forms of amino acids into proteins (Fig. 22.1A5). Thus, two cell populations are grown in culture media that are identical, except that one of them contains a particular light amino acid and the other the heavy form of this amino acid (Ong et al., 2002). This method allows quantitative measurements of relative changes in the abundance of thousands of proteins in complex samples. When this is combined with phosphoprotein enrichment, the phosphorylation changes of thousands of proteins can be determined, including site-specific changes.

The most used techniques to enrich these samples in phosphopeptides are antibody-based affinity chromatography (Lee & Lee, 2004), immobilized metal–ion chromatography (Porath, 1992; Yip, Tam, Kong, Belew, & Porath, 1985), and metal–oxide affinity chromatography (Wolschin, Wienkoop, & Weckwerth, 2005), to target phosphorylated peptides. Recently, good data for phospho-peptide enrichment have been obtained using metal-oxides columns, such as with TiO_2 and ZrO_2 (Engholm-Keller & Larsen, 2011; Thingholm, Jorgensen, Jensen, & Larsen, 2006). The method described below is based on affinity chromatography phosphoprotein enrichment (Fig. 22.1A6).

SPECIFIC MATERIAL

1. SILAC (Invitrogen, Cat. No. SP10001)
2. Urea (Applichem, Cat. No. 57-13-6)
3. Dynabeads (Invitrogen, Cat. No. 10007D)
4. Ammonium Bicarbonate (Fluka, Cat. No. 40867)
5. Anti-Phospho-(Ser/Thr) PKA Substrate (Cell Signaling, Cat. No. #9621)
6. Trypsin (Promega, Cat. No. V5111).

We selected SILAC from Invitrogen, which uses L-lysine that is isotopically labeled with nonradioactive heavy carbon (^{13}C) and light carbon (^{12}C) isotopes. In this way, the light and heavy amino acids are incorporated into all of the newly synthesized proteins in two separate cell populations. After a number of cell divisions, there is the need to enrich the incorporation of the amino acid into the cell proteins. Traffic-pulse experiments are performed as described in Section 22.1.1, with the protein lysates collected from cells at 40 and 32 °C, separately, as previously described in Section 22.2.2. Then the lysates from the light and heavy cell incubations are mixed, with equal protein quantities, and subjected to affinity chromatography using phospho-PKA substrate and dynabeads (this method is suitable for any phospho-antibody). We incubate antibody (10 μg) with the dynabeads for 10 min at Room Temperature, then we washed three times with the washing buffer and we incubate 3 mg of total protein lysate (H + L) with the dynabeads 10 min at 4 °C, we washed three times and we eluted protein with 8 M Urea plus 0,4 M ammonium bicarbonate 30 min at RT. The proteins are then proteolytically digested using trypsin and subjected to mass spectrometry analysis. Identification of the proteins from the primary sequences in the mass spectrometry database is carried out using the MASCOT software (Perkins, Pappin, Creasy, & Cottrell, 1999). The relative peptide abundances in the samples at 40 and 32 °C (as the heavy to light carbon [H/L] ratios) are calculated from raw data using the MAX-QUANT free software (Cox & Mann, 2008). An H/L ratio of 1.0 means that the relative protein abundance is not changed between the treated/ stimulated cells over time. The selection is thus for the proteins with an H/L ratio <1.0, which represents proteins that were phosphorylated by PKA during the traffic-pulse.

22.5 DISCUSSION

Over the last few years in particular, the Golgi complex has emerged as even more "complex" than previously thought, with the finding that it is implicated in cellular processes other than the traditional trafficking and biosynthetic pathways (Wilson et al., 2010). As examples, lipid homoeostasis (McMaster, 2001; van Meer & Sprong, 2004), the unfolded protein response (Nadanaka, Yoshida, Sato, & Mori, 2006; Schindler & Schekman, 2009), nutrient sensing (Zeng et al., 2004), mitotic checkpoints (Colanzi & Corda, 2007; Wei & Seemann, 2009), oxidative stress (Jiang et al., 2011), and microtubule nucleation (Kodani & Sutterlin, 2009) have all been shown to be Golgi-dependent to some extent. Moreover, the Golgi complex can serve as a relay station for plasma-membrane-initiated signaling, and for positive (Chiu et al., 2002) or negative (Feng et al., 2007) regulation of some signaling pathways. Taking this all into consideration, our present vision of the Golgi complex is far from that of a simple "processor/distributor" of cargo. From now on, the Golgi complex can be considered as a stage, where biosynthetic and cellular homeostasis processes co-exist, and are probably interconnected by signaling (Cancino & Luini, 2013). The secretory machinery control mechanism should be able to react to changes on demand; for example, changes in secretion levels. For this purpose, the amount of cargo must be sensed to regulate the balance of the membrane fluxes, the changes in the lipid composition, the turnover of the molecular machinery, and the rate of energy production and cargo synthesis. This regulation must be the result of interactions between several organelles (interorganelle signaling), and their interplay should rely on common mechanisms (or molecules) as well.

Interestingly, secretory genes are coregulated by specific transcriptional factors (Fox, Hanlon, & Andrew, 2010), and probably along with that, metabolic genes (Fox et al., 2010; Zhang et al., 2005). What the molecular mechanisms for this coregulation are, and whether the secretory compartments, specifically the Golgi complex, have roles in this, need to be addressed.

In short, the Golgi complex could be considered as a hub for cellular signaling.

SUMMARY

The secretory pathway needs a system to monitor its function; moreover, it should react to changes in secretion levels. The presence of a variety of signaling molecules, including G-proteins, kinases, phosphatases, and phospholipases, which are physically associated with secretory endomembranes, and in particular with the Golgi complex, might help to prevent structural and functional disruption (Cancino & Luini, 2013; Farhan & Rabouille, 2011). How these signaling pathways work in a coordinated fashion to keep the necessary precise balance is poorly understood. The amount of information collected over the last few decades is large, but it is mainly made up of a myriad of scattered data on the phosphorylation of molecular components of the transport machinery and sometimes on its effects on various trafficking steps. Thus, the precise functional "meaning" of this body of information remains for the most part mysterious (Cancino & Luini, 2013). To analyze the flow of information through the signaling pathways that control membrane trafficking, we need to be able to establish how and when Golgi-based signaling pathways are activated.

The experimental approaches herein described will help in the study of these Golgi-based signaling cascades.

Acknowledgment

We would like to thank C.P. Berrie for critical reading of the chapter.

References

Beckers, C. J., Keller, D. S., & Balch, W. E. (1987). Semiintact cells permeable to macromolecules: use in reconstitution of protein transport from the endoplasmic reticulum to the Golgi complex. *Cell*, *50*(4), 523–534.

Bergmann, J. E. (1989). Using temperature-sensitive mutants of VSV to study membrane protein biogenesis. *Methods Cell Biol*, *32*, 85–110.

Bond, L. M., Peden, A. A., Kendrick-Jones, J., Sellers, J. R., & Buss, F. (2011). Myosin VI and its binding partner optineurin are involved in secretory vesicle fusion at the plasma membrane. *Molecular Biology of the Cell*, *22*(1), 54–65.

Bonfanti, L., Mironov, A. A., Jr., Martinez-Menarguez, J. A., Martella, O., Fusella, A., Baldassarre, M., et al. (1998). Procollagen traverses the Golgi stack without leaving the lumen of cisternae: Evidence for cisternal maturation. *Cell*, *95*(7), 993–1003.

Bruckner, P., & Eikenberry, E. F. (1984). Procollagen is more stable in cellulo than in vitro. *European Journal of Biochemistry*, *140*(2), 397–399.

Cancino, J., & Luini, A. (2013). Signaling circuits on the Golgi complex. *Traffic*, *14*(2), 121–134.

Chiu, V. K., Bivona, T., Hach, A., Sajous, J. B., Silletti, J., Wiener, H., et al. (2002). Ras signalling on the endoplasmic reticulum and the Golgi. *Nature Cell Biology*, *4*(11988737), 343–350.

Colanzi, A., & Corda, D. (2007). Mitosis controls the Golgi and the Golgi controls mitosis. *Current Opinion in Cell Biology*, *19*(17689238), 386–393.

Connolly, C. N., Futter, C. E., Gibson, A., Hopkins, C. R., & Cutler, D. F. (1994). Transport into and out of the Golgi complex studied by transfecting cells with cDNAs encoding horseradish peroxidase. *J Cell Biol*, *127*(3), 641–652.

Cox, J., & Mann, M. (2008). MaxQuant enables high peptide identification rates, individualized p.p.b.-range mass accuracies and proteome-wide protein quantification. *Nature Biotechnology*, *26*(12), 1367–1372.

Doms, R. W., Keller, D. S., Helenius, A., & Balch, W. E. (1987). Role for adenosine triphosphate in regulating the assembly and transport of vesicular stomatitis virus G protein trimers. *Journal of Cell Biology*, *105*(5), 1957–1969.

Engholm-Keller, K., & Larsen, M. R. (2011). Titanium dioxide as chemo-affinity chromatographic sorbent of biomolecular compounds–applications in acidic modification-specific proteomics. *Journal of Proteomics*, *75*(2), 317–328.

Farhan, H., & Rabouille, C. (2011). Signalling to and from the secretory pathway. *Journal of Cell Science*, *124*(Pt 2), 171–180.

Feng, L., Xie, X., Ding, Q., Luo, X., He, J., Fan, F., et al. (2007). Spatial regulation of Raf kinase signaling by RKTG. *Proceedings of the National Academy of Sciences of the United States of America*, *104*(17724343), 14348–14353.

Fox, R. M., Hanlon, C. D., & Andrew, D. J. (2010). The CrebA/Creb3-like transcription factors are major and direct regulators of secretory capacity. *Journal of Cell Biology*, *191*(3), 479–492.

Gallione, C. J., & Rose, J. K. (1985). A single amino acid substitution in a hydrophobic domain causes temperature-sensitive cell-surface transport of a mutant viral glycoprotein. *Journal of Virology*, *54*(2), 374–382.

Giannotta, M., Ruggiero, C., Grossi, M., Cancino, J., Capitani, M., Pulvirenti, T., et al. (2012). The KDEL receptor couples to Galpha(q/11) to activate Src kinases and regulate transport through the Golgi. *The EMBO Journal*, *31*(13), 2869–2881.

Hirschberg, K., Miller, C. M., Ellenberg, J., Presley, J. F., Siggia, E. D., Phair, R. D., et al. (1998). Kinetic analysis of secretory protein traffic and characterization of golgi to plasma membrane transport intermediates in living cells. *Journal of Cell Biology*, *143*(6), 1485–1503.

Ivanov, S. S., Chung, A. S., Yuan, Z. L., Guan, Y. J., Sachs, K. V., Reichner, J. S., et al. (2004). Antibodies immobilized as arrays to profile protein post-translational modifications in mammalian cells. *Molecular & Cellular Proteomics*, *3*(8), 788–795.

Jiang, Z., Hu, Z., Zeng, L., Lu, W., Zhang, H., Li, T., et al. (2011). The role of the Golgi apparatus in oxidative stress: Is this organelle less significant than mitochondria? *Free Radical Biology & Medicine*, *50*(8), 907–917.

Kodani, A., & Sutterlin, C. (2009). A new function for an old organelle: Microtubule nucleation at the Golgi apparatus. *The EMBO Journal*, *28*(8), 995–996.

Kreis, T. E., & Lodish, H. F. (1986). Oligomerization is essential for transport of vesicular stomatitis viral glycoprotein to the cell surface. *Cell*, *46*(6), 929–937.

Lee, W. C., & Lee, K. H. (2004). Applications of affinity chromatography in proteomics. *Analytical Biochemistry*, *324*(1), 1–10.

Lewis, M. J., & Pelham, H. R. (1990). A human homologue of the yeast HDEL receptor. *Nature*, *348*(6297), 162–163.

Lowe, M. (2011). Structural organization of the Golgi apparatus. *Current Opinion in Cell Biology*, *23*(1), 85–93.

McMaster, C. R. (2001). Lipid metabolism and vesicle trafficking: More than just greasing the transport machinery. *Biochemistry and Cell Biology*, *79*(11800009), 681–692.

Mironov, A. A., Beznoussenko, G. V., Nicoziani, P., Martella, O., Trucco, A., Kweon, H. S., et al. (2001). Small cargo proteins and large aggregates can traverse the Golgi by a common mechanism without leaving the lumen of cisternae. *Journal of Cell Biology*, *155*(7), 1225–1238.

Nadanaka, S., Yoshida, H., Sato, R., & Mori, K. (2006). Analysis of ATF6 activation in Site-2 protease-deficient Chinese hamster ovary cells. *Cell Structure and Function*, *31*(17110786), 109–116.

Ong, S. E., Blagoev, B., Kratchmarova, I., Kristensen, D. B., Steen, H., Pandey, A., et al. (2002). Stable isotope labeling by amino acids in cell culture, SILAC, as a simple and accurate approach to expression proteomics. *Molecular & Cellular Proteomics*, *1*(5), 376–386.

Pap, E. H., Dansen, T. B., & Wirtz, K. W. (2001). Peptide-based targeting of fluorophores to peroxisomes in living cells. *Trends in Cell Biology*, *11*(1), 10–12.

Perkins, D. N., Pappin, D. J., Creasy, D. M., & Cottrell, J. S. (1999). Probability-based protein identification by searching sequence databases using mass spectrometry data. *Electrophoresis*, *20*(18), 3551–3567.

Peterkofsky, B., & Udenfriend, S. (1965). Enzymatic hydroxylation of proline in microsomal polypeptide leading to formation of collagen. *Proceedings of the National Academy of Sciences of the United States of America*, *53*, 335–342.

Porath, J. (1992). Immobilized metal ion affinity chromatography. *Protein Expression and Purification*, *3*(4), 263–281.

Presley, J. F., Cole, N. B., Schroer, T. A., Hirschberg, K., Zaal, K. J., & Lippincott-Schwartz, J. (1997). ER-to-Golgi transport visualized in living cells. *Nature*, *389*(6646), 81–85.

Pulvirenti, T., Giannotta, M., Capestrano, M., Capitani, M., Pisanu, A., Polishchuk, R. S., et al. (2008). A traffic-activated Golgi-based signalling circuit coordinates the secretory pathway. *Nature Cell Biology*, *10*(8), 912–922.

Rivera, V. M., Wang, X., Wardwell, S., Courage, N. L., Volchuk, A., Keenan, T., et al. (2000). Regulation of protein secretion through controlled aggregation in the endoplasmic reticulum. *Science*, *287*(5454), 826–830.

Sallese, M., Pulvirenti, T., & Luini, A. (2006). The physiology of membrane transport and endomembrane-based signalling. *The EMBO Journal*, *25*(16763561), 2663–2673.

Scheel, A. A., & Pelham, H. R. (1998). Identification of amino acids in the binding pocket of the human KDEL receptor. *Journal of Biological Chemistry*, *273*(4), 2467–2472.

Schindler, A. J., & Schekman, R. (2009). In vitro reconstitution of ER-stress induced ATF6 transport in COPII vesicles. *Proceedings of the National Academy of Sciences of the United States of America*, *106*(19822759), 17775–17780.

Semenza, J. C., Hardwick, K. G., Dean, N., & Pelham, H. R. (1990). ERD2, a yeast gene required for the receptor-mediated retrieval of luminal ER proteins from the secretory pathway. *Cell*, *61*(7), 1349–1357.

Sengupta, D., & Linstedt, A. D. (2011). Control of organelle size: The Golgi complex. *Annual Review of Cell and Developmental Biology*, *27*, 57–77.

Singh, P., Tang, B. L., Wong, S. H., & Hong, W. (1993). Transmembrane topology of the mammalian KDEL receptor. *Molecular and Cellular Biology*, *13*(10), 6435–6441.

Thingholm, T. E., Jorgensen, T. J., Jensen, O. N., & Larsen, M. R. (2006). Highly selective enrichment of phosphorylated peptides using titanium dioxide. *Nature Protocols*, *1*(4), 1929–1935.

Trucco, A., Polishchuk, R. S., Martella, O., Di Pentima, A., Fusella, A., Di Giandomenico, D., et al. (2004). Secretory traffic triggers the formation of tubular continuities across Golgi sub-compartments. *Nature Cell Biology*, *6*(11), 1071–1081.

van Meer, G., & Sprong, H. (2004). Membrane lipids and vesicular traffic. *Current Opinion in Cell Biology*, *16*(4), 373–378.

Volchuk, A., Amherdt, M., Ravazzola, M., Brugger, B., Rivera, V. M., Clackson, T., et al. (2000). Megavesicles implicated in the rapid transport of intracisternal aggregates across the Golgi stack. *Cell*, *102*(3), 335–348.

Wei, J. H., & Seemann, J. (2009). Mitotic division of the mammalian Golgi apparatus. *Seminars in Cell & Developmental Biology*, *20*(7), 810–816.

Wilson, C., Venditti, R., Rega, L. R., Colanzi, A., D'Angelo, G., & De Matteis, M. A. (2010). The Golgi apparatus: An organelle with multiple complex functions. *The Biochemical Journal*, *433*(1), 1–9.

Wolschin, F., Wienkoop, S., & Weckwerth, W. (2005). Enrichment of phosphorylated proteins and peptides from complex mixtures using metal oxide/hydroxide affinity chromatography (MOAC). *Proteomics*, *5*(17), 4389–4397.

Yip, T. T., Tam, Y. Y., Kong, Y. C., Belew, M. C., & Porath, J. (1985). Isolation of the epidermal growth factor from the shrew submaxillary gland. *FEBS Letters*, *187*(2), 345–348.

Zeng, L., Lu, M., Mori, K., Luo, S., Lee, A. S., Zhu, Y., et al. (2004). ATF6 modulates SREBP2-mediated lipogenesis. *The EMBO Journal*, *23*(4), 950–958.

Zhang, X., Odom, D. T., Koo, S. H., Conkright, M. D., Canettieri, G., Best, J., et al. (2005). Genome-wide analysis of cAMP-response element binding protein occupancy, phosphorylation, and target gene activation in human tissues. *Proceedings of the National Academy of Sciences of the United States of America*, *102*(12), 4459–4464.

CHAPTER

Signaling at the Golgi During Mitosis

23

Antonino Colanzi* and Christine Sütterlin†

**Institute of Protein Biochemistry, CNR, Naples, Italy*

†*Department of Developmental and Cell Biology, University of California, Irvine, California, USA*

CHAPTER OUTLINE

Introduction **384**
23.1 The Organization of Mammalian Golgi Membranes in Interphase **385**
23.2 Mechanism of Mitotic Golgi Disassembly **386**
23.2.1 Severing the Golgi Ribbon Into Stacks 386
23.2.1.1 BARS 387
23.2.1.2 MEK1 387
23.2.1.3 Golgi Reassembly and Stacking Proteins 388
23.2.1.4 Plk1 389
23.2.2 Disassembly of Golgi Stacks Into "Blobs" and "Haze" 389
23.3 Mechanism of Postmitotic Golgi Reassembly **390**
23.4 The Significance of Golgi Reorganization in Mitosis **390**
23.4.1 Golgi Ribbon Severing 390
23.4.2 Golgi Stack Fragmentation 391
23.5 Assays to Study Mitotic Golgi Reorganization **393**
23.5.1 Semi-intact Cell Assay to Study Mitotic Golgi Fragmentation 393
23.5.1.1 Mitotic Cytosol 393
23.5.1.2 Interphase Cytosol 393
23.5.1.3 Selective Permeabilization of Tissue Culture Cells 393
23.5.1.4 Semi-intact Cell Assay 394
23.5.2 Cell-free Assays to Study Mitotic Golgi Fragmentation 394
23.5.3 Microinjection Assay in Intact Cells to Study Mitotic Golgi Fragmentation 394
23.5.3.1 Cell Cycle Synchronization of NRK Cells and Microinjection 394

Methods in Cell Biology, Volume 118

ISSN 0091-679X
http://dx.doi.org/10.1016/B978-0-12-417164-0.00023-9

Concluding Remarks 395
Acknowledgments 395
References 395

Abstract

The Golgi complex of mammalian cells is composed of interconnected stacks of flattened cisternae that form a continuous membrane system in the pericentriolar region of the cell. At the onset of mitosis, this so-called Golgi ribbon is converted into small tubular–vesicular clusters in a tightly regulated fragmentation process, which leads to a temporary loss of the physical Golgi–centrosome proximity. Mitotic Golgi breakdown is required for Golgi partitioning into the two daughter cells, cell cycle progression and may contribute to the dispersal of Golgi-associated signaling molecules. Here, we review our current understanding of the mechanisms that control mitotic Golgi reorganization, its biological significance, and assays that are used to study this process.

INTRODUCTION

The Golgi apparatus fulfills numerous critical functions. As a central organelle of the secretory pathway, it is necessary for the modification, sorting and transport of proteins and lipids, and for the organization of a subset of polarized microtubules (Wilson et al., 2011). Furthermore, Golgi membranes form a centrally localized platform that facilitates the integration of various signaling pathways (Cancino & Luini, 2013; Wilson et al., 2011).

Although many basic Golgi functions are conserved throughout evolution, the structural organization of this organelle varies greatly between species. For example, in the yeast *Saccharomyces cerevisiae* or the protozoan parasite *Toxoplasma gondii*, individual Golgi cisternae or stacks of cisternae, respectively, are dispersed throughout the cytosol (Pelletier et al., 2002; Preuss, Mulholland, Franzusoff, Segev, & Botstein, 1992). In mammalian cells, in contrast, the Golgi complex is arranged as a interconnected pericentriolar ribbon in interphase and dispersed tubular–reticular and vesicular elements in mitosis (Misteli & Warren, 1995). This reversible conversion between two very different three-dimensional (3D) Golgi structures is achieved in tightly regulated steps (Fig. 23.1). First, the Golgi ribbon is disconnected to yield separate Golgi stacks (Colanzi et al., 2007). Then, stacks are converted into tubular–reticular membranes, termed "Golgi blobs," which are dispersed throughout the cytosol (Colanzi, Suetterlin, & Malhotra, 2003; Nelson, 2000). Finally, "Golgi blobs" are broken down by vesiculation to yield the "Golgi haze," which corresponds to small dispersed vesicles (Colanzi, Suetterlin, et al., 2003). This cell cycle-dependent reorganization leads to a temporary block in Golgi function. For example, in mitosis, protein transport does not occur and the ability of Golgi membranes to nucleate microtubules is significantly downregulated

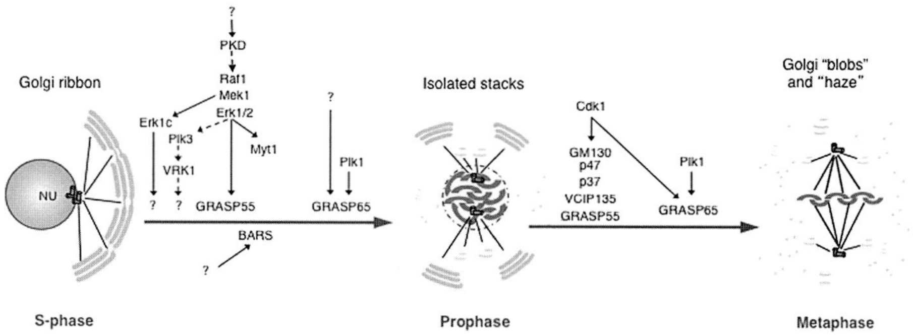

FIGURE 23.1

The fragmentation of the Golgi ribbon during mitosis is tightly controlled. Schematic representation of the mechanisms leading to breakdown of the Golgi complex in mitosis. The mammalian Golgi complex in S-phase forms a ribbon next to the centrosome (red cylinders) and the nucleus (NU). In G2, BARS, MEK1, the GRASPs, and Plk1 activities are required to convert this interconnected ribbon into isolated stacks. This step is necessary and sufficient for entry into mitosis (prophase). In prophase, the isolated Golgi stacks undergo further disassembly into the Golgi "blobs" and "haze" through Plk1 and Cdk1-mediated phosphorylation of golgins (GM130), and components of the fusion machinery (p47, p37, and VCIP135). The relative fraction of Golgi membranes dispersed as "blobs" or "haze" is cell type specific. (See color plate.)

(Maia et al., 2013; Yeong, 2013). Postmitotic Golgi reassembly begins in telophase and involves the formation of two smaller Golgi ribbons that eventually coalesce (Gaietta et al., 2006).

In this review, we will focus on the regulation of cell cycle-dependent Golgi dynamics in mammalian cells. We will describe how this organelle is organized at each stage of the cell cycle and how signaling pathways regulate the conversion of the Golgi ribbon into dispersed tubular–vesicular clusters. In addition, we will discuss why Golgi membranes may fragment during mitosis and describe some assays that have been used to monitor mitotic Golgi reorganization and to identify regulatory components.

23.1 THE ORGANIZATION OF MAMMALIAN GOLGI MEMBRANES IN INTERPHASE

The characteristic ribbon organization of the mammalian Golgi apparatus depends on a diverse set of proteins. Up to 100 Golgi stacks, each composed of five to eight flattened cisternae are connected by membranous tubular bridges, called the noncompact zones, and are positioned next to the centrosome (Sütterlin & Colanzi, 2010). Golgins are peripheral membrane proteins that are proposed to maintain this 3D arrangement of Golgi membranes by providing a structural skeleton (Munro, 2011). Their involvement in Golgi organization is supported by the finding that the depletion of a golgin, such as Golgin-160, resulted in the conversion of the Golgi ribbon into dispersed, but functional mini-stacks (Yadav, Puri, & Linstedt, 2009). The peripheral Golgi proteins GRASP65 and GRASP55 also contribute to Golgi ribbon formation because RNAi-mediated depletion of either protein resulted in the unlinking of Golgi stacks (Feinstein & Linstedt, 2008; Puthenveedu, Bachert, Puri, Lanni, & Linstedt, 2006). Golgi ribbon organization also depends on an intact microtubule and actin cytoskeleton, specialized cytoskeleton-based motors and membrane input from the endoplasmic reticulum (ER) (Marra et al., 2007; Rios & Bornens, 2003; Thyberg & Moskalewski, 1999), but these factors will not be discussed here in detail.

It is not completely understood why mammalian Golgi membranes in interphase are organized into stacks of cisternae and why these interconnected Golgi stacks are positioned in the pericentriolar region of the cell. It is unlikely that Golgi cisternae stacking is important for protein transport because the dispersed Golgi cisternae of *S. cerevisiae* are able to secrete efficiently (Papanikou & Glick, 2009). Interestingly, a recent study showed that loss of Golgi stacking leads to an increased rate of protein transport and defects in glycosylation (Xiang et al., 2013). Thus, cisternae stacking may increase the accuracy of protein glycosylation.

There is increasing support for the functional significance of the pericentriolar positioning of the Golgi apparatus. For example, pericentriolar Golgi membranes are proposed to control cell polarization and migration. Interfering with Golgi positioning by depleting Golgin-160 or GMAP210 had no effect on general protein transport to the cell surface, but there were defects in directional protein secretion, which is necessary for cell migration (Yadav et al., 2009). In addition, this specific Golgi position may facilitate functional Golgi–centrosome interactions, such as the regulation of centrosome organization and function from the Golgi (Kodani, Kristensen, Huang, & Sütterlin, 2009). Finally, Golgi fragmentation, which leads to a temporary loss of pericentriolar Golgi positioning, is required for entry into mitosis (Sütterlin, Hsu, Mallabiabarrena, & Malhotra, 2002). In conclusion, ribbon organization, cisternae stacking, and pericentriolar positioning of the Golgi complex may add another level of regulation for mammalian cell-specific processes.

23.2 MECHANISM OF MITOTIC GOLGI DISASSEMBLY

23.2.1 Severing the Golgi ribbon into stacks

In the first step of mitotic Golgi disassembly, the interconnected Golgi ribbon is converted into isolated stacks, which requires the cleavage of the membrane tubules of the noncompact zones (Fig. 23.1). While these morphological changes appear minor and can only be detected with sensitive approaches, such as FRAP and electron microscopy (Colanzi et al., 2007), they are of functional importance because they are necessary and sufficient for entry of cells into mitosis (Colanzi et al.,

2007; Feinstein & Linstedt, 2007). The severing of Golgi stacks depends on at least four factors, which include the fission-inducing protein CtBP1-S/BARS (referred to as BARS) (Hidalgo Carcedo et al., 2004), the protein kinase MEK1 (Acharya, Mallabiabarrena, Acharya, & Malhotra, 1998; Feinstein & Linstedt, 2007), the peripheral Golgi proteins GRASP65 and GRASP55 (Sütterlin et al., 2002; Xiang & Wang, 2010), and the mitotic kinase Plk1 (Sütterlin et al., 2001). Each of these proteins has a specialized role in Golgi ribbon breakdown and their combined activities produce isolated stacks in G2. It is possible that these same factors also control the breakdown of isolated Golgi stacks into the "Golgi blobs," which have been detected in metaphase (Fig. 23.1).

23.2.1.1 BARS

This membrane fission factor controls the disassembly of Golgi stacks by severing the tubular network of the noncompact zones (Colanzi et al., 2007). Interfering with BARS activity in a semi-intact Golgi fragmentation assay resulted in groups of large tubular–vesicular–saccular networks of Golgi membranes, which were continuous and localized in the pericentriolar region (Colanzi et al., 2007). Interestingly, BARS is required for several membrane-trafficking steps (Bonazzi et al., 2005; Valente et al., 2012; Yang et al., 2011), but Golgi membranes are only fragmented late in G2 indicating that BARS is specifically activated in G2 to promote Golgi ribbon severing. It is likely that this regulation involves the phosphorylation of BARS, which is a known substrate for several protein kinases (Liberali et al., 2008; Valente et al., 2012). BARS phosphorylation may facilitate the recruitment of specific, still unknown binding partners, which then mediate the cutting of the Golgi ribbon.

23.2.1.2 MEK1

The dual specificity kinase MEK1 controls Golgi disassembly by activating several regulatory kinases (Fig. 23.1). MEK1 is a central player of the MAP kinase signal transduction pathway, which normally functions as a tripartite signaling module (Robinson & Cobb, 1997). Loss of MEK1 activity prevented Golgi fragmentation in intact and semi-intact cells (Acharya et al., 1998; Feinstein & Linstedt, 2007).

Two MEK1 activators have been identified that function upstream of MEK1 in Golgi fragmentation. Raf1 is known to directly activate MEK1 and its activity was necessary for mitotic Golgi fragmentation and entry into mitosis (Colanzi, Sutterlin, & Malhotra, 2003). There is also an indirect role for PKD1 and PKD2 in the activation of Raf1/MEK1 in mitosis (Kienzle et al., 2013). Interfering with PKD activity blocked Raf1/MEK1 activation and prevented entry into mitosis. Loss of PKD activity was rescued by expression of active MEK1, indicating that these proteins function in the same regulatory pathway.

MEK1 regulates Golgi breakdown through several effectors. ERK1/2, the best characterized MEK1 substrate, controls mitotic Golgi breakdown through phosphorylation of GRASP55 (Feinstein & Linstedt, 2008). Plk3, another MEK1 substrate localizes to the spindle and the Golgi complex (Ruan et al., 2004), where it activates VRK1 through phosphorylation (Lopez-Sanchez, Sanz-Garcia, & Lazo, 2009).

Interfering with VRK1 function prevented MEK1-induced Golgi breakdown (Lopez-Sanchez et al., 2009). The ERK1 splice variant ERK1c also associates with the Golgi complex, and its depletion reduced the efficiency of mitotic Golgi fragmentation and entry into mitosis (Shaul & Seger, 2006). As Plk3, VRK1, and ERK1c function downstream of MEK1, each of these components could be involved in the severing of the Golgi ribbon or the conversion of mini-stacks into "Golgi blobs" (Fig. 23.1); however, their targets and specific contributions to Golgi fragmentation remain to be determined.

Myt1 is an MEK1 effector that has been implicated in the control of mitotic Golgi fragmentation. This protein kinase, a known negative regulator of Cdk1 activity, associates with the Golgi complex and the endoplasmic reticulum (Liu, Stanton, Wu, & Piwnica-Worms, 1997). RNAi studies in *Drosophila* and HeLa cells have revealed a requirement for Myt1 in mitotic Golgi dynamics (Cornwell, Kaminski, & Jackson, 2002; Nakajima et al., 2008). In addition, a recent study reported that Myt1 depletion caused accelerated entry into mitosis, and there was a small but significant increase in cells that were in G2 and that had fragmented Golgi membranes (Villeneuve, Scarpa, Ortega-Bellido, & Malhotra, 2013). The idea that Myt1 function downstream of MEK1 was confirmed by the result that Myt1 is inactivated by MEK1-mediated phosphorylation and that its depletion was able to bypass a requirement for MEK1.

23.2.1.3 Golgi reassembly and stacking proteins

The two related Golgi reassembly and stacking proteins (GRASPs), GRASP65 and GRASP55, which localize to the *cis*- or *medial/trans* Golgi, respectively, are critical regulators of Golgi organization. Their depletion caused Golgi ribbon breakdown and a reduction in the number of cisternae per stack (Feinstein & Linstedt, 2008; Puthenveedu et al., 2006; Sütterlin, Polishchuk, Pecot, & Malhotra, 2005). Interfering with their functions in semi-intact cell assays or in intact cells affected mitotic Golgi fragmentation and mitotic entry (Duran et al., 2008; Preisinger et al., 2005; Sütterlin et al., 2002).

GRASPs have unique structural features with two N-terminal PDZ-like domains and a C-terminal serine/proline-rich regulatory domain (Vinke, Grieve, & Rabouille, 2011). Their Golgi localization is mediated by myristoylation of an N-terminal glycine and interaction with specific golgins (Vinke et al., 2011). GRASPs are phosphorylated in a cell cycle-controlled manner, with at least eight phosphorylation sites on GRASP65 and at least five sites on GRASP55 (Duran et al., 2008; Preisinger et al., 2005; Tang, Yuan, Vielemeyer, Perez, & Wang, 2012). These sites are modified by Cdk1 and Plk1, or ERK1/2 in the case of GRASP65 or GRASP55, respectively (Duran et al., 2008; Preisinger et al., 2005; Sengupta & Linstedt, 2010; Tang et al., 2012).

Cell cycle-regulated *trans*-oligomerization of GRASPs provides a possible mechanism for the control of Golgi reorganization during mitosis. *In vitro* binding and bead aggregation assays showed that GRASPs form homo-oligomers in *trans* through their N-terminal GRASP domain and that phosphorylation of specific residues within the C-terminal serine/proline-rich regulatory domain prevents these interactions (Tang, Yuan, & Wang, 2010). In addition, microinjection of GRASP65-specific antibodies into mitotic cells blocked Golgi stacks reformation after cell division (Wang, Seemann, Pypaert, Shorter, & Warren, 2003). These findings suggest that the *trans*-oligomerization of GRASPs promotes Golgi stacking and ribbon assembly in interphase, and that GRASP phosphorylation in mitosis disrupts these associations, inducing Golgi ribbon disassembly. While this oligomerization-based model can explain how GRASPs may regulate Golgi stack assembly and disassembly, it not clear whether GRASPs are long enough to link individual Golgi stacks into a ribbon.

Live imaging studies with fluorescently tagged GRASP55 support an additional role of GRASP55 in the stabilization of contacts between mini-stacks (Feinstein & Linstedt, 2008). Monitoring Golgi reassembly after nocodazole-induced Golgi fragmentation revealed GRASP55-GFP-positive membrane tubules that formed stable bridges between adjacent mini-stacks. Such bridges were rarely seen in the absence of GRASP55, indicating that GRASP55 may contribute to the formation of these stable contacts between mini-stacks. Thus, the assembly of the Golgi ribbon from dispersed Golgi fragments may require GRASP *trans*-oligomerization to form mini-stacks and GRASP55-mediated stabilization of tubular contacts to form the interconnected ribbon.

23.2.1.4 Plk1

Several studies have identified a requirement for mitotic kinase Plk1 for mitotic Golgi fragmentation (Sengupta & Linstedt, 2010; Sütterlin et al., 2001). Plk1 phosphorylates GRASP65 on serine 189 in the N-terminal domain that mediates *trans*-oligomerization (Truschel, Zhang, Bachert, Macbeth, & Linstedt, 2012). Expression of phospho-mimetic GRASP65 mutants prevented cisternae tethering and caused Golgi ribbon disassembly, while expression of nonphosphorylatable mutants prevented Golgi ribbon cleavage during mitosis (Sengupta & Linstedt, 2010). This study led to a model in which Cdk1 first phosphorylates the GRASP65 C-terminal domain, creating a docking site for Plk1. Plk1 then binds to GRASP65 and phosphorylates Ser189, which inhibits the *trans*-oligomerization properties of GRASP65. However, since the activity of Cdk1 in G2 is low, it is possible that a kinase other than Cdk1 phosphorylates GRASP65 to prime it for Plk1 binding. A likely candidate is the Raf1/MEK1/ERK1/2 pathway, which is active in G2 and which recognizes and modifies the same phosphorylation consensus site as Cdk1 (Yoshimura et al., 2005).

23.2.2 Disassembly of Golgi stacks into "blobs" and "haze"

In prophase, the isolated Golgi stacks that result from Golgi ribbon cleavage are further fragmented into "Golgi blobs" and "Golgi haze." This additional fragmentation step depends on unstacking of cisternae to disassemble Golgi mini-stacks and vesiculation to fragment Golgi cisternae. Unstacking is mediated by Cdk1 and Plk1, which phosphorylate GRASP65 and thereby disrupt its ability to form *trans*-oligomers. Once unstacked, Golgi cisternae are broken down by extensive vesiculation through the continuous budding of COPI-coated vesicles (Misteli & Warren, 1994b), and this process is facilitated by the reported block in transport from the ER and inhibition of vesicle fusion with the Golgi complex. The mechanism behind this transport block is incompletely understood, but it may involve the phosphorylation of components of the ER-to-Golgi transport machinery, such as the ARF exchange factor GBF1 (Morohashi, Balklava, Ball, Hughes, & Lowe, 2010). In summary, cell cycle-dependent Golgi organization appears to be controlled by equilibrium between disassembly and reassembly factors. How the functions of these maintenance factors and BARS are coordinated remains to be addressed in future studies.

23.3 MECHANISM OF POSTMITOTIC GOLGI REASSEMBLY

In vitro reconstitution assays have provided great insight into the mechanism of postmitotic Golgi reassembly. Two AAA ATPases, NSF and p97 (also called VCP for valosin-containing protein), were found to be necessary for the reassembly of small vesicular Golgi elements into short cisternae (Acharya, McCaffery, Jacobs, & Malhotra, 1995; Rabouille, Levine, Peters, & Warren, 1995). Both ATPases utilize specific co-factors, with SNAPs and p115 promoting NSF-mediate fusion, and complexes of p97 with p47 and VCIP135, or with p37 controlling p97 function in Golgi reformation (Kondo et al., 1997; Uchiyama et al., 2006). NSF-mediated fusion is proposed to produce larger vesicles and tubular–reticular elements, which are then fused by a p97-mediated process to generate cisternae. In mitosis, p47, p37, and VCIP135 are phosphorylated by Cdk1, which blocks p97-mediated membrane fusion so that Golgi membranes remain disassembled (Kaneko, Tamura, Totsukawa, & Kondo, 2010; Totsukawa, Matsuo, Kubota, Taguchi, & Kondo, 2013; Uchiyama et al., 2003). Thus, mitotic phosphorylation of the membrane fusion machinery can explain the mitotic Golgi phenotype of dispersed tubular–reticular membranes and vesicles in the cytosol (Fig. 23.1).

In addition to phosphorylation, ubiquitination plays an important role in mitotic Golgi reassembly. The p97 co-factor p47 was found to contain a UBA ubiquitin-binding domain. Upon association with p97, p47 was able to bind ubiquitinated proteins via this UBA domain, and this domain was necessary for Golgi reassembly (Meyer, Shorter, Seemann, Pappin, & Warren, 2000). In addition, VCIP135 was shown to possess deubiquitinase activity, which was necessary for Golgi reassembly *in vitro* and in intact cells (Wang, Satoh, Warren, & Meyer, 2004). Further support for a role of a ubiquitination–deubiquination cycle in mitotic Golgi dynamics comes from the identification of HACE1, a Golgi-associated ubiquitin ligase, whose activity is required for mitotic Golgi dynamics (Tang et al., 2011). It will be interesting to identify the proteins that are ubiquitinated during Golgi disassembly and deubiquitinated during reassembly.

23.4 THE SIGNIFICANCE OF GOLGI REORGANIZATION IN MITOSIS

23.4.1 Golgi ribbon severing

Although many mechanistic aspects of mitotic Golgi disassembly remain to be defined, it has become clear that entry of cells into mitosis depends on the ability of the

Golgi complex to fragment (Hidalgo Carcedo et al., 2004; Sütterlin et al., 2002). This conclusion is based on microinjection and overexpression studies in synchronized cells and implies that Golgi organization and cell cycle progression are tightly linked.

It is incompletely understood how a block in Golgi fragmentation prevents cell cycle progression. As the block in Golgi fragmentation induces a G2 arrest, it is likely to involve cyclin B-dependent kinase 1 (CycB-Cdk1) because this protein kinase is the major regulator of the G2/M transition (Nigg, 2001). The activity of CycB1-Cdk1 activity is known to be controlled by the recruitment and activation of the Ser/Thr kinase Aurora-A at the centrosome (Marumoto et al., 2002). Interestingly, a block in Golgi fragmentation interfered with Aurora-A recruitment to the centrosome in G2 and prevented its activation (Persico, Cervigni, Barretta, Corda, & Colanzi, 2010). This link between Golgi organization in G2 and Aurora-A recruitment and activation provides a first mechanistic insight into how Golgi dynamics may be coordinated with cell-cycle progression.

However, a second cyclin B-dependent, but Aurora-independent mechanism may also contribute to the Golgi-mediated control of mitotic entry. Mammalian cells express two B-type cyclins, CycB1 and CycB2, of which CycB2 associates with the Golgi complex and the ER (Jackman, Firth, & Pines, 1995). Both isoforms cooperate to promote mitotic entry. It is possible that the activation of the Golgi-associated CycB2-Cdk1 complex is regulated by the organization state of the Golgi complex so that impaired Golgi fragmentation could prevent the activation of this central mitotic regulator. In conclusion, Golgi fragmentation may control mitotic entry by simultaneously activating the Aurora-A/Cdk1 pathway as well as the Golgi-associated CycB2-Cdk1 complex.

23.4.2 Golgi stack fragmentation

The reason for Golgi stack disassembly is only beginning to being understood. This process does not seem to be necessary for mitotic progression because cell types that lack a ribbon-like Golgi structure, such plant cells or *Drosophila* embryos, undergo cell division without breakdown of their stack. In addition, accurate partitioning of an organelle can be achieved without complete fragmentation of this organelle, as has been seen for the ER (Terasaki, 2000). Therefore, the disassembly of isolated stacks into the Golgi haze may achieve goals that are not directly linked to mitotic entry. For example, disassembly of Golgi stacks has been found to lead to inactivation of the small GTPase Arf1 and release of a set of peripheral Golgi proteins (Altan-Bonnet et al., 2006). When Arf1 was kept in an active state, stack disassembly was blocked and cells entered mitosis, but there were severe defects in chromosome segregation and cytokinetic furrow ingression (Altan-Bonnet et al., 2006). Additional Golgi proteins dissociate from the Golgi complex during mitosis to complete Golgi-independent functions (Table 23.1). For instance, the Golgi-associated protein p115 was found to dissociate from the Golgi during mitosis and to partition with the poles of the mitotic spindle, where it controlled spindle formation, chromosome segregation and cytokinesis (Radulescu et al., 2011). Similarly, Miki loses its Golgi association during mitosis and localizes to centrosome to control the recruitment of

Table 23.1 Golgi-associated proteins that acquire new location and function during mitosis

Protein name	Mitotic localization	Mitotic function	Stimulus for relocalization	Upstream regulator	References
Arf1	Cytosol	Unknown	Unknown	Unknown	Altan-Bonnet et al. (2006)
p115	Spindle	Spindle formation/ chromosome segregation	Unknown	Unknown	Radulescu, Mukherjee, and Shields (2011)
Miki	Centrosome	Spindle formation/ chromosome segregation	PARsylation	Tankyrase1	Ozaki et al. (2012)
KBTBD8	Spindle	Unknown	Unknown	Unknown	Luhrig, Kolb, Mellies, and Nolte (2013)
CtBP	Centrosome	Unknown	Unknown	Unknown	Spyer and Allday (2006)
Rab6A	Unknown	Dynein/dynactin complex at kinetochores	Unknown	Unknown	Miserey-Lenkei et al. (2006)
Nir2	Cleavage forrow and midbody	Plk1 recruitment	Phosphorylation	Cdk1	Litvak et al. (2004)
ACBD3	Cytosol	Regulation of Numb signaling	Unknown	Unknown	Zhou et al. (2007)
Zw10	Kinetochores	Mitotic checkpoint	Unknown	Unknown	Vallee, Varma, and Dujardin (2006)

CG-NAP, a scaffold protein of the γ-tubulin ring complex (Ozaki et al., 2012). While Miki relocalization is induced by tankyrase-mediate PARsylation, it cannot be excluded that Golgi fragmentation contributes to Miki dispersal. Thus, complete Golgi vesiculation may function as a regulatory factor in the distribution of these proteins, and it will be interesting to examine whether blocking Golgi "blob" or "haze" formation interferes with the function of these proteins

23.5 ASSAYS TO STUDY MITOTIC GOLGI REORGANIZATION

23.5.1 Semi-intact cell assay to study mitotic Golgi fragmentation

(Acharya et al., 1998)

23.5.1.1 Mitotic cytosol (Nakagawa, Kitten, & Nigg, 1989)

1. Grow NRK or HeLa cells in 15 cm petri dishes (70% confluency).
2. Arrest cells in S-phase with 2 mM thymidine for 10–12 h.
3. Wash cells to remove thymidine and incubate with 500 ng/ml nocodazole for 10–12 h to arrest cells in metaphase.
4. Harvest mitotic cells by "shake off."
5. Wash mitotic cells with cold PBS and with mitotic extract buffer (MEB: 15 mM PIPES (pH 7.4), 50 mM KCl, 10 mM $MgCl_2$, 20 mM β-mercaptoethanol, 20 mM β-glycerophosphate, 15 mM EGTA, 0.5 mM spermidine, 0.2 mM spermine, 1 mM DTT, 0.1 mM PMSF, 0.2 μg/ml aprotinin, 0.2 μg/ml leupeptin, and 0.2 μg/ml pepstatin).
6. Resuspend cell pellet in twice the packed cell volume in MEB and allow to swell for 10 min on ice.
7. Homogenize by passage through an 18-gauge needle and lyse by repeated passages through a 24-gauge needle. Monitor lysis state under the light microscope.
8. Centrifuge lysate in a table top ultracentrifuge at 48,000 rpm for 45 min using a TLS55 rotor.
9. Separate high-speed supernatant (approximate protein concentration of 10–12 mg/ml) and freeze as "mitotic extract" in aliquots in liquid nitrogen.
10. Store at −80 °C.

23.5.1.2 Interphase cytosol

As for mitotic cells, but asynchronous cells were harvested by scraping.

23.5.1.3 Selective permeabilization of tissue culture cells

1. Grow cells (NRK or HeLa) on coverslips (60% confluency) and incubate with 2 mM thymidine for 12–16 h.
2. Wash with KHM buffer (25 mM HEPES-KOH [pH 7.2], 125 mM potassium acetate, 2.5 mM magnesium acetate) at room temperature.
3. Move dish to ice and wash with cold KHM buffer.

4. Permeabilize plasma membrane by treating cells with 30 μg/ml digitonin in cold KHM for 3–5 min—The exact timing of digitonin treatment has to be determined using antibody staining in the absence of detergent.
5. Stop permeabilization by washing with cold KHM.
6. Incubate permeabilized cells with cold KHM, 1 M KCL to remove peripheral membrane proteins for 5 min.
7. Allow permeabilized cells to warm up during the last 2 min of the salt wash (step 6).
8. Wash with KHM buffer at RT.

23.5.1.4 Semi-intact cell assay

1. Prepare 50 μl incubation mixture: 7 mg/ml cytosol, ATP-regeneration system (final concentration 0.5 mM ATP, 0.3 mM UTP, 10 mM creatine phosphate, and 12 IU/ml creatine kinase), and purified recombinant proteins to be tested.
2. Place on parafilm in a 10 cm tissue culture dish.
3. Inverse coverslip with permeabilized cells onto reaction mixture and incubate for 60 min at 32 °C.
4. Fix and analyze by immunofluorescence.

23.5.2 Cell-free assays to study mitotic Golgi fragmentation

(Misteli & Warren, 1994a; Rabouille et al., 1995)

Golgi disassembly and reassembly have been reconstituted in an *in vitro* assay using Golgi membranes, and mitotic cytosol or purified components. The extent of Golgi fragmentation is evaluated by electron microscopy. As excellent reviews describing this assay have been published, we refer to these for more details (Tang, Xiang, & Wang, 2010; Warren, Levine, & Misteli, 1995).

23.5.3 Microinjection assay in intact cells to study mitotic Golgi fragmentation (Persico et al., 2010; Sütterlin et al., 2002)

23.5.3.1 Cell cycle synchronization of NRK cells and microinjection

1. Grow NRK cells on fibronectin-coated glass coverslips (60% confluency).
2. Synchronize cells at the G1/S transition by treating them with 2.5 μg/ml aphidicolin in complete medium for 16 h.
3. Remove aphidicolin by washing three times 3 min with warm complete medium.
4. 30 min after aphidicolin removal, inject 200 cells with purified recombinant proteins or antibodies (8–12 mg/ml) or Dextran-FITC as a negative control, using an Eppendorf 5246 transjector (Eppendorf, Milan, Italy).
5. Incubate for 7–8 h, fix, and process for immunofluorescence with antibodies to phospho-Histone H3 to determine mitotic index and with antibodies to detect the injected antibodies or proteins.

CONCLUDING REMARKS

The initial observation that Golgi membranes in mammalian cells undergo regulated reorganization in mitosis was made more than three decades ago and has stimulated many studies on the mechanisms of Golgi biogenesis and maintenance. About 10 years ago, these investigations led to the discovery that mitotic Golgi fragmentation is required for entry of cells into mitosis, implying that cell-cycle progression depends not only on the state of the DNA, but also on the organization of an organelle. However, many questions on the link between organelle structure and cell-cycle regulation have remained unanswered. For example, we are only beginning to understand why Golgi membranes undergo such extensive reorganization during mitosis, and why they are positioned next to the centrosome in interphase. This incomplete insight into this process is due to the overall complexity of the process, but also to a lack of appropriate methods. An effective protocol to block the cleavage of the Golgi ribbon is the microinjection of "blocking reagents" (antibodies, recombinant proteins, and peptides), which target factors involved in the process. As a result, there is an acute block of Golgi fragmentation, which then induces a potent and prolonged G2 arrest. However, this approach is limited to the observation of single cells by immunofluorescence and is not amenable to unbiased screens for additional components. Other treatments (kinase inhibitors, RNAi, or protein overexpression) reach every cell within a population, but appear to produce effects that are in general smaller so that results may be difficult to interpret. Thus, it will be essential to develop methods to rapidly inactivate the function of Golgi organization regulators in an entire cell population. Such approaches will help understand the timing and temporal order in which regulatory kinases modify their targets to induce a fragmented Golgi, and to identify regulatory networks involved in this process through global investigations (proteomics, phosphoproteomic).

Acknowledgments

We apologize to those colleagues whose work we were not able to discuss due to space limitation. A. C. acknowledges the Italian Association for Cancer Research (AIRC, Milan, Italy; IG6074) for financial support. C. S. is supported by a grant from the National Institutes of Health (R01 GM089913).

References

Acharya, U., Mallabiabarrena, A., Acharya, J. K., & Malhotra, V. (1998). Signaling via mitogen-activated protein kinase kinase (MEK1) is required for Golgi fragmentation during mitosis. *Cell*, *92*, 183–192.

Acharya, U., McCaffery, J. M., Jacobs, R., & Malhotra, V. (1995). Reconstitution of vesiculated Golgi membranes into stacks of cisternae: Requirement of NSF in stack formation. *Journal of Cell Biology*, *129*, 577–589.

Altan-Bonnet, N., Sougrat, R., Liu, W., Snapp, E. L., Ward, T., & Lippincott-Schwartz, J. (2006). Golgi inheritance in mammalian cells is mediated through endoplasmic reticulum export activities. *Molecular Biology of the Cell*, *17*, 990–1005.

Bonazzi, M., Spano, S., Turacchio, G., Cericola, C., Valente, C., Colanzi, A., et al. (2005). CtBP3/BARS drives membrane fission in dynamin-independent transport pathways. *Nature Cell Biology*, *7*, 570–580.

Cancino, J., & Luini, A. (2013). Signaling circuits on the Golgi complex. *Traffic*, *14*, 121–134.

Colanzi, A., Hidalgo Carcedo, C., Persico, A., Cericola, C., Turacchio, G., Bonazzi, M., et al. (2007). The Golgi mitotic checkpoint is controlled by BARS-dependent fission of the Golgi ribbon into separate stacks in G2. *The EMBO Journal*, *26*, 2465–2476.

Colanzi, A., Suetterlin, C., & Malhotra, V. (2003). Cell-cycle-specific Golgi fragmentation: How and why? *Current Opinion in Cell Biology*, *15*, 462–467.

Colanzi, A., Sutterlin, C., & Malhotra, V. (2003). RAF1-activated MEK1 is found on the Golgi apparatus in late prophase and is required for Golgi complex fragmentation in mitosis. *Journal of Cell Biology*, *161*, 27–32.

Cornwell, W. D., Kaminski, P. J., & Jackson, J. R. (2002). Identification of Drosophila Myt1 kinase and its role in Golgi during mitosis. *Cellular Signalling*, *14*, 467–476.

Duran, J. M., Kinseth, M., Bossard, C., Rose, D. W., Polishchuk, R., Wu, C. C., et al. (2008). The role of GRASP55 in Golgi fragmentation and entry of cells into mitosis. *Molecular Biology of the Cell*, *19*, 2579–2587.

Feinstein, T. N., & Linstedt, A. D. (2007). Mitogen-activated protein kinase kinase 1-dependent Golgi unlinking occurs in G2 phase and promotes the G2/M cell cycle transition. *Molecular Biology of the Cell*, *18*, 594–604.

Feinstein, T. N., & Linstedt, A. D. (2008). GRASP55 regulates Golgi ribbon formation. *Molecular Biology of the Cell*, *19*, 2696–2707.

Gaietta, G. M., Giepmans, B. N., Deerinck, T. J., Smith, W. B., Ngan, L., Llopis, J., et al. (2006). Golgi twins in late mitosis revealed by genetically encoded tags for live cell imaging and correlated electron microscopy. *Proceedings of the National Academy of Sciences of the United States of America*, *103*, 17777–17782.

Hidalgo Carcedo, C., Bonazzi, M., Spano, S., Turacchio, G., Colanzi, A., Luini, A., et al. (2004). Mitotic Golgi partitioning is driven by the membrane-fissioning protein CtBP3/BARS. *Science*, *305*, 93–96.

Jackman, M., Firth, M., & Pines, J. (1995). Human cyclins B1 and B2 are localized to strikingly different structures: B1 to microtubules, B2 primarily to the Golgi apparatus. *The EMBO Journal*, *14*, 1646–1654.

Kaneko, Y., Tamura, K., Totsukawa, G., & Kondo, H. (2010). Phosphorylation of p37 is important for Golgi disassembly at mitosis. *Biochemical and Biophysical Research Communications*, *402*, 37–41.

Kienzle, C., Eisler, S. A., Villeneuve, J., Brummer, T., Olayioye, M. A., & Hausser, A. (2013). PKD controls mitotic Golgi complex fragmentation through a Raf-MEK1 pathway. *Molecular Biology of the Cell*, *24*, 222–233.

Kodani, A., Kristensen, I., Huang, L., & Sütterlin, C. (2009). GM130-dependent control of Cdc42 activity at the Golgi regulates centrosome organization. *Molecular Biology of the Cell*, *20*, 1192–1200.

Kondo, H., Rabouille, C., Newman, R., Levine, T. P., Pappin, D., Freemont, P., et al. (1997). p47 is a cofactor for p97-mediated membrane fusion. *Nature*, *388*, 75–78.

Liberali, P., Kakkonen, E., Turacchio, G., Valente, C., Spaar, A., Perinetti, G., et al. (2008). The closure of Pak1-dependent macropinosomes requires the phosphorylation of CtBP1/BARS. *The EMBO Journal*, *27*, 970–981.

Litvak, V., Argov, R., Dahan, N., Ramachandran, S., Amarilio, R., Shainskaya, A., et al. (2004). Mitotic phosphorylation of the peripheral Golgi protein Nir2 by Cdk1 provides a docking mechanism for Plk1 and affects cytokinesis completion. *Molecular Cell*, *14*, 319–330.

Liu, F., Stanton, J. J., Wu, Z., & Piwnica-Worms, H. (1997). The human Myt1 kinase preferentially phosphorylates Cdc2 on threonine 14 and localizes to the endoplasmic reticulum and Golgi complex. *Molecular and Cellular Biology*, *17*, 571–583.

Lopez-Sanchez, I., Sanz-Garcia, M., & Lazo, P. A. (2009). Plk3 interacts with and specifically phosphorylates VRK1 in Ser342, a downstream target in a pathway that induces Golgi fragmentation. *Molecular and Cellular Biology*, *29*, 1189–1201.

Luhrig, S., Kolb, S., Mellies, N., & Nolte, J. (2013). The novel BTB-kelch protein, KBTBD8, is located in the Golgi apparatus and translocates to the spindle apparatus during mitosis. *Cell Division*, *8*, 3.

Maia, A. R., Zhu, X., Miller, P., Gu, G., Maiato, H., & Kaverina, I. (2013). Modulation of Golgi-associated microtubule nucleation throughout the cell cycle. *Cytoskeleton*, *70*, 32–43.

Marra, P., Salvatore, L., Mironov, A., Jr., Di Campli, A., Di Tullio, G., Trucco, A., et al. (2007). The biogenesis of the Golgi ribbon: The roles of membrane input from the ER and of GM130. *Molecular Biology of the Cell*, *18*, 1595–1608.

Marumoto, T., Hirota, T., Morisaki, T., Kunitoku, N., Zhang, D., Ichikawa, Y., et al. (2002). Roles of aurora-A kinase in mitotic entry and G2 checkpoint in mammalian cells. *Genes to Cells*, *7*, 1173–1182.

Meyer, H. H., Shorter, J. G., Seemann, J., Pappin, D., & Warren, G. (2000). A complex of mammalian ufd1 and npl4 links the AAA-ATPase, p97, to ubiquitin and nuclear transport pathways. *The EMBO Journal*, *19*, 2181–2192.

Miserey-Lenkei, S., Couedel-Courteille, A., Del Nery, E., Bardin, S., Piel, M., Racine, V., et al. (2006). A role for the Rab6A' GTPase in the inactivation of the Mad2-spindle checkpoint. *The EMBO Journal*, *25*, 278–289.

Misteli, T., & Warren, G. (1994). COP-coated vesicles are involved in the mitotic fragmentation of Golgi stacks in a cell-free system. *Journal of Cell Biology*, *125*, 269–282.

Misteli, T., & Warren, G. (1995). Mitotic disassembly of the Golgi apparatus in vivo. *Journal of Cell Science*, *108*(Pt. 7), 2715–2727.

Morohashi, Y., Balklava, Z., Ball, M., Hughes, H., & Lowe, M. (2010). Phosphorylation and membrane dissociation of the ARF exchange factor GBF1 in mitosis. *Biochemical Journal*, *427*, 401–412.

Munro, S. (2011). The golgin coiled-coil proteins of the Golgi apparatus. *Cold Spring Harbor Perspectives in Biology*, *3*, a005256.

Nakagawa, J., Kitten, G. T., & Nigg, E. A. (1989). A somatic cell-derived system for studying both early and late mitotic events in vitro. *Journal of Cell Science*, *94*(Pt. 3), 449–462.

Nakajima, H., Yonemura, S., Murata, M., Nakamura, N., Piwnica-Worms, H., & Nishida, E. (2008). Myt1 protein kinase is essential for Golgi and ER assembly during mitotic exit. *Journal of Cell Biology*, *181*, 89–103.

Nelson, W. J. (2000). W(h)ither the Golgi during mitosis? *Journal of Cell Biology*, *149*, 243–248.

Nigg, E. A. (2001). Mitotic kinases as regulators of cell division and its checkpoints. *Nature Reviews. Molecular Cell Biology*, *2*, 21–32.

Ozaki, Y., Matsui, H., Asou, H., Nagamachi, A., Aki, D., Honda, H., et al. (2012). Poly-ADP ribosylation of Miki by tankyrase-1 promotes centrosome maturation. *Molecular Cell*, *47*, 694–706.

Papanikou, E., & Glick, B. S. (2009). The yeast Golgi apparatus: Insights and mysteries. *FEBS Letters*, *583*, 3746–3751.

Pelletier, L., Stern, C. A., Pypaert, M., Sheff, D., Ngo, H. M., Roper, N., et al. (2002). Golgi biogenesis in Toxoplasma gondii. *Nature*, *418*, 548–552.

Persico, A., Cervigni, R. I., Barretta, M. L., Corda, D., & Colanzi, A. (2010). Golgi partitioning controls mitotic entry through Aurora-A kinase. *Molecular Biology of the Cell*, *21*, 3708–3721.

Preisinger, C., Korner, R., Wind, M., Lehmann, W. D., Kopajtich, R., & Barr, F. A. (2005). Plk1 docking to GRASP65 phosphorylated by Cdk1 suggests a mechanism for Golgi checkpoint signalling. *The EMBO Journal*, *24*, 753–765.

Preuss, D., Mulholland, J., Franzusoff, A., Segev, N., & Botstein, D. (1992). Characterization of the Saccharomyces Golgi complex through the cell cycle by immunoelectron microscopy. *Molecular Biology of the Cell*, *3*, 789–803.

Puthenveedu, M. A., Bachert, C., Puri, S., Lanni, F., & Linstedt, A. D. (2006). GM130 and GRASP65-dependent lateral cisternal fusion allows uniform Golgi-enzyme distribution. *Nature Cell Biology*, *8*, 238–248.

Rabouille, C., Levine, T. P., Peters, J. M., & Warren, G. (1995). An NSF-like ATPase, p97, and NSF mediate cisternal regrowth from mitotic Golgi fragments. *Cell*, *82*, 905–914.

Radulescu, A. E., Mukherjee, S., & Shields, D. (2011). The Golgi protein p115 associates with gamma-tubulin and plays a role in Golgi structure and mitosis progression. *Journal of Biological Chemistry*, *286*, 21915–21926.

Rios, R. M., & Bornens, M. (2003). The Golgi apparatus at the cell centre. *Current Opinion in Cell Biology*, *15*, 60–66.

Robinson, M. J., & Cobb, M. H. (1997). Mitogen-activated protein kinase pathways. *Current Opinion in Cell Biology*, *9*, 180–186.

Ruan, Q., Wang, Q., Xie, S., Fang, Y., Darzynkiewicz, Z., Guan, K., et al. (2004). Polo-like kinase 3 is Golgi localized and involved in regulating Golgi fragmentation during the cell cycle. *Experimental Cell Research*, *294*, 51–59.

Sengupta, D., & Linstedt, A. D. (2010). Mitotic inhibition of GRASP65 organelle tethering involves Polo-like kinase 1 (PLK1) phosphorylation proximate to an internal PDZ ligand. *Journal of Biological Chemistry*, *285*, 39994–40003.

Shaul, Y. D., & Seger, R. (2006). ERK1c regulates Golgi fragmentation during mitosis. *Journal of Cell Biology*, *172*, 885–897.

Spyer, M., & Allday, M. J. (2006). The transcriptional co-repressor C-terminal binding protein (CtBP) associates with centrosomes during mitosis. *Cell Cycle*, *5*, 530–537.

Sütterlin, C., & Colanzi, A. (2010). The Golgi and the centrosome: Building a functional partnership. *Journal of Cell Biology*, *188*, 621–628.

Sütterlin, C., Hsu, P., Mallabiabarrena, A., & Malhotra, V. (2002). Fragmentation and dispersal of the pericentriolar Golgi complex is required for entry into mitosis in mammalian cells. *Cell*, *109*, 359–369.

Sütterlin, C., Lin, C. Y., Feng, Y., Ferris, D. K., Erikson, R. L., & Malhotra, V. (2001). Polo-like kinase is required for the fragmentation of pericentriolar Golgi stacks during mitosis.

Proceedings of the National Academy of Sciences of the United States of America, *98*, 9128–9132.

Sütterlin, C., Polishchuk, R., Pecot, M., & Malhotra, V. (2005). The Golgi-associated protein GRASP65 regulates spindle dynamics and is essential for cell division. *Molecular Biology of the Cell*, *16*, 3211–3222.

Tang, D., Xiang, Y., De Renzis, S., Rink, J., Zheng, G., Zerial, M., et al. (2011). The ubiquitin ligase HACE1 regulates Golgi membrane dynamics during the cell cycle. *Nature Communications*, *2*, 501.

Tang, D., Xiang, Y., & Wang, Y. (2010). Reconstitution of the cell cycle-regulated Golgi disassembly and reassembly in a cell-free system. *Nature Protocols*, *5*, 758–772.

Tang, D., Yuan, H., Vielemeyer, O., Perez, F., & Wang, Y. (2012). Sequential phosphorylation of GRASP65 during mitotic Golgi disassembly. *Biology Open*, *1*, 1204–1214.

Tang, D., Yuan, H., & Wang, Y. (2010). The role of GRASP65 in Golgi cisternal stacking and cell cycle progression. *Traffic*, *11*, 827–842.

Terasaki, M. (2000). Dynamics of the endoplasmic reticulum and golgi apparatus during early sea urchin development. *Molecular Biology of the Cell*, *11*, 897–914.

Thyberg, J., & Moskalewski, S. (1999). Role of microtubules in the organization of the Golgi complex. *Experimental Cell Research*, *246*, 263–279.

Totsukawa, G., Matsuo, A., Kubota, A., Taguchi, Y., & Kondo, H. (2013). Mitotic phosphorylation of VCIP135 blocks p97ATPase-mediated Golgi membrane fusion. *Biochemical and Biophysical Research Communications*, *433*, 237–242.

Truschel, S. T., Zhang, M., Bachert, C., Macbeth, M. R., & Linstedt, A. D. (2012). Allosteric regulation of GRASP protein-dependent Golgi membrane tethering by mitotic phosphorylation. *Journal of Biological Chemistry*, *287*, 19870–19875.

Uchiyama, K., Jokitalo, E., Lindman, M., Jackman, M., Kano, F., Murata, M., et al. (2003). The localization and phosphorylation of p47 are important for Golgi disassembly-assembly during the cell cycle. *Journal of Cell Biology*, *161*, 1067–1079.

Uchiyama, K., Totsukawa, G., Puhka, M., Kaneko, Y., Jokitalo, E., Dreveny, I., et al. (2006). p37 is a p97 adaptor required for Golgi and ER biogenesis in interphase and at the end of mitosis. *Developmental Cell*, *11*, 803–816.

Valente, C., Turacchio, G., Mariggio, S., Pagliuso, A., Gaibisso, R., Di Tullio, G., et al. (2012). A 14-3-3gamma dimer-based scaffold bridges CtBP1-S/BARS to PI(4)KIIIbeta to regulate post-Golgi carrier formation. *Nature Cell Biology*, *14*, 343–354.

Vallee, R. B., Varma, D., & Dujardin, D. L. (2006). ZW10 function in mitotic checkpoint control, dynein targeting and membrane trafficking: Is dynein the unifying theme? *Cell Cycle*, *5*, 2447–2451.

Villeneuve, J., Scarpa, M., Ortega-Bellido, M., & Malhotra, V. (2013). MEK1 inactivates Myt1 to regulate Golgi membrane fragmentation and mitotic entry in mammalian cells. *The EMBO Journal*, *32*, 72–85.

Vinke, F. P., Grieve, A. G., & Rabouille, C. (2011). The multiple facets of the Golgi reassembly stacking proteins. *Biochemical Journal*, *433*, 423–433.

Wang, Y., Satoh, A., Warren, G., & Meyer, H. H. (2004). VCIP135 acts as a deubiquitinating enzyme during p97-p47-mediated reassembly of mitotic Golgi fragments. *Journal of Cell Biology*, *164*, 973–978.

Wang, Y., Seemann, J., Pypaert, M., Shorter, J., & Warren, G. (2003). A direct role for GRASP65 as a mitotically regulated Golgi stacking factor. *The EMBO Journal*, *22*, 3279–3290.

Warren, G., Levine, T., & Misteli, T. (1995). Mitotic disassembly of the mammalian Golgi apparatus. *Trends in Cell Biology*, *5*, 413–416.

Wilson, C., Venditti, R., Rega, L. R., Colanzi, A., D'Angelo, G., & De Matteis, M. A. (2011). The Golgi apparatus: An organelle with multiple complex functions. *Biochemical Journal*, *433*, 1–9.

Xiang, Y., & Wang, Y. (2010). GRASP55 and GRASP65 play complementary and essential roles in Golgi cisternal stacking. *Journal of Cell Biology*, *188*, 237–251.

Xiang, Y., Zhang, X., Nix, D. B., Katoh, T., Aoki, K., Tiemeyer, M., et al. (2013). Regulation of protein glycosylation and sorting by the Golgi matrix proteins GRASP55/65. *Nature Communications*, *4*, 1659.

Yadav, S., Puri, S., & Linstedt, A. D. (2009). A primary role for Golgi positioning in directed secretion, cell polarity, and wound healing. *Molecular Biology of the Cell*, *20*, 1728–1736.

Yang, J. S., Valente, C., Polishchuk, R. S., Turacchio, G., Layre, E., Moody, D. B., et al. (2011). COPI acts in both vesicular and tubular transport. *Nature Cell Biology*, *13*, 996–1003.

Yeong, F. M. (2013). Multi-step down-regulation of the secretory pathway in mitosis: A fresh perspective on protein trafficking. *BioEssays: News and Reviews in Molecular, Cellular and Developmental Biology*, *35*, 462–471.

Yoshimura, S., Yoshioka, K., Barr, F. A., Lowe, M., Nakayama, K., Ohkuma, S., et al. (2005). Convergence of cell cycle regulation and growth factor signals on GRASP65. *Journal of Biological Chemistry*, *280*, 23048–23056.

Zhou, Y., Atkins, J. B., Rompani, S. B., Bancescu, D. L., Petersen, P. H., Tang, H., et al. (2007). The mammalian Golgi regulates numb signaling in asymmetric cell division by releasing ACBD3 during mitosis. *Cell*, *129*, 163–178.

CHAPTER

Ice Recovery Assay for Detection of Golgi-Derived Microtubules

24

Ashley D. Grimaldi*, Maria Fomicheva*,† and Irina Kaverina*

**Department of Cell and Developmental Biology, Vanderbilt University Medical Center, Nashville, Tennessee, USA*

†Department of Biology, Lomonosov Moscow State University, Moscow, Russia

CHAPTER OUTLINE

Introduction and Rationale 402
24.1 Materials 403
24.1.1 Cell Culture and Transfection 403
24.1.2 Ice Treatment and Ice Recovery 404
24.1.2.1 Required Materials for Ice Treatment 404
24.1.2.2 Optional Materials for Recovery At 37 °C and Free Tubulin Extraction Step 404
24.1.3 Immunofluorescence 404
24.1.4 Confocal Microscope and Analysis 404
24.2 Methods 405
24.2.1 Cell Maintenance and Preparation for Transfection 405
24.2.2 Ice Treatment and Ice Recovery 407
24.2.3 Immunofluorescence 408
24.2.4 Image Acquisition 409
24.2.5 Analysis of MT Nucleation and MT Subpopulations 410
24.3 Discussion 412
Summary 413
Acknowledgments 413
References 414

Abstract

Proper organization of the microtubule cytoskeleton is essential for many cellular processes including maintenance of Golgi organization and cell polarity. Traditionally, the centrosome is considered to be the major microtubule organizing center (MTOC)

ISSN 0091-679X
http://dx.doi.org/10.1016/B978-0-12-417164-0.00024-0

of the cell; however, microtubule nucleation can also occur through centrosome-independent mechanisms. Recently, the Golgi has been described as an additional, centrosome-independent, MTOC with distinct cellular functions. Golgi-derived microtubules contribute to the formation of an asymmetric microtubule network, control Golgi organization, and support polarized trafficking and directed migration in motile cells. In this chapter, we present an assay using recovery from ice treatment to evaluate the potential of the Golgi, or other MTOCs, to nucleate microtubules. This technique allows for clear separation of distinct MTOCs and observation of newly nucleated microtubules at these locations, which are normally obscured by the dense microtubule network present at steady-state conditions. This type of analysis is important for discovery and characterization of noncentrosomal MTOCs and, ultimately, understanding of their unique cellular functions.

INTRODUCTION AND RATIONALE

Microtubules (MTs) function as tracks within the cell to organize the intracellular space by distributing organelles and proteins. The inherent polarity of MTs allows for directionality in intracellular transport. In addition to the innate polarity provided by MTs, it is important for the cell to establish a structurally and functionally asymmetric MT network for many important processes, such as cell migration and cell division. Conventionally, centrosomes serve as the principal microtubule organizing center (MTOC) from which the majority of MTs in the cell are nucleated; however, MTs nucleated at the centrosome have been shown to be symmetric in nature, both *in vitro* (Bergen, Kuriyama, & Borisy, 1980) and in polarized cells (Salaycik, Fagerstrom, Murthy, Tulu, & Wadsworth, 2005). Therefore, the cell must have other mechanisms in place to allow for the establishment of asymmetric MT networks.

Multiple mechanisms exist that allow for the cell to establish an asymmetric MT network. For example, repositioning of the centrosome away from the cell center, local stabilization of MTs at a specific region in the cell, or severing and releasing of MTs from the centrosome (reviewed in Vinogradova, Miller, & Kaverina, 2009). In addition, an alternative, noncentrosomal MTOC can serve as a site for nucleation of an asymmetric MT array. Several recent studies have demonstrated the role of the Golgi complex as a major MTOC in mammalian cells (Chabin-Brion et al., 2001; Efimov et al., 2007; Rivero, Cardenas, Bornens, & Rios, 2009). Interestingly, the Golgi complex itself is intrinsically asymmetric: the Golgi is comprised of membrane cisternae stacks with distinct *cis*-to-*trans* polarity (Ladinsky, Wu, McIntosh, McIntosh, & Howell, 2002). In addition, Golgi-derived MTs are oriented predominately toward the leading edge of a motile cell, allowing for directional post-Golgi trafficking and cell migration (Efimov et al., 2007; Miller et al., 2009). Noncentrosomal, Golgi-derived MTs have also been shown to be essential in maintaining the cellular architecture of complex MT-dependent cells, such as neurons (Ori-McKenney, Jan, & Jan, 2012).

MTs are required for proper Golgi organization. Our recent analyses of Golgi-derived MTs have shown that this specific MT subpopulation is essential for proper

Golgi assembly, morphology, and integrity (Miller et al., 2009). MT depolymerization causes disruption of the Golgi complex into individual mini-stacks; while, in the presence of a normal MT network, the Golgi complex accumulates near the centrosome, due to the function of the minus-end-directed motor Dynein (Vaughan, 2005). Initially it was unclear to what degree the centrosomal and Golgi-derived MTOCs were involved in the reassembly of Golgi mini-stacks after MT depolymerization. Further studies in our lab revealed that both centrosomal and Golgi-derived MT populations are required for proper Golgi complex assembly; however, Golgi-derived MTs are necessary and sufficient for Golgi maintenance once formed (Vinogradova et al., 2012).

In addition to MTs, the actin cytoskeleton also plays a role in Golgi organization. Actin and many actin-binding proteins have been localized to the Golgi complex and have been shown to be important in the function of this organelle (reviewed in Egea, Lazaro-Dieguez, & Viella, 2006). In plants and fungi, actin and the actin-based motor, myosin, are required for the movement and organization of Golgi stacks; however, it has recently been shown that myosin also plays a role in Golgi organization in mammalian cells (reviewed in Lowe, 2011).

Frequently within the cell, the Golgi complex is located near the centrosome resulting in the maintenance of two independent MTOCs, one symmetric and one asymmetric, in close proximity to each other. This close proximity can make it technically difficult to distinguish between MTs nucleated at either structure and it can be nearly impossible to observe newly nucleated Golgi-derived MTs at steady state due to the dense MT network existing around the MTOCs. There are multiple tricks and techniques, which have been developed to circumvent these dilemmas. Here we present a protocol for clear detection of Golgi-derived MTs using an ice recovery assay, first applied for this purpose by Rivero et al. (2009). This technique can be easily adapted to characterize and study MT nucleation at other centrosomal and noncentrosomal MTOCs as well.

24.1 MATERIALS

24.1.1 Cell culture and transfection

1. *Cells*: Immortalized human pigment epithelial cells hTert-RPE1 (Clontech).
2. *Complete media*: Dulbecco's Modification of Eagle's Medium/Ham's F-12 50/50 Mix (DMEM F-12K, Mediatech) supplemented with 10% fetal bovine serum (Atlanta Biologicals), and 1 mM L-glutamine (Mediatech).
3. *Trypsin solution*: 0.25% trypsin with EDTA, without sodium bicarbonate, calcium, and magnesium (Mediatech).
4. *Transfection reagent*: FuGENE 6 (Roche).
5. *Plasmid*: Centrosome marker (e.g., Centrin2-EGFP, Supplied by Jeffrey Salisbury on Addgene).
6. Plastic tissue culture flasks and dishes (Santa Cruz).

24.1.2 Ice treatment and ice recovery

24.1.2.1 Required Materials for Ice Treatment

1. Ice bucket and Ice

24.1.2.2 Optional Materials for Recovery at 37 °C and Free Tubulin Extraction Step

1. 37 °C water bath.
2. Nocodazole (Sigma) stock: 16.6 mM in DMSO. Store at −20 °C.
3. Paclitaxel (Sigma) stock: 5 μM in DMSO. Store at −20 °C.
4. Extraction buffer: 60 mM PIPES, 25 mM HEPES, 10 mM EGTA, 2 mM $MgCl_2$, 0.1% Saponin; Adjust pH to 6.9 and store at 4 °C. Immediately before use, supplement with 0.25 nM nocodazole and 0.25 nM paclitaxel (Zhai & Borisy, 1994).

24.1.3 Immunofluorescence

1. No. 1.5 glass microscope coverslips (circular, 12 mm width) and microscope slides for immunofluorescence (Fisher Scientific).
2. *Human fibronectin (BD Biosciences) stock*: Prepare a 2.27 μM (1 mg/mL) stock in 1 M urea. For long-term storage keep at −80 °C, working solution can be kept at 4 °C.
3. *Phosphate-buffered saline (PBS)*: Prepare a 10 × stock solution by mixing 1.37 M NaCl, 27 mM KCl, 80 mM Na_2HPO_4, 20 mM KH_2PO_4 then autoclave. Alternatively, use pre-made 10 × PBS (Corning). Dilute to 1 × working solution before use, store at room temperature.
4. *Fixation*: Methanol (EMD Chemicals Inc.) and −20 °C freezer for incubation.
5. *Blocking and antibody dilution buffer*: Prepare a 1% bovine serum albumin (Fisher) in PBS supplemented with 5% donor horse serum (Atlanta Biologicals). For long-term storage, keep at −20 °C; working solution can be kept at 4 °C.
6. *Primary antibodies*: Mouse monoclonal anti-GM130 antibody (used at 1:300 dilution, BD Biosciences), rabbit polyclonal anti-alpha tubulin (used at 1:1000 dilution, Abcam).
7. *Secondary antibodies*: Highly cross-absorbed Alexa Fluor 568 anti-mouse (Invitrogen), highly cross-absorbed Alexa Fluor 647 anti-rabbit (Invitrogen).
8. *Mounting medium*: ProLong Gold antifade reagent (Invitrogen). Store at −20 °C.
9. Clear nail polish.

24.1.4 Confocal microscope and analysis

1. *Microscope*: Leica TCS SP5 laser-scanning confocal system with 488, 543, and 633 excitation laser lines for triple-color imaging.
2. *Optics*: 100 ×, 1.47 numerical aperture (NA) Plan Apo lens.
3. Software packages, such as NIH ImageJ (http://rsbweb.nih.gov/ij/), can be used for image analysis.

24.2 METHODS

We suggest the following method for identification and analysis of noncentrosomal MT nucleation sites, using the Golgi as an example. Because the centrosome and Golgi are often located in close proximity to each other within the cell, it can be difficult to study MT nucleation at these independent sites during steady state. In some cells, this dilemma can be overcome if a significant distance spatially separates the centrosome and the Golgi; however, this is not always possible. Alternatively, MT nucleation can be monitored after ice treatment recovery to clearly visualize multiple MTOCs. When cells are subjected to ice treatment, the entire MT network is disassembled; subsequent warming up allows for MT regrowth. Without obstruction from the entire MT network, individual MTOCs and MT subpopulations can now be resolved and examined (Fig. 24.1).

24.2.1 Cell maintenance and preparation for transfection

RPE1 cells are maintained in 5-mL tissue culture flasks and passaged when approximately 80% confluent. To passage, cells are treated with trypsin solution for 1–2 min at room temperature or until cells begin to round up and detach from the bottom of the flask; cells are then split into a new 5-mL flask for maintenance at a 1:3 or 1:6 dilution. For transfection and ice recovery followed by immunofluorescence, cells are plated at a 1:6 dilution in 35-mm tissue culture dishes containing prepared glass microscope coverslips.

To prepare coverslips for immunofluorescence, empty a pack of coverslips into a glass petri dish and autoclave to sterilize. Sterilized coverslips are kept in the tissue culture hood, under UV, until ready to use. Using forceps, transfer coverslips from the petri dish to a 35-mm tissue culture dish. Before plating cells onto the coverslips, we first coat the glass surface with fibronectin to promote cell attachment and cell spreading. Fibronectin stock (see Step 2 in Section 24.1.3) is diluted 1:100 with 1 × PBS, then a 50–100 μL drop of diluted fibronectin is added onto the center of each coverslip. Finally, the coverslips are incubated with fibronectin for a minimum of 1 h at room temperature, or 30 min at 37 °C, to allow coating of the glass surface. Prior to plating your cells, remove excess fibronectin solution from the coverslip surface by aspiration; be careful to only aspirate at the top of the droplet to prevent disruption on the fibronectin that has settled onto the coverslip.

To plate your cells onto the coverslips, trypsinize and split them into the 35 mm dish with fibronectin-coated coverslips at the desired dilution. Cells are then incubated at 37 °C to allow for attachment to the fibronectin-coated surface and cell spreading.

For transfection, the cells are transfected with 0.5 μg of Centrin-GFP using the FuGENE transfection reagent according to the manufacture's protocol: In a 1.5-mL Eppendorf tube containing 100 μL of plain DMEM/F12K medium, add 3 μL of FuGENE and mix by gently flicking the tube several times. Then, add 0.5 μg of Centrin-GFP to the transfection mixture and gently flick the tube to mix

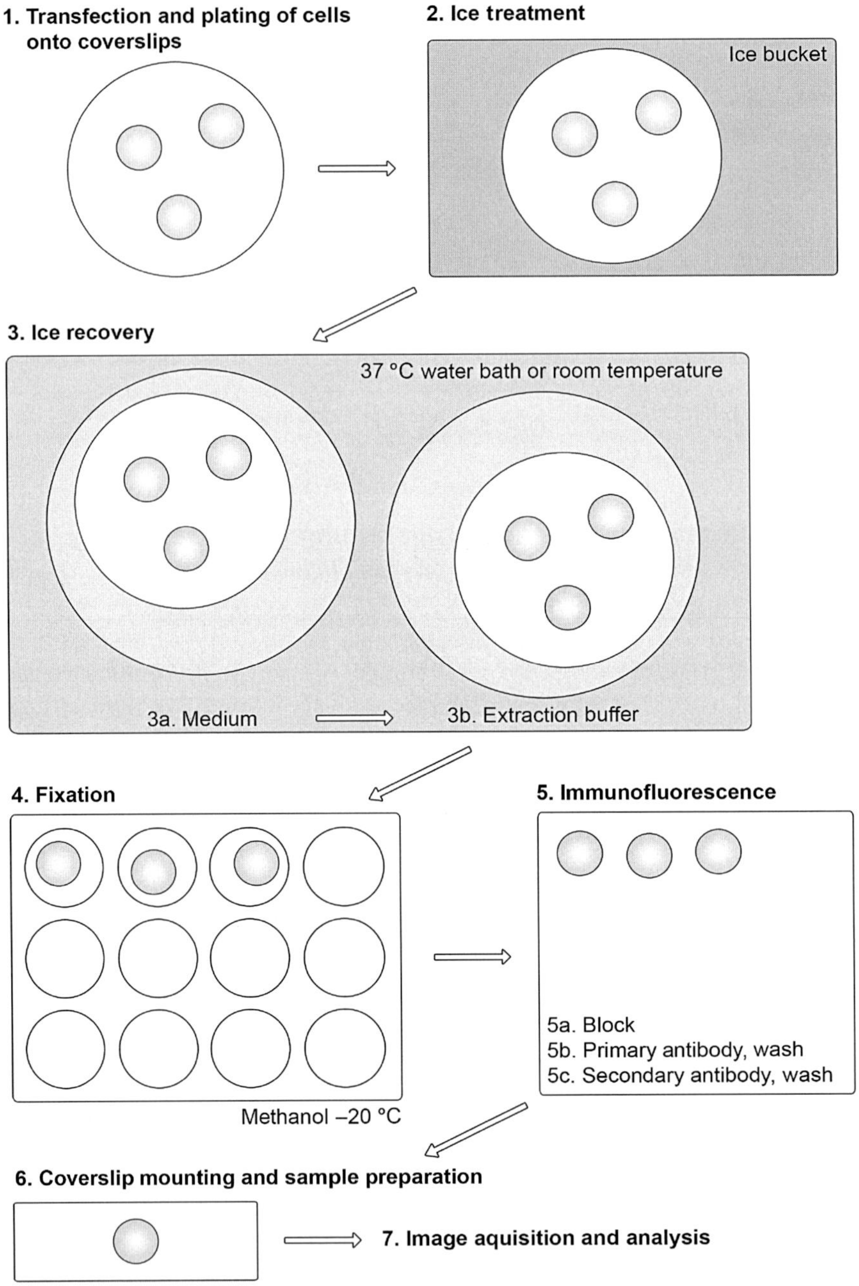

FIGURE 24.1

Steps of the ice recovery assay. The sequence of the main steps during the ice recovery assay. Small gray circles represent coverslips and white arrows indicate the transition from one step in the procedure to the next. For details of the ice recovery procedure, see main text. (See color plate.)

once again. Incubate the mixture at room temperature for a minimum of 30 min then add drop-wise to the cells you have already plated onto the fibronectin-coated coverslips. Gently swirl the dish to mix the transfection mixture with the medium already in the dish. Incubate the cells with the transfection mixture for 1–2 days to allow expression of the protein of interest prior to the ice recovery experiment. We recommend that you try various transfection conditions (such as amount of DNA, amount of FuGENE, and expression time) to find the optimal Centrin-GFP expression conditions for your cell type.

24.2.2 Ice treatment and ice recovery

For the ice treatment, fill a bucket with ice and keep on your bench at room temperature. To allow for disassembly of the MT cytoskeleton, the 35-mm dish containing cells adhered to coverslips is placed directly onto ice and incubated for 40 min. Shorter incubation times will result in partial disassembly of the MT network with stable populations, such as acetylated-tubulin, still remaining within the cell.

During the ice incubation, prepare the fixative by adding 1–2 mL of 100% methanol to a 12-well plate. Chill the 12-well plate at −20 °C for at least 15 min to ensure the fixative reaches the required temperature. Be careful not to contaminate the methanol with any ice or water, as this will interfere with optimal fixation.

Once all MTs are disassembled, ice recovery can be achieved through two methods: the first is recovery at 37 °C and the second is recovery at room temperature. Recovery from ice treatment at room temperature is ideal because MT nucleation and regrowth occurs more slowly than at 37 °C, which allows for longer recovery periods and the ability to capture the fine details of the process.

To prepare for ice recovery at 37 °C, first float two 60 mm tissue culture dishes on a Styrofoam raft in a 37 °C water bath. Fill one dish with 5 mL of fresh complete media and one dish with 5 mL of freshly prepared extraction buffer; allow the solutions to warm to 37 °C for several minutes. To recover MT nucleation after ice treatment, quickly lift the coverslip from the 35-mm dish on ice and place it directly into the pre-warmed complete media. Gently swirl the dish for 45 s to allow MT regrowth. Next, immediately remove the coverslip from the media and immerse in the pre-warmed extraction buffer. Gently swirl the dish for another 40 s to allow extraction of free tubulin from the cytoplasm. After recovery at 37 °C, coverslips are immediately subjected to fixation. For recovery at room temperature, simply remove the 35-mm dish containing your coverslips from the ice bucket and place on your bench at room temperature. Incubate the coverslips at room temperature for 3–5 min to allow MT regrowth. If desired, the coverslips can then be incubated in extraction buffer to remove excess free tubulin from the cytoplasm as described above.

Please note that the extraction buffer containing low amounts of nocodazole and paclitaxel can be used as an optional step after ice recovery and prior to fixation to decrease free tubulin in the cytoplasm. Because the majority of the MT network remains depolymerized after short ice recovery times, a large amount of free tubulin exists in the cytoplasm, which upon immunofluorescence can decrease the quality of your images. If desired, brief incubation of your coverslips in extraction buffer can help to eliminate background fluorescence created by free tubulin, and can enhance the contrast in your images.

It is important to use short recovery times because MT nucleation and regrowth are quite rapid; various recovery times should be tested to optimize visualization of desired dynamic processes. Ideally, for the best detection of MT regrowth at MTOCs, we recommend allowing the MTs to grow between 1 and 5 μm in length prior to fixation (e.g., images, see Fig. 24.3).

The ice incubation time necessary to fully disassemble the entire MT network may vary between cell types. Note that it is important to find the minimal time on ice required to disassemble all MTs without exposing the cells to unnecessarily long periods on ice. Therefore, before conducting the full ice recovery experiment, it is recommended that you test various ice incubation times and stain for alpha-tubulin to confirm that the MT network is completely disassembled. For RPE1 cells, we and others have determined that 40 min is sufficient to disassemble the entire MT network (Rivero et al., 2009; Vinogradova et al., 2012).

24.2.3 Immunofluorescence

Using the pre-chilled methanol (see Section 24.2.2), immerse the coverslip into the fixative immediately after completion of ice recovery. The coverslip should be carefully lifted from the dish (at either room temperature or 37 °C), with forceps, and placed cell-side facing up into the fixation solution. Ensure that the coverslip is completely covered in methanol (i.e., not floating) by gently pressing on the edge of the coverslip with forceps. Incubate the coverslip in methanol at −20 °C for 5 min, then immediately wash three times with 1 × PBS. During the wash steps, be sure not to expose the cells to air by keeping a small amount of PBS in the 12-well plate at all times. Following methanol fixation, it is important to keep the coverslip in PBS for a minimum of 1 h to allow for rehydration of the cells after fixation.

To prepare for immunofluorescence, adhere a piece of parafilm to the bottom of a plastic box. Then, place a slightly damp Kimwipe in the lid of the box to create a humid chamber. Be sure to check that the Kimwipe is moist throughout the immunofluorescence procedure and remoisten if necessary.

For the blocking step, place 18 μL drops of blocking solution (see Step 5 in Section 24.1.3) onto the parafilm. Using forceps, carefully transfer the coverslip from the PBS wash (in the 12-well plate) onto the blocking solution with the cell-side facing down; this will allow for the cells to be in contact with the blocking solution. Incubate 40 min to 1 h at room temperature.

For all steps of immunofluorescence carried out within the parafilm-lined plastic box, it is important that the coverslip be floating on top of the droplet of solution (e.g., block or antibody) so the glass surface with cells is completely exposed to the solution. The cell-side of the coverslip should not touch the parafilm or be exposed to air and should never be pressed down, which may damage the cells.

Next, place 18 μL drops of primary antibody solution onto clean parafilm and transfer the coverslip onto the primary antibody solution using the same procedure described above. Incubate for 1 h at room temperature. For longer incubation times, place the plastic box at 4 °C. To wash, immerse the coverslip in a 12-well plate containing approximately 3 mL of 1 × PBS with the cell-side facing up; wash three times for 5 min each, using fresh PBS each time.

For the secondary antibody incubation, repeat the same procedure and wash steps described for the primary antibody and protect the coverslip from light by covering the plastic box with aluminum foil or placing it in a closed desk drawer.

To mount the coverslip, place a drop of mounting medium (see Step 8 in Section 24.1.3) onto a clean glass microscope slide. Then, using forceps, invert the coverslip onto the mounting medium with the cell-side face down. Use a Kimwipe to wick away any excess mounting media from the sides of the coverslip. Keep the newly prepared slide at room temperature over night or 37 °C for 1 h, protected from light, to allow for the mounting medium to set. Lastly, seal the edge of the coverslip with clear nail polish to prevent it from moving during image acquisition. It is important to use clear nail polish as some colored nail polishes can fluoresce under imaging conditions, which will result in poor image quality.

After the nail polish has dried, to prepare the sample for image acquisition it is important to carefully clean the surface of the coverslip with ethanol to remove residual dried PBS. The samples are now ready for examination on the microscope. For long-term storage, keep samples at −20 °C protected from light.

Please note that the volumes outlines in this immunofluorescence procedure (18 μL drops of the appropriate solution) have been optimized in our lab for use with 12 mm coverslips (see Step 1 in Section 24.1.3). This procedure can also be used with coverslips of varying sizes; however, the volumes listed above must be adjusted accordingly. Importantly, this "drop" procedure for immunofluorescence should be considered as an alternative to traditional methods because it requires only small volumes of each reagent and is, therefore, an extremely cost-effective method that provides excellent sample quality.

24.2.4 Image acquisition

Because you will be looking at structures that are below optical resolution, it is important to use the highest resolution possible when acquiring data for analysis of MT nucleation after ice recovery. We recommend the use of a high-resolution 100 × objective lens with an NA of at least 1.4 (see Step 2 in Section 24.1.4). In addition, we recommend that you acquire confocal slices from individual channels sequentially to allow for the best separation of fluorescence in neighboring channels and to enhance image quality.

In order to monitor all MTs nucleated in 3D, it is important to acquire Z-stack confocal images of the entire cell. As a starting point, we recommend using 0.2 μm slices for optimal resolution in Z; however, the Z interval may need to be adjusted based on cell type and microscope system.

24.2.5 Analysis of MT nucleation and MT subpopulations

ImageJ, or any other appropriate software, can be used for image processing. To analyze MT nucleation at MTOCs following ice recovery experiments, colocalization between MT minus ends and the MTOC of interest must be examined in 3D. This can be achieved by examining a stack of confocal slices or by creating a 3D reconstruction from the stack. Quantification of colocalization is done manually by counting the instances where one end of an MT can be seen touching either a centrosome or a Golgi stack (for examples of colocalization, see Fig. 24.3). If desired, colocalization can be detected by software, although algorithms that combine detection of colocalization and automated detection of single MTs are not readily available.

Z-stack images are projected, individually for each channel acquired, using a maximum intensity projection algorithm. Individual channels can then be contrasted and overlaid to analyze colocalization between MTs and your subcellular structure of interest.

For the example of centrosomal and Golgi-derived MTs: The position of the centrosome can be identified as a single spot with high fluorescence intensity (visualized with Centrin-GFP) and a diameter of less than 1 μm (Zhu & Kaverina, 2011). After ice recovery, MTs clearly emanating from this location can be considered nucleated from the centrosome and are hence centrosomal MTs. Similarly, a Golgi marker can be used to determine MTs nucleating from the Golgi (visualized with an anti-GM130 antibody), which are then considered to be Golgi-derived MTs. Other subcellular structures can also be examined using specific markers, respectively (Figs. 24.2 and 24.3).

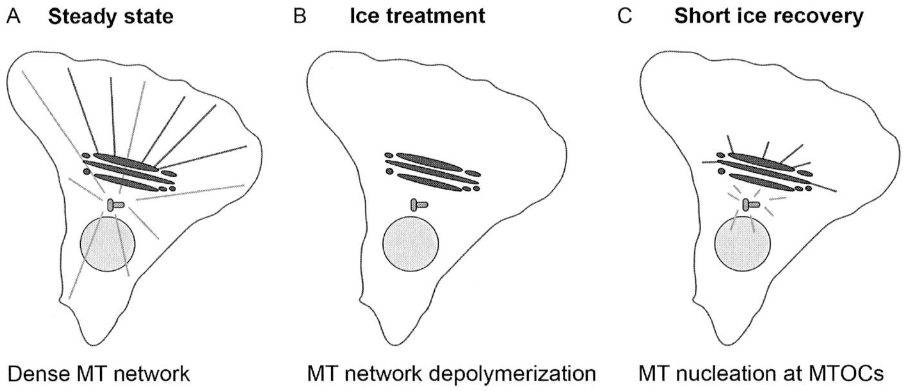

FIGURE 24.2

Microtubule network organization during the steps of the ice recovery assay. (A–C) Schematic showing organization of the microtubule network during main steps of the ice recovery assay. (A) At steady state, the dense MT network obscures MT nucleation events at MTOCs (centrosome and Golgi). (B) Upon ice treatment, the entire MT network undergoes depolymerization. (C) Short ice recovery times permit MT nucleation and repolymerization allowing for the clear detection of MT nucleation events at multiple MTOCs. (A–C) MTs nucleated at the centrosome (gray) are shown in green, while Golgi-derived MTs are shown in magenta emanating from the Golgi (purple). (See color plate.)

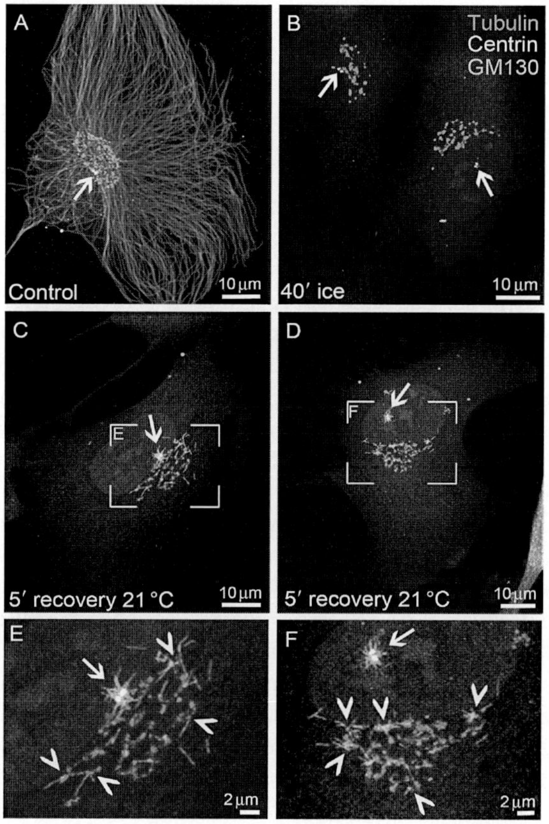

FIGURE 24.3

Detection of Golgi-derived MTs using the ice recovery assay. (A–F) Immunofluorescence images of RPE cells expressing Centrin-GFP (green) stained for alpha-tubulin (red) and the Golgi marker GM130 (blue). Maximum intensity projection of a confocal stack. White arrow indicates the location of the centrosome. (A) Untreated control cell at steady state with intact MT network. (B) Cell after 40 min ice treatment to depolymerize the entire MT network. (C–F) Examples of MT nucleation at MTOCs after 5 min ice recovery at 21 °C (room temperature). Short, newly nucleated MTs can be seen colocalizing with both Centrin-GFP (centrosomal MTOC) and GM130 (noncentrosomal MTOC, Golgi complex). (C) Example of ice recovery in a cell in which the centrosome and Golgi are in close proximity to each other. (D) Example of ice recovery in a cell in which the centrosome and Golgi are physically separated allowing for clearer distinction between MT subpopulations. (E) Zoomed region from C. (F) Zoomed region from D. (E–F) White arrow indicates MTs nucleated at the centrosome, alpha-tubulin colocalization with Centrin-GFP. Yellow arrowhead indicates MTs nucleated at the Golgi, alpha-tubulin colocalization with GM130. (See color plate.)

24.3 DISCUSSION

The above outlined procedure, for detection of Golgi-derived MTs using an ice recovery assay, is a simple and invaluable tool for discovery and characterization of noncentrosomal MTOCs. Ice recovery, along with other techniques, has been monumental in the establishment of the Golgi complex as a major noncentrosomal MTOC and has allowed the scientific community to begin to elucidate the unique roles that such nontraditional MTOCs play within the cell. Our assay allows for the clear detection of individual MT nucleation events, which can be difficult to detect by other means because they occur both transiently and on a scale below optical resolution.

An additional advantage of this ice recovery protocol is that it does not require the use of MT drugs, such as nocodazole. Nocodazole can have unknown off-target effects within the cell and, in comparison, ice recovery is a more physiological treatment. Also, the use of nocodazole alone makes it difficult to study MT nucleation in mitotic cells, which form aberrant spindles and exit mitosis upon exposure to the drug (Jordan, Thrower, & Wilson, 1992). Cold-treated cells, in contrast, are preserved at distinct stages of mitosis so that comparison of Golgi-derived MT numbers can be performed (Maia et al., 2013). Moreover, recovery after ice, or a combination of ice and nocodazole treatments, can be used to detect other noncentrosomal MTs formed at distinct mitotic stages (e.g., augmin-dependent central spindle MTs, Uehara & Goshima, 2010).

On the other hand, manual analysis of the colocalization between the ends of newly nucleated MTs and MTOCs is tedious and time-consuming. Development of easily accessible image analysis software would be beneficial to those hoping to utilize this method. Such software would require an automated detection algorithm that recognizes MTs as single entities of distinct shape and detects colocalization of their ends with a Golgi marker. Automated analysis would likely require shorter recovery times in order to avoid MT overlapping.

Another negative of this assay is that it requires a fairly advanced confocal imaging system for the 3D detection of colocalization between MTs and the MTOC of interest. Because MT nucleation at an MTOC occurs below optical resolution, the best imaging equipment available is necessary to properly detect these events. In addition, at least three laser excitation lines are required to capture data on MTs, the centrosome, and your noncentrosomal MTOC, in our example, the Golgi complex.

An alternative method to the simple ice recovery assay is live cell imaging of cells expressing a +TIP marker and a Golgi marker. Microtubule plus-end tracking proteins (+TIPs) are a diverse group of proteins that track MT plus-ends as they grow (Akhmanova & Steinmetz, 2008). Near the MTOC, +TIPs only localize to the tips of newly nucleated MTs and move away from the MTOC as the MT polymerizes outwards. Expression of a +TIP protein, for example, EB3-GFP, in any cell type allows tracking of MTs as they are nucleated from either the centrosome or noncentrosomal MTOCs. This can be done either at steady state or over the course of either ice recovery or a nocodazole washout experiment.

At steady state, this alternative method allows for the identification of MT nucleation sites within the cell without modifying the MT network organization. However, as mentioned previously, the close proximity of the Golgi complex to the centrosome can make it difficult to monitor nucleation at the Golgi even when utilizing a clear marker of nucleation, such as a +TIP protein. For this type of experiment, the best way to improve detection of Golgi-derived MT nucleation events is to image cells in which the centrosome and the Golgi complex are physically separated within the intracellular space. Alternatively, when this is not possible, MTs can be depolymerized by your chosen method and allowed to undergo regrowth for the detection of nucleation in the absence of obstruction from the MT network. The use of a +TIP marker has been a common approach to track steady-state MT nucleation at centrosomes (e.g., Piehl, Tulu, Wadsworth, & Cassimeris, 2004) and in our lab to detect noncentrosomal MT nucleation at the Golgi in both steady state and upon nocodazole washout (Efimov et al., 2007). Our lab's detailed protocol of how to use a +TIP marker in conjunction with live cell imaging to study nucleation of Golgi-derived MTs can be found in *Microtubule Dynamics: Methods and Protocols* (Zhu & Kaverina, 2011).

We hope that our ice recovery assay, as well as our recommendations for optimal image acquisition and analysis of MT nucleation events, proves useful in guiding your detection of Golgi-derived MTs or in your studies of additional noncentrosomal MTOCs. We believe that this method can be employed throughout the cell biology community for the examination of nucleation at MTOCs and serve as a platform for the development of improved techniques in the future.

SUMMARY

Ice recovery followed by immunofluorescence is a unique method to detect nucleation of Golgi-derived microtubules at the nonconventional MTOC, the Golgi complex. Disassembly of the microtubule network during ice treatment allows for the clear separation of MTs nucleated at the centrosomes in comparison to alternative MTOCs, which cannot be readily observed under steady-state conditions. The protocol outlined here provides a detailed guide for sample preparation, ice treatment, and recovery, as well as immunofluorescence, image acquisition, and analysis. Our method for detection of MT nucleation at the Golgi is fairly straightforward, inexpensive, and reliable. Therefore, we believe that this method can be applied for the novel discovery of additional noncentrosomal MTOCs and their functional characterization.

Acknowledgments

This work was supported by National Institutes of Health grant R01-GM078373 and by American Heart Association grant-in-aid 10GRNT4230026 (to I. K.) as well as American Heart association predoctoral fellowship 12PRE12040153 (to A. D. G.). M. F. was supported by the Vanderbilt International Scholar Program.

Author Contributions: A. D. G. cowrote the chapter. M. F. performed ice recovery and immunofluorescence experiments for Fig. 24.3. I. K. directed all of the work and project planning and cowrote the chapter.

References

Akhmanova, A., & Steinmetz, M. O. (2008). Tracking the ends: A dynamic protein network controls the fate of microtubule tips. *Nature Reviews. Molecular Cell Biology*, *9*, 309–322.

Bergen, L. G., Kuriyama, R., & Borisy, G. G. (1980). Polarity of microtubules nucleated by centrosomes and chromosomes of Chinese hamster ovary cells *in vitro*. *Journal of Cell Biology*, *84*, 151–159.

Chabin-Brion, K., Marceiller, J., Perez, F., Settegrana, C., Drechou, A., & Poüs, C. (2001). The Golgi complex is a microtubule-organizing organelle. *Molecular Biology of the Cell*, *12*, 2047–2060.

Efimov, A., Kharitonov, A., Efimova, N., Loncarek, J., Miller, P. M., Andreyeva, N., et al. (2007). Asymmetric CLASP-dependent nucleation of noncentrosomal microtubules at the *trans*-Golgi network. *Developmental Cell*, *12*, 917–930.

Egea, G., Lazaro-Dieguez, F., & Viella, M. (2006). Actin dynamics at the Golgi complex in mammalian cells. *Current Opinion in Cell Biology*, *18*, 168–178.

Jordan, M. A., Thrower, D., & Wilson, L. (1992). Effects of vinblastine, podophyllotoxin and nocodazole on mitotic spindles. Implications for the role of microtubule dynamics in mitosis. *Journal of Cell Science*, *102*, 401–416.

Ladinsky, M. S., Wu, C. C., McIntosh, S., McIntosh, J. R., & Howell, K. E. (2002). Structure of the Golgi and distribution of reporter molecules at 20 degrees C reveals the complexity of the exit compartments. *Molecular Biology of the Cell*, *13*, 2810–2825.

Lowe, M. (2011). Structural organization of the Golgi apparatus. *Current Opinion in Cell Biology*, *23*, 85–93.

Maia, A. R., Zhu, X., Miller, P., Gu, G., Maiato, H., & Kaverina, I. (2013). Modulation of Golgi-associated microtubule nucleation throughout the cell cycle. *Cytoskeleton*, *70*, 32–43.

Miller, P. M., Folkmann, A. W., Maia, A. R. R., Efimova, N., Efimov, A., & Kaverina, I. (2009). Golgi-derived CLASP-dependent microtubules control Golgi organization and polarized trafficking in motile cells. *Nature Cell Biology*, *11*, 1069–1080.

Ori-McKenney, K. M., Jan, L. Y., & Jan, Y. N. (2012). Golgi outposts shape dendrite morphology by functioning as sites of acentrosomal microtubule nucleation in neurons. *Neuron*, *76*, 921–930.

Piehl, M., Tulu, U. S., Wadsworth, P., & Cassimeris, L. (2004). Centrosome maturation: Measurement of microtubule nucleation throughout the cell cycle by using GFP-tagged EB1. *Proceedings of the National Academy of Sciences of the United States of America*, *101*, 1584–1588.

Rivero, S., Cardenas, J., Bornens, M., & Rios, R. M. (2009). Microtubule nucleation at the cis-side of the Golgi apparatus requires AKAP450 and GM130. *The EMBO Journal*, *28*, 1016–1028.

Salaycik, K. J., Fagerstrom, C. J., Murthy, K., Tulu, U. S., & Wadsworth, P. (2005). Quantification of microtubule nucleation, growth and dynamics in wound-edge cells. *Journal of Cell Science*, *118*, 4113–4122.

Uehara, R., & Goshima, G. (2010). Functional central spindle assembly requires de novo microtubule generation in the interchromosomal region during anaphase. *Journal of Cell Biology*, *191*, 259–267.

Vaughan, K. T. (2005). Microtubule plus ends, motors and traffic of Golgi membranes. *Biochimica et Biophysica Acta*, *1744*, 316–324.

Vinogradova, T., Miller, P. M., & Kaverina, I. (2009). Microtubule network asymmetry in motile cells. *Cell Cycle*, *8*, 2168–2174.

Vinogradova, T., Paul, R., Grimaldi, A. D., Loncarek, J., Miller, P. M., Yampolsky, D., et al. (2012). Concerted effort of centrosomal and Golgi-derived microtubules is required for proper Golgi complex assembly but not for maintenance. *Molecular Biology of the Cell*, *23*, 820–833.

Zhai, Y., & Borisy, G. G. (1994). Quantitative determination of the proportion of microtubule polymer present during mitosis-interphase transition. *Journal of Cell Science*, *107*, 881–890.

Zhu, X., & Kaverina, I. (2011). Quantification of asymmetric microtubule nucleation at subcellular structures. In J. M. Walker (Series Ed.) & A. Straube (Vol. Ed.), *Methods in molecular biology*: Vol. 777. *Microtubule dynamics: Methods and protocols* (pp. 235–244). New York: Humana Press.

Index

Note: Page numbers followed by *f* indicate figures and *t* indicate tables.

A

Acetone fixation, 113
Active fission-fusion, 307
Anterograde transport assay, 42, 43*f*
Apical Lumen trafficking
  3D culture, MDCK cysts (*see* 3D-MDCK cyst culture)
  ECM components, 106
  epithelial cyst formation *in vitro*, 107
  epithelial organs, 106
  Transwells® filters, 106
Arabidopsis, 72–73

B

Benzylguanine (BG), retrograde transport. *See* SNAP-tag, retrograde transport
Bicinchonic acid (BCA) protein assay kit, 91, 92–93
Bradford assay, 5–6
Brefeldin A (BFA), 87*f*, 89, 95, 96*f*, 98*f*
B-subunit of Shiga toxin (STxB)
  brefeldin-A, 147
  DMEM medium, 147
  GalT-GFP-SNAP fusion protein, 145–146, 146*f*
  glycosphingolipid Gb3, 145
  immunoprecipitation, 147
  peptide identification, 149, 150*t*
  purification, 145
  saponin, 145

C

Caenorhabditis elegans. *See* Nematode *Caenorhabditis elegans*
Calcium phosphate-mediated delivery
  materials, 20
  method
    CO_2 level, 21
    HEPES, 21
    PBS, 20–21
    protocol, 20–21
    transfection, 20–21
Cation-independent mannose 6-phosphate receptor (CI-MPR), 150, 151, 151*f*, 152*f*
CDG. *See* Congenital disorders of glycosylation (CDG)
Cell confinement, micropatterning techniques
  in vitro cultured mammalian cells, 327–328
  mesenchymal stem cells, 327–328
  normalization, adhesion, 328–329, 328*f*
  population size, local cell density and position, 327–328
Cell fixation
  acetone, 113
  methanol, 112–113
  PFA/formalin, 112
Cell polarity
  and lumen formation rates, 108
  matrigel, 109
  vectorial orientation, 108
Cell-surface biotinylation
  description, 92
  Golgi, 92
  ice-cold PBS, 94
  immunoprecipitation, 94–95
  incubating cells, 88
  measurement rates, protein trafficking, 94–95
  vs. total VSVG, 94
  ts045-VSVG-GFP, 93–94, 93*f*
CI-MPR. *See* Cation-independent mannose 6-phosphate receptor (CI-MPR)
Clathrin-coated vesicles, 4, 11
CLEM. *See* Correlative light-electron microscopy (CLEM)
Coarse-grained models
  active fission-fusion, 307
  coat proteins, 305–306
  description, 308
  kinetics, coat-assisted budding, 306
  mammalian and plant cells, 300
  molecular processes, 300
  organelle biogenesis, 300–302
  pancakes, 305–306
  plasma membrane (PM), 300
  protein transport and cisternal identity
    anterograde and retrograde, 303, 303*f*
    bulk Golgi flow and enzyme localization, 304–305
    cisternal progression, 303–304
    FRAP experiments, 304
    "maturation velocity", 303–304
    regulation, intercisternal exchange, 305
    stack and Golgi resident enzymes, 302
  top-down and bottom-up approach, 307–308

Confocal microscopy
epithelial cells
image filter-grown cells, 134–135
Microsytem LSM510 and Carl Zeiss Microimaging, Inc., 135
Volocity software, 134
Lab-Tek chamber, 202
laser power, 202
Oil, water and glycerol objectives, 202–203
parameters, 203
quantitative FRAP, 203
saturated pixels, 202–203
set up, 202
software program, 203
start up, 202
Congenital disorders of glycosylation (CDG)
alkyne-tagged sialic acid, 160–162, 163*f*
HPLC separation, 170, 173*f*
N- and *O*-glycoproteins, 160
N-Glycome fingerprinting, 172–174
types, 172–174
COPI vesicles
Arf1 and coatomer, 4
cargo proteins, 10–11
characterization, 4
clathrin-coated vesicles, 11
coatomer promotes, 11
coat proteins, 5
electron microscopic analysis, 4
formation, semi-intact cells
ATP-regenerating system, 6–7
buoyant density, 8
characterization, 8, 9*f*
donor membranes, 7–8, 7*f*
HeLa cells, 6–7
nucleotide-dependent signals, 8
vesicle-containing pellets, 6–7
ultarstructural analysis, 8–9, 10*f*
COPII vesicles
donor membranes, 9
ER and Golgi apparatus, 4
purified Sar1, 4–5
secretory pathway, 17
Correlative light-electron microscopy (CLEM)
alpha-mannosidase II (Man-II), 246–247
cells
fixation process, 250
observations, 249–250, 249*f*
seeding, 249–250
transfection, 250
"classical vesicles", 244
3D reconstruction and serial sections
fiduciary markers, 253
fluorescent and phase-contrast images, 253–255
image acquisition, 255
tomography approach, 255
efficient and fascinating tool, 247
embedding, 251–252
fluorescence Golgi, 244
genetic, neurodegenerative and metabolic disorders, 255
GFP-tagged Golgi proteins, 255–256
immunolabeling
anti-GFP IgG, 251
gold-enhancement reaction, 251
horseradish peroxidase (HRP), 250–251
lysosomal membrane protein LAMP1, 246
mannose-6-phosphate receptors, 245–246
materials, 247–248
metabolism, signaling and cell-fate decision, 255
N-acetylgalactosaminyltransferase-2-GFP, 246
procollagen III aggregates, 247
sectioning procedure
EPON block and MatTek dish, 252–253
glass-knife stage, 249*f*, 252–253
top and bottom sides, pyramid, 253, 254*f*
TGN and GFP, 244
VSVG and PC-I, 244–245
VSVG–CFP and GPI–YFP-myc, 245
Cycloheximide chase, 92–94

D

DAG. *See* Diacylglycerol (DAG)
Density mapping
automatic testing procedure, 339
density estimation
commands, 335
document, 335
packages, 335
Rab6-labeled structures and R script, 335
program installation, 334–339
Rab6-labeled compartment, normalized cells, 339, 340*f*
Diacylglycerol (DAG), 347, 354
3D-MDCK cyst culture
acetone fixation, 113
advantages, 120
blocking and antibody incubations, 113–114
cell death, 117
cell strains, 110
collagen-I, 107–108
in vivo experiments, 119–120

lumen numbers, 115–117
matrigel
BD :356234, 109
BMEs, 108
cell polarization and lumen formation rates, 108
functions, 109
matrix overlay method, 111–112
MEM, 108
methanol fixation, 112–113
micropatterns
chips, 110
CYTOOchips, 109
disk-shaped laminin-coating, 109–110
laminin-coating, 109–110
round shapes, 109–110
mounting and imaging, 114
permeabilization, 113
PFA/formalin fixation, 112
preparation, 118–119
pressure and stretching, 107
secondary antibodies and staining reagents, 114
spindle orientation, 117–118
trypsinization, resuspension and seeding cells, 110–111
tubular epithelia, 114
Drosophila cell lines and tissues, secretory pathway
COPI in lipid droplet biogenesis, 37
ER, 36
ERES, 36
genetic screens, 37
genome encodes proteins, 36
link to development, 37
mammalian cells, 36
materials, 37–38
methods
protein localization (*see* Protein localization)
S2 cells (*see* S2 cells)
tER-Golgi units, 47

E

Electron microscopy (EM) and IEM, 40, 41*f*, 46–47
Electron tomography (ET)
cis- and *trans*-Golgi
cellular trafficking, 269
3D imaging approaches, 274, 275*f*
lipid exchange/modification, 269
membrane modifications, TGN, 269, 273*f*
microtubule cytoskeleton and cisternae, 269–274
retrograde pathways, 274
Golgi apparatus
computer hardware and software, 265–267
"dual-axis" tomography, 265–267
"fiduciary model", 265–267
"intermediate voltage", microscopes, 265–267, 270*f*
membrane trafficking, 265
plastic sections, 265–267
serial section tomography, 267–268
TGN, 265
stacks *vs*. ribbons
high-resolution tomograms, 268
insulin-secreting cell, 268–269, 271*f*, 272*f*
protein Rab6, 268
structure–function diversity, 268–269
ultrastructural investigations, 268
EM. *See* Electron microscopy (EM)
Embedding, CLEM, 251–252
EMS. *See* Ethyl methane sulfonate (EMS)
Endocytosis
inhibition, 62*f*
intestine and oocytes, 55
and secretion, 54
Endo H digestion
CHX-chase assays, 93–94, 93*f*
denatured samples, 92–93
endogenous transferrin receptor, 94–95
glycosylated protein, 92
medial-Golgi apparatus, 100
pulse-chase assays, 94–95
trans-Golgi/TGN, 88
treatment, 92–93
ts045-VSVG-GFP-infected cells, 92
Endoplasmic reticulum (ER)-to-Golgi traffic synchronization protocols
ER-retention and cargo proteins
FKBP-domain mutant, 363*f*, 367
HeLa cells, 367–368
human skin fibroblasts, PC-I, 365–366
materials, 364–365
polymerization, 363*f*, 364–365
VSVG traffic synchronization, 366–367
Epithelial cells, replication defective lentiviruses
ARH depletion, 127–128, 128*f*
cell cultures, 130
clathrin-mediated endocytosis, 135
confocal analysis (*see* Confocal microscopy)
description, 128
gene silencing, 131
immunofluorescence staining, 133–134

Epithelial cells, replication defective lentiviruses *(Continued)*
materials, 129–130
MDCK and LDLR, 126
pGIPZ system, 127–128
Rab13 mutant protein Rab13Q67L, 135
recycling endosomes, 135
reporter genes
clathrin adaptor AP-1B, 132
filter-grown cells, 132
gene expression, MDCK and LLC-PK1 cells, 131–132
tubes, 132
siRNA-mediated knockdown, genes, 127–128
tissue culture plates, 130
VSV-G, 127–128
Epithelial morphogenesis
coating preactivated micropatterns, 118–119
ECM molecule, 108
lumen formation defects, 116*f*
MDCK cells, 108
microenvironment role, 120
micropattern shape and size, 107–108
ER N-glycosylation efficiency
materials, 162–166
metabolic labeling, 166
oligosaccharides purification
HPLC separation, 170, 173*f*
ion-exchange chromatography, fOSs, 168
monophosphorylated and pyrophosphorylated species, 169
oligosaccharide-PP-Dol, 169–170
phosphorylated, 169
synthesized glycoproteins, 170
sequential extraction procedure
$CHCl_3/MeOH/H_2O$, 167
CM fraction, 167
CMW fraction, 167
flowchart, 161*f*, 167
fOS fraction, 167
protein pellet, 167
ET. *See* Electron tomography (ET)
Ethyl methane sulfonate (EMS)
CaMV 35S promoter, 74
confocal *vs.* conventional fluorescence microscopy, 74
high-density oligonucleotide arrays, 74–75
mapping, 73–75
M1 generation, 74
mutagenized plants, 73–74
potential Golgi-ER tethers, 73–74
seedlings, 74
treatment, 73–74

F

Fetal bovine serum (FBS), 89
FLIM
analysis software, 80
data-interactions, 78–79, 79*f*
GFP and mRFP, 80
time-correlated SPC module, 79–80
FLIP. *See* Fluorescence loss in photobleaching (FLIP)
Fluorescence loss in photobleaching (FLIP)
cell identification, 207
COP molecules, 207
depletion, 207–210
FP reporter and treatment, 209
GalT-GFP, HeLa cell, 207, 208*f*
imaging and photobleaching conditions, 208
microscope and cells, 207
pinhole and detector gain, 207
prebleach fluorescence intensity, 209
shielding, 209
Fluorescence recovery after photobleaching (FRAP)
cytoplasmic GFP, 204–205
data sets, 205
FLIP (*see* Fluorescence loss in photobleaching (FLIP))
laser illumination, 205
prebleach images, collection, 205
ROI fluorescence intensity, 203–204, 204*f*
sensitive emCCD cameras, 205–206
software packages, 204–205
Fluorescent proteins (FPs)
bleach, 75–76
brightness, photostability and folding time, 198–199
cellular proteins, 200
cysteines, 198–199
donor and acceptor, 77–78
and EMS, 73–74
FLIP, 206–207
fusions, 71–72
GC size and organization, 199–200, 199*f*
leaf epidermal cells, 81
Lipofectamine 2000, 201
maturation time, 73
mobile fraction, 210
N-linked sugar, 198–199
oligomerization, 198–199
recovery rates, 76
SEC-green, 73

superfolder GFP (sfGFP), 200, 202, 203
vacuole and apoplast, 73
Fluorescent reporters, 348–349
Förster resonance energy transfer (FRET)
advantages, 78
CFP/YFP/GFP/mRFP, 77–78
description, 77–78
emission signals, 79–80
FLIM data-interactions, 78–79, 79*f*
fusion protein interactions, 77–78
measurement techniques, 78
methods
Chi-square, 80
data, 80
drug-mediated cytoskeleton depolymerization, 80
GFP, 80
photon counts, 80
save images and import, 80
percentage efficiency (E%), 78
SPC modules, 78
FPs. *See* Fluorescent proteins (FPs)
FRAP. *See* Fluorescence recovery after photobleaching (FRAP)
FRET. *See* Förster resonance energy transfer (FRET)

G

Galactosyl transferase (GalT)-EGFP
dynamics, ER and Golgi pool, 227–229
fluorescence
ROI_1, 222*f*, 230
ROI_3, 222*f*, 230
localization, 229–230
retrograde and anterograde movement, 222*f*, 230
GalT-EGFP. *See* Galactosyl transferase (GalT)-EGFP
Gene depletion methods
Caenorhabditis elegans, 16
"off-target" effects, 16
organelle function analysis, 17–18
rescue experiment, 30–31
RNAi (*see* RNA interference (RNAi))
shRNAs (*see* Short hairpin RNAs (shRNAs))
siRNAs (*see* Short-interfering RNAs (siRNAs))
Gene suppression
description, 24
immunoblotting, 24–25
immunofluorescence, 26–27
qPCR, 27–30
GFP. *See* Green fluorescent protein (GFP)
GFP fluorescence, 40, 41*f*
Golgi and secretory pathway, 71–73
Golgi apparatus
architecture, 262–263
cellular processes, 260–261
CLEM, 260–261
complex internal mechanisms, 260–261
3D imaging techniques, 261–262
Endo H, 92
functional priorities, 261–262
and plasma membrane, 93*f*
pulse-chase, 96*f*
pulse-chase labeling, 88
secretory cells, 86
stacks, HepG2 hepatoma cell, 263–265, 264*f*
"state-of-the-art" approach, 261–262
structure–function relationships, 260–261
Golgi complex, CLEM. *See* Correlative light-electron microscopy (CLEM)
Golgi-dependent signaling
cellular processes, 361
description, 378–379
ER-to-Golgi traffic synchronization protocols, 364–368
intracellular membrane transport, 361
KDELR-dependent signaling, 371–374
KDELR transduction machinery, 375–378
posttranslational modifications and phosphorylation, 362, 363*f*
regulatory mechanism, 361
traffic-generated Golgi-based signaling, 375–378
transport synchronization assays, 362
Golgi depletion, living cells
equipment, 313–315
laser nanosurgery, 319–322
materials, 315
microcontact printing, 315–319
molecular mechanisms, 312
Golgi N-glycosylation efficiency
immunolabeling, 171
materials, 164–166
metabolic labeling, alkyne-tagged sialic acid and fixation, 171
microscopic analysis, fluorescent labeling, 172
probe labeling reactions, 171
Golgi reassembly and stacking proteins (GRASPs)
cell cycle-regulated *trans*-oligomerization, 388–389
GRASP65 and GRASP55, 385–386, 388
myristoylation, N-terminal glycine, 388
Plk1, 389

Golgi reassembly and stacking proteins (GRASPs) *(Continued)*
stabilization, GRASP55, 389
Golgi structure, HCS and high-content analysis. *See* High-content screening (HCS)
GPCRs. *See* G-protein-coupled receptors (GPCRs)
G-protein-coupled receptors (GPCRs), 362
G-protein of vesicular stomatitis virus (VSVG) traffic synchronization
anti-VSVG P5D4 antibody, 367
cell confluence, 367
DMEM–HEPES, 367
FBS removal, 367
"restrictive" and "permissive" temperatures, 366–367
tsO45 (V) mutant of VSV, 367
GRASPs. *See* Golgi reassembly and stacking proteins (GRASPs)
Green fluorescent protein (GFP)
alpha-mannosidase II (Man-II), 246–247
ATP7A-GFP distribution, 252–253, 254*f*
cell transfection, 250
distribution, *N*-acetylgalactosaminyltransferase-2-GFP, 246
jellyfish *Aequorea victoria*, 244
GTPase signaling
Ras activity, 350–351
Rho family
bleaching, acceptor, 352–353
Cdc42 activation, 352, 353*f*
C-terminal-CAAx motif, 351–352
FRET-based sensors, 351–352
Raichu sensors, 353

H

HCS. *See* High-content screening (HCS)
HeLa cells, 5–6
High-content analysis, Golgi structure. *See* High-content screening (HCS)
High-content screening (HCS)
biochemical assays and genetic methods, 282
cell biology tool, 284
CellProfiler, 282–283
cell seeding, fixing and staining, 290
design and planning
antibody staining, 288
depletion, protein, 288
GFP-tagged Golgi marker, 285–286
Golgi complex morphology, 285–286, 287*f*
human HeLa cell line, 285–286, 286*f*
materials, 285
negative and positive control siRNAs, 288
endoplasmic reticulum (ER), 283
fluorescence microcopy, 292
high-content analysis, 292–293
and HTS, 282–283
human glucocorticoid receptor, 282–283
image acquisition, 290–291
image analysis and data management
Golgi fragmentation, 287*f*, 291–292
Hoechst 33342 and GalNAc-T2-GFP, 291
morphological and intensity measurements, 291–292
software availability, 291
kinases and phosphatases, RNAi screen, 284
materials, 284–285
membranes and proteins, 283
RNAi strategies, 283–284
solid-phase reverse transfection, 289–290
High-performance liquid chromatography (HPLC)
and CDG, 172–174
identification, glycosylation ER, 172–174
separation, oligosaccharides purification, 170, 173*f*
transferrin, 172–174
High-throughput screening (HTS), 282–283
HPLC. *See* High-performance liquid chromatography (HPLC)
HTS. *See* High-throughput screening (HTS)

I

Ice recovery, Golgi-derived microtubules
actin cytoskeleton, 403
cell maintenance and transfection
fibronectin, coverslips, 405
FuGENE, 405–407
RPE1 cells, 405
trypsinize, 405
depolymerization, 402–403
expression, +TIP protein, 412–413
ice treatment, 407–408
image acquisition, 409
immunofluorescence
blocking solution, 408
Kimwipe, 408
methanol fixation, 408
nail polish, 409
parafilm-lined plastic box, 408
primary antibody solution, 409
secondary antibody incubation, 409
size variation, coverslips, 409
intracellular space, 402
materials
cell culture and transfection, 403

confocal microscope and analysis, 404
ice bucket and ice, 404
immunofluorescence, 404
MT nucleation and subpopulations
centrosome, identification, 410
colocalization, 410, 412
subcellular structures, 410, 410*f*, 411*f*
Z-stack images, 410
MTOC (*see* Microtubule organizing center (MTOC))
nocodazole, 412
repositioning, centrosome, 402
steps, 405, 406*f*
Ice treatment and ice recovery
bucket, 407
incubation time, 408
methanol, 407
methods, 407
MT nucleation and regrowth, 408
nocodazole and paclitaxel, 407–408
tissue culture dishes, 407
IEM. *See* Immunoelectron microscopy (IEM)
iFRAP protocol. *See* Inverse FRAP (iFRAP) protocol
Immunoblotting
definition, 24
employ quantitative imaging techniques, 24
materials, 24
method
acrylamide gel, 25
Blunt-ended tweezers, 25
nitrocellulose/PVDF membranes, 25
NuPAGE® gels, 25
Odyssey system, 25, 26*f*
place membrane, 25
set transfer device to blot, 25
wash and scrape cells, 24–25
Odyssey imaging, 24
Immunoelectron microscopy (IEM)
Drosophila tissues, 46–47
salivary gland fixed, 44*f*
Immunofluorescence
description, 26–27
materials, 27
method
aspirate medium, 27
imageJ, 27
methanol permeability, 27
PBS-BSA, 27
RNAi knockdown validation, 27, 28*f*
thick sections, 46
whole mount tissues (*see* Whole mount tissues)
Immunofluorescence staining, epithelial cells
Cy5-labeled goat anti-rat antibodies, 128*f*, 133
rabbit anti-TGN38 antibodies, 133–134
Rab13-EGFP, 133–134
rat monoclonal antibody 2.4G2, 128*f*, 133
tissue culture hood, procedures, 133
Immunolabeling protocol, 40, 41*f*
Immunoprecipitation (IP)
anti-saposin D antibody, 98*f*
cell lysates and secreted material, 99
Endo H and N-glycans, 94–95
VSVG (P5D4), 96*f*
Intracellular trafficking pathways
cargo proteins, 339–340
cell confinement, micropatterning techniques, 327–329
cell seeding, treatment and fixation, 332
density mapping, 334–339
eukaryotic cells, 326
experimental procedure, 330, 331*f*
Golgi apparatus, mammalian cells, 327
image acquisition, 333
immunolabeling, 333
malignant development, 339–340
materials, 331
nocodazole, Rab6-labeled compartment, 339, 340*f*
preparation, photolithography, 331–332
probabilistic density-mapping procedure, 327, 328*f*, 329–330
Rab6-labeled transport carriers, 330
Rab proteins, 326–327
secretory and endocytic pathways, 326
segregation, cellular processes, 326–327
Inverse FRAP (iFRAP) protocol, 206–207
In vitro reconstitution systems, Golgi complex functions
antibodies, 5
ArfGAP1, 4
COPI (*see* COPI vesicles)
description, 4–5
liposomal systems, 4
reagents, 5
semi-intact cells (*see* Semi-intact cells)
transport vesicles, 4
In vivo experiments, 119–120

K

KDELR-dependent signaling
fluorescent membrane-permeant peptides, BODIPY
FRET measurements, 373
HPLC chromatography, 373–374

KDELR-dependent signaling *(Continued)*
KDEL and KDEA peptides, 373–374
materials, 373
GPCRs, 362
materials, 371–372
receptor overexpression, 372
SFK activation, 363*f*, 371–372
ssHRP (control), 363*f*, 371–373
KDELR transduction machinery
phosphoproteomics, 375
traffic-induced phosphorylation
antibody microarrays, 375–377
immunophosphoproteomic mass spectrometry, 377–378

L

Laser nanosurgery
advantages, 322–323
biological structures, living cells, 312–313
blebbing, 321
cell dissection and glass etching, 319, 320*f*, 323
cell preparation, 319
centrosome, 313
cutting pattern, 321
limitations, 323
microtubules and actin stress fibers, 313, 314*f*
protein and lipid components, 312
seeding BSC1 cells, 313, 314*f*
signal transduction, 312
time-lapse imaging, 321–322
Leaves
Arabidopsis thaliana, 71–72
epidermal cells, 71
FP-fusion expressions, 71
imaging leaf material, 73
tobacco and arabidopsis, 71–72
transient expression, 71–72
transient transformation
in *Arabidopsis*, 72–73
in tobacco, 72
Lentiviral transduction systems
advantages, 185
DNA size, 185–186
HEK-293T cells
expression level, population, 187
harvest and filtrate, 186
puromycin, 186
seed, 186
split and expansion, 187
transfection, plasmids, 186
variety, 186
IRES, 185–186, 191
materials, 186–187
pCDH vector, 186
production and infection, cells, 186–187
Lentivirus production
biosafety classification, 22
description, 22
infect human cells, 22
materials, 23
method, 23
Lenti-X system, 23
Light microscopy, 39–40
Lipid-based transfection
materials, 21
method
advantages, 22
incubation, 21–22
knockdown efficiency, 21–22
life technologies, Lipofectamine® 2000, 21
nucleic acids, 21
plasmid DNA and siRNA, 22
plate cells, 21–22
Lipid signaling
acute depletion, PI4P, 349–350, 351*f*
fluorescent reporters, 348–349
immunofluorescence, PI4P, 347–348
phosphoinositides and metabolites, 347
Live imaging. *See also* Super-resolution confocal live imaging microscopy
agarose pads, 61
assessing gene function in transgenic *C. elegans*, 61, 62*f*
control and RNAi worms, 61
fluorescent microscope, 61
transgenic worms, 60
YP170-GFP and RME-2-GFP, 61
Lumen numbers
apical membrane markers, 115
basolateral marker, 115
ILK, 115–117
inverted polarity/apical bipolarity, 115–117
per cyst, 115, 116*f*
spindle orientation controls, 115–117
tight-junction markers, 115
tubular epithelium, 115–117

M

Madin–Darby canine kidney (MDCK) cells
gene expression, 131–132
LDLR mutant protein, 126
Rab13-EGFP, 126–127, 127*f*
VSV-G, 127–128
Mammalian cells
advantages and disadvantages, 100

biochemical analysis, protein trafficking, 86, 87*f*, 89, 90*f*
biotinylated proteins, 88
DNA vector constructs, 99
endogenous cargoes, 100
ERGIC, 86, 88
exogenous cargoes, 99
F_M tag, 100–101
fusion protein, horseradish peroxidase, 100
glycoprotein, 88–89
materials
cell culture, 89–90
sample preparation and analysis, 91
methods
cell-surface biotinylation, 92–94
cycloheximide chase, 92–94
Endo H digestion, 92–94
IP, 94–97
pulse-chase (*see* Pulse-chase)
recombinant adenovirus-encoding ts045-VSVG-GFP, 91–92
neuroaminidases, 88
procollagen, 88–89
pulse-chase labeling, 88
RNAi/mutagenesis, 101
RUSH system, 101
TGN, 86, 100
transferrin receptor, 88–89
ts045-VSVG, 88–89
Western blotting, 86
Matrix overlay method, 111–112
MDCK cells. *See* Madin–Darby canine kidney (MDCK) cells
Membrane trafficking. *See* Nematode *Caenorhabditis elegans*
Methanol fixation, 112–113
Microcontact printing
description, 316–319
glass coverslips, 316, 318*f*
stamp preparation, 316, 317*f*
Microscopy
double transgenic approach, 63–64
marker distribution in *C. elegans* oocytes, 61–62, 63*f*
RME-2-GFP, 62–64
secretory/endocytic pathways, 61–62
SP12-GFP marker, 62–63
Microtubule organizing center (MTOC)
colocalization, MT nucleation, 410, 411*f*, 412
detection, MT regrowth, 408
expression, +TIP protein, 412
maintenance, 403
noncentrosomal, 402, 412
Mitosis, Golgi ribbon fragmentation
B-type cyclins, CycB1 and CycB2, 391
cell-free assays, 394
CycB1-Cdk1 activity, 391
description, 385
disassembly
BARS, 387
"blobs" and "haze", 389–390
FRAP and electron microscopy, 386–387
GRASPs (*see* Golgi reassembly and stacking proteins (GRASPs))
MEK1, 387–388
Plk1, 389
Drosophila embryos, 391–393
Golgi-associated proteins, 391–393, 392*t*
"Golgi blobs", 384–385
interphase cytosol, 393
mammalian Golgi membranes
characteristics, 385–386
cisternae stacking, 386
Golgin-160/GMAP210 depletion, 386
Golgi proteins GRASP65 and GRASP55, 385–386
noncompact zones, 385–386
mechanism, postmitotic Golgi reassembly, 390
mitotic cytosol, 393
NRK cells and microinjection, 394
permeabilization, tissue culture cells, 393–394
secretory pathway, central organelle, 384
semi-intact cell assay, 394
three-dimensional (3D) Golgi structures, 384–385, 385*f*
yeast *Saccharomyces cerevisiae*, 384–385
MTOC. *See* Microtubule organizing center (MTOC)
Mutant screening
EMS (*see* Ethyl methane sulfonate (EMS))
FPs, 73
microscopy-based screens, 73
SEC-GFP, 73
trafficking proteins, 73–75

N

Nematode *Caenorhabditis elegans*
forward and reverse genetics approaches, 64–65
genetic analyses in model animals, 64
intracellular and organismal process, 52
materials, 55–56
methods
live imaging, 60–61
microscopy and analysis, 61–64
RNAi (*see* RNA interference (RNAi))
worm maintenance, 56–57
pathways and molecular machineries, 52

Nematode *Caenorhabditis elegans (Continued)*
RISC, 53
RNAi (*see* RNA interference (RNAi))
siRNAs, 53
N-glycosylation, Golgi apparatus
CDG (*see* Congenital disorders of glycosylation (CDG))
ER N-glycosylation efficiency (*see* ER N-glycosylation efficiency)
fluorescent probe, 174
Glc, Man and GlcNAc, 159–160
Golgi N-glycosylation efficiency (*see* Golgi N-glycosylation efficiency)
Huisgen azide–alkyne cycloaddition reaction, 160–162, 162*f*
N-glycans, 159–160
N-glycoproteins, 160–162
posttranslational modification, 158
sequential extraction procedure, 160–162, 161*f*
SialNAl, 160–162, 163*f*
types, 158–159

O

Organelle biogenesis
computational models
Brownian dynamics simulation models, 301
"compartments", 301
discrete-event models, 301
protein concentrations, 301
de novo generation, 300
transport models, 302
Organelle function analysis
classical RT-PCR approaches, 17
definition, 17
Golgi morphology, 18
in vitro, 17
metazoans, 17
mouse knockout, Sec24A, 17
Rab protein family, 17
RNAi, 17
trafficking components, 17
Ovaries, 43–45

P

PBS. *See* Phosphate-buffered saline (PBS)
Permeabilization, 113
PFA/formalin fixation, 112
PH. *See* Pleckstrin homology (PH)
Phenylmethylsulfonyl fluoride (PMSF), 91, 95
Phosphate-buffered saline (PBS), 89, 91–92, 94, 95, 97–99
Phosphatidylinositol-4-phosphate (PI4P)
acute depletion, 349–350, 351*f*
cell permeabilization, 348
colocalization, 348
FAPP-PH domain, 349
formaldehyde and paraformaldehyde, 347
HeLa cells, 347
mouse IgM antibody, 348
PI4 kinases, 347
polyvinyl alcohol, 348
RNAi transfection, 347
Photoactivation, ManII-PAmCh
brightness, photostability and contrast ratio, 231–232
experiment flow, 231
irradiation, wavelength, 230
laser power and gain, 232
mEos2, 232
PAGFP, 231–232
retrograde movement, 232
spatial and temporal evolution, fluorescent signal, 230
time-lapse imaging, 230–231
Photobleaching methods
biochemistry experiments, 196
bleach protocol
actin depolymerizing agents, 76
excitation and emission wavelengths, 76
FRAP experiment, 76
laser beam intensity, 76
record full-field scans, 77
ROI, 76–77
cell culture and transfection
adherent cells, 200
expression levels, 201
fluorescent signal, 200
Lab-Tek chambers, 201–202
Lipofectamine 2000, 201
poly-L-lysine, 200–201
confocal microscope (*see* Confocal microscopy)
description, 75, 213–214
diffusion analysis, 210–212
FLIP, 207–209
FPs (*see* Fluorescent proteins (FPs))
fractional fluorescence, 213
FRAP protocol, 203–206
GFP-labeled proteins, 196–197
iFRAP protocol, 206–207
latrunculin B/cytochalasin D, 75–76
materials
cell culture medium, 197
confocal laser scanning microscope, 197
eukaryotic cells, 197

formaldehyde/PBS, 197
Nunc Lab-Tek chambered coverglass, 197, 198*f*
phenol red free medium, 197
stage heater, 197
mobile fraction, 210
secretory proteins, 196
$t_{1/2}$ analysis, 212
Photohighlighting techniques
cell imaging, 220
cell plating and transfection, 220
description, 218–219
development, 218
ER and PM, 218
FRAP and photoactivation, 219
Golgi enzymes
cycling back, ER, 226–227
cycloheximide, 229
fluorescence recovery, 226–227
GALT-EGFP (*see* Galactosyl transferase (GalT)-EGFP)
half-life, 229
movement, proteins, 229
population, 227
image protocols and acquisition
calibration, laser power and exposure time, 223–224
experimental strategies, 221–223, 221*f*, 222*f*, 223*f*
pixel intensities, 224
time-lapse imaging, 224
instrument, 220
measurement of cargo export
cisternal maturation model, 224
classes, transmembrane protein, 225
data collection, 224–225
experiment, 225–226
fluorescence intensity, 226
FP-tagged cargo molecule, 226
Golgi export rate, 226
Golgi ROI, 221*f*, 224–225
photoactivation, Golgi enzymes (*see* Photoactivation, ManII-PAmCh)
protein populations, 218
reagents, 219–220
types, proteins, 219
PI4P. *See* Phosphatidylinositol-4-phosphate (PI4P)
Plant cells
endomembranes, 70–71
FRET, 77–80
Golgi apparatus, 70
leaves, 71–73
mutant screens, 73–75
organelles, 70
photobleaching, 75–77
secretory pathway (*see* Secretory pathway)
Pleckstrin homology (PH)
advantage and disadvantage, 348–349
adverse effects, 349
FAPP1 and OSBP1, 349
PI4P-sequestering effect, 349
PMSF. *See* Phenylmethylsulfonyl fluoride (PMSF)
Probabilistic density-mapping
endomembranes, 330
kernel estimators, 329–330
procedure, single cell and cell population, 327, 328*f*
smoothing parameter, 329–330, 329*f*
statistical tools, 329–330
Protein localization
in *Drosophila* tissues
EM and IEM, 46–47
immunofluorescence (*see* Immunofluorescence)
in early secretory pathway, S2 cells
antibodies, 38
confocal microscopy, 40
cycloheximide, 38–39
EM and IEM, 40, 41*f*
fixation, light microscopy, 39–40
GFP fluorescence, 40, 41*f*
Golgi, 38–39
immunolabeling protocol, 40, 41*f*
Kc cells and S2 cell, 38
metallothionein promoter, 38
pulsing, 38–39
SMSr, 39
stable transfection protocol, 39
transient transfection protocol, 39
V5 tagging, 38
Protein phosphorylation, 351–353
Pulse-chase
analysis
cellular proteins, 97–99
culture medium, 99
pellets, 99
protein secretion, 97–99, 98*f*
surface biotinylation, 97–99
assays
cell surface and cellular proteins, 94–95
Endo H digestion/surface biotinylation, 94–95
ice-cold PBS and lysed, 95
membrane glycoproteins, 95–97, 96*f*
methionine and cysteine, 95

Pulse-chase *(Continued)*
RIPA buffer, 95
SDS-PAGE and dried gels, 95–97
siRNAs, 94–95
ts045-VSVG-GFP adenoviral, 95

Q

qPCR. *See* Quantitative PCR (qPCR)
Quantitative PCR (qPCR)
description, 27
materials, 28–29
real-time, 30
reverse transcription, 29
RNA isolation, 29

R

Recombinant adenovirus-encoding ts045-VSVG-GFP, 91–92
Region of interest (ROI)
"bagel bleach", 206, 206*f*
and FLIP (*see* Fluorescence loss in photobleaching (FLIP))
fluorescence intensity, 203–204
laser illumination, 205
mobile fraction, 210
$t_{1/2}$ analysis, 212–213
Rescue experiment
cDNA, 30–31
EM/advanced light microscopy techniques, 31
GFP-tagging, 31
lipid-based transfection, 31
RNAi, 30–31
siRNA/shRNA, 30–31
small molecule inhibitors, 30–31
Retention Using Selective Hooks (RUSH) system
advantages, 182
biotin, SBP, 182, 183*f*
donor compartment, 182
generation, stable cell lines (*see* Lentiviral transduction systems)
infection, *Shigella*, 191–192
and multi-color imaging, 191, 192*f*
protein of interest, 187–188, 190
reporter
cloning cassettes, 184–185, 184*f*
dephosphorylate, 185
E. coli, ampicillin, 185
living cells, 185–187
materials, 184–185
NucleoSpin Quickpure, 185
synchronization, trafficking, 184
SBP and fluorescent proteins, 182–183
synchronization, biotin, 182
time-lapse imaging, 188–190
UniProt database, 183
Reverse transcription, 29
RNAi. *See* RNA interference (RNAi)
RNA interference (RNAi)
advantages, 64
Ahringer library, 58
animal models, 32
bacterial RNAi clones, 58
C. elegans, 53–54
components, 54
and dsRNA, 57
feeding libraries, 57
gene silencing, 53
genomic DNA, 58
knockdown, 31–32
L4440, 57
libraries targeting, 52–53
metazoans, 52–53
molecular mechanism, 53
off-target effects, 32
plants and fungi, 52–53
positive and negative controls, 60
and qPCR, 18
reagent sources, 19
RME-1 and SNAP-29, 60
and RME-2-GFP, 54–55
and RNase III, 57
rrf-3 and *eri-1*, 60
in S2 cells, 41–42
secretory and endocytic trafficking pathways, 54
and shRNAs, 18
and siRNAs, 18
standard cloning techniques, 58
synchronized/semi-synchronized worms, 58–60
vertebrate model, 52–53
Vidal lab RNAi library, 58
YP170-GFP assay, 54–55, 60
RNAi screen, Golgi complex. *See* High-content screening (HCS)
RNA isolation, 29
ROI. *See* Region of interest (ROI)
RUSH system. *See* Retention Using Selective Hooks (RUSH) system

S

Salivary glands, 45
SBP. *See* Streptavidin-binding peptide (SBP)
S2 cells
live cell imaging, 43

protein knockdown, 36
protein localization, 38–40
recombinant eukaryotic proteins, 36
RNAi, 41–42
transport/secretion assays, 42, 43*f*
Sec16
in *Drosophila* tissues, 44*f*
endogenous visualization, 41*f*
and Sec23, 38
Sec16-DNC1, 43
Secretion
and endocytosis, 54
RNAi, SNAP-29, 60
Secretory pathway
aggregation systems, 181–182
glycosylation, 180
Golgins, 180
RUSH system (*see* Retention Using Selective Hooks (RUSH) system)
synthesized proteins, 180
temperature blocks and reversible inhibitory molecules, 181
VSVGtsO45, 181, 191
Secretory trafficking. *See* Mammalian cells
Semi-intact cells
advantages, 10–11
Arf1 and coatomer, 9–10
COPI vesicles, 6–9
donor membranes, 9–10
intracellular membranes, 10–11
preparation, 5–6
Short hairpin RNAs (shRNAs)
lentivirus production, 22–23
plasmid backbone, 18
small-scale screen, 18–19
stable cell line production
materials, 23
polybrene, 23
protein knockdown, 23–24
puromycin/antibiotics, 23–24
suppression validation (*see* Gene suppression)
and TRC, 22
Short-interfering RNAs (siRNAs)
calcium phosphate-mediated delivery, 20–21
cheap and highly effective, 18
fluorescent and immunofluorescence labeling, 20
GL2, 19
glyceraldehyde-3-phosphate dehydrogenase, 20
lamin A/C, 19
lipid-based transfection, 21–22
nontargeting sequences *vs.* specific targeting, 19
reverse transfection methods, 20
targeting and nontargeting controls, 20
transient suppression, 18–19
shRNAs. *See* Short hairpin RNAs (shRNAs)
Signaling, Golgi apparatus
description, 356
FRET reporter
DAG activity, 354
PKC activity, 354
GTPases, 350–353
lipids (*see* Lipid signaling)
materials, 346–347
protein phosphorylation, 351–353
secretory pathway, 346
SILAC method. *See* Stable isotope labeling by amino acids in cell culture (SILAC) method
siRNAs. *See* Short-interfering RNAs (siRNAs)
SNAP-tag, retrograde transport
advantages, 140
BG tagging, antibody uptake assay
CI-MPR, 150
GalT-GFP-SNAP capture reagent, 150, 151, 151*f*
pGalTGFP-SNAP, 150–151
SDS-PAGE gels, 150–151, 152*f*
sodium phosphate, 150
capture reagent, 144
cell surface modification reagent
BG-PEG$_9$-NHS, 142–144, 143*f*
NHS, amino acceptors, 143*f*, 144
endoplasmic reticulum and cytosol, 152
Golgi membranes
GalT-GFP-SNAP, 145, 146*f*
green GFP fluorescence, 145
HeLa-GGS cells, glass coverslips, 144–145
intracellular trafficking, 152
materials, 141–142
modification, plasma membrane, 152–153
proteins and lipids, trafficking, 140
proteomics format
BG-PEG$_9$-NHS, 147–148, 149*f*
glycolipid receptor Gb3, 148
peptide identification, STxB, 149, 150*t*
SDS-PAGE gel, 148–149
SNAP-Cell Block reagent, 148
STxB, 145–147
transferrin and mannose 6-phosphate receptors, 140
vectorial proteomics approach, 140–141, 141*f*
Spindle orientation
angle measurement, 117–118
lumen phenotypes, 117–118
MDCK cysts, 117–118

Spindle orientation *(Continued)*
planar cell division, 117–118
quantifications, 117–118, 118*f*
Stable isotope labeling by amino acids in cell culture (SILAC) method, 363*f*, 368, 375, 377–378
Streptavidin-binding peptide (SBP), 182–183, 183*f*, 184–185, 192*f*
STxB. *See* B-subunit of Shiga toxin (STxB)
Super-resolution confocal live imaging microscopy
color separation, 239–241
data collection, 236
demonstration, cisternal maturation, 236
dichroic mirrors and band-pass filters, 239–241, 241*f*
fluorescent markers, 239
GFP-Gos1 and Sec7-mRFP, 239, 240*f*
Golgi apparatus, 236
instrumentation
camera, 237
confocal scanner, 237
equipment, 237
image intensifier, 237
lasers, 237
microscope, 237
objective lens, 237
piezo actuator, 237
setup, 237, 238*f*
software, 237
spectroscopic unit, 237
materials
coverslips, Concanavalin A, 238
fluorescent proteins, 238
yeast cells, 237
methods, 238–239
time and space resolutions, 236–237
yeast *Saccharomyces cerevisiae*, 236

T

Temperature block, 86, 87*f*, 88, 101
TGN. *See* Trans-Golgi network (TGN)
Time-lapse imaging, RUSH system
advantages, fluorescent protein, 188
biotin, 189–190
L-shaped tubing, Chamlide, 189–190, 189*f*
materials
adherent cells, 190
Chamlide, L-shaped tubing, 189*f*, 190
fluorescence microscope, 190
imaging medium, 190
immersion oil, 190
25-mm glass coverslips, 190
plasmid encoding, 190
syringes and needles, 190
transfection reagent, 190
method
biotin, 190
coverslips, Chamlide chamber, 189*f*, 190
immersion oil, 190
seed, cells, 190
software, 190
wavelength acquisition, 190
SbfI and *FseI* restriction enzyme, 188, 189*f*
Tobacco, 72
Traffic-induced phosphorylation
immunofluorescence (IF) microscopy
image acquisition, 369
materials, 368–369
MetaMorph/ImageJ software, 368–369
PBS-CM, 369
total intensity, 369
materials, 368
SILAC method, 363*f*, 368
Western blotting
cell plate, 370–371
cellular processes, 370
fat-free milk saturation, 370–371
preparation, total lysates, 363*f*, 370
protease and phosphatase inhibitors, 370
protein concentration, 370–371
Trafficking, living cells
aggregation systems, 181–182
proteins, interest, 187–188
RUSH system (*see* Retention Using Selective Hooks (RUSH) system)
secretory pathway, 180
temperature blocks and reversible inhibitory molecules, 181
VSVGtsO45, 181
Transfection
stable transfection protocol, 39
transient transfection protocol, 39
Trans-Golgi network (TGN)
Golgi stack, 244
lysosomal membrane protein LAMP1, 246
mannose-6-phosphate receptors, 245–246
VSVG, 244–245
Transient transformation, leaves
in *Arabidopsis*, 72–73
in tobacco, 72
Trichloroacetic acid (TCA), 91, 98*f*, 99

U
Ultarstructural analysis, COPI vesicles, 8–9, 10*f*

V
Vesicular stomatitis virus glycoprotein (VSV-G), 127–128
VSVG traffic synchronization. *See* G-protein of vesicular stomatitis virus (VSVG) traffic synchronization

W
Whole mount tissues
  imaginal discs, 45–46
  ovaries, 43–45
  overexpressed protein visualizations, 46
  salivary glands, 45
Worms, 56–57

Volumes in Series

Founding Series Editor
DAVID M. PRESCOTT

Volume 1 (1964)
Methods in Cell Physiology
Edited by David M. Prescott

Volume 2 (1966)
Methods in Cell Physiology
Edited by David M. Prescott

Volume 3 (1968)
Methods in Cell Physiology
Edited by David M. Prescott

Volume 4 (1970)
Methods in Cell Physiology
Edited by David M. Prescott

Volume 5 (1972)
Methods in Cell Physiology
Edited by David M. Prescott

Volume 6 (1973)
Methods in Cell Physiology
Edited by David M. Prescott

Volume 7 (1973)
Methods in Cell Biology
Edited by David M. Prescott

Volume 8 (1974)
Methods in Cell Biology
Edited by David M. Prescott

Volume 9 (1975)
Methods in Cell Biology
Edited by David M. Prescott

Volume 10 (1975)
Methods in Cell Biology
Edited by David M. Prescott

Volume 11 (1975)
Yeast Cells
Edited by David M. Prescott

Volume 12 (1975)
Yeast Cells
Edited by David M. Prescott

Volume 13 (1976)
Methods in Cell Biology
Edited by David M. Prescott

Volume 14 (1976)
Methods in Cell Biology
Edited by David M. Prescott

Volume 15 (1977)
Methods in Cell Biology
Edited by David M. Prescott

Volume 16 (1977)
Chromatin and Chromosomal Protein Research I
Edited by Gary Stein, Janet Stein, and Lewis J. Kleinsmith

Volume 17 (1978)
Chromatin and Chromosomal Protein Research II
Edited by Gary Stein, Janet Stein, and Lewis J. Kleinsmith

Volume 18 (1978)
Chromatin and Chromosomal Protein Research III
Edited by Gary Stein, Janet Stein, and Lewis J. Kleinsmith

Volume 19 (1978)
Chromatin and Chromosomal Protein Research IV
Edited by Gary Stein, Janet Stein, and Lewis J. Kleinsmith

Volume 20 (1978)
Methods in Cell Biology
Edited by David M. Prescott

Advisory Board Chairman
KEITH R. PORTER

Volume 21A (1980)
Normal Human Tissue and Cell Culture, Part A: Respiratory, Cardiovascular, and Integumentary Systems
Edited by Curtis C. Harris, Benjamin F. Trump, and Gary D. Stoner

Volume 21B (1980)
Normal Human Tissue and Cell Culture, Part B: Endocrine, Urogenital, and Gastrointestinal Systems
Edited by Curtis C. Harris, Benjamin F. Trump, and Gray D. Stoner

Volume 22 (1981)
Three-Dimensional Ultrastructure in Biology
Edited by James N. Turner

Volume 23 (1981)
Basic Mechanisms of Cellular Secretion
Edited by Arthur R. Hand and Constance Oliver

Volume 24 (1982)
The Cytoskeleton, Part A: Cytoskeletal Proteins, Isolation and Characterization
Edited by Leslie Wilson

Volume 25 (1982)
The Cytoskeleton, Part B: Biological Systems and *In Vitro* Models
Edited by Leslie Wilson

Volume 26 (1982)
Prenatal Diagnosis: Cell Biological Approaches
Edited by Samuel A. Latt and Gretchen J. Darlington

Series Editor
LESLIE WILSON

Volume 27 (1986)
Echinoderm Gametes and Embryos
Edited by Thomas E. Schroeder

Volume 28 (1987)
***Dictyostelium discoideum:* Molecular Approaches to Cell Biology**
Edited by James A. Spudich

Volume 29 (1989)
Fluorescence Microscopy of Living Cells in Culture, Part A: Fluorescent Analogs, Labeling Cells, and Basic Microscopy
Edited by Yu-Li Wang and D. Lansing Taylor

Volume 30 (1989)
Fluorescence Microscopy of Living Cells in Culture, Part B: Quantitative Fluorescence Microscopy—Imaging and Spectroscopy
Edited by D. Lansing Taylor and Yu-Li Wang

Volume 31 (1989)
Vesicular Transport, Part A
Edited by Alan M. Tartakoff

Volume 32 (1989)
Vesicular Transport, Part B
Edited by Alan M. Tartakoff

Volume 33 (1990)
Flow Cytometry
Edited by Zbigniew Darzynkiewicz and Harry A. Crissman

Volume 34 (1991)
Vectorial Transport of Proteins into and across Membranes
Edited by Alan M. Tartakoff

Selected from Volumes 31, 32, and 34 (1991)
Laboratory Methods for Vesicular and Vectorial Transport
Edited by Alan M. Tartakoff

Volume 35 (1991)
Functional Organization of the Nucleus: A Laboratory Guide
Edited by Barbara A. Hamkalo and Sarah C. R. Elgin

Volume 36 (1991)
***Xenopus laevis:* Practical Uses in Cell and Molecular Biology**
Edited by Brian K. Kay and H. Benjamin Peng

Series Editors
LESLIE WILSON AND PAUL MATSUDAIRA

Volume 37 (1993)
Antibodies in Cell Biology
Edited by David J. Asai

Volume 38 (1993)
Cell Biological Applications of Confocal Microscopy
Edited by Brian Matsumoto

Volume 39 (1993)
Motility Assays for Motor Proteins
Edited by Jonathan M. Scholey

Volume 40 (1994)
A Practical Guide to the Study of Calcium in Living Cells
Edited by Richard Nuccitelli

Volume 41 (1994)
Flow Cytometry, Second Edition, Part A
Edited by Zbigniew Darzynkiewicz, J. Paul Robinson, and Harry A. Crissman

Volume 42 (1994)
Flow Cytometry, Second Edition, Part B
Edited by Zbigniew Darzynkiewicz, J. Paul Robinson, and Harry A. Crissman

Volume 43 (1994)
Protein Expression in Animal Cells
Edited by Michael G. Roth

Volume 44 (1994)
***Drosophila melanogaster:* Practical Uses in Cell and Molecular Biology**
Edited by Lawrence S. B. Goldstein and Eric A. Fyrberg

Volume 45 (1994)
Microbes as Tools for Cell Biology
Edited by David G. Russell

Volume 46 (1995)
Cell Death
Edited by Lawrence M. Schwartz and Barbara A. Osborne

Volume 47 (1995)
Cilia and Flagella
Edited by William Dentler and George Witman

Volume 48 (1995)
***Caenorhabditis elegans:* Modern Biological Analysis of an Organism**
Edited by Henry F. Epstein and Diane C. Shakes

Volume 49 (1995)
Methods in Plant Cell Biology, Part A
Edited by David W. Galbraith, Hans J. Bohnert, and Don P. Bourque

Volume 50 (1995)
Methods in Plant Cell Biology, Part B
Edited by David W. Galbraith, Don P. Bourque, and Hans J. Bohnert

Volume 51 (1996)
Methods in Avian Embryology
Edited by Marianne Bronner-Fraser

Volume 52 (1997)
Methods in Muscle Biology
Edited by Charles P. Emerson, Jr. and H. Lee Sweeney

Volume 53 (1997)
Nuclear Structure and Function
Edited by Miguel Berrios

Volume 54 (1997)
Cumulative Index

Volume 55 (1997)
Laser Tweezers in Cell Biology
Edited by Michael P. Sheetz

Volume 56 (1998)
Video Microscopy
Edited by Greenfield Sluder and David E. Wolf

Volume 57 (1998)
Animal Cell Culture Methods
Edited by Jennie P. Mather and David Barnes

Volume 58 (1998)
Green Fluorescent Protein
Edited by Kevin F. Sullivan and Steve A. Kay

Volume 59 (1998)
The Zebrafish: Biology
Edited by H. William Detrich III, Monte Westerfield, and Leonard I. Zon

Volume 60 (1998)
The Zebrafish: Genetics and Genomics
Edited by H. William Detrich III, Monte Westerfield, and Leonard I. Zon

Volume 61 (1998)
Mitosis and Meiosis
Edited by Conly L. Rieder

Volume 62 (1999)
Tetrahymena thermophila
Edited by David J. Asai and James D. Forney

Volume 63 (2000)
Cytometry, Third Edition, Part A
Edited by Zbigniew Darzynkiewicz, J. Paul Robinson, and Harry Crissman

Volume 64 (2000)
Cytometry, Third Edition, Part B
Edited by Zbigniew Darzynkiewicz, J. Paul Robinson, and Harry Crissman

Volume 65 (2001)
Mitochondria
Edited by Liza A. Pon and Eric A. Schon

Volume 66 (2001)
Apoptosis
Edited by Lawrence M. Schwartz and Jonathan D. Ashwell

Volume 67 (2001)
Centrosomes and Spindle Pole Bodies
Edited by Robert E. Palazzo and Trisha N. Davis

Volume 68 (2002)
Atomic Force Microscopy in Cell Biology
Edited by Bhanu P. Jena and J. K. Heinrich Hörber

Volume 69 (2002)
Methods in Cell–Matrix Adhesion
Edited by Josephine C. Adams

Volume 70 (2002)
Cell Biological Applications of Confocal Microscopy
Edited by Brian Matsumoto

Volume 71 (2003)
Neurons: Methods and Applications for Cell Biologist
Edited by Peter J. Hollenbeck and James R. Bamburg

Volume 72 (2003)
Digital Microscopy: A Second Edition of Video Microscopy
Edited by Greenfield Sluder and David E. Wolf

Volume 73 (2003)
Cumulative Index

Volume 74 (2004)
Development of Sea Urchins, Ascidians, and Other Invertebrate Deuterostomes: Experimental Approaches
Edited by Charles A. Ettensohn, Gary M. Wessel, and Gregory A. Wray

Volume 75 (2004)
Cytometry, 4th Edition: New Developments
Edited by Zbigniew Darzynkiewicz, Mario Roederer, and Hans Tanke

Volume 76 (2004)
The Zebrafish: Cellular and Developmental Biology
Edited by H. William Detrich, III, Monte Westerfield, and Leonard I. Zon

Volume 77 (2004)
The Zebrafish: Genetics, Genomics, and Informatics
Edited by William H. Detrich, III, Monte Westerfield, and Leonard I. Zon

Volume 78 (2004)
Intermediate Filament Cytoskeleton
Edited by M. Bishr Omary and Pierre A. Coulombe

Volume 79 (2007)
Cellular Electron Microscopy
Edited by J. Richard McIntosh

Volume 80 (2007)
Mitochondria, 2nd Edition
Edited by Liza A. Pon and Eric A. Schon

Volume 81 (2007)
Digital Microscopy, 3rd Edition
Edited by Greenfield Sluder and David E. Wolf

Volume 82 (2007)
Laser Manipulation of Cells and Tissues
Edited by Michael W. Berns and Karl Otto Greulich

Volume 83 (2007)
Cell Mechanics
Edited by Yu-Li Wang and Dennis E. Discher

Volume 84 (2007)
Biophysical Tools for Biologists, Volume One: In Vitro Techniques
Edited by John J. Correia and H. William Detrich, III

Volume 85 (2008)
Fluorescent Proteins
Edited by Kevin F. Sullivan

Volume 86 (2008)
Stem Cell Culture
Edited by Dr. Jennie P. Mather

Volume 87 (2008)
Avian Embryology, 2nd Edition
Edited by Dr. Marianne Bronner-Fraser

Volume 88 (2008)
Introduction to Electron Microscopy for Biologists
Edited by Prof. Terence D. Allen

Volume 89 (2008)
Biophysical Tools for Biologists, Volume Two: In Vivo Techniques
Edited by Dr. John J. Correia and Dr. H. William Detrich, III

Volume 90 (2008)
Methods in Nano Cell Biology
Edited by Bhanu P. Jena

Volume 91 (2009)
Cilia: Structure and Motility
Edited by Stephen M. King and Gregory J. Pazour

Volume 92 (2009)
Cilia: Motors and Regulation
Edited by Stephen M. King and Gregory J. Pazour

Volume 93 (2009)
Cilia: Model Organisms and Intraflagellar Transport
Edited by Stephen M. King and Gregory J. Pazour

Volume 94 (2009)
Primary Cilia
Edited by Roger D. Sloboda

Volume 95 (2010)
Microtubules, in vitro
Edited by Leslie Wilson and John J. Correia

Volume 96 (2010)
Electron Microscopy of Model Systems
Edited by Thomas Müeller-Reichert

Volume 97 (2010)
Microtubules: In Vivo
Edited by Lynne Cassimeris and Phong Tran

Volume 98 (2010)
Nuclear Mechanics & Genome Regulation
Edited by G.V. Shivashankar

Volume 99 (2010)
Calcium in Living Cells
Edited by Michael Whitaker

Volume 100 (2010)
The Zebrafish: Cellular and Developmental Biology, Part A
Edited by: H. William Detrich III, Monte Westerfield and Leonard I. Zon

Volume 101 (2011)
The Zebrafish: Cellular and Developmental Biology, Part B
Edited by: H. William Detrich III, Monte Westerfield and Leonard I. Zon

Volume 102 (2011)
Recent Advances in Cytometry, Part A: Instrumentation, Methods
Edited by Zbigniew Darzynkiewicz, Elena Holden, Alberto Orfao, William Telford and Donald Wlodkowic

Volume 103 (2011)
Recent Advances in Cytometry, Part B: Advances in Applications
Edited by Zbigniew Darzynkiewicz, Elena Holden, Alberto Orfao, Alberto Orfao and Donald Wlodkowic

Volume 104 (2011)
The Zebrafish: Genetics, Genomics and Informatics 3rd Edition
Edited by H. William Detrich III, Monte Westerfield, and Leonard I. Zon

Volume 105 (2011)
The Zebrafish: Disease Models and Chemical Screens 3rd Edition
Edited by H. William Detrich III, Monte Westerfield, and Leonard I. Zon

Volume 106 (2011)
Caenorhabditis elegans: Molecular Genetics and Development 2nd Edition
Edited by Joel H. Rothman and Andrew Singson

Volume 107 (2011)
Caenorhabditis elegans: Cell Biology and Physiology 2nd Edition
Edited by Joel H. Rothman and Andrew Singson

Volume 108 (2012)
Lipids
Edited by Gilbert Di Paolo and Markus R Wenk

Volume 109 (2012)
Tetrahymena thermophila
Edited by Kathleen Collins

Volume 110 (2012)
Methods in Cell Biology
Edited by Anand R. Asthagiri and Adam P. Arkin

Volume 111 (2012)
Methods in Cell Biology
Edited by Thomas Müler Reichart and Paul Verkade

Volume 112 (2012)
Laboratory Methods in Cell Biology
Edited by P. Michael Conn

Volume 113 (2013)
Laboratory Methods in Cell Biology
Edited by P. Michael Conn

Volume 114 (2013)
Digital Microscopy, 4th Edition
Edited by Greenfield Sluder and David E. Wolf

Volume 115 (2013)
Microtubules, *in Vitro*, 2nd Edition
Edited by John J. Correia and Leslie Wilson

Volume 116 (2013)
Lipid Droplets
Edited by H. Robert Yang and Peng Li

Volume 117 (2013)
Receptor-Receptor Interactions
Edited by P. Michael Conn

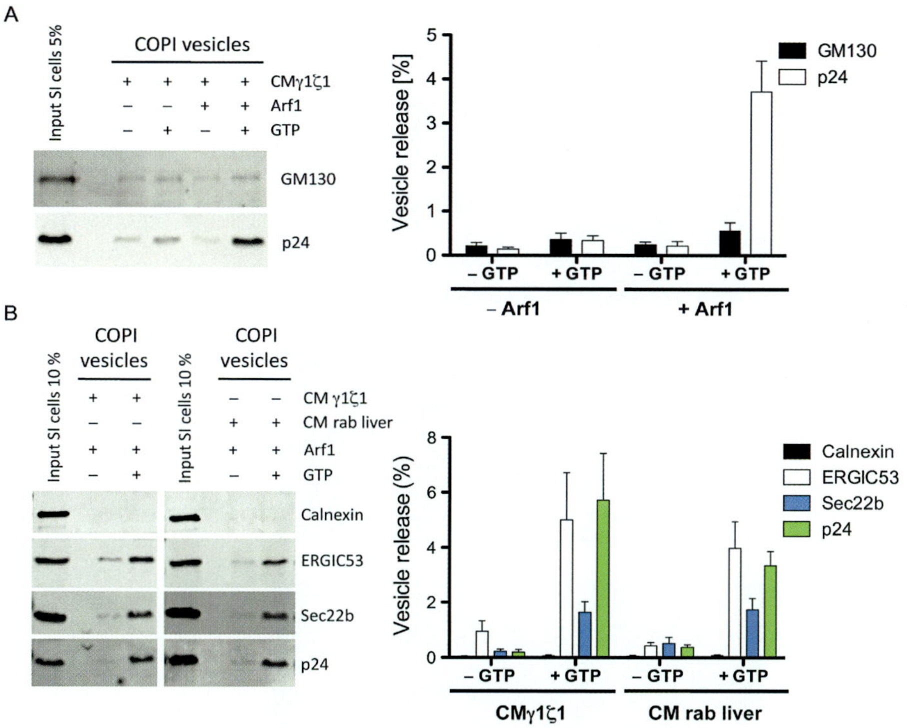

PLATE 1 (Fig. 1.1 on page 7 of this volume).

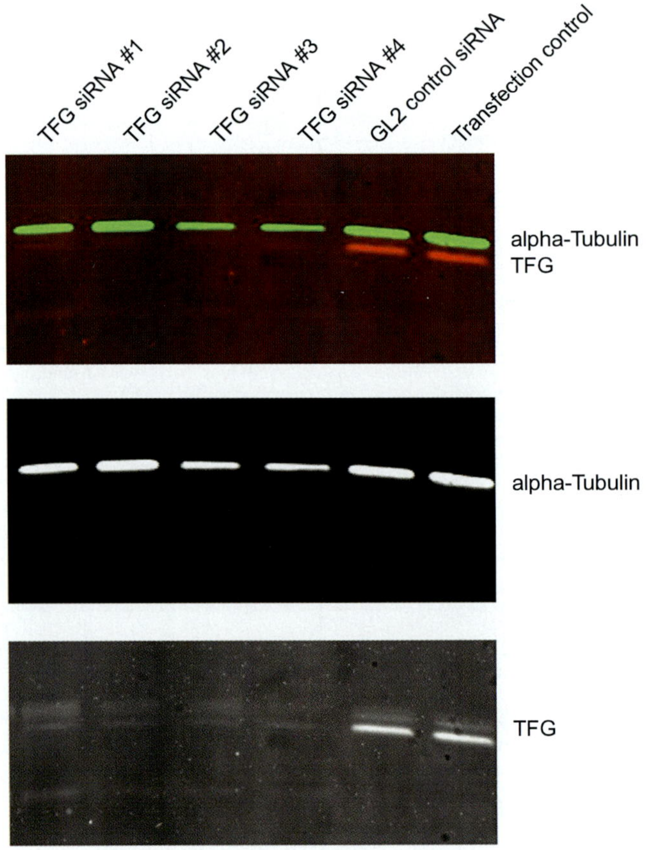

PLATE 2 (Fig. 2.1 on page 26 of this volume).

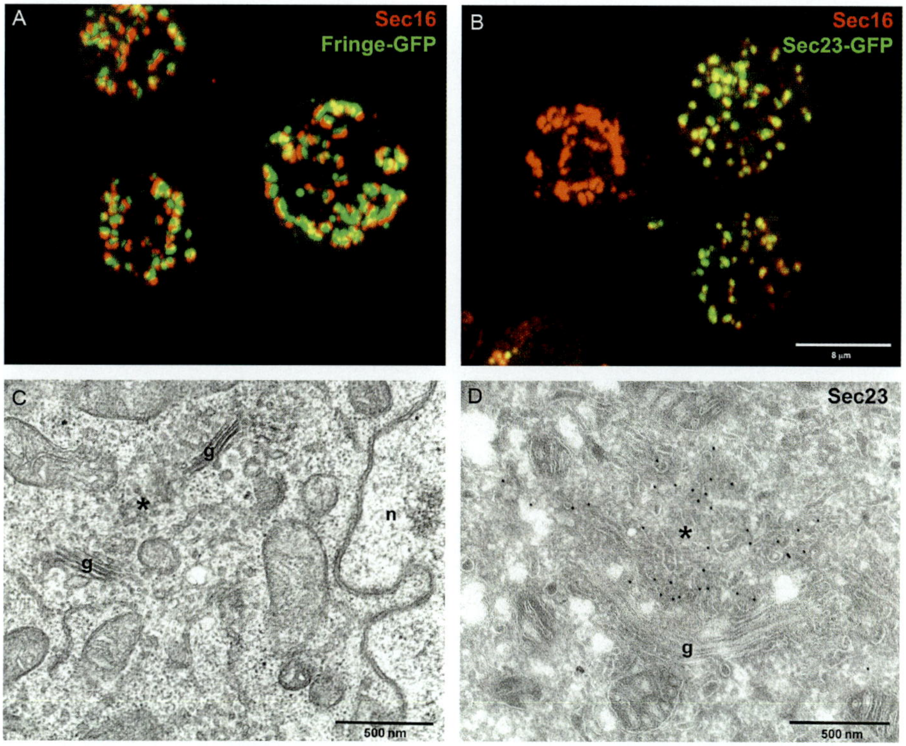

PLATE 3 (Fig. 3.1 on page 40 of this volume).

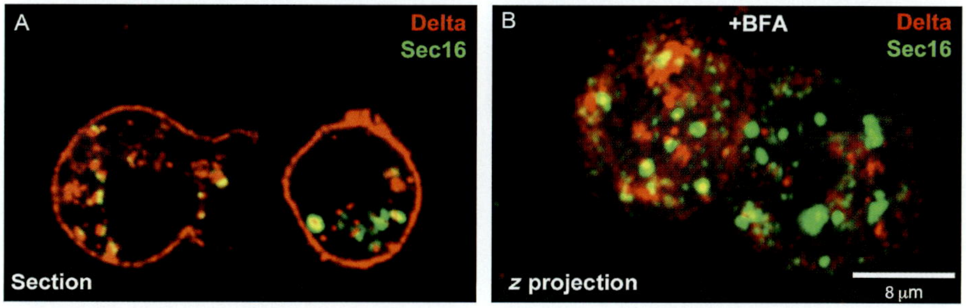

PLATE 4 (Fig. 3.2 on page 43 of this volume).

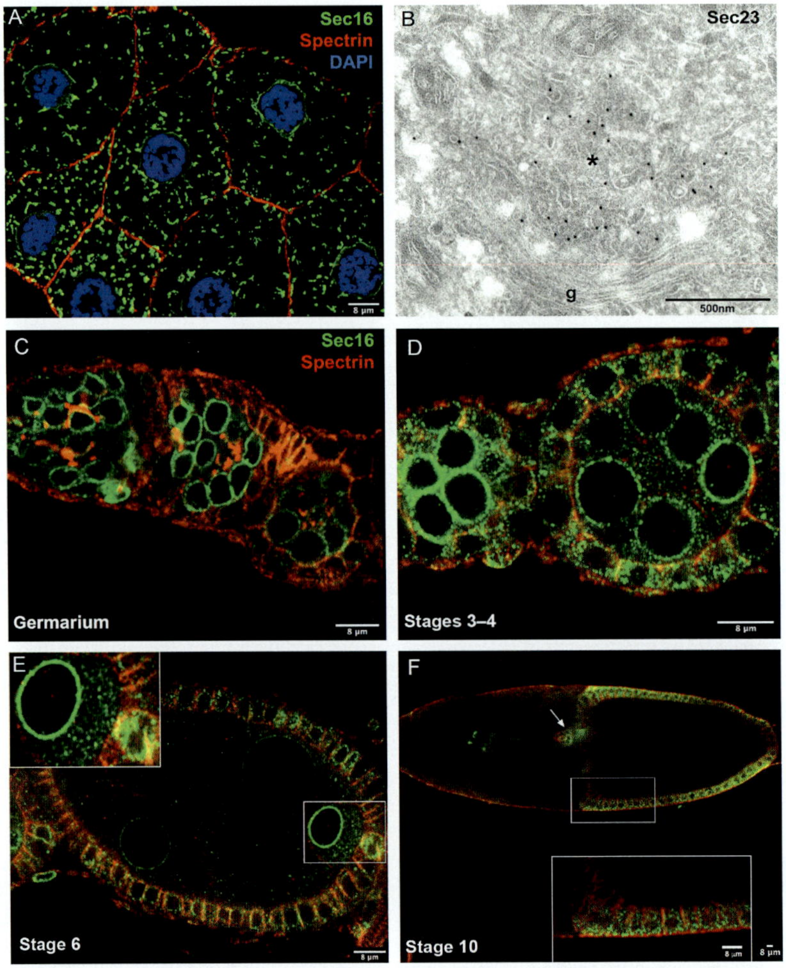

PLATE 5 (Fig. 3.3 on page 44 of this volume).

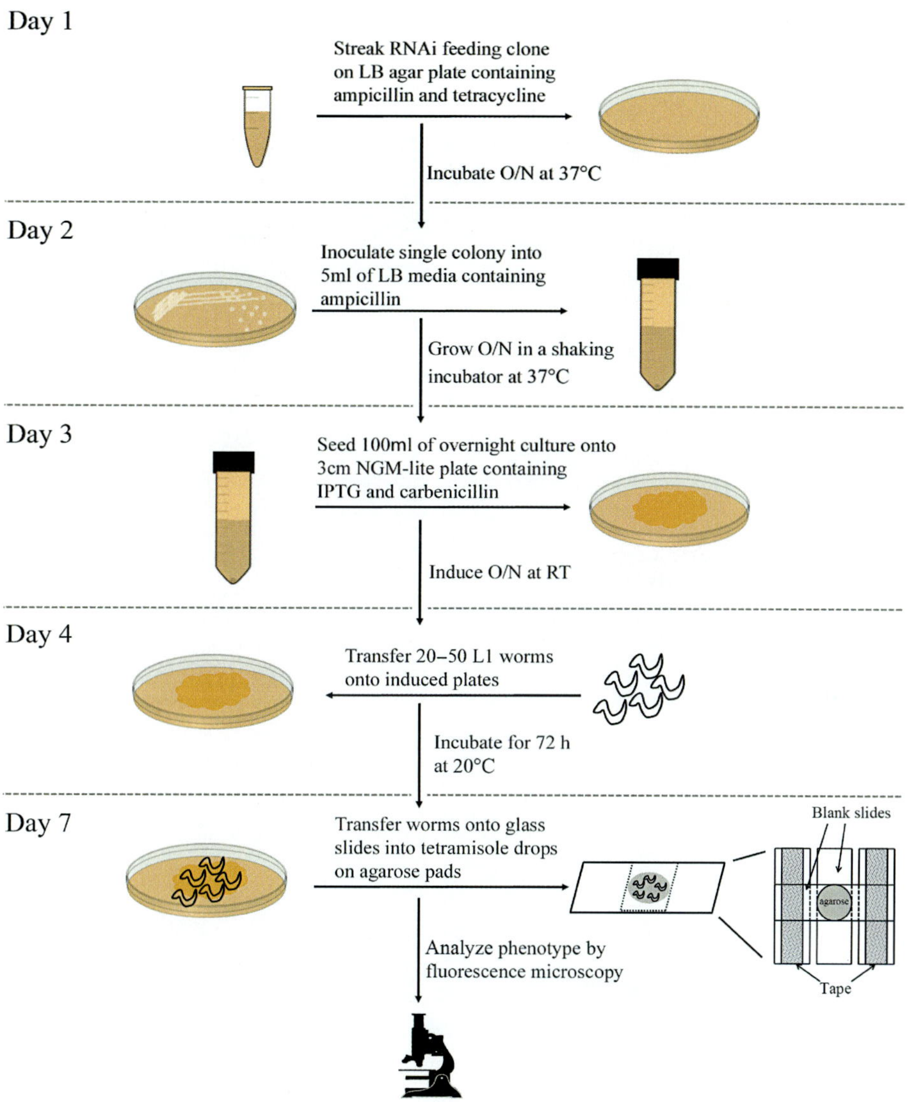

PLATE 6 (Fig. 4.1 on page 59 of this volume).

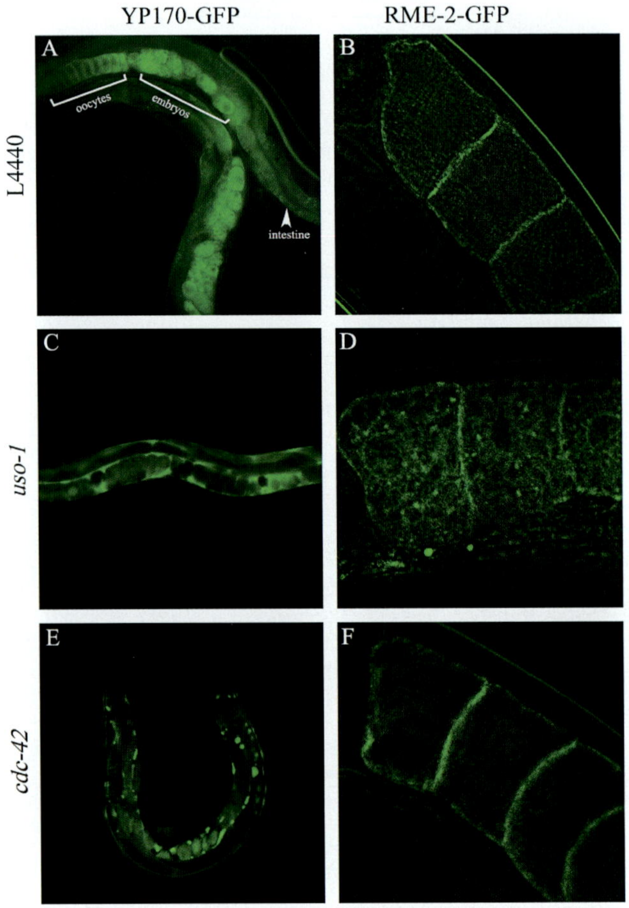

PLATE 7 (Fig. 4.2 on page 62 of this volume).

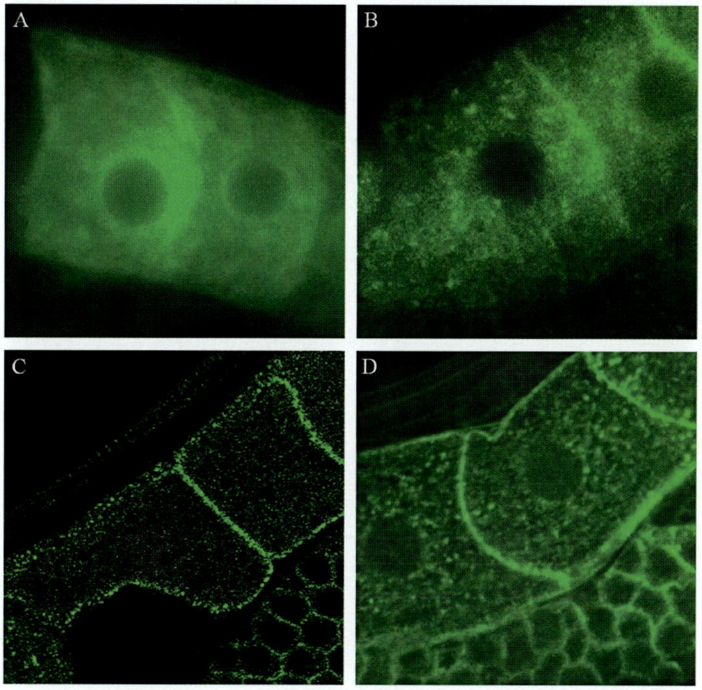

PLATE 8 (Fig. 4.3 on page 63 of this volume).

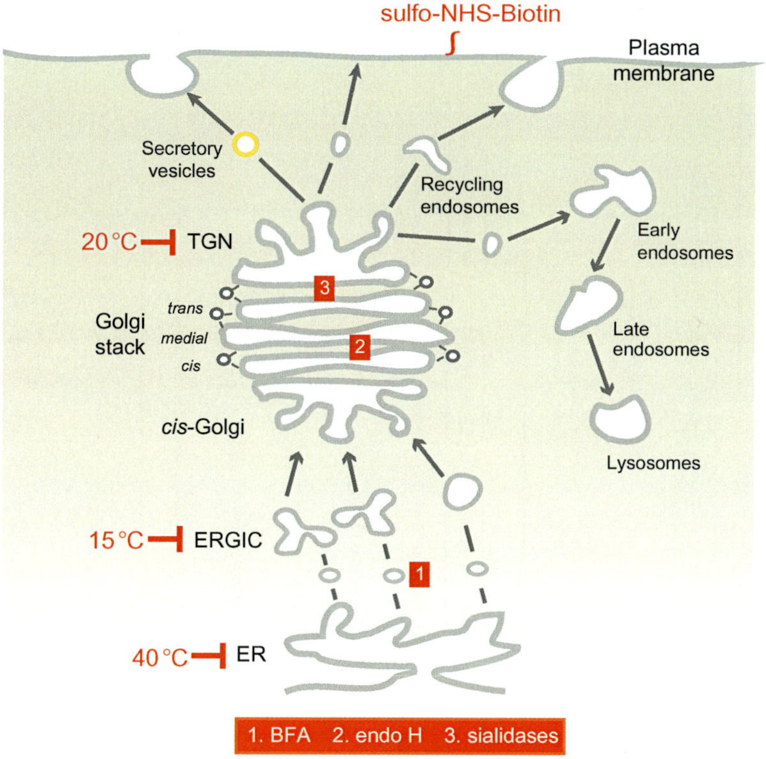

PLATE 9 (Fig. 6.1 on page 87 of this volume).

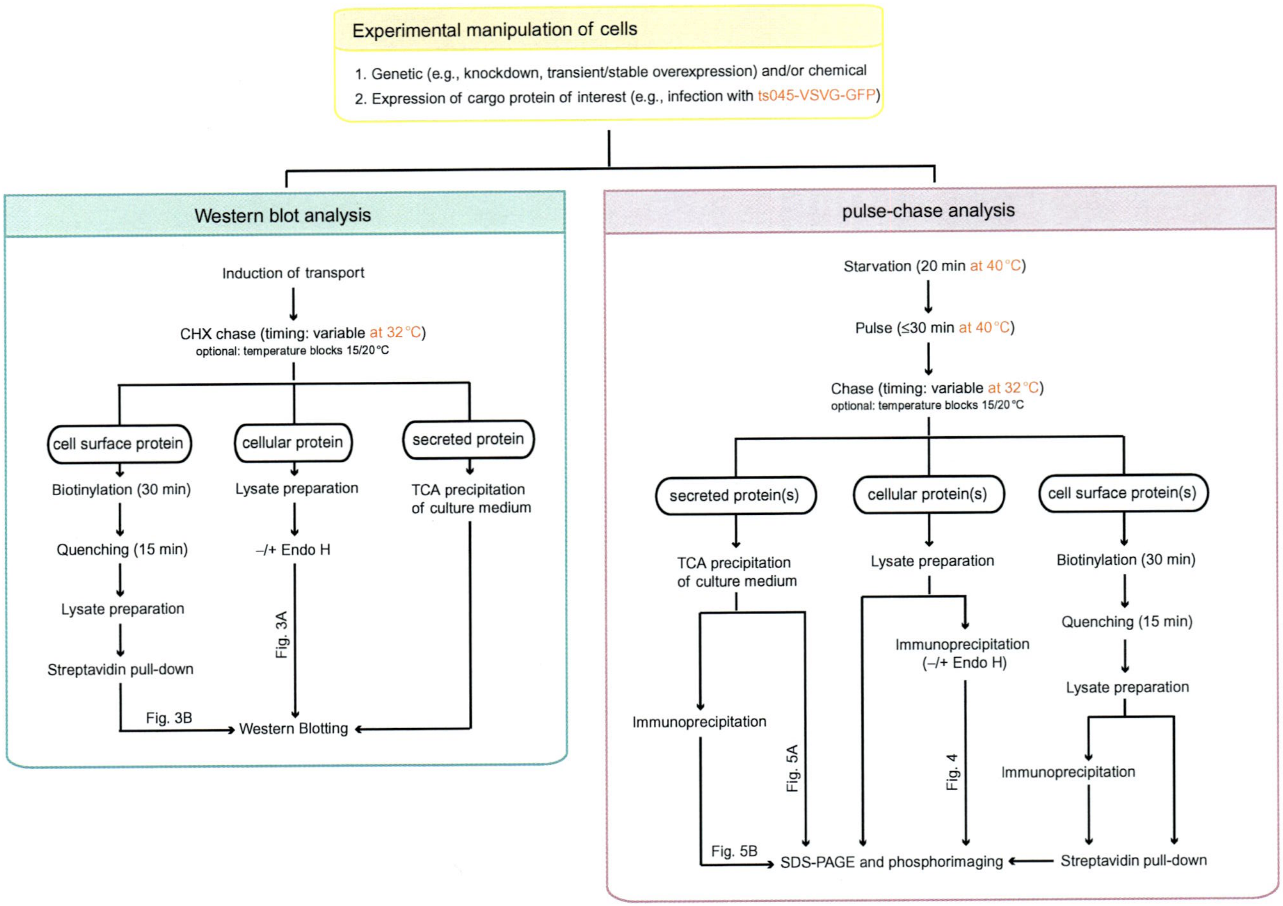

PLATE 10 (Fig. 6.2 on page 90 of this volume).

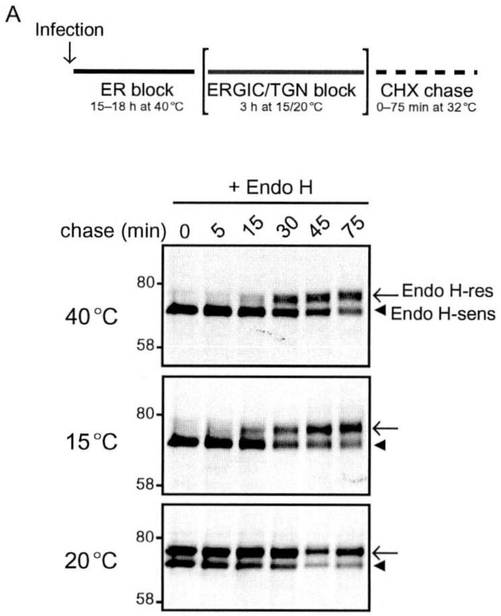

PLATE 11 (Fig. 6.3 on page 93 of this volume).

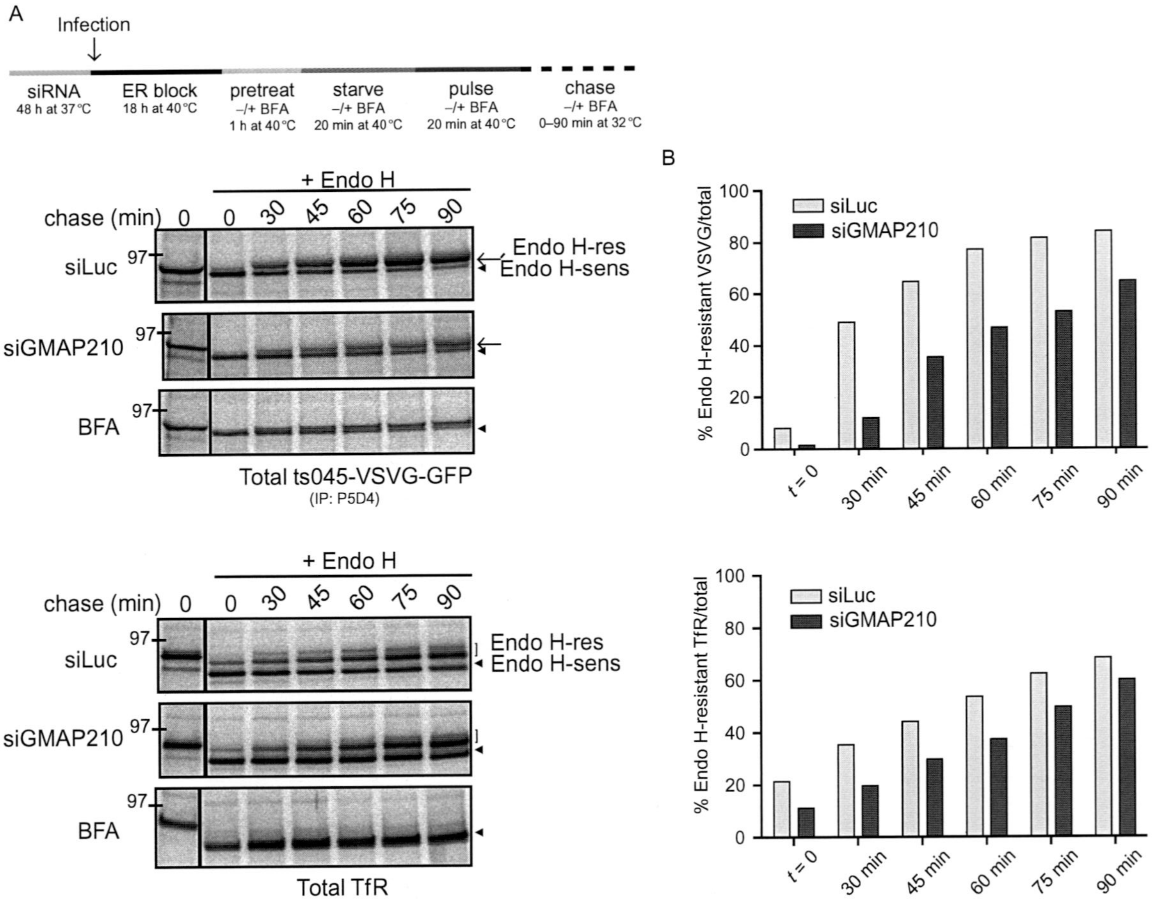

PLATE 12 (Fig. 6.4 on page 96 of this volume).

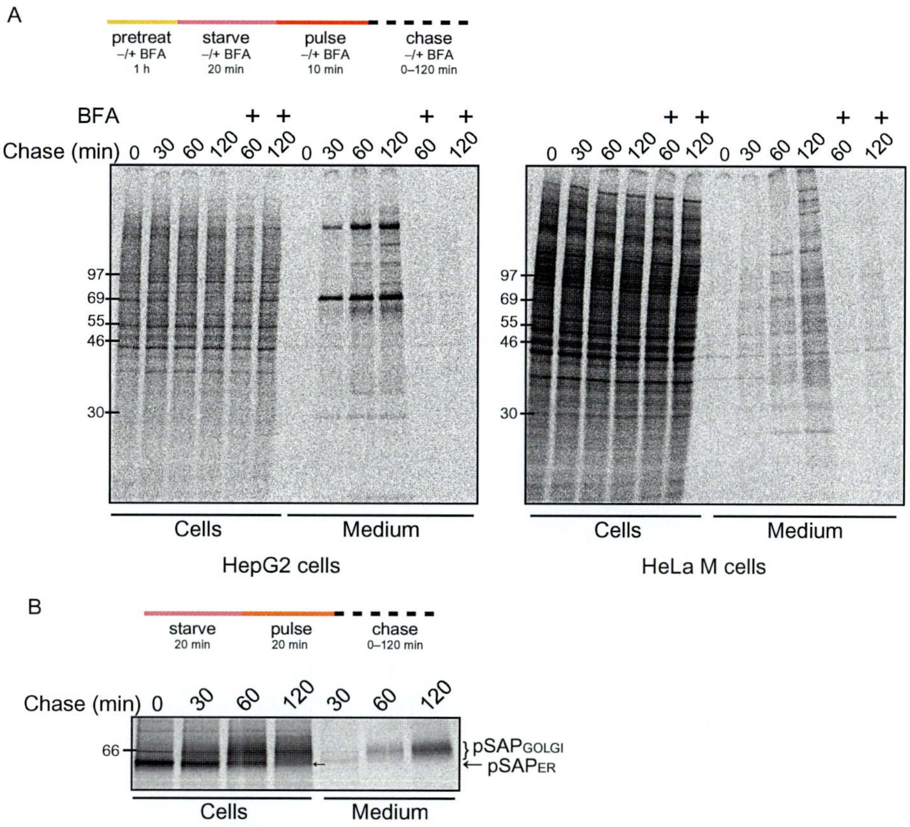

PLATE 13 (Fig. 6.5 on page 98 of this volume).

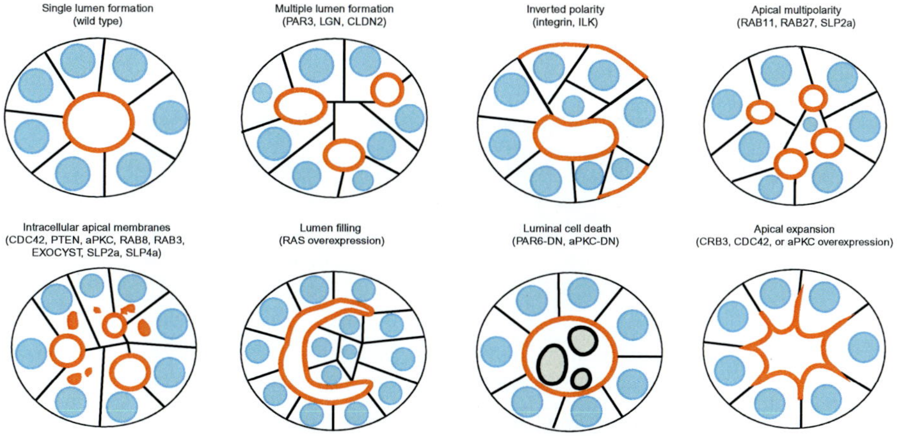

PLATE 14 (Fig. 7.1 on page 116 of this volume).

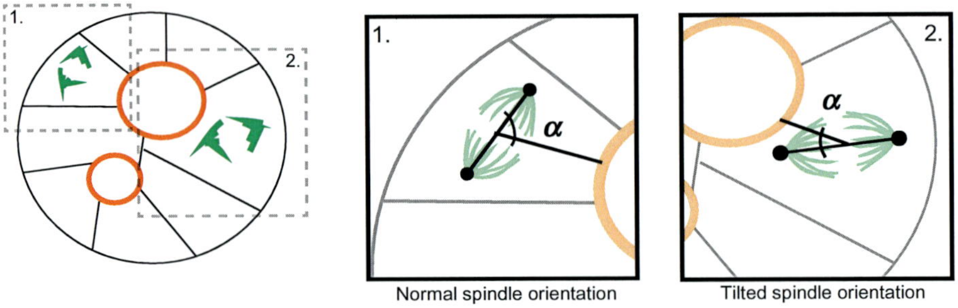

PLATE 15 (Fig. 7.2 on page 118 of this volume).

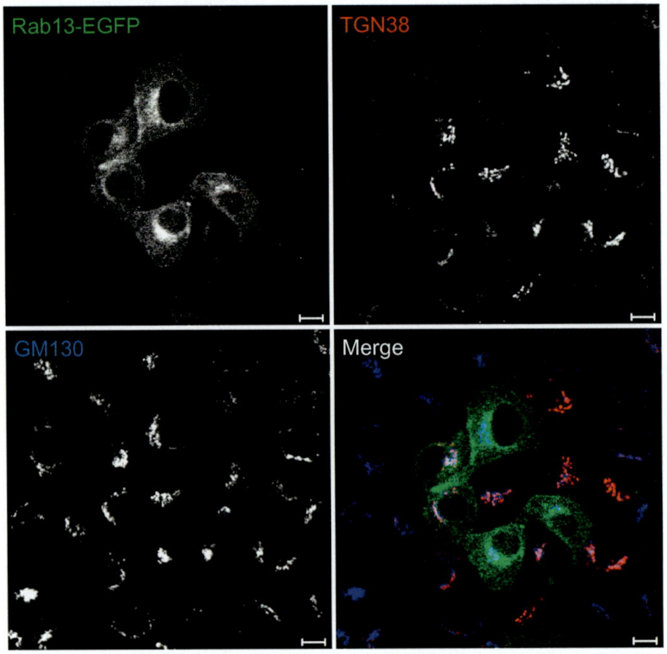

PLATE 16 (Fig. 8.1 on page 127 of this volume).

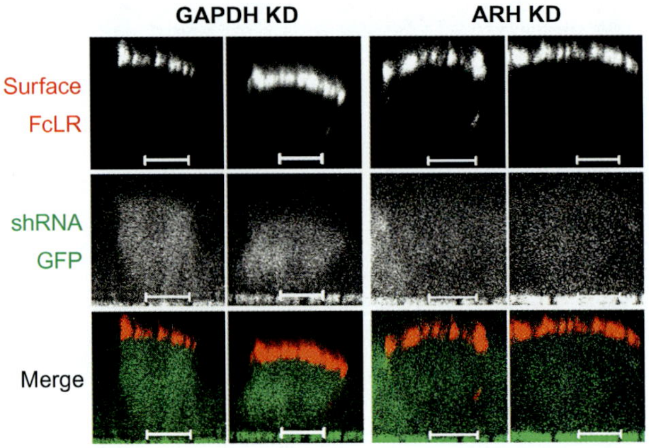

PLATE 17 (Fig. 8.2 on page 128 of this volume).

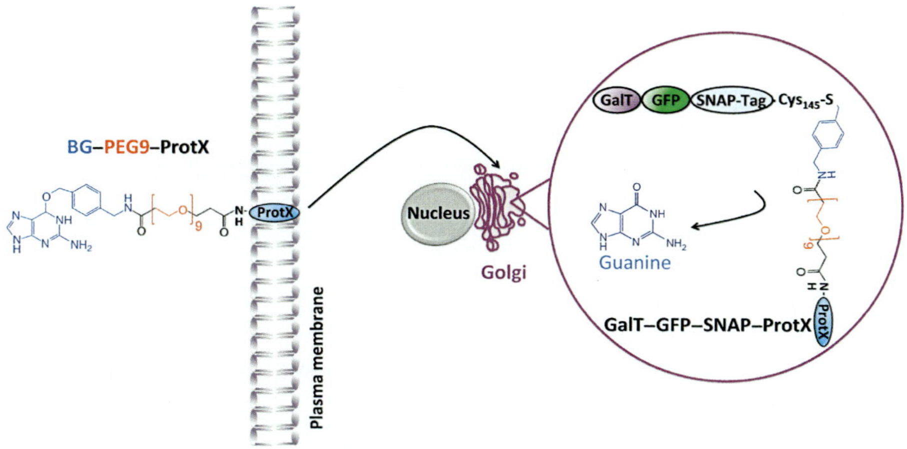

PLATE 18 (Fig. 9.1 on page 141 of this volume).

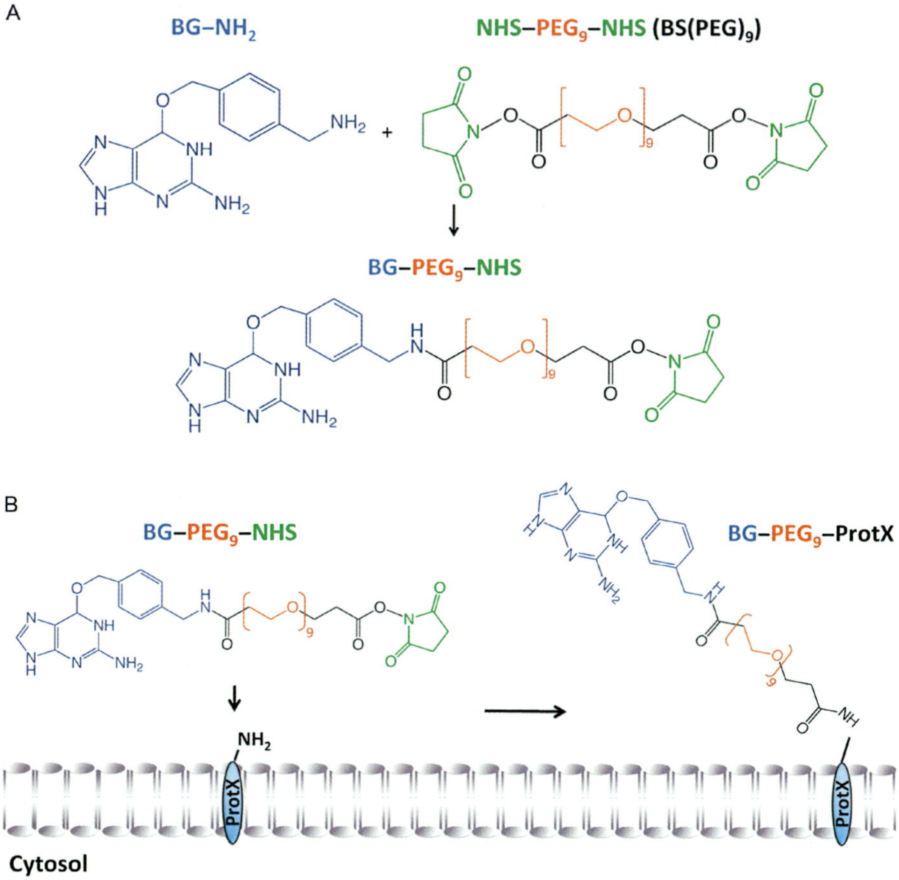

PLATE 19 (Fig. 9.2 on page 143 of this volume).

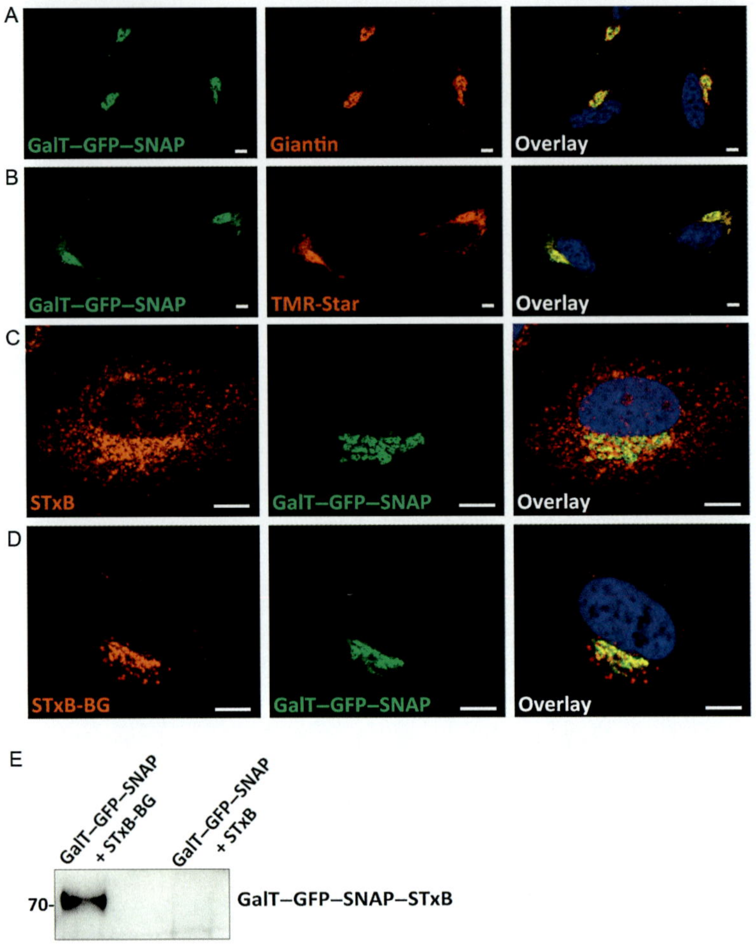

PLATE 20 (Fig. 9.3 on page 146 of this volume).

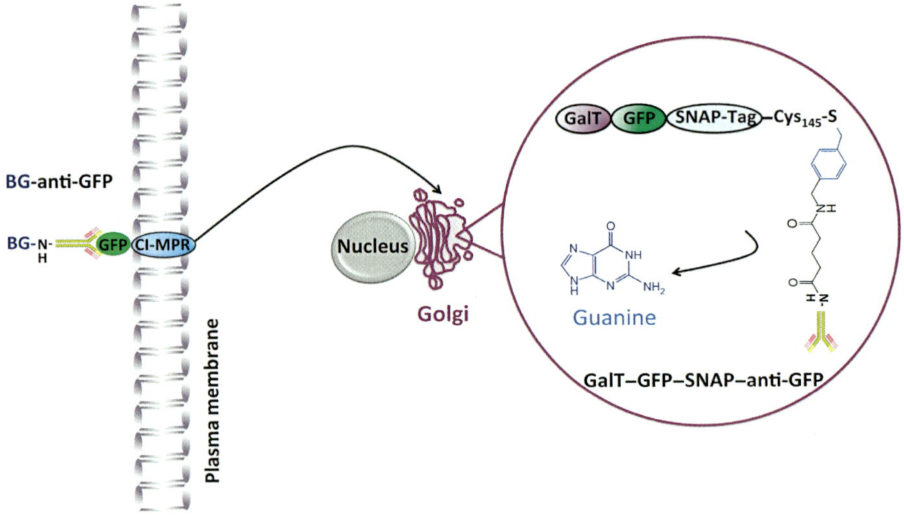

PLATE 21 (Fig. 9.5 on page 151 of this volume).

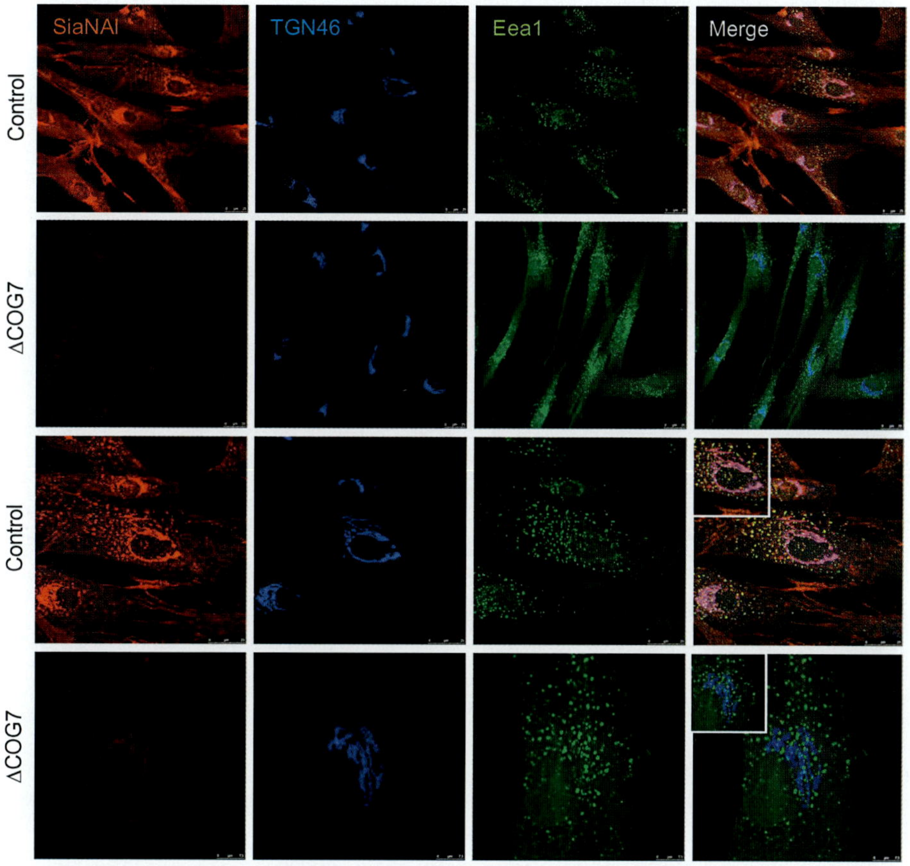

PLATE 22 (Fig. 10.3 on page 163 of this volume).

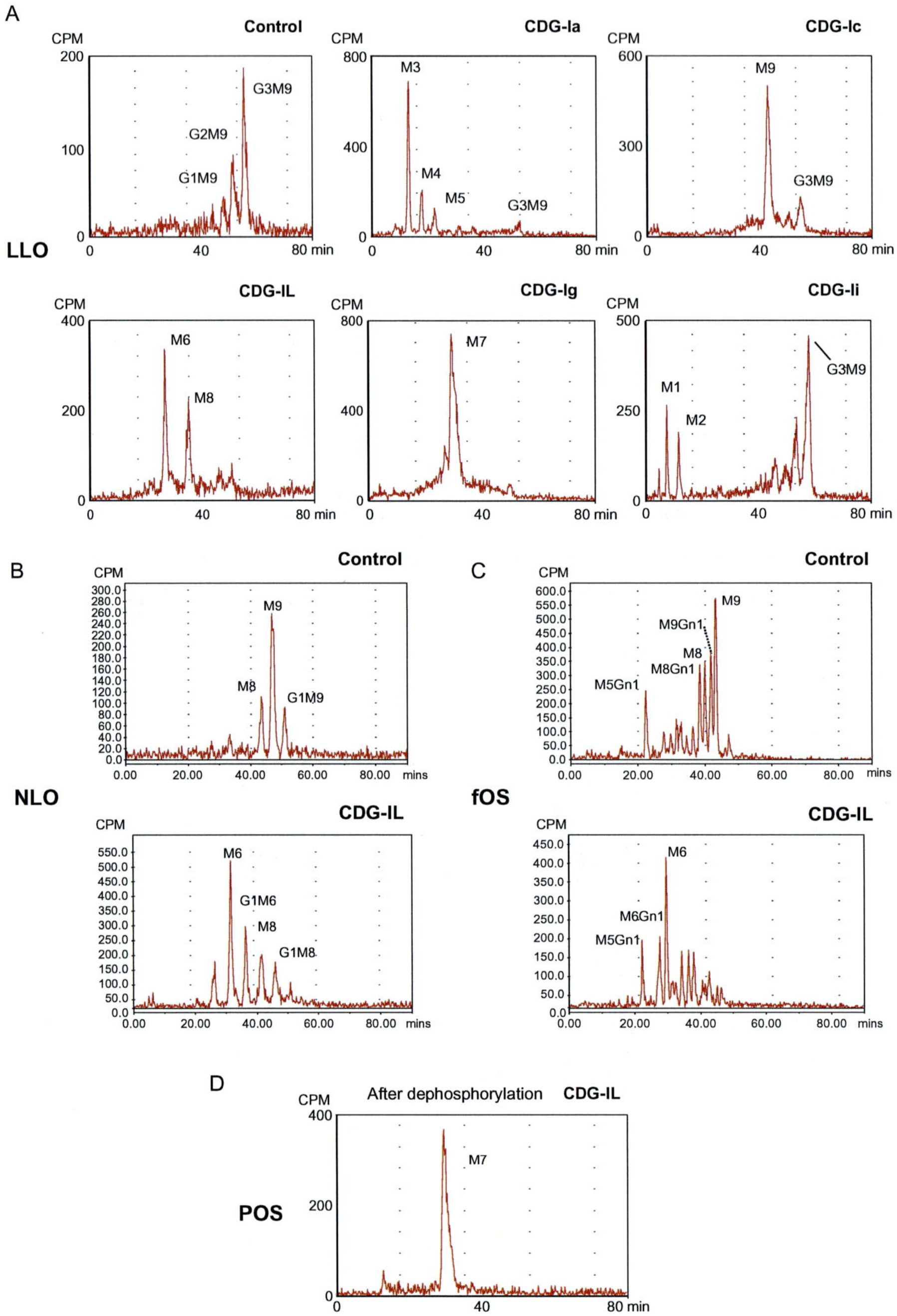

PLATE 23 (Fig. 10.4 on page 173 of this volume).

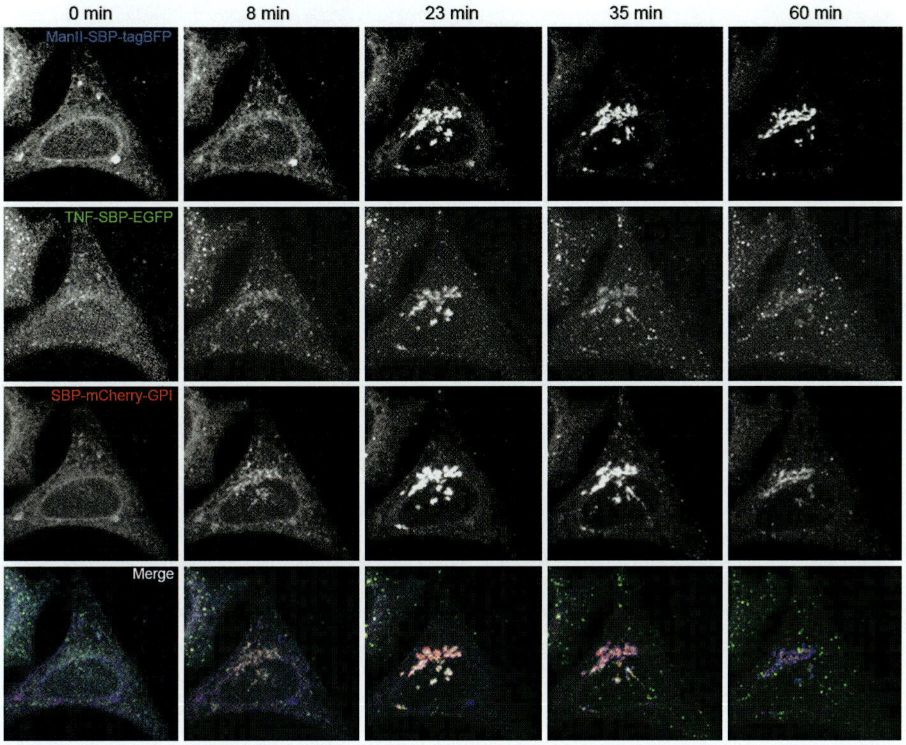

PLATE 24 (Fig. 11.4 on page 192 of this volume).

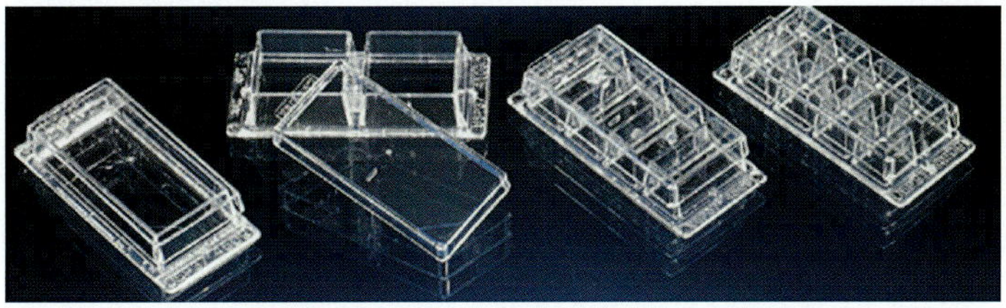

PLATE 25 (Fig. 12.1 on page 198 of this volume).

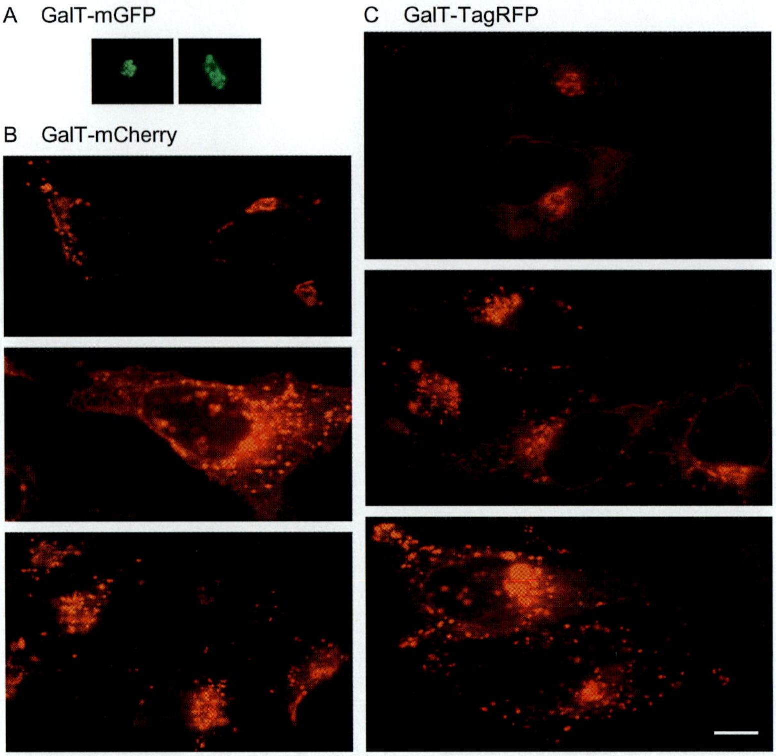

PLATE 26 (Fig. 12.2 on page 199 of this volume).

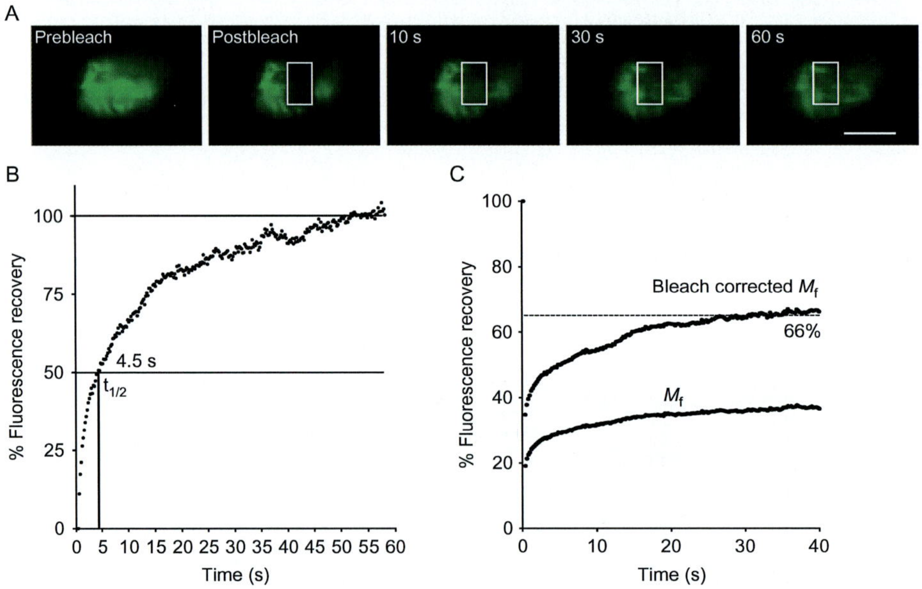

PLATE 27 (Fig. 12.3 on page 204 of this volume).

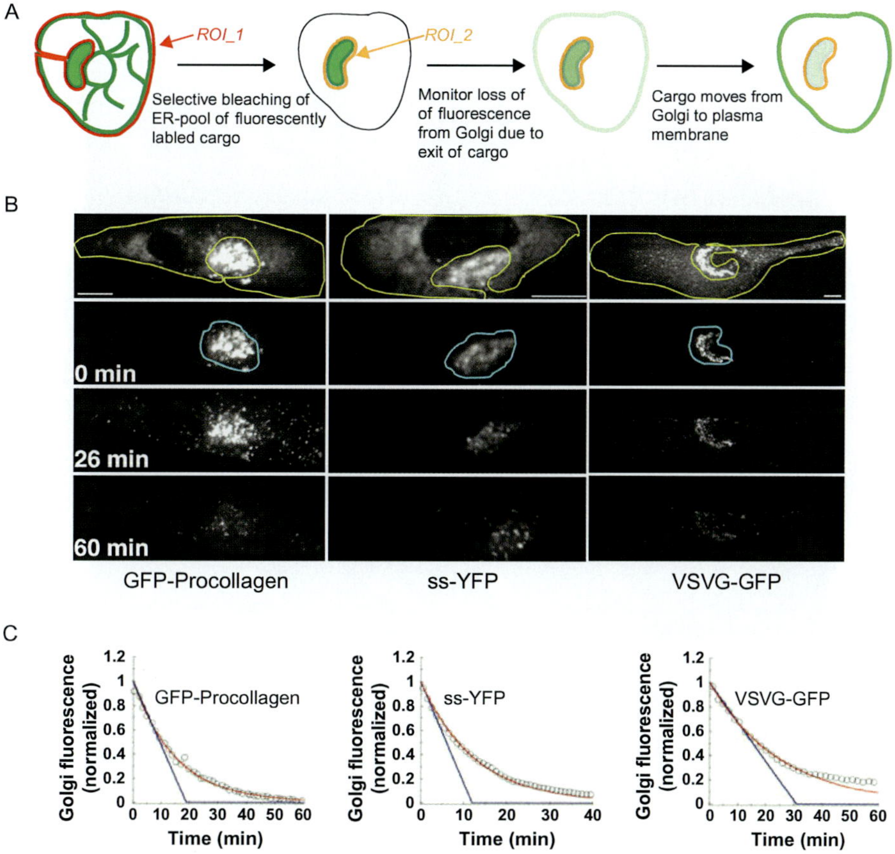

PLATE 28 (Fig. 13.1 on page 221 of this volume).

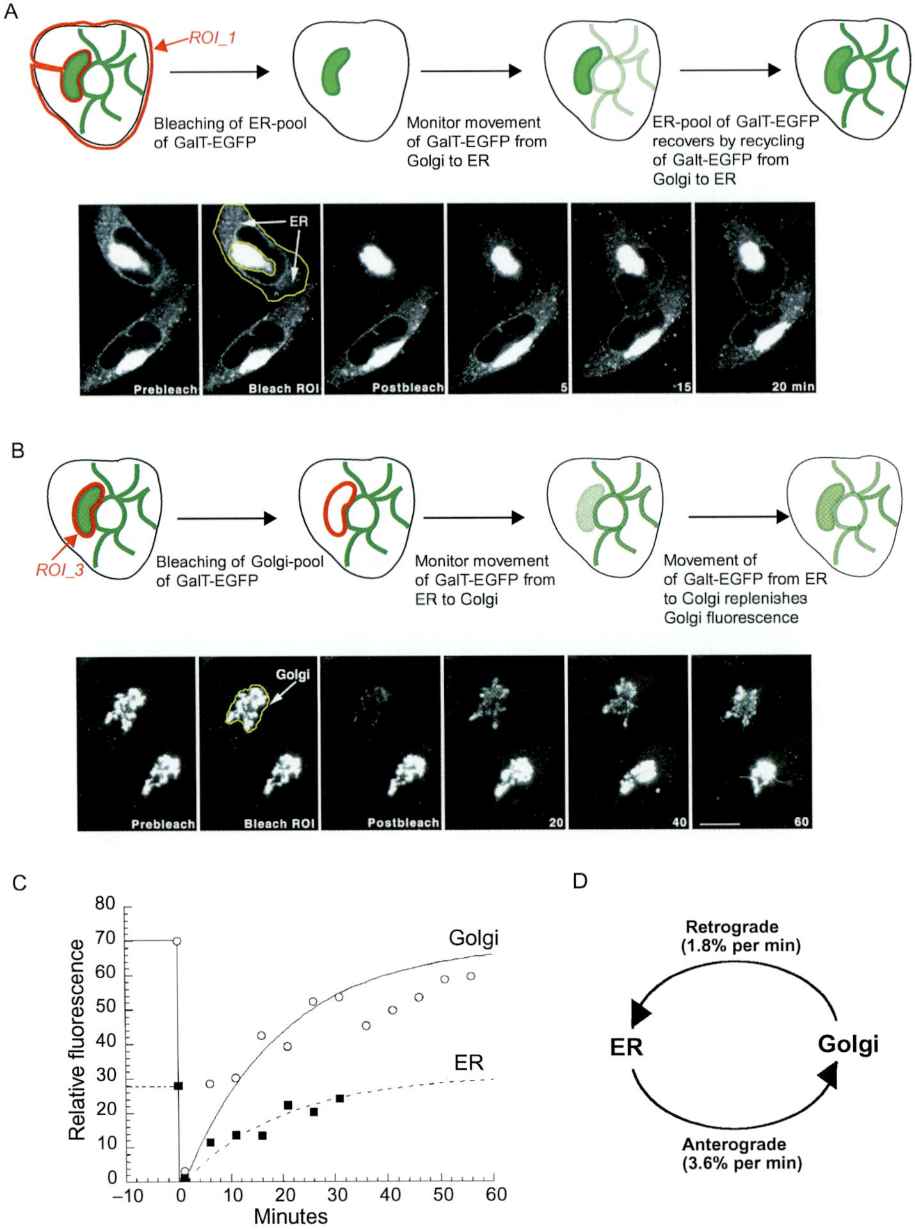

PLATE 29 (Fig. 13.2 on page 222 of this volume).

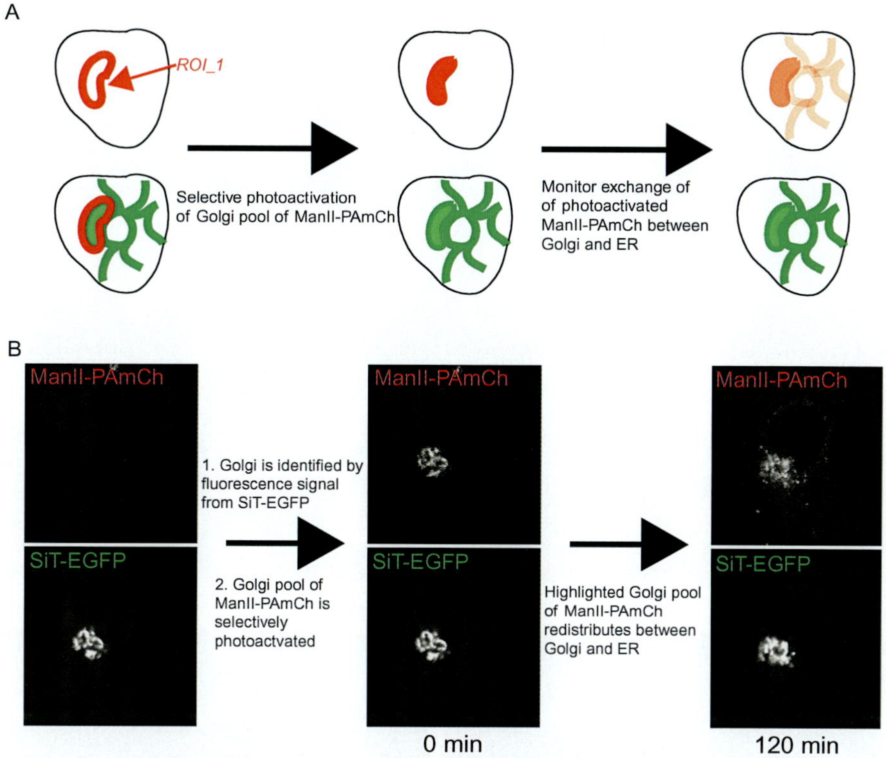

PLATE 30 (Fig. 13.3 on page 223 of this volume).

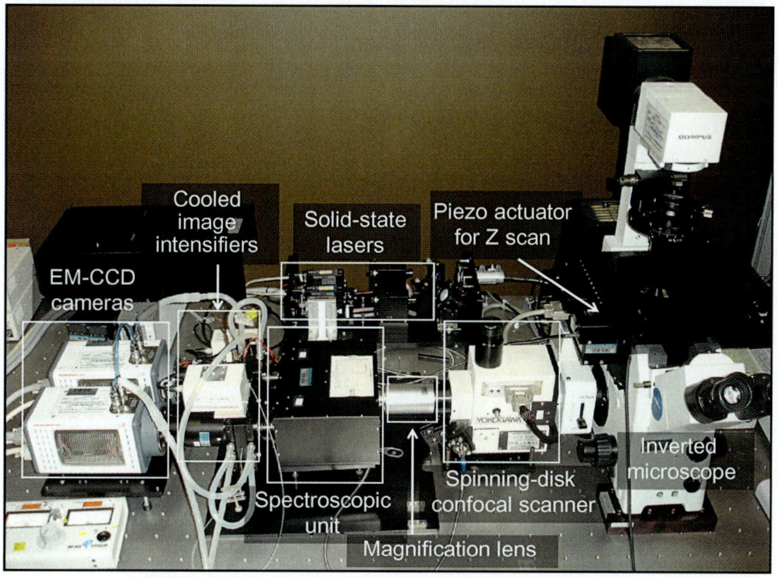

PLATE 31 (Fig. 14.1 on page 238 of this volume).

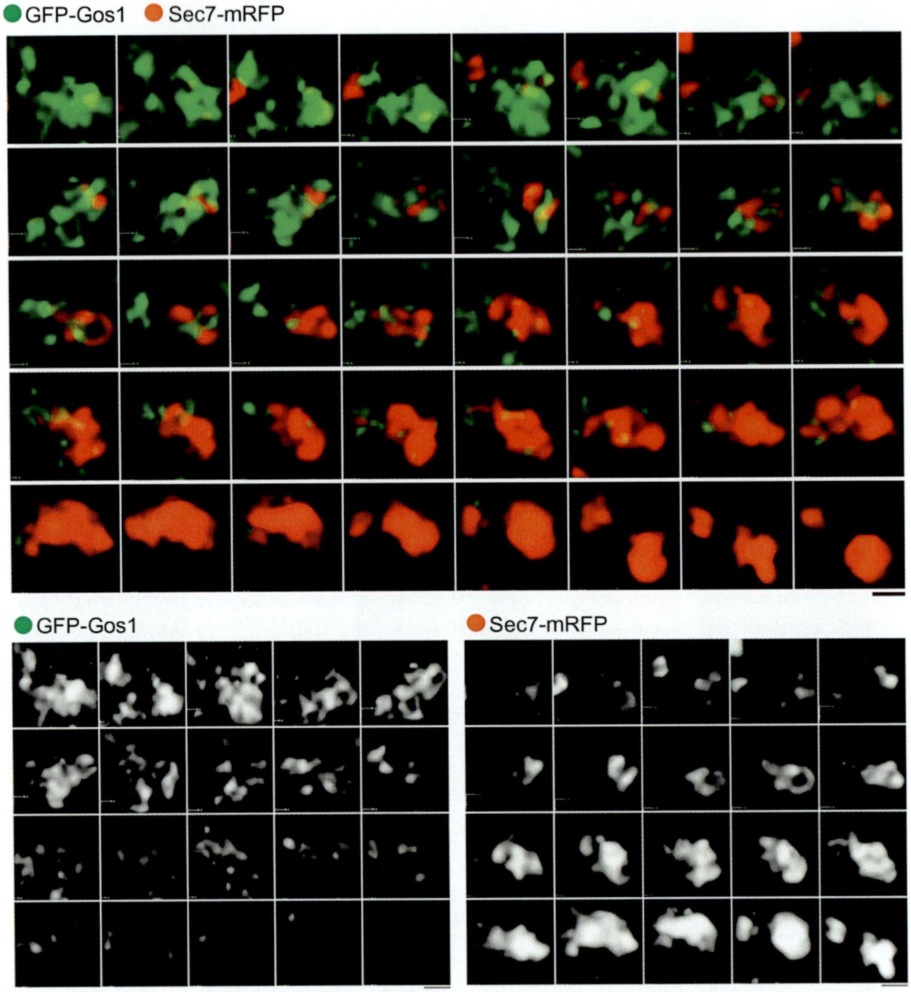

PLATE 32 (Fig. 14.2 on page 240 of this volume).

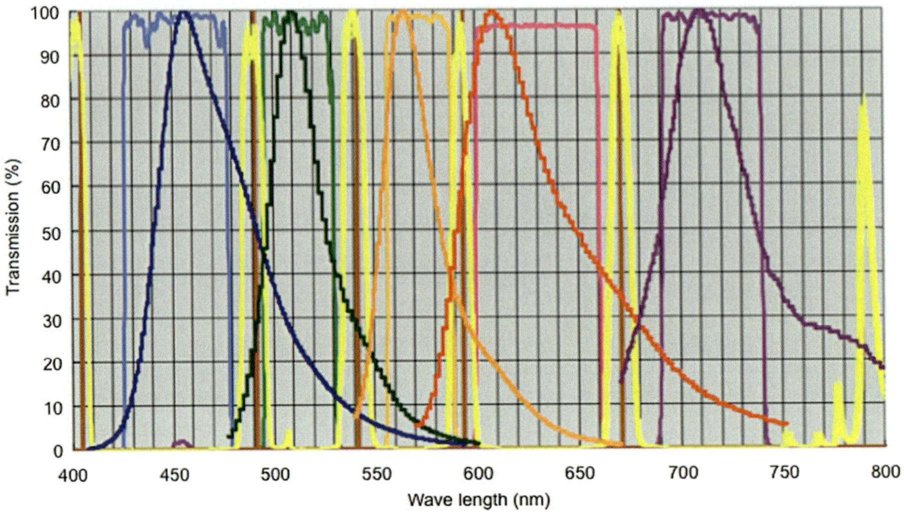

PLATE 33 (Fig. 14.3 on page 241 of this volume).

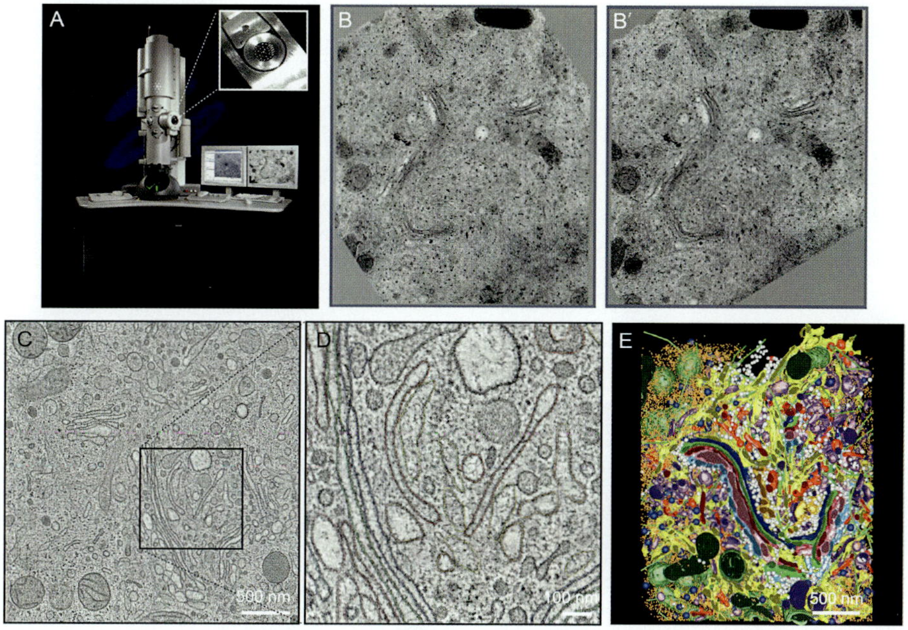

PLATE 34 (Fig. 16.2 on page 267 of this volume).

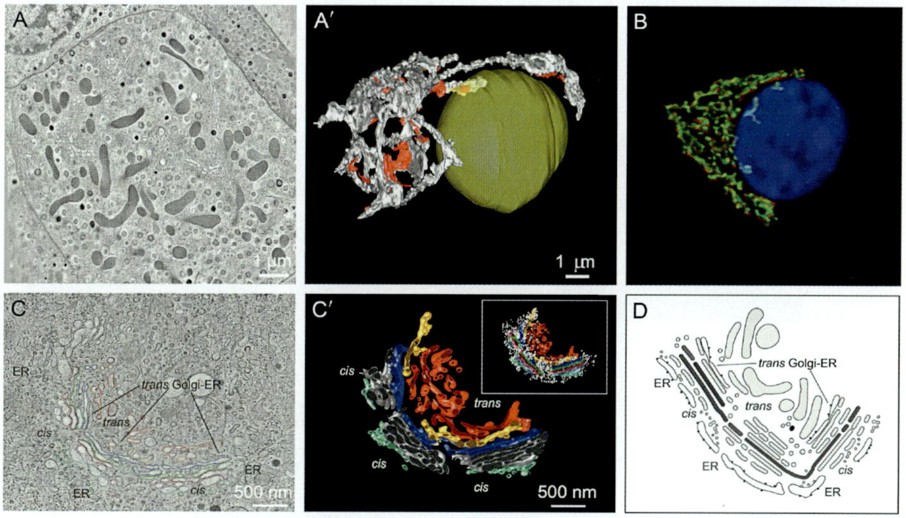

PLATE 35 (Fig. 16.3 on page 271 of this volume).

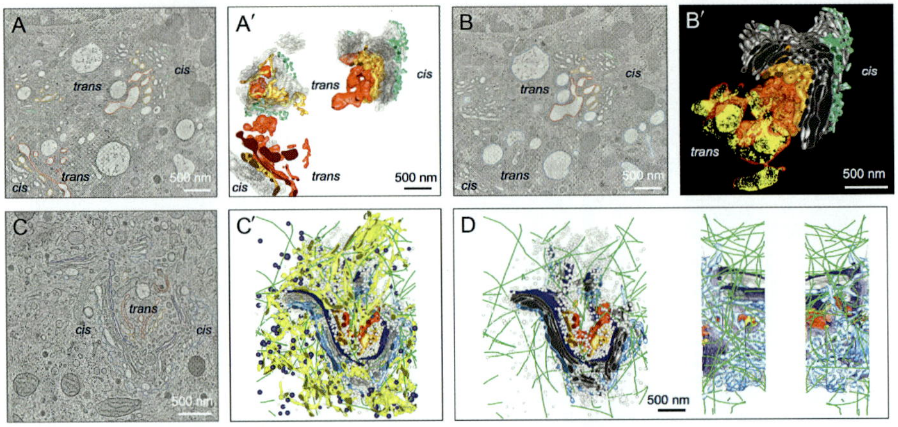

PLATE 36 (Fig. 16.4 on page 272 of this volume).

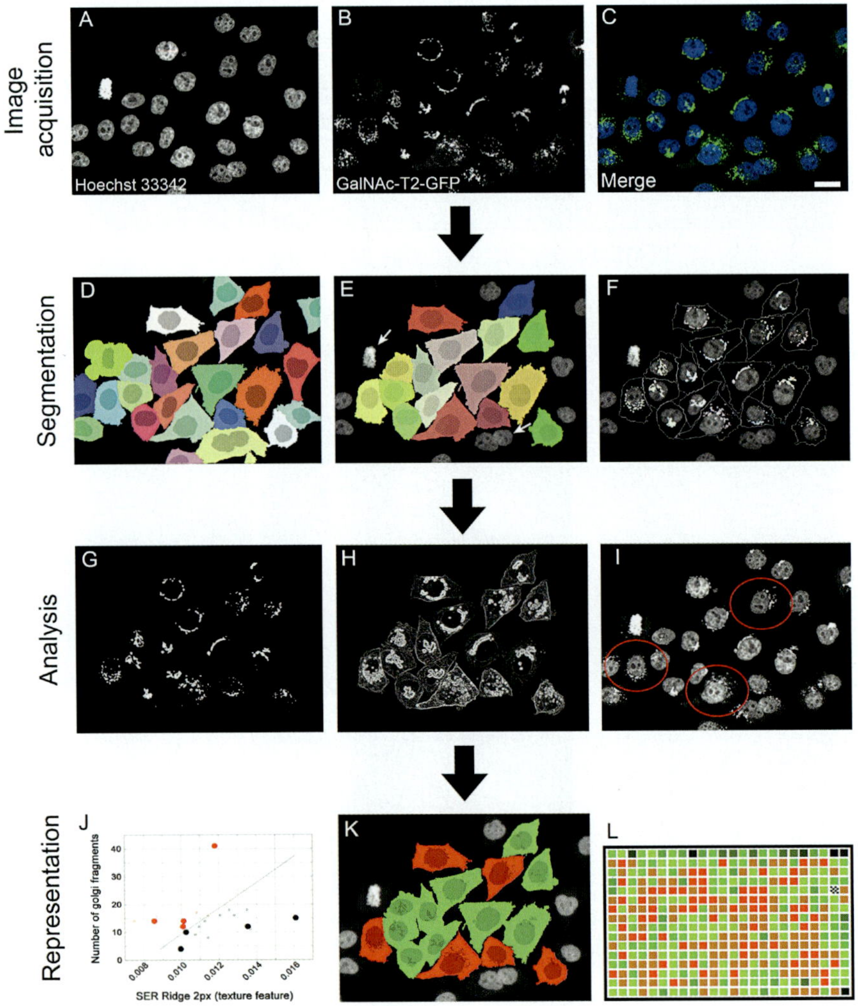

PLATE 37 (Fig. 17.2 on page 287 of this volume).

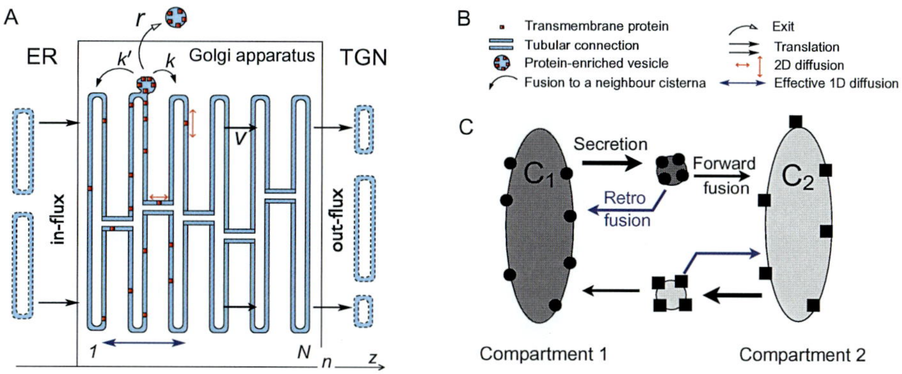

PLATE 38 (Fig. 18.1 on page 303 of this volume).

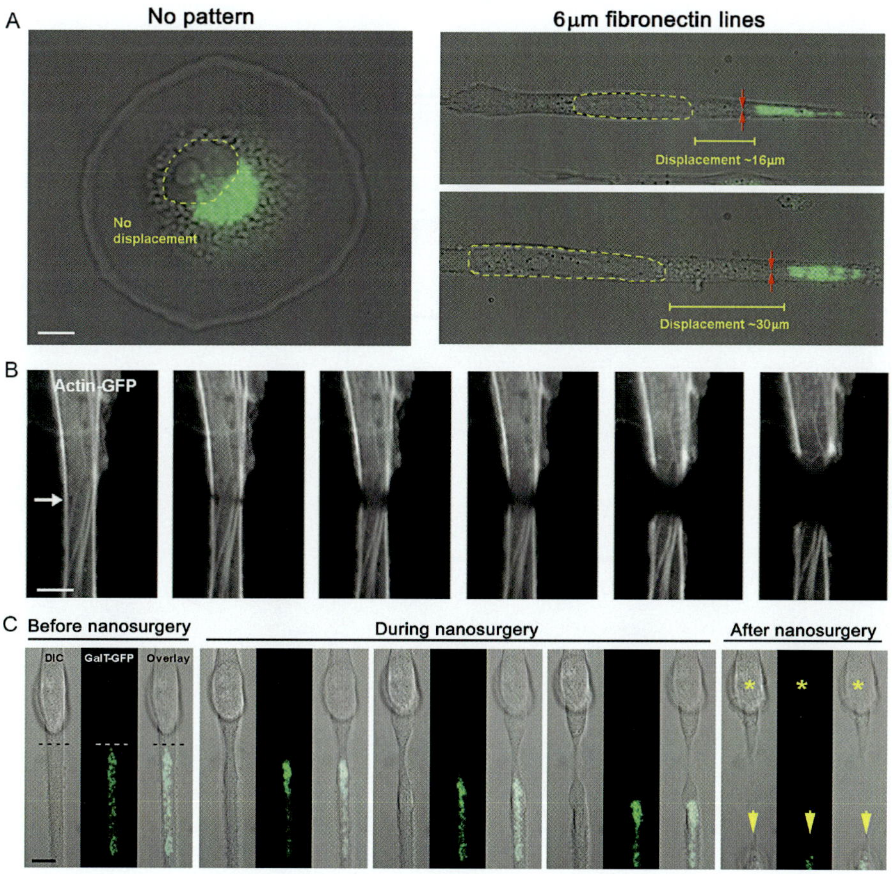

PLATE 39 (Fig. 19.1 on page 314 of this volume).

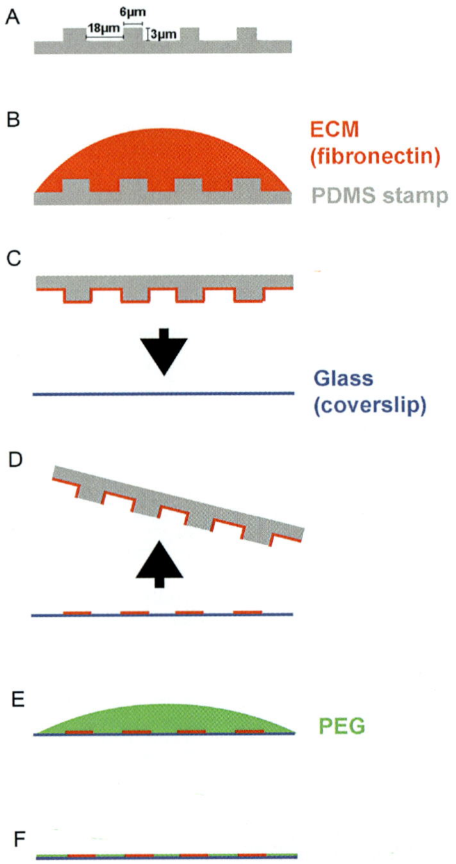

PLATE 40 (Fig. 19.2 on page 317 of this volume).

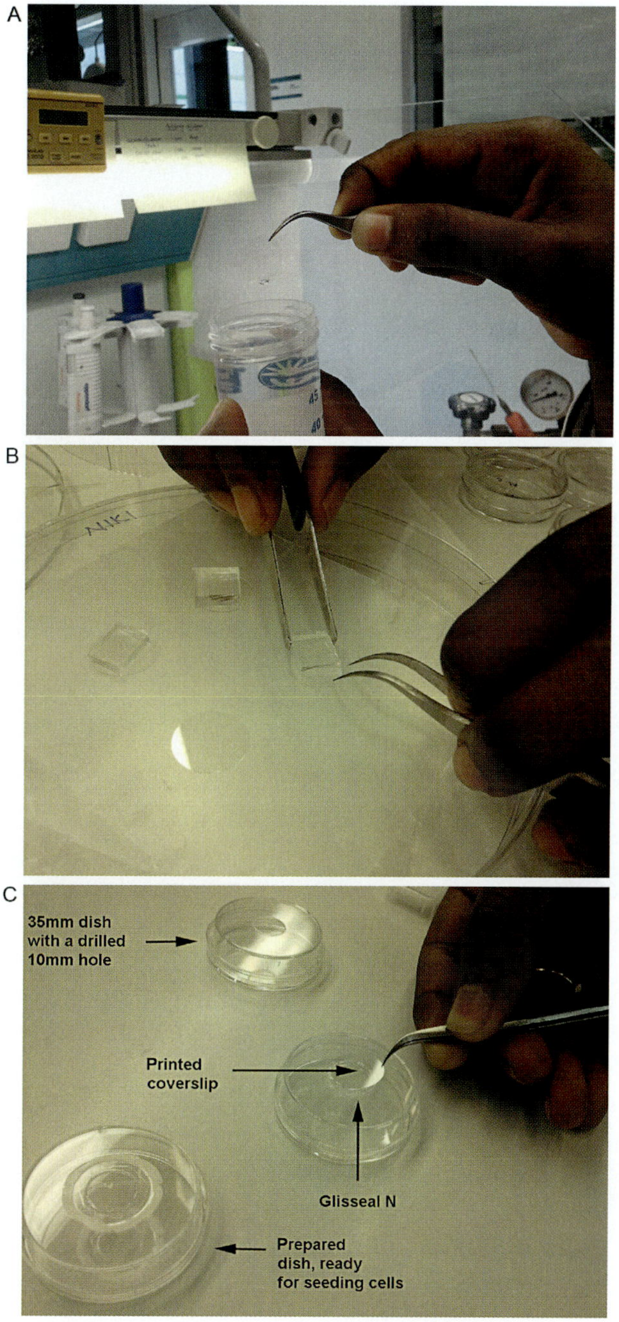

PLATE 41 (Fig. 19.3 on page 318 of this volume).

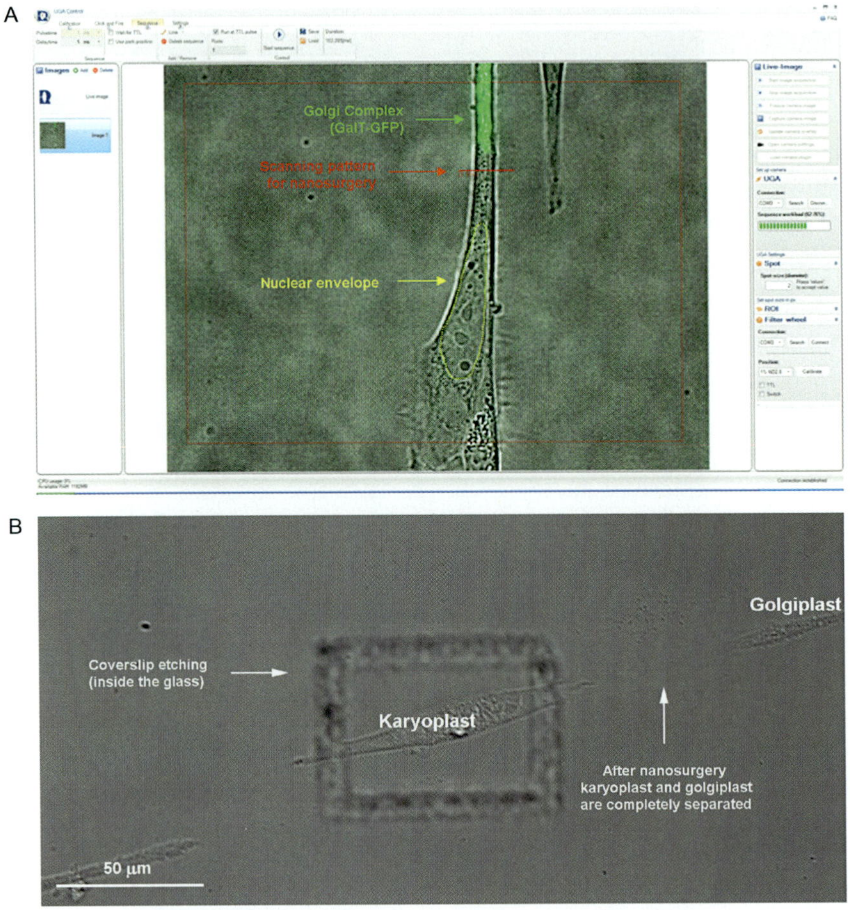

PLATE 42 (Fig. 19.4 on page 320 of this volume).

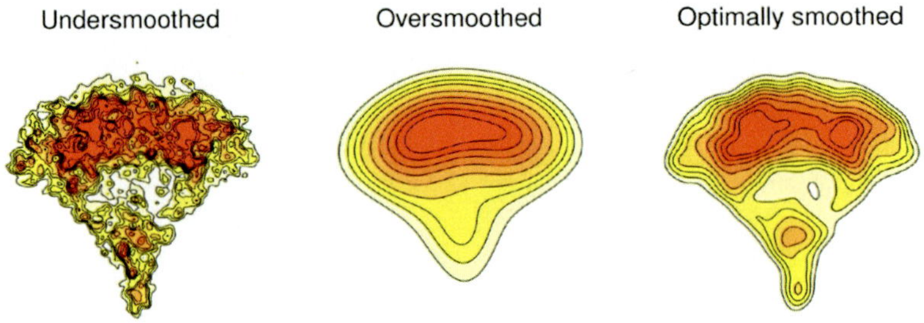

PLATE 43 (Fig. 20.2 on page 329 of this volume).

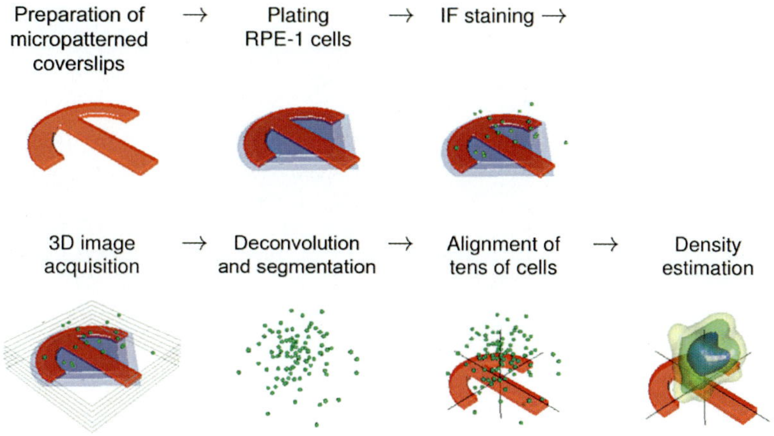

PLATE 44 (Fig. 20.3 on page 331 of this volume).

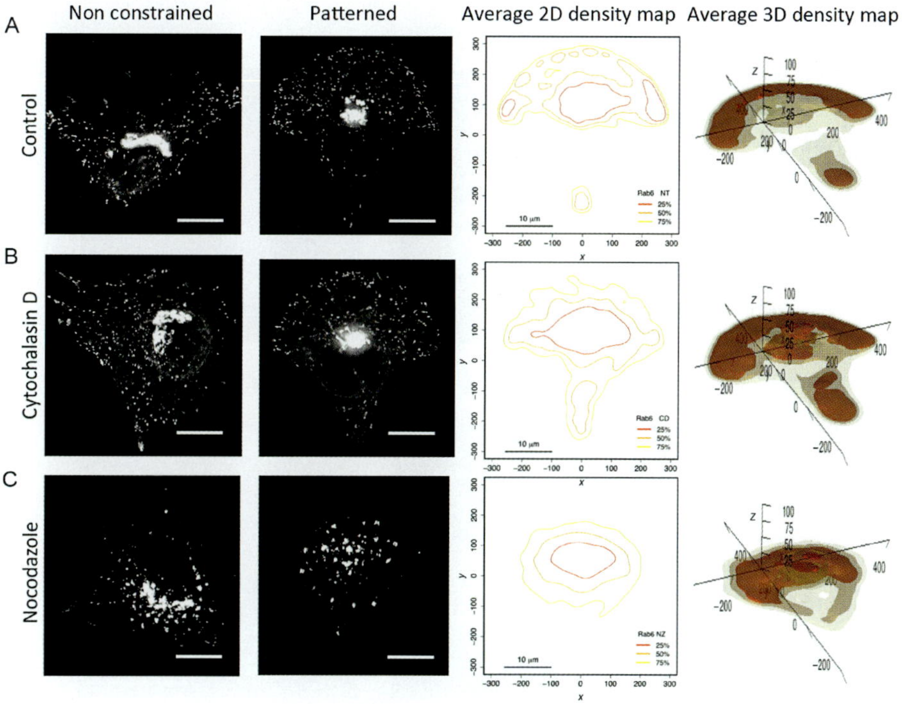

PLATE 45 (Fig. 20.4 on page 340 of this volume).

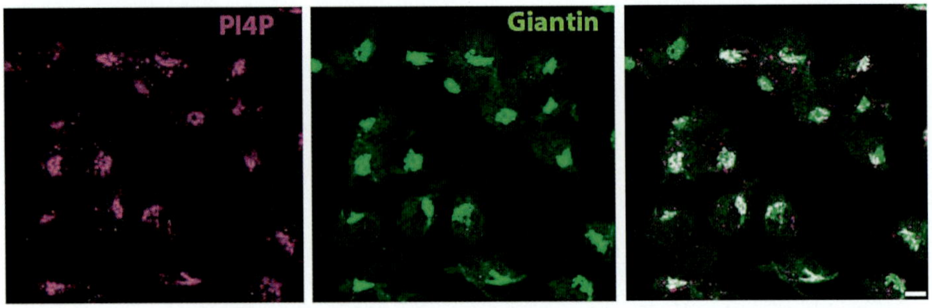

PLATE 46 (Fig. 21.1 on page 348 of this volume).

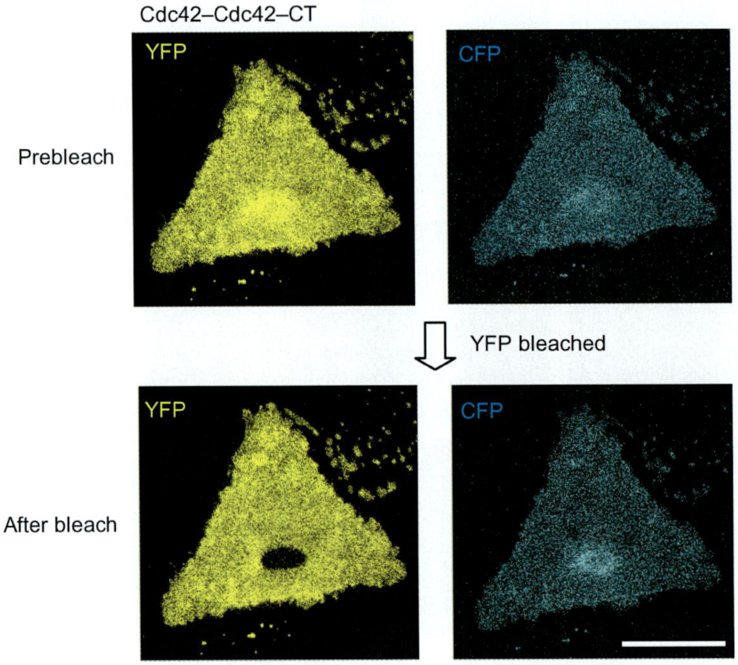

PLATE 47 (Fig. 21.3 on page 353 of this volume).

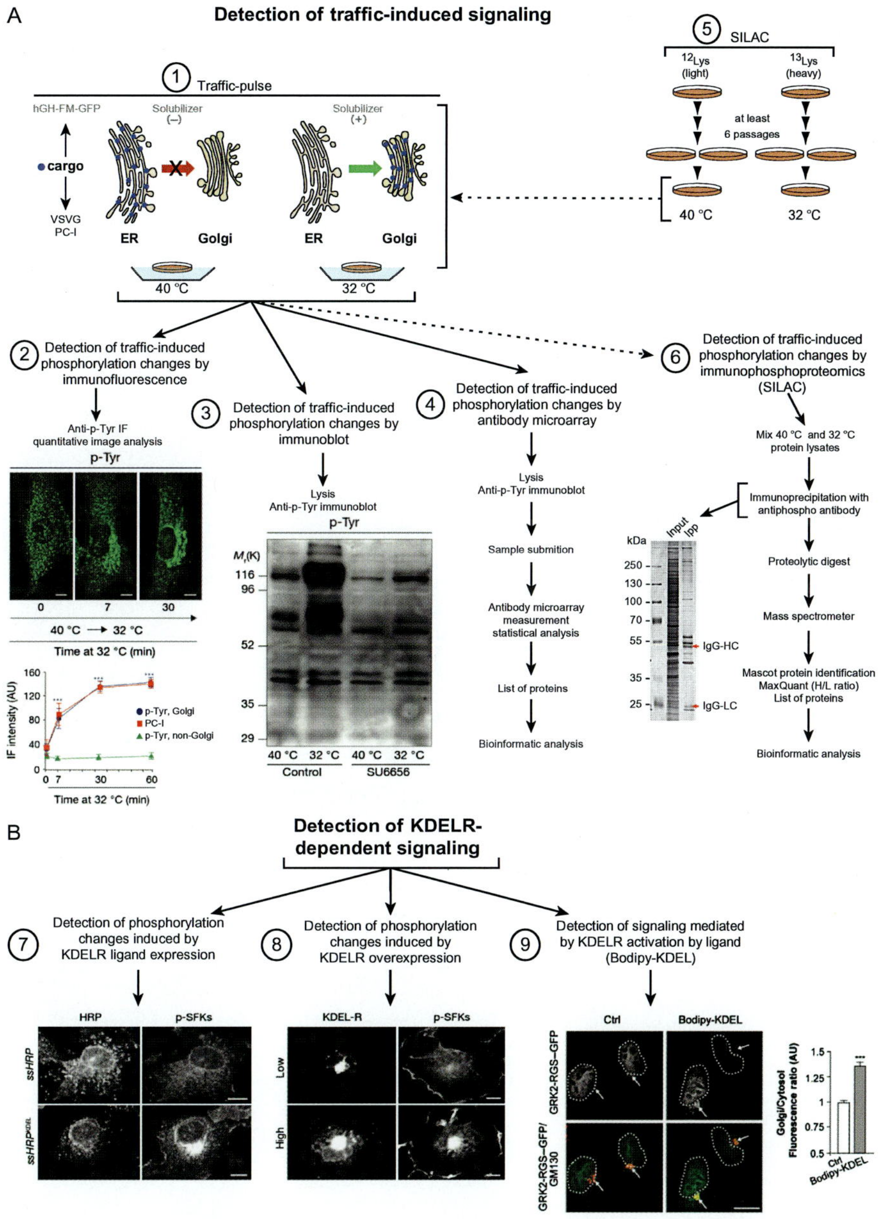

PLATE 48 (Fig. 22.1 on page 363 of this volume).

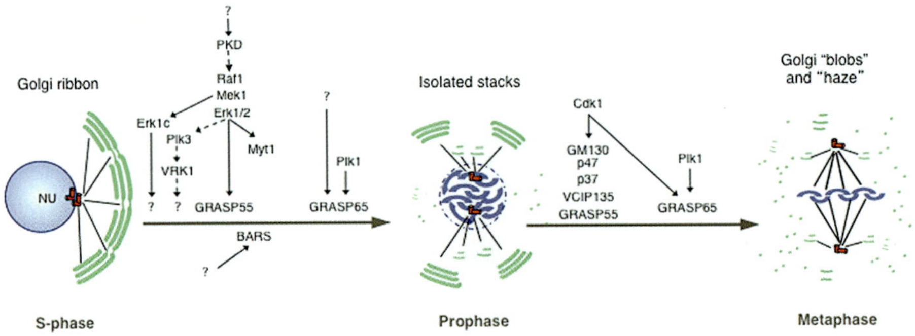

PLATE 49 (Fig. 23.1 on page 385 of this volume).

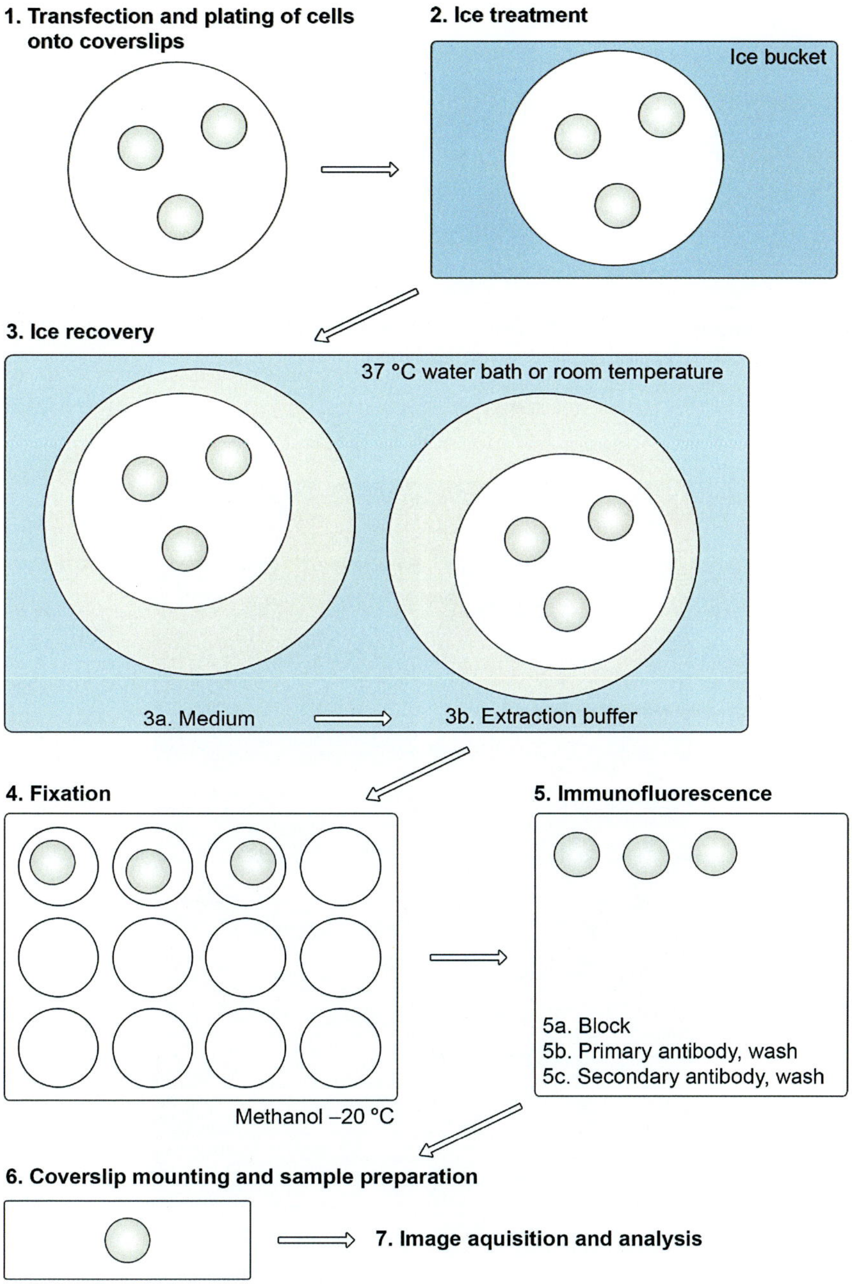

PLATE 50 (Fig. 24.1 on page 406 of this volume).

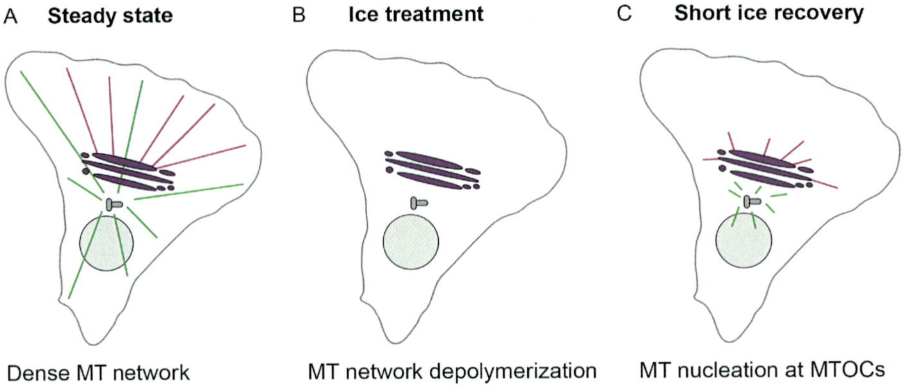

PLATE 51 (Fig. 24.2 on page 410 of this volume).

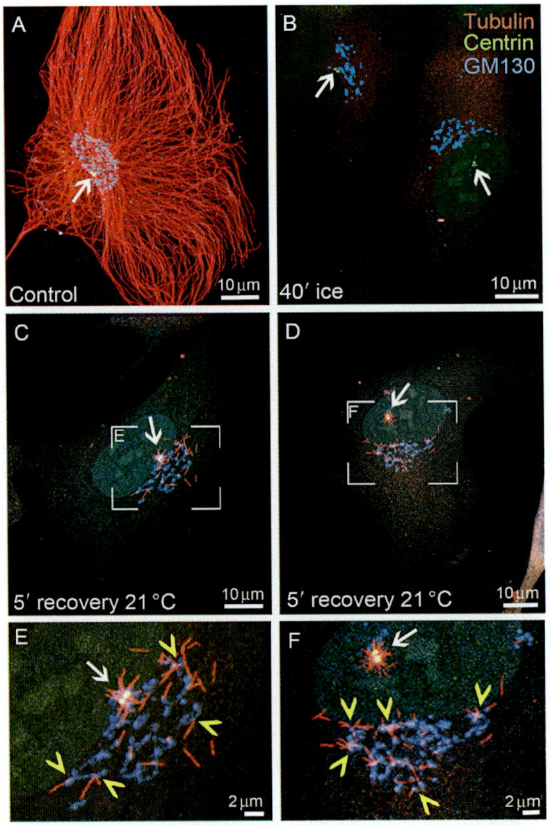

PLATE 52 (Fig. 24.3 on page 411 of this volume).

Edwards Brothers Malloy
Ann Arbor MI. USA
November 22, 2013